Novel Plant Natural Product Skeletons

Yongxian Cheng
School of Pharmaceutical Sciences
Shenzhen University Health Science Center
Shenzhen, Guangdong, China

Dapeng Qin
School of Pharmaceutical Sciences
Shenzhen University Health Science Center
Shenzhen, Guangdong, China

ISBN 978-981-99-7328-6 ISBN 978-981-99-7329-3 (eBook)
https://doi.org/10.1007/978-981-99-7329-3

This work was supported by grants from the National Natural Science Foundation of China (81903508, 81525026, U1702287).

This Springer imprint is published by the registered company Springer Nature Singapore Pte Ltd.
The registered company address is: 152 Beach Road, #21-01/04 Gateway East, Singapore 189721, Singapore

Conflicts of Interest

The authors have no conflicts of interest to declare.

Preface

New skeleton natural products (NSNPs) are usually defined to be those structures which are formed by carbon-carbon bond cleavage followed by the subsequent rearrangement and have structural frameworks that are different from those of previously known NPs. The discovery of NSNPs will add new members for corresponding structure types, bring new opportunities for biological/pharmaceutical areas, and provide new structure templates for synthetic chemists. However, dozens of works dealing with NPs were organized according to a certain structure type, a certain biological activity, or chemical substances in a certain family or genus beyond NSNPs from plant kingdoms. Therefore, we have created a comprehensive and systematic book of NSNPs derived from plants between 1999 and 2021. This book categorizes NSNPs from plants by providing their names, source distributions, family and genus classifications, structural types, structure characteristics, and bioactivities. A total of 1365 NSNPs in 99 families were summarized, which cover more than 30 structure types with NSNPs in the Hypericaceae family and in alkaloids' structures being the most. The summary of biological activities reveals the biological profiling of NSNPs in the last 23 years and exposes disadvantages in this aspect. This book will provide a handbook like tools for chemical, biological, pharmaceutical, or biosynthetic communities.

This book is also dedicated to the memory of Dr. Yongxian Cheng's mother.

Shenzhen, China

Yongxian Cheng
Dapeng Qin

Contents

Chapter 1
Introduction

Since antiquity, humans have utilized substances derived from natural sources to treat diseases. Early books describing medicinal agents date back to between 1550 and 1292 BC when "Ebers Papyrus" was published in Egypt. This book describes 700 substances used as drug. "Shen Nong's Herbal Classic", written around the first century A.D. by ancient Chinese authors, includes descriptions of 365 medicinal herbs. The therapeutic effects of herbs and other natural materials are thought to be associated with their component natural products (NPs), usually defined as substances having molecular weights less than 1500 Dalton that are created as secondary metabolites through biosynthetic pathways that rely on enzyme catalyzed processes. In 1806, morphine, was the first NP isolated by Friedrich Sertürner, and many alkaloids such asstrychnine, quinine, atropine, cocaine and emetine were identified in the nineteenth century, some of which are currently still utilized as medicinal agents.

Assignment of the structures of NPs typically utilized chemical synthesis and was enormously challenging prior to the middle of the twentieth century. For example, it took 148 years from the time of its isolation to determine the complete structure of morphine in 1952. With the advent of NMR spectroscopy and other sophisticated instrumental methods, structural assignment of NPs has advanced at a surprising speed. So far, more than 300,000 NPs have been isolated and identified, and some of these substances such as aspirin, penicillin, morphine, quinine, paclitaxel and artemisinin have become drugs that have had a major impact on world health and the longevity of the human population (Fig. 1.1). It is hard to predict the precise numbers of NPs that exist in Nature and it is not easy to identify the biological roles that some of them play in their natural sources. Despite these limitations, the value of plant derived NPs as drugs is high. Data show that 60% anti-cancer drugs and 75% antibiotics approved by the FDA during the 1981–2002 period are either themselves or close relatives of NPs [1]. A recent review by Gordon M. Cragg summarized the roles played by NPs as new drugs from 1981 to 2019, emphasizing that NPs still hold great potential in the arena of drug development [2].

Y. Cheng and D. Qin, *Novel Plant Natural Product Skeletons*,
https://doi.org/10.1007/978-981-99-7329-3_1

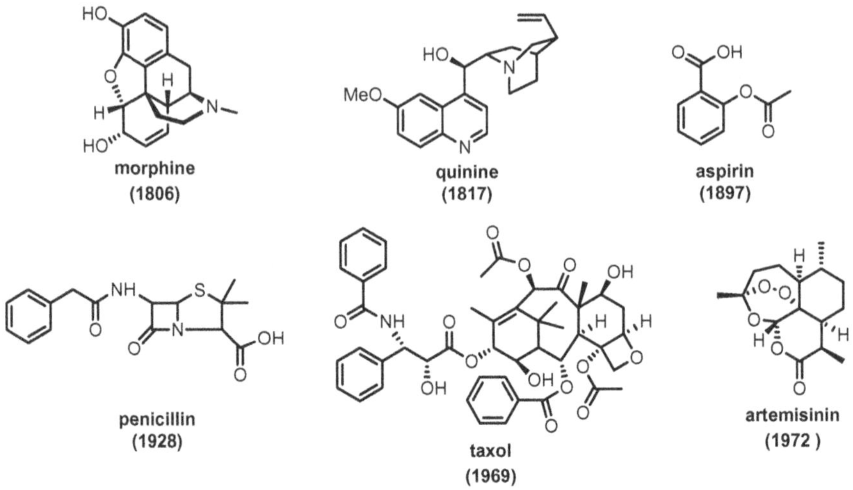

Fig. 1.1 Representative NPs that serve as drugs

The most interesting members of the large family of NPs are those that come from the plant kingdom and that have novel structural architectures. Consequently, members of this group, which have structural frameworks that are different from those of previously known NPs are the focus of the current Review. Plant NPs have been extensively reviewed in the past using formats based on either structure types or plant family/genus [3–12]. In contrast, this Review covers NSNPs from plant sources that have unique molecular structures and that should attract the attention different readers in related scientific communities. For example, numerous fascinating NSNPs bearing multiple stereogenic centers and a variety of complex structures are produced in plants and have served as a stimulus for the development of new synthetic strategies. In the process of accomplishing total syntheses of target NSNPs, organic chemists have uncovered new and highly selective methods for creating structures and manipulating functionality. Moreover, these synthetic efforts have encouraged the development of new theories about the regiochemical and stereochemical control of organic reactions, an accomplishment probably best exemplified by the well-known Woodward-Hoffmann rules for control of pericyclic reactions arising from proposed by Woodward's and Eschenmoser's studies of the total synthesis of vitamin B_{12} [13]. Other examples of achievements made in organic chemistry that were inspired by studies of NPs include Haworth projection notation, [14] the concept of biomimetic synthesis, [15] the biogenetic isoprene rule, [16] and the concepts of conformational, [17] and retrosynthetic analysis (Fig. 1.2) [18–20].

Plant derived NPs, possessing new structures and biological functions, have led to advances in pharmacology and drug discovery. For instance, the concept of receptor was proposed in 1878 to understand the action of the NPs atropine and pilocarpine against cat saliva secretion [21]. Other examples of well-known NPs that aided the development of pharmacology are morphine, quinine, aspirin, atropine, reserpine

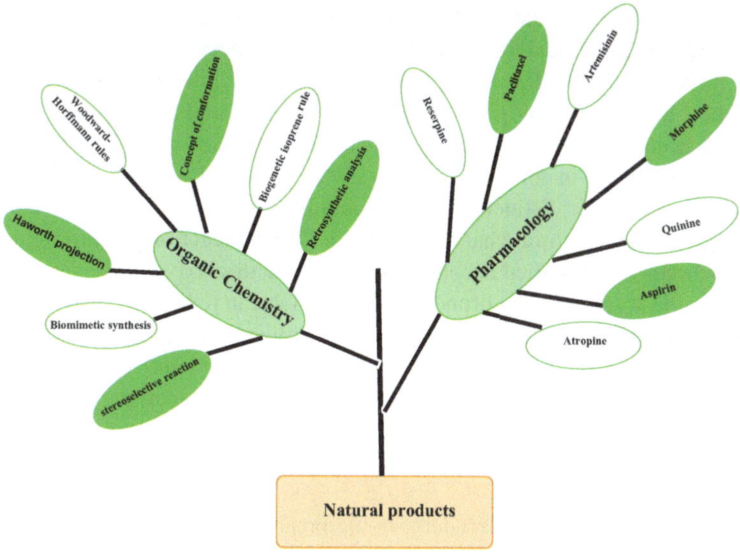

Fig. 1.2 Development of related fields inspired by natural products

and paclitaxel [22]. Thus, plant NSNPs, having new molecular structures, possess the potential for unique pharmacological activities and, as a result, they serve as unique substances for use in studies aimed at unveiling the pathogenesis of a wide array of human diseases.

Synthetic biology is a new emerging multidisciplinary field that is benefited by the discovery of new plant NPs that have unique molecular structures. For the most part, earlier studies of the biosynthesis of NPs mainly focused on microorganism derived substances. More recently, the greater attention being given to NPs from plant origins has benefited synthetic biology, and led to breakthroughs in the biosynthesis of plant NPs such as artemisinic acid (artemisinin precursor), [23, 24] ginkgolides, [25] paclitaxel, [26] tanshinone, [27] morphine[28] and cannabinoids [29].

A summary of plant NPs with novel organic structures should benefit workers in the fields of organic synthesis, pharmacology and synthetic biology. As a result, we have created a comprehensive and systematic Review of structurally NSNPs derived from plants, which have been identified in the time period between January 1999 and December 2021. This Review categorizes the NSNPs by providing their names, source distributions, family and genus classifications, structural types, structure characteristics, and bioactivities. The literature search that uncovered these substances covered journals including *Angewandte Chemie International Edition*, *Chemical Communications*, *Chemistry-A European Journal*, *European Journal of Medicinal Chemistry*, *Journal of the American Chemical Society*, *Journal of Medicinal Chemistry*, *Journal of Natural Products*, *Journal of Organic Chemistry*, *Organic Chemistry Frontiers*, *Organic Letters*, *Phytochemistry*, *Planta Medica*, *Tetrahedron*, *Tetrahedron Letters*, and others. The search uncovered more than 1,300 substances derived from plants, which are organized into 12 classifications including sesquiterpenoids, diterpenoids, sesterterpenoids, triterpenes, limonoids, phloroglucinols, meroterpenoids, flavonoid hybrids, lignans, alkaloids, steroids, and a few other types. Sesquiterpenoids, diterpenoids, sesterterpenoids, triterpenes, flavonoid hybrids, lignans, alkaloids and steroides are classified according to their plant family of origin, and those in the same family are organized based on their structural types. Because of their distinct nature, limonoids and phloroglucinols are classified according to their structural characteristics. With the exception of phloroglucinols, meroterpenoids are classified according to their structural types. In general, NSNPs in the same class are ordered chronologically, which can reflect the structure change rule of the new skeleton compounds. We have utilized 8 categories to characterize biological activities including cytotoxic, anti-inflammatory, antiviral, immunomodulatory, antioxidant, neuroprotective, antibacteria and anti-malarial, which are the most common biological assays performed on NSNPs. Of course, the biological activities of the NSNPs are more extensive and, consequently, other activities are also listed.

By consulting this Review, readers should be able to gain information about the family and genus distribution of NSNPs in plants, and the structural characteristics and biological activities of these new compounds. Organization of new skeleton containing NPs according to their plant families and genera should improve the efficiency of finding new structural types by chemists interested in the synthesis of NPs possessing novel structural characteristics. The summary of biological activities should assist those interested in assessing the potential biological function of the new groups of NSNPs. In addition, the Review can be employed to trace the structural chronological evolution of NSNPs. Moreover, it can be used to gain an understanding of contributions made by famous scientists to the discovery of NSNPs, some of whom have focused on one or several families of plants, while some have focused on one or several new types of NSNPs.

Furthermore, we tried to clarify the confusing trivial names which were not proposed by their discoverers (**446–450**, **471—475**, **656—658**, **697**, **698**, **878**, **879**, **1147**, **1148** and **1319**), and they were presented only with numbers in the Tables.

References

1. Newman DJ, Cragg GM, Snader KM. Natural products as sources of new drugs over the period 1981–2002. J Nat Prod. 2003;66:1022–37.
2. Newman DJ, Cragg GM. Natural products as sources of new drugs over the nearly four decades from 01/1981 to 09/2019. J Nat Prod. 2020;83:770–803.
3. Ciochina R, Grossman RB. Polycyclic Polyprenylated acylphloroglucinols. Chem Rev. 2006;106:3963–86.
4. Wu YB, Ni ZY, Shi QW, Dong M, Kiyota H, Gu YC, Cong B. Constituents from *Salvia* species and their biological activities. Chem Rev. 2012;112:5967–6026.
5. Vasas A, Hohmann J. Euphorbia diterpenes: isolation, structure, biological activity, and synthesis (2008–2012). Chem Rev. 2014;114:8579–612.

6. Wang HB, Wang XY, Liu LP, Qin GW, Kang TG. Tigliane diterpenoids from the *Euphorbiaceae* and *Thymelaeaceae* families. Chem Rev. 2015;115:2975–3011.
7. Tan QG, Luo XD. Meliaceous limonoids: chemistry and biological activities. Chem Rev. 2011;111:7437–522.
8. Silva LN, Zimmer KR, Macedo AJ, Trentin DS. Plant natural products targeting bacterial virulence factors. Chem Rev. 2016;116:9162–236.
9. Yang XW, Grossman RB, Xu G. Research progress of polycyclic polyprenylated acylphloroglucinols. Chem Rev. 2018;118:3508–58.
10. Almeida A, Dong L, Appendino G, Bak S. Plant triterpenoids with bond-missing skeletons: biogenesis, distribution and bioactivity. Nat Prod Rep. 2020;37:1207–28.
11. Shen Y, Liang WJ, Shi YN, Kennelly EJ, Zhao DK. Structural diversity, bioactivities, and biosynthesis of natural diterpenoid alkaloids. Nat Prod Rep. 2020;37:763–96.
12. Zhang, H, Liu HB, Yue JM. Organic carbonates from natural sources. Chem Rev 2014;114:883–898.
13. Woodward RB, Edited Zagalak B, Friedrich W. Synthetic vitamin B12. Proc Eur Symp 1979;3rd:37–87.
14. Gooch JW. Encyclopedic dictionary of polymers. Springer 2011;Haworth Projection.
15. de la Torre MC, Sierra MA. Comments on recent achievements in biomimetic organic synthesis. Angew Chem Int Ed 2004;43:160–181.
16. Hillier SG, Lathe R. Terpenes, hormones and life: isoprene rule revisited. J Endocrinol. 2019;242:R9–22.
17. Moss GP. Basic terminology of stereochemistry. Pure Appl Chem. 1996;68:2193–222.
18. Corey EJ, Cheng XM. The logic of chemical synthesis. Wiley; 1989.
19. Corey EJ. Retrosynthetic thinking–essentials and examples. Chem Soc Rev. 1988;17:111–33.
20. Corey EJ. The logic of chemical synthesis: multistep synthesis of complex carbogenic molecules (Nobel Lecture). Angew Chem Int Ed. 1991;30:455–65.
21. Kenakin T. Principles: receptor theory in pharmacology. Trends Pharmacol Sci. 2004;25:186–92.
22. Gilani AH, Rahman AU. Trends in ethnopharmocology. J Ethnopharmacol. 2005;100:43–9.
23. Ro DK, Paradise EM, Ouellet M, Fisher KJ, Newman KL, Ndungu JM, Ho KA, Eachus RA, Ham TS, Kirby J, Chang MCY, Withers ST, Shiba Y, Sarpong R, Keasling JD. Production of the antimalarial drug precursor artemisinic acid in engineered yeast. Nature. 2006;440:940–3.
24. Paddon CJ, Westfall PJ, Pitera DJ, et al. High-level semi-synthetic production of the potent antimalarial artemisinin. Nature. 2013;496:528–32.
25. Leonard E, Ajikumar PK, Thayer K, Xiao WH, Mo JD, Tidor B, Stephanopoulos G, Prather KLJ. Combining metabolic and protein engineering of a terpenoid biosynthetic pathway for overproduction and selectivity control. P Natl Acad Sci USA. 2010;107:13654–9.
26. Ajikumar PK, Xiao WH, Tyo KEJ, Wang Y, Simeon F, Leonard E, Mucha O, Phon TH, Pfeifer B, Stephanopoulos G. Isoprenoid pathway optimization for taxol precursor overproduction in *Escherichia Coli*. Science. 2010;330:70–4.
27. Zhou YJ, Gao W, Rong Q, Jin G, Chu H, Liu W, Yang W, Zhu Z, Li G, Zhu G, Huang L, Zhao ZK. Modular pathway engineering of diterpenoid synthases and the mevalonic acid pathway for miltiradiene production. J Am Chem Soc. 2012;134:3234–41.
28. Galanie S, Thodey K, Trenchard IJ, Interrante MF, Smolke CD. Complete biosynthesis of opioids in yeast. Science. 2015;349:1095–100.
29. Luo X, Reiter MA, d'Espaux L, Wong J, Denby CM, Lechner A, Zhang Y, Grzybowski AT, Harth S, Lin W, Lee H, Yu C, Shin J, Deng K, Benites VT, Wang G, Baidoo EEK, Chen Y, Dev I, Petzold CJ, Keasling JD. Complete biosynthesis of cannabinoids and their unnatural analogues in Yeast. Nature. 2019;567:123–6.

Chapter 2
Classification of Diverse Novel Sesquiterpenoids

Sesquiterpenoids are a class of structurally and biologically diverse natural products derived from farnesyl pyrophosphate (FPP) and generated by carbocation cascade reactions programmed in sesquiterpene synthases. 125 skeletal sesquiterpenoids were summarized in the past 23 years (**1–9** Fig. 2.1, **10–22** Fig. 2.2, **23–40** Fig. 2.3, **41–68** Fig. 2.4, **69–78** Fig. 2.5, **79–90** Fig. 2.6, **91–107** Fig. 2.7, **108–125** Fig. 2.8; Table A1). Based on the degree of polymerization, they are divided into two types: sesquiterpenoid monomers (hereafter solo-sesquiterpenoids) (**1–40**) and sesquitepenoid oligomers (**41–125**). 40 skeletal solo-sesquiterpenoids were found in the past 23 years, they are formed by carbon–carbon bond cleavage followed by the subsequent rearrangement and therefore different from their original skeletons. There are two types of the uncommon sesquitepenoid oligomers mainly dealing with dimers: homodimers and heterodimers. They are the same or different classes of sesquiterpenoids furnished by forming an unusual rings or cage structures. The oligomers are mainly formed by Diels–Alder, hetero-Diels–Alder, [2 + 2] cycloaddition, or free-radical coupling reactions. It is apparent that plants belonging to the family Compositae provide a considerable amount of uncommon sesquitepenoids.

2.1 Solo-Sesquiterpenoids

2.1.1 Asteraceae/Compositae

To date, seven unprecedented solo-sesquiterpenoids involving five references were obtained from Asteraceae. Ligulolide A (**1**) was isolated from *Ligularia virgaurea*, [1] heliespirones B (**2**) and C (**3**) were obtained from *Helianthus annuus* [2]. **1–3** are all spiro sesquiterpenoids. Interestingly, **1** possesses a sesquiterpene carbon backbone, **2** and **3** bear six- and five-membered spiroheterocyclic skeletons, respectively.

Y. Cheng and D. Qin, *Novel Plant Natural Product Skeletons*,
https://doi.org/10.1007/978-981-99-7329-3_2

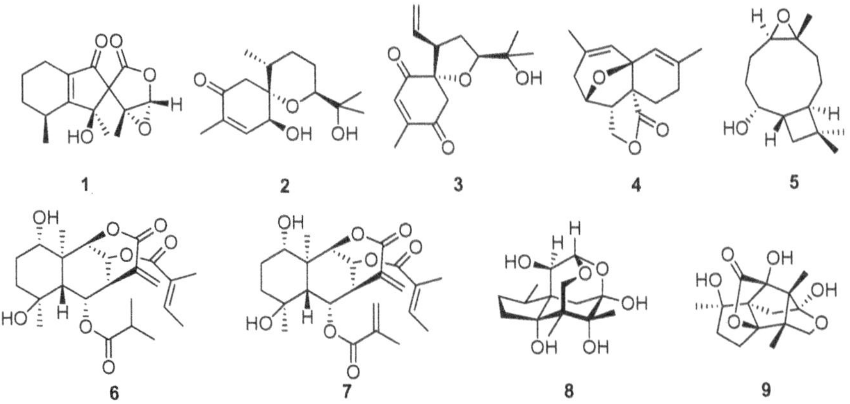

Fig. 2.1 Structures of new skeleton sesquiterpenoids **1–9**

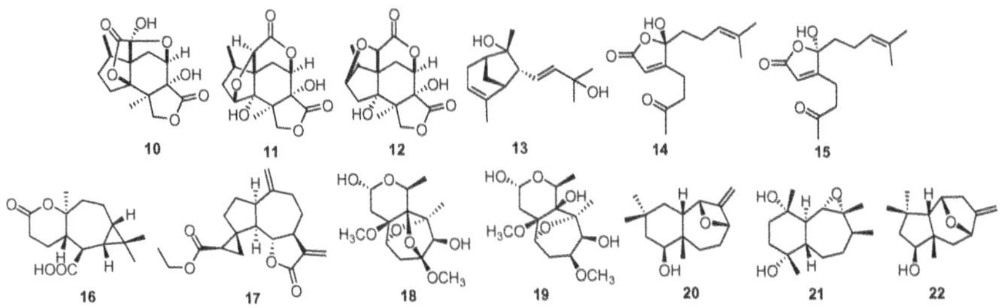

Fig. 2.2 Structures of new skeleton sesquiterpenoids **10–22**

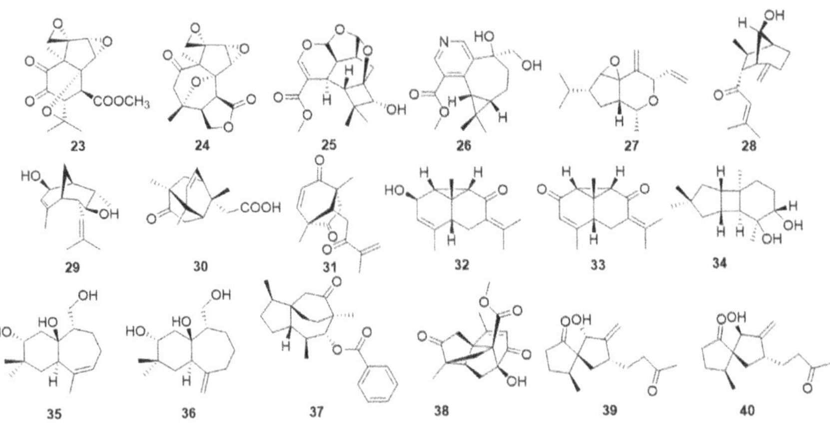

Fig. 2.3 Structures of new skeleton sesquiterpenoids **23–40**

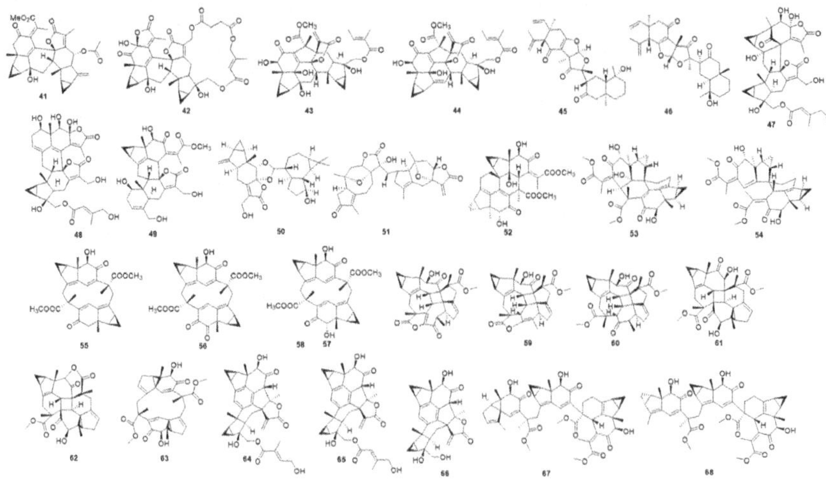

Fig. 2.4 Structures of new skeleton sesquiterpenoids **41–68**

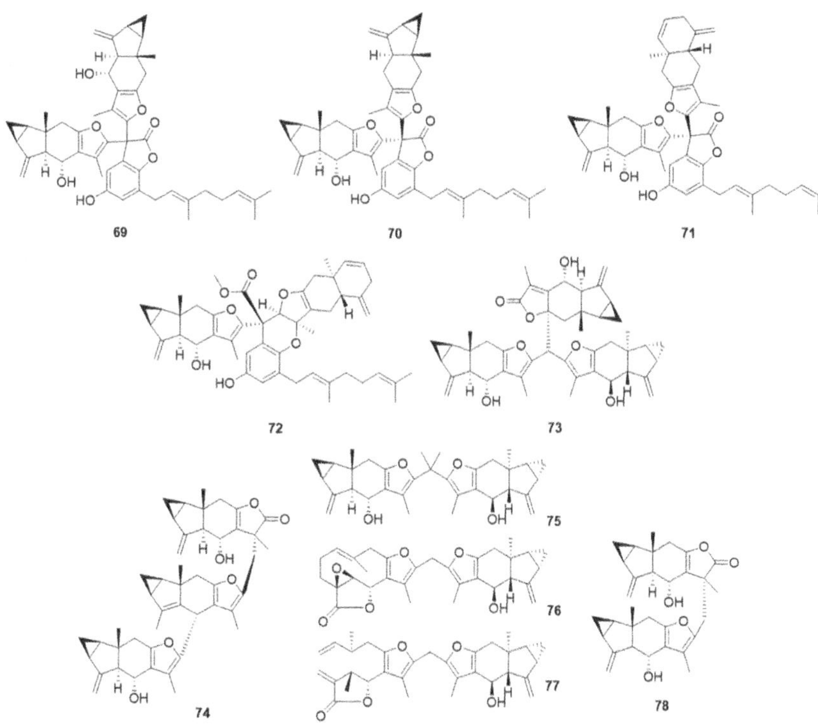

Fig. 2.5 Structures of new skeleton sesquiterpenoids **69–78**

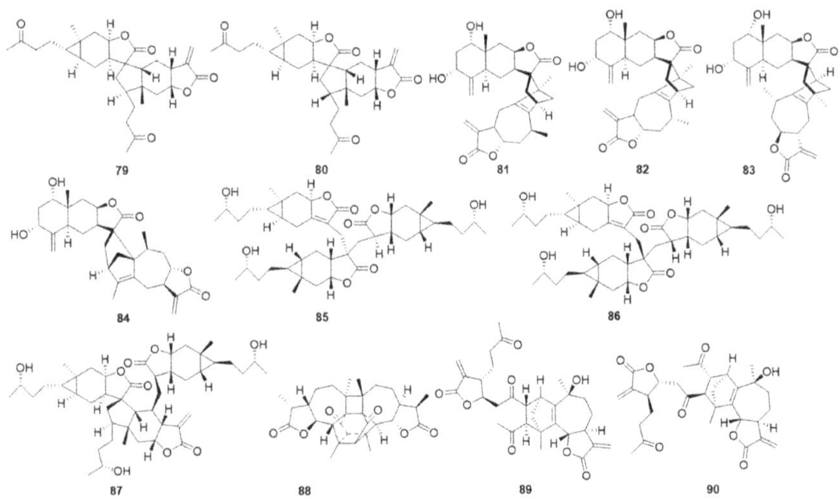

Fig. 2.6 Structures of new skeleton sesquiterpenoids **79–90**

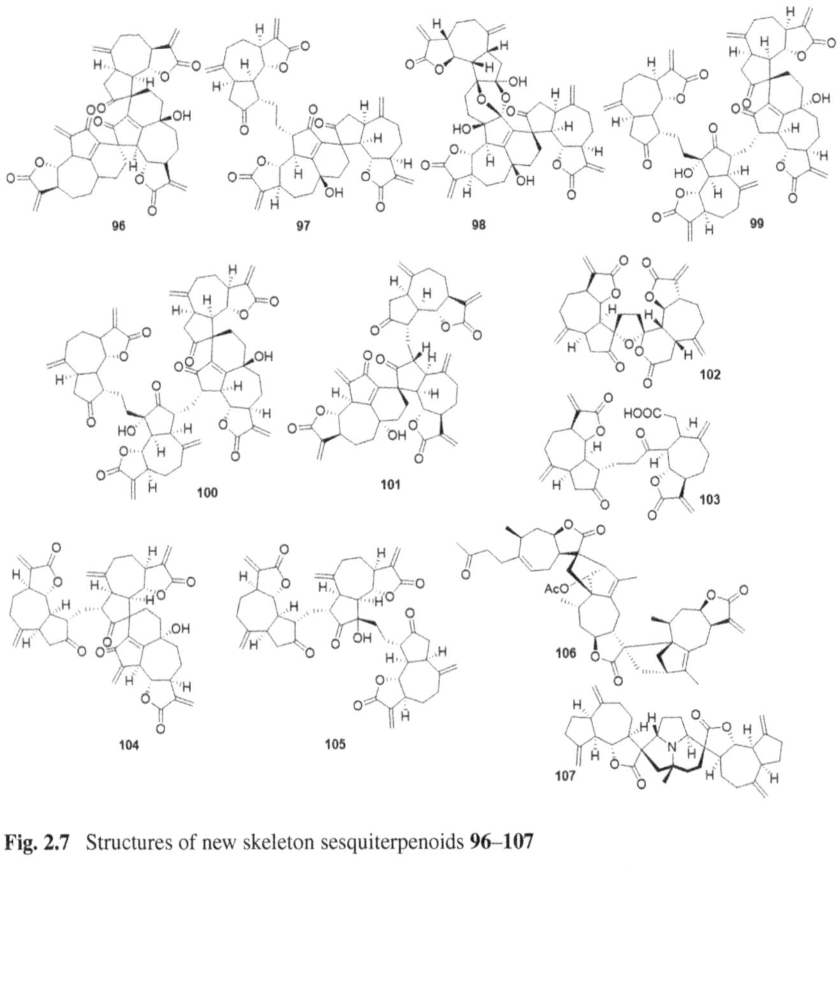

Fig. 2.7 Structures of new skeleton sesquiterpenoids **96–107**

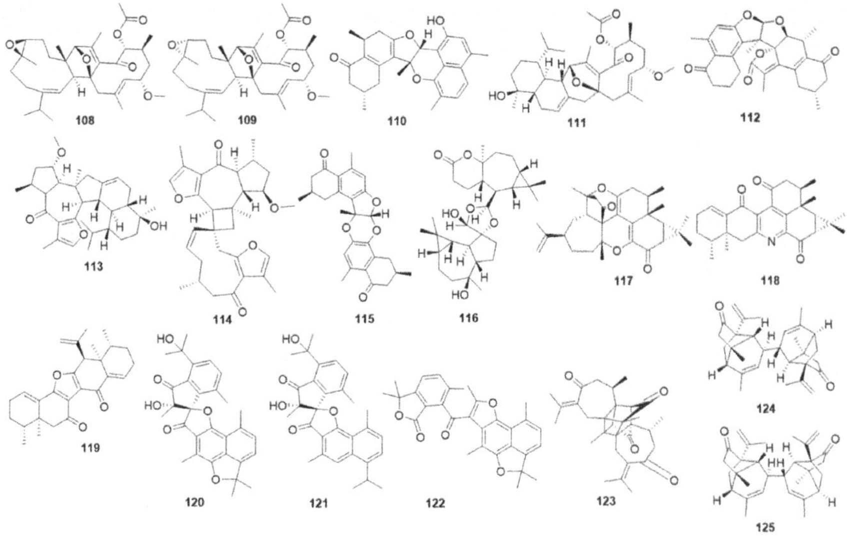

Fig. 2.8 Structures of new skeleton sesquiterpenoids **108**–**125**

Anthecularin (**4**), possessing a novel ring system, was isolated from *Anthemis auriculate* [3]. A major nonlactonic constituent, artarborol (**5**), featuring the presence of fused cyclobutane and cyclononane rings, was isolated from *Artemisia arborescens* [4]. Another two novel sesquiterpene δ-lactones, wedelolides A (**6**) and B (**7**) which are the first examples of an unprecedented framework characteristic of a sesquiterpene δ-lactone, (9*R*)-eudesman-9,12-olide, were obtained from *Wedelia trilobata* [5].

2.1.2 Magnoliaceae

The solo-sesquiterpenoids from the family Magnoliaceae are mostly caged architectures and were all isolated from the genus *Illicium.* Cycloparvifloralone (**8**) were isolated from the leaves of *Illicium parviflorum* (swamp star anise, yellow star anise), it features a hitherto unknown ring system with a cagelike acetal/hemiketal structure [6]. Illisimonin A (**9**), a novel sesquiterpenoid having a tricyclo[5.2.1.$0^{1,6}$]decane skeleton, was isolated from the fruits of *Illicium simonsii.* This compound is also characteristic of a caged 2-oxatricyclo[3,3,0,$1^{4,7}$]nonane ring system fused to a five-membered carbocyclic ring and a five-membered lactone ring [7]. Jiadifenolide (**10**) and jiadifenoxolanes A (**11**) and B (**12**), three novel *seco*-prezizaane-type sesquiterpenoids, were isolated from *Illicium jiadifengpi* pericarps. Their pentacyclic caged structures were identified by the combination of 2D NMR methods, chemical conversion, and single crystal X-ray diffraction. Compound **10** represents the first example of a *seco*-prezizaane-type sesquiterpenoid bearing a γ-lactone resulting from the construction between C-11 (carbonyl) and a hydroxy group attached to C-4 [8].

2.1.3 Annonaceae

Artaboterpenoids A, (+)-, and (−)-artaboterpenoids B (**13–15**) are novel bisabolene-derived sesquiterpenoids isolated from *Artabotrys hexapetalus*, their structures including absolute configurations were identified by spectroscopic methods and electronic circular dichroism calculations. Notably, **13** features a novel carbon skeleton with a new C-2–C-10 linkage. Whereas enantiomeric **14** and **15** represent the first examples of 1,2-*seco*-bisabolene-type sesquiterpene lactones [9].

2.1.4 Valerianaceae

Volvalerelactone A (**16**), a new sesquiterpenoid lactone, was isolated from *Valeriana officinalis*. **16** possesses an unprecedented 3/7/6 tricyclic ring system which makes it unusual, a biogenetic pathway for **16** was proposed with methyl migration and oxidation are the key steps of biosynthesis [10]. Narjatamolide (**17**), an unusual homoguaiane sesquiterpene lactone, was isolated from the roots and rhizomes of *Nardostachys jatamansi* DC. It represents the new carbon skeleton of a homoguaiane sesquiterpenoid possessing an additional acetate unit spiro-fused with C-4 and C-15 to form a cyclopropane ring [11].

2.1.5 Apocynaceae

Urechitols A (**18**) and B (**19**) were isolated from *Pentalinon andrieuxii*, which is a commonly used Yucatecan traditional medicine to treat leishmaniasis. **18** and **19** are the first examples with a novel "campechane" trinorsesquiterpenoid skeleton [12].

2.1.6 Anacardiaceae

Toxicodenanes A–C (**20–22**), representing three new carbon skeletons, were isolated from the resin of *Toxicodendron vernicifluum*. **20** and **21** possess a 6/7 ring system and **22** bears a 5/8 ring system. In particularly, both **20** and **22** have a tetrahydrofuran ring moiety in the molecules which make them cagelike structures [13].

2.1.7 *Coriariaceae*

In this family, two new skeletal sesquiterpenoids, coriatone (**23**), a highly oxygenated picrotoxane-type sesquiterpene, and coriane (**24**), a corianlactone possessing an unprecedented sesquiterpene basic skeleton, were isolated from *Coriaria nepalensis* [14].

2.1.8 *Plagiochilaceae*

Plagiochianin A (**25**) features the presence of an unprecedented 2,3:6,7-di-seco-6,8-cycloaromadendrane carbon scaffold conjugated with three cyclic acetals, and plagiochianin B (**26**) is an exceptional pyridine-type aromadendrane alkaloid. They are two novel *ent*-2,3-*seco*-aromadendrane derivatives isolated from the liverwort *Plagiochila duthiana.* Their absolute configurations were assigned by single-crystal X-ray analysis and electronic circular dichroism calculations [15].

2.1.9 *Lepidolaenaceae*

Hodgsonox (**27**) is another skeletal sesquiterpenoid isolated from liverwort, representing a new class of sesquiterpenoid with a cyclopenta[5,1-*c*]pyran ring system fused to an oxirane ring in the structure. The combination of a mono- and a 1,1-disubstituted double bond flanking the oxygenated carbon of a the pyran ring is a unique structural feature [16].

2.1.10 *Zingiberaceae*

Curcumanes A (**28**) and B (**29**) were isolated from traditional Chinese medicine *Curcuma longa.* They are structurally unique sesquiterpenoids possessing unprecedented skeletons with a dicyclo[3.2.1]octane and a dicyclo[3.3.1]nonane moiety in the molecules [17].

2.1.11 *Myrtaceae*

(±)-Eugenunilones C–F (**30–33**) were isolated from the fruits of *Eugenia uniflora.* Compounds **30** and **31** represent the first examples of sesquiterpenoids possessing a caged tricyclo[4.3.1.$0^{3,7}$]decane core and an isopentyl substituted bicyclo[3.2.1]octane backbone, respectively, while compounds **32** and **33** share a tricyclo[4.4.0.0^{2,10}]decane scaffold which is found in nature for the first time [18].

2.1.12 *Oleaceae*

(±)-syringanoid A (**34**), (±)-pinnatanoids A (**35**) and B (**36**), that represent an unprecedented 5/4/6 tricyclic backbone and a rare 6/7 bicyclic backbone, respectively, were isolated from the peeled stems of *Syringa pinnatifolia* [19].

2.1.13 *Illiciaceae*

Illihenin A (**37**), a novel sesquiterpenoid, was isolated from the roots of *Illicium henryi*. Compound **37** represents a class of novel 5/7/6 tricyclic sesquiterpenoids featuring a rare cage-like tricyclo[6.2.2.0^{1,5}]dodecane core [20].

2.1.14 *Thymelaeaceae*

Daphnenoids A–C (**38**–**40**), three unusual sesquiterpenes with distinctive ring skeletons, were obtained from the herb of *Daphne penicillata* by molecular networking strategies. Daphnenoid A (**38**) possesses a unique caged tetracyclo [5.3.2.0^{1,6}.0^{4,11}] dodecane scaffold by unexpected cyclizations of C-1/C-11 and C-2/C-14. Daphnenoids B and C (**39** and **40**) were the first discovered natural sesquiterpenes with unique 5/5 spirocyclic systems in nature [21].

2.2 Sesquitepenoid Oligomers

2.2.1 *Chloranthaceae*

In this family, a total of 28 new skeletal dimeric sesquiterpenoids belonging to 10 references were summarized. Of note, Prof. Jianmin Yue, a scholar from China, contributed most to skeletal compounds of this family. Chlorahololides A (**41**) and B (**42**) are both highly complex sesquiterpenoid dimers isolated from *Chloranthus holostegius*. Especially, **42** is a decacyclic lindenane-type sesquiterpenoid dimer having a unique 18-membered macrocyclic trilactone ring in the structure [22]. Sarcanolides A (**43**) and B (**44**) are two sesquiterpenoid dimers from *Sarcandra hainanensis*. They feature a new nonacyclic scaffold in which the bond formation of

C-11–C-7’ makes the five-membered lactone ring in a full β-direction [23]. Serratustones A (**45**) and B (**46**), isolated from *Chloranthus serratus*, are the first examples of nonlindenane-type sesquiterpenoid dimers from the Chloranthaceae family. In a biogenic point of view, they are formed from a novel dimerizationpattern of anelemane and an eudesmane by involving aldol condensation as the key step [24]. Three dimeric sesquiterpenoids (**47–49**) were isolated from *Chloranthus fortunei*. Fortunoid A (**47**) is a new carbon skeleton characteristic of a rearranged lindenane dimer, whereas, fortunoids B (**48**) and C (**49**) are the first examples of heterodimeric backbones composed of a lindenane and an eudesmane sesquiterpenoid, whose structures including stereochemistry were established by spectroscopic methods and electric circular dichroism calculations [25]. Two sesquiterpenoid dimers hedyorienoids A (**50**) and B (**51**), featuring different polycyclic skeletons, were described from *Hedyosmum orientale*. It is noteworthy that compound **50** possesses an unprecedented heterodimeric framework of a lindenane and an aromadendrane sesquiterpenoid via forming an unusual 1,3-dioxolane ring, and **51** features a new dimerization pattern of two guaiane-type sesquiterpenoids by forming a new carbon–carbon bond [26]. Chloraserrtone A (**52**) is the first lindenane-type sesquiterpenoid dimer with extremely unique connections of C-15–C15′, C-4–C-6′, and C-6–C-11′ to form two six-membered rings between the monomeric units. X-ray diffraction data were used to elucidate the absolute configuration of **52**.[27] Two highly fused lindenane sesquiterpenoid dimers, chlotrichenes A (**53**) and B (**54**), were obtained from *Chloranthus holostegius*. Both **53** and **54** possess a new type of spirocarboncyclic dimeric framework formed by *endo*-Diels–Alder reaction. **53** possesses a unique 3/5/6/6/6/6/5/3-fused octacyclic skeleton arising from subsequently plausible epoxidation-cyclization reactions of **54** [28]. In addition, three dimeric lindenane-type sesquiterpenoids, japonicones A–C (**55–57**), characteristic of a rare 12-membered ring framework, were isolated from *Chloranthus japonicus*. Their absolute configurations were unequivocally established by computational methods and X-ray crystallography [29]. Chlorahupetones A–I (**58–66**), nine sesquiterpenoid dimers, were isolated from *Chloranthus henryi* var. *hupehensis*. Compound **58** features a unique 3/5/6/5/4/7/7/5-fused octacyclic carbon skeleton of lindenane-type and guaiane-type sesquiterpenoid monomers bridged by a seven-four-five-membered ring system. Compounds **59** and **60**, possessing an unusual 3/5/6/5/4/7/6/5-fused carbon skeleton, are two structural analogues of 1. Compounds **61** and **62** possess an unprecedented 3/5/6/6/4/6/6/3/5-fused carbon framework. Compound **63** is a unique C_2-symmetric cyclic rearranged lindenane-type sesquiterpenoid dimer. Compounds **64–66** are lindenane-type sesquiterpenoid dimers featuring a carbon framework via the formation of a C-11–C-7’ bond with an aromatized ring D and a five-membered lactone ring fused at C-11 and C-7’ [30]. Two new lindenane sesquiterpenoid trimers, designated as trichloranoids A and B (**67**and **68**) were isolated from *Chloranthus spicatus*. **67** and **68** incorporated one unprecedented skeletal unit respectively in their western hemisphere, and each of them was likely formed biogenetically via a biradical rearrangement of the vinylcyclopropane [31].

2.2.2 Lauraceae

10 unprecedented sesquitepenoid oligomers were characterized from *Lindera aggregata* by Prof. Guoyuan Zhu. Among them, linderalides A–C (**69–71**) are the first examples of disesquiterpenoid-geranylbenzofuranone hybrids which are directly linked by two C–C bonds, linderalide D (**72**), featuring the presence of an unusual linearly 6/6/5/6/6 pentacyclic system fused by a sesquiterpenoid unit and a geranylbenzofuranone moiety, represents another unprecedented carbon skeleton [32]. Aggreganoids A–F (**73–78**) are a new class of oligomeric sesquiterpenoids featuring the conjugation via different or identical sesquiterpenoid monomers through a carbon bridge [33]. **73** and **74** are two methine- or methylene-bridged sesquiterpenoid trimers possessing a unique C_{46} skeleton, **75–78** represent the first examples of carbon-bridged disesquiterpenoids with a C_{33} or C_{31} skeleton.

2.2.3 Asteraceae/Compositae

In addition to unusual solo-sesquiterpenoids, there are plenty of unique sesquitepenoid oligomers in Asteraceae. Dicarabrones A (**79**) and B (**80**) are skeletal epimers from *Carpesium abrotanoides* featuring a cyclopentane ring connecting with two sesquiterpenoid lactone units. Such an unusual carbon skeleton was presumably generated via a [3 + 2] cycloaddition reaction [34]. Carpedilactones A–D (**81–84**), isolated from *Carpesium faberi*, are four isomeric *exo*-Diels–Alder adducts conjugated via a eudesmanolide dienophile and a guaianolide diene [35]. Among them, **81–83** are the first three 2,4-linked *exo*-Diels–Alder adducts between a eudesmanolide dienophile and a guaianolide diene. It is a remarkable fact that either the 2,4-linkage of **81–83** or the 1,3-linkage of **84**, the four dimers possessing the same exostereochemistry, is different from the endoproducts of the Diels–Alder addition according to the "Endo-Rule". Tricarabrols A–C (**85–87**) are three sesquiterpenoid lactone trimers from *Carpesium faberi* [36]. Of which **85** and **86** are a pair of stereoisomers possessing a novel C_{44} skeleton via a methylene-tethered linkage among the sesquiterpenoid scaffolds. Besides the unique linkage of the cyclopentane ring, **87** also manifests a methylene bridge. Artesin A (**88**) was isolated from the aerial parts of *Artemisia sieversiana* as a rare symmetrical dimeric guaianolide, which was speculated to be biosynthesized from two monomeric sesquiterpenoid lactones through two [2 + 2] cycloaddition reactions. However, the fact that the two four-membered rings are present in **88** is very rare in plants due to high ring tension [37]. Artemisians A–D (**89–92**) were afforded from a well-known traditional Chinese medicine *Artemisia annua* (Qinghao) [38]. **89–92**, consisting of a $\Delta^{1(2), 4(5)}$ or a $\Delta^{1(5), 3(4)}$ guaianolide diene, represent the first examples of [4 + 2] Diels–Alder type adducts, which was presumably biosynthesized from a rare 1,10–4,5-diseco-guaianolide and a guaianolide diene. Arteannoide A (**93**), isolated from *Artemisia annua*, is the first sesquiterpenoid dimer composed of two cadinene sesquiterpenoid units linked by 6,8-dioxabicyclo[3.2.1]octan-7-one, an unusual fused ring system. Notably, artemisinin, possessing a peroxide bridge and potent antimalarial activity, is also an unusual sesquiterpenoid purified from Qinghao. The 2015 Nobel Prize in Physiology or Medicine was shared by Professor Youyou Tu for her impact research work on the development of artemisinin. To our knowledge, few isolation studies on Qinghao have been reported since the beginning of the new century, while the discovery of unprecedented **93** makes the research more distinct and precious [39]. Two Chinese scientists Profs. Weidong Zhang and Yang Ye contributed much to the discovery of unusual sesquiterpenoid oligomers in genus *Ainsliaea*. Ainsliadimer A (**94**), isolated from *Ainsliaea macrocephala*, represents a skeletal sesquiterpenoid dimer having a cyclopentane system connected by two monomeric sesquiterpenoid lactone units, [40] while ainsliatrimers A (**95**) and B (**96**) are the first two guaianolide-type sesquiterpenoid lactone trimers described from *Ainsliaea fulvioides* [41]. Investigation on *Ainsliaea fragrans* afforded four unprecedented sesquiterpenoid oligomers. Of which ainsliatriolides A (**97**) and B (**98**) are the first examples trimerized from guaianolide sesquiterpenoids through two different C–C linkages, [42] whereas ainsliatetramers A (**99**) and B (**100**), featuring a complex skeleton constructed by four sesquiterpenoid units via three different linkages, are the first examples of uncommon sesquiterpenoid tetramers, whose generation was proposed to be accomplished by a Michael addition, a regular and a hetero-Diels–Alder cycloaddition [43]. Ainsfragolide (**101**) is an unusual guaianolide sesquiterpene trimer generated with a novel C–C linkage at C2'–C15″, which may be biosynthesized prospectively through a further Michael addition [44]. Macrocephadiolides A (**102**) and B (**103**), featuring a rare 5,6-spirocyclic ketal lactone core and a C-15/C-15' linkage between a guaianolide and a 3,4-*seco*-guaianolide monomer, were isolated from *Ainsliaea macrocephala* [45]. Their absolute configurations were secured by single-crystal X-ray crystallography. Macrocephatriolides A and B (**104** and **105**), two novel guaiane-type sesquiterpene lactone trimers possessing unique linkage patterns, were identified from the whole plant of *Ainsliaea macrocephala* [46]. The trimeric architecture of **104** features a cyclohexene linkage and a methylene bridge, which were presumably constructed from three constitutive monomers via a Diels–Alder cycloaddition and a Michael addition, respectively. The three monomers of **105** were tethered by a 1,2-ethanediyl and a methylene linkage at the same timeInulajaponicolide A (**106**) is an undescribed unusual sesquiterpene trimer from *Inula japonica*, which is comprised by one xanthanolide and two guaianolide units via the linkage mode of C-11/C-3′ and C-11′/C-1″ characteristic of a Diels–Alder cycloaddition reaction [47]. Vlasoulamine A (**107**) is an unprecedented *N*-containing sesquiterpenoid lactone dimer from *Vladimiria souliei*, featuring a fully hydrogenated pyrrolo[2,1,5-*cd*]indolizine core and two sesquiterpenoid lactone units symmetrically located at each side [48]. The absolute configuration of **107** was secured by single-crystal X-ray diffraction analysis.

2.2.4 *Burseraceae*

Plant resins are defined primarily as lipophilic mixtures of volatile and nonvolatile terpenoids and/or phenolic compounds. Plants secrete resins usually for their protective benefits such as defending against herbivores, insects, and pathogens. *Resina Commiphora* is a plant resin secreted by the genus *Commiphora*, which has been used as a Chinese medicine for the treatment of blood stagnation.

Recently, Prof. Yongxian Cheng's group from China has focused on the biologically active compounds from *Resina Commiphora*, and about eight novel dimeric sesquiterpenoids (**108–115**) were successfully isolated. Commiphoroids A (**108**) and B (**109**) are stereoisomers of putative [2 + 4]-cycloaddition reactions. Commiphoroid C **110** is a trinorsesquiterpenoid dimer containing a 6/6/5/6/6/6 hexacyclic framework, while **111** features a 8-oxabicyclo[3.2.1]oct-6-ene skeletal core [49]. Commiphoratones A (**112**) and B (**113**) both contain a unique 6/6/5/5/6(5)/6 heptacyclic architecture and represent an unusual pattern of dimerization between two types of sesquiterpenoids [50]. Moreover, **112** has a saddle shape. Commiphorine A (**114**) is an unprecedented heterodimer conjugated by a guaiane- and a germacrane-type sesquiterpenoid via the formation of a four-membered spirocycle to create a 5/7(5)/4/10/5 ring system. There exists seven continuous chiral centers and two middle rings (seven- and ten-membered) in **114** which brings a great challenge for its total chemical synthesis. In contrast to **114**, the fusion of two cadinane-type sesquiterpenoid units via an unusual 1,4-dioxin core makes commiphorine B (**115**) a 6/6/5/6/6/6 backbone [51].

2.2.5 *Valerianaceae*

Volvalerelactone B (**116**) is a dimeric sesquiterpenoid connected via a unique five-member ring system and features an unprecedented 3/7/6 tricyclic ring system [10]. Another three unusual sesquiterpenoid dimers, nardochinoids A–C (**117–119**), were isolated from *Nardostachys chinensis* [52]. Their absolute configurations were all characterized by X-ray single-crystal diffraction analysis. Among them, **117** represents an unprecedented pattern of dimerization composed of two types of sesquiterpenes, uniquely defined by the presence of a new fused 3,8-dioxatricyclo[7.2.1.0^{1,6}]dodecane-11-one ring system. **118**, a unique sesquiterpene dimer with a pyridine core, is the first nitrogen containing nornardosinane-aristolane sesquiterpenoid conjugate. As for **119**, it is the first sesquiterpene dimer fusing a nornardosinane sesquiterpene to a nardosinane sesquiterpene via a five-membered *O*-heterocyclic ring to generate a surprising 6/6/5/6/6 polycyclic ring system.

2.2.6 *Zingiberaceae*

Involucratustones A–C (**120–122**) are three cadinane dimers from *Stahlianthus involucratus*, which feature multiple contiguous quaternary carbons [53]. **120** and **121** are rearranged homodimers of cadinane sesquiterpene fused with a unique fully substituted 1-oxaspiro[4.4]nonane core. This was observed for the first time in natural products. **122** is characteristic of a novel 3',4'-*seco*-cadinane-dimer.

2.2.7 *Annonaceae*

Xylopiana A (**123**), a unique dimeric guaiane, was isolated from *Xylopia vielana*. It possesses an unusual case-shaped pentacyclo[$5.2.1.0^{1,2}.0^{4,5'}0.0^{5,4'}$]decane-3,2'-dione core in the structure. Interestingly, a biomimetic conversion from vielanin F to **123** fulfilled by a photochemical [2 + 2] cycloaddition reaction [54].

2.2.8 *Myrtaceae*

meso-Eugenunilone A (**124**) and (±)-eugenunilone B (**125**) were isolated from the fruits of *Eugenia uniflora*. Compounds **124** and **125** were two unprecedented dimers with caged tricyclo[4.4.0.02,8]decane units [18].

References

1. Wu QX, Shi YP, Yang L. Unusual sesquiterpene lactones from *Ligularia virgaurea* spp. *oligocephala*. Org Lett. 2004;6:2313–2316.
2. Macías FA, Galindo JLG, Varela RM, Torres A, Molinillo JMG, Fronczek FR. Heliespirones B and C: two new plant heliespiranes with a novel spiro heterocyclic sesquiterpene skeleton. Org Lett. 2006;8:4513–6.
3. Karioti A, Skaltsa H, Linden A, Perozzo R, Brun R, Tasdemir D. Anthecularin: a novel sesquiterpene lactone from *anthemis auriculata* with antiprotozoal activity. J Org Chem. 2007;72:8103–6.
4. Fattorusso C, Stendardo E, Appendino G, Fattorusso E, Luciano P, Romano A, Taglialatela-Scafati O. Artarborol, a nor-caryophyllane sesquiterpene alcohol from *artemisia arborescens*. stereostructure assignment through concurrence of NMR data and computational analysis. Org Lett. 2007;9:2377–2380.
5. Ton That Q, Jossang J, Jossang A, Nguyen Kim PP, Jaureguiberry G. Wedelolides A and B: novel sesquiterpene δ-lactones, (9*R*)-eudesman-9,12-olides, from *Wedelia trilobata*. J Org Chem. 2007;72:7102–5.
6. Schmidt TJ. Novel *seco*-prezizaane sesquiterpenes from north *American Illicium* species. J Nat Prod. 1999;62:684–7.

7. Ma SG, Li M, Lin MB, Li L, Liu YB, Qu J, Li Y, Wang XJ, Wang RB, Xu S, Hou Q, Yu SS. Illisimonin A, a caged sesquiterpenoid with a tricyclo[5.2.1.$0^{1,6}$]decane skeleton from the fruits of *Illicium simonsii*. Org Lett. 2017;19:6160–6163.
8. Kubo M, Okada C, Huang JM, Harada K, Hioki H, Fukuyama Y. Novel pentacyclic *seco*-prezizaane-type sesquiterpenoids with neurotrophic properties from *Illicium jiadifengpi*. Org Lett. 2009;11:5190–3.
9. Xi FM, Ma SG, Liu YB, Li L, Yu SS. Artaboterpenoids A and B, bisabolene-derived sesquiterpenoids from *Artabotrys hexapetalus*. Org Lett. 2016;18:3374–7.
10. Wang PC, Ran XH, Luo HR, Hu JM, Chen R, Ma QY, Dai HF, Liu YQ, Xie MJ, Zhou J, Zhao YX. Volvalerelactones A and B, two new sesquiterpenoid lactones with an unprecedented skeleton from *Valeriana officinalis* Var. *latifolia*. Org Lett. 2011;13:3036–3039.
11. Chai T, Meng XH, Wang CB, Wang K, Ma LM, Shi YP, Yang JL. Narjatamolide, an unusual homoguaiane sesquiterpene lactone from *Nardostachys jatamansi*. J Org Chem. 2021;86:11006–10.
12. Yam-Puc A, Escalante-Erosa F, Pech-López M, Chan-Bacab MJ, Arunachalampillai A, Wendt OF, Sterner O, Peña-Rodríguez LM. Trinorsesquiterpenoids from the root extract of *Pentalinon andrieuxii*. J Nat Prod. 2009;72:745–8.
13. He JB, Luo J, Zhang L, Yan YM, Cheng YX. Sesquiterpenoids with New Carbon Skeletons from the Resin of *Toxicodendron vernicifluum* as new types of extracellular matrix inhibitors. Org Lett. 2013;15:3602–5.
14. Shen YH, Li SH, Li RT, Han QB, Zhao QS, Liang L, Sun HD, Lu Y, Cao P, Zheng QT. Coriatone and Corianlactone, Two novel sesquiterpenes from *Coriaria nepalensis*. Org Lett. 2004;6:1593–5.
15. Han JJ, Zhang JZ, Zhu RX, Li Y, Qiao YN, Gao Y, Jin XY, Chen W, Zhou JC, Lou HX. Plagiochianins A and B, two *ent*-2,3-*seco*-aromadendrane derivatives from the Liverwort *Plagiochila duthiana*. Org Lett. 2018;20:6550–3.
16. Ainge GD, Gerard PJ, Hinkley SFR, Lorimer SD, Weavers RT. Hodgsonox, a new class of sesquiterpene from the Liverwort *Lepidolaena hodgsoniae*. Isolation directed by insecticidal activity. J Org Chem. 2001;66:2818–2821.
17. Liu Y, Liu F, Qiao MM, Guo L, Chen MH, Peng C, Xiong L. Curcumanes A and B, two bicyclic sesquiterpenoids with significant vasorelaxant activity from *Curcuma longa*. Org Lett. 2019;21:1197–201.
18. Chen M, Cao JQ, Ang S, Zeng TN, Li NP, Yang TJ, Liu JS, Wu Y, Ye WC, Wang L. Eugenunilones A-H: rearranged sesquiterpenoids from *Eugenia uniflora*. Org Chem Front. 2022;9:667–75.
19. Jiao SG, Su GZ, Zhou XC, Ge FX, Liu CX, Zhang RF, Peng B, Chen S, Huang LQ, Tu PF, Chai XY. Three pairs of enantiomeric sesquiterpenoids from *Syringa pinnatifolia*. J Org Chem. 2021;86:7263–70.
20. Yong JY, Li WR, Wang XJ, Su GZ, Li M, Zhang JP, Jia HL, Li YH, Wang RB, Gan ML, Ma SG. Illihenin A: an antiviral sesquiterpenoid with a cage-like tricyclo[6.2.2.$0^{1,5}$]dodecane skeleton from *Illicium henryi*. J Org Chem. 2021;86:2017–2022.
21. Zhao P, Li ZY, Qin SY, Xin BS, Liu YY, Lin B, Yao GD, Huang XX, Song SJ. Three unusual sesquiterpenes with distinctive ring skeletons from *Daphne penicillata* uncovered by molecular networking strategies. J Org Chem. 2021;86:15298–306.
22. Yang SP, Gao ZB, Wang FD, Liao SG, Chen HD, Zhang CR, Hu GY, Yue JM. Chloraholides A and B, two potent and selective blockers of the potassium channel isolated from *Chloranthus holostegius*. Org Lett. 2007;9:903–6.
23. He XF, Zhang S, Zhu RX, Yang SP, Yuan T, Yue JM. Sarcanolides A and B: two sesquiterpenoid dimers with a nonacyclic scaffold from *Sarcandra hainanensis*. Tetrahedron. 2011;67:3170–4.
24. Yuan T, Zhu RX, Yang SP, Zhang H, Zhang CR, Yue JM. Serratustones A and B representing a new dimerization pattern of two types of sesquiterpenoids from *Chloranthus serratus*. Org Lett. 2012;14:3198–201.
25. Zhou B, Liu QF, Dalal S, Cassera MB, Yue JM. Fortunoids A-C, three sesquiterpenoid dimers with different carbon skeletons from *Chloranthus fortunei*. Org Lett. 2017;19:734–7.

26. Fan YY, Sun YL, Zhou B, Zhao JX, Sheng L, Li JY, Yue JM. Hedyorienoids A and B, two sesquiterpenoid dimers featuring different polycyclic skeletons from *Hedyosmum orientale*. Org Lett. 2018;20:5435–8.
27. Bai B, Ye SX, Yang DP, Zhu LP, Tang GH, Chen YY, Li GQ, Zhao ZM. Chloraserrtone A, a sesquiterpenoid dimer from *Chloranthus serratus*. J Nat Prod. 2019;82:407–11.
28. Chi J, Xu W, Wei S, Wang X, Li J, Gao H, Kong L, Luo J. Chlotrichenes A and B, two lindenane sesquiterpene dimers with highly fused carbon skeletons from *Chloranthus holostegius*. Org Lett. 2019;21:789–92.
29. Yan H, Qin XJ, Li XH, Yu Q, Ni W, He L, Liu HY. Japonicones A-C: three lindenane sesquiterpenoid dimers with a 12-membered ring core from *Chloranthus japonicus*. Tetrahedron Lett. 2019;60:713–7.
30. Zhang DY, Wang XX, Wang YN, Wang M, Zhuang PY, Jin Y, Liu H. Nine sesquiterpenoid dimers with four unprecedented types of carbon skeleton from *Chloranthus henryi* var. *hupehensis*. Org Chem Front. 2021;8:4374–4386.
31. Zhou JS, Liu QF, Zimbres Flavia M, Butler Joshua H, Cassera Maria B, Zhou B, Yue JM. Trichloranoids A-D, antimalarial sesquiterpenoid trimers from *Chloranthus spicatus*. Org Chem Front. 2021;8:1795–801.
32. Liu X, Yang J, Yao XJ, Yang X, Fu J, Bai LP, Liu L, Jiang ZH, and Zhu GY. Linderalides A–D, disesquiterpenoid– geranylbenzofuranone conjugates from Lindera aggregata. J Org Chem 2019;84:8242–8247.
33. Liu X, Yang J, Fu J, Yao XJ, Wang JR, Liu L, Jiang ZH, Zhu GY. Aggreganoids A-F, carbon-bridged sesquiterpenoid dimers and trimers from *Lindera aggregata*. Org Lett. 2019;21:5753–6.
34. Wu J, Tang C, Chen L, Qiao Y, Geng M, Ye Y. Dicarabrones A and B, a pair of new epimers dimerized from sesquiterpene lactones via a [3 + 2] cycloaddition from *Carpesium abrotanoides*. Org Lett. 2015;17:1656–9.
35. Yang YX, Shan L, Liu QX, Shen YH, Zhang JP, Ye J, Xu XK, Li HL, Zhang WD. Carpedilactones A-D, four new isomeric sesquiterpene lactone dimers with potent cytotoxicity from *Carpesium faberi*. Org Lett. 2014;16:4216–9.
36. Yuan J, Wen X, Ke CQ, Zhang T, Lin L, Yao S, Goodpaster JD, Tang C, Ye Y. Tricarabrols A-C, three anti-inflammatory sesquiterpene lactone trimers featuring a methylene-tethered linkage from *Carpesium faberi*. Org Chem Front. 2020;7:1374–82.
37. Zhou XD, Chai XY, Zeng KW, Zhao MB, Jiang Y, Tu PF. Artesin A, a new cage-shaped dimeric guaianolide from *Artemisia sieversiana*. Tetrahedron Lett. 2015;56:1141–3.
38. Xue GM, Han C, Chen C, Li LN, Wang XB, Yang MH, Gu YC, Luo JG, Kong LY. Artemisians A-D, diseco-guaianolide involved heterodimeric [4 + 2] adducts from *Artemisia argyi*. Org Lett. 2017;19:5410–3.
39. Qin DP, Pan DB, Xiao W, Li HB, Yang B, Yao XJ, Dai Y, Yu Y, Yao XS. Dimeric cadinane sesquiterpenoid derivatives from *Artemisia annua*. Org Lett. 2018;20:453–6.
40. Wu ZJ, Xu XK, Shen YH, Su J, Tian JM, Liang S, Li HL, Liu RH, Zhang WD. Ainsliadimer A, a new sesquiterpene lactone dimer with an unusual carbon skeleton from *Ainsliaea macrocephala*. Org Lett. 2008;10:2397–400.
41. Wang Y, Shen YH, Jin HZ, Fu JJ, Hu XJ, Qin JJ, Liu JH, Chen M, Yan SK, Zhang WD. Ainsliatrimers A and B, the first two guaianolide trimers from *Ainsliaea fulvioides*. Org Lett. 2008;10:5517–20.
42. Zhang R, Tang C, Liu HC, Ren Y, Xu C, Ke CQ, Yao S, Huang X, Ye Y. Ainsliatriolides A and B, two guaianolide trimers from *Ainsliaea fragrans* and their cytotoxic activities. J Org Chem. 2018;83:14175–80.
43. Zhang R, Tang C, Liu HC, Ren Y, Ke CQ, Yao S, Cai Y, Zhang N, Ye Y. Tetramerized sesquiterpenoid ainsliatetramers A and B from *Ainsliaea fragrans* and their cytotoxic activities. Org Lett. 2019;21:8211–4.
44. Ding N, Wang JY, Liu J, Zhu YJ, Hou SR, Zhao HM, Yang YS, Chen XB, Hu LH, Wang XC. Cytotoxic guaianolide sesquiterpenoids from *Ainsliaea fragrans*. J Nat Prod. 2021;84:2568–74.
45. Ren YM, Zhou SZ, Zhang T, Qian M, Zhang R, Yao S, Zhu H, Tang C, Lin L, Ye Y. Targeted isolation of two disesquiterpenoid macrocephadiolides A and B from *Ainsliaea macrocephala* using a molecular networking-based dereplication strategy. Org Chem Front. 2020;7:1481–9.

46. Ren YM, Zhang R, Feng ZL, Ke CQ, Yao S, Tang CP, Lin LG, Ye Y. Macrocephatriolides A and B: two guaianolide trimers from *Ainsliaea macrocephala* as PTP1B inhibitors and insulin sensitizers. J Org Chem. 2021;86:17782–9.
47. Jin Q, Lee JW, Jang H, Lee HL, Kim JG, Wu W, Lee D, Kim EH, Kim Y, Hong JT, Lee MK, Hwang BY. Dimeric- and trimeric sesquiterpenes from the flower of *Inula japonica*. Phytochemistry. 2018;155:107–13.
48. Wu Z, Wang Q, Wang J, Dong H, Xu X, Shen Y, Li H, Zhang W. Vlasoulamine A, a neuroprotective [3.2.2]cyclazine sesquiterpene lactone dimer from the roots of *Vladimiria souliei*. Org Lett 2018;20:7567–7570.
49. Liu JW, Zhang MY, Yan YM, Wei XY, Dong L, Zhu YX, Cheng YX. Characterization of sesquiterpene dimers from *Resina commiphora* that promote adipose-derived stem cell proliferation and differentiation. J Org Chem. 2018;83:2725–33.
50. Liu JW, Liu Y, Yan YM, Yang J, Lu XF, Cheng YX. Commiphoratones A and B, two sesquiterpene dimers from *Resina commiphora*. Org Lett. 2018;20:2220–3.
51. Dong L, Qin DP, Di QQ, Liu Y, Chen WL, Wang SM, Cheng YX. Commiphorines A and B, unprecedented sesquiterpenoid dimers from *Resina commiphora* with striking activities on anti-inflammation and lipogenesis inhibition. Org Chem Front. 2019;6:3825–33.
52. Shen XY, Qin DP, Zhou H, Luo JF, Yao YD, Lio CK, Li HB, Dai Y, Yu Y, Yao XS. Nardochinoids A-C, three dimeric sesquiterpenoids with specific fused-ring skeletons from *Nardostachys chinensis*. Org Lett. 2018;20:5813–6.
53. Li QM, Luo JG, Zhang YM, Li ZR, Wang XB, Yang MH, Luo J, Sun HB, Chen YJ, Kong LY. Involucratustones A-C: unprecedented sesquiterpene dimers containing multiple contiguous quaternary carbons from *Stahlianthus involucratus*. Chem Eur J. 2015;21:13206–9.
54. Zhang YL, Zhou XW, Wang XB, Wu L, Yang MH, Luo J, Yin Y, Luo JG, Kong LY. Xylopiana A, a dimeric guaiane with a case-shaped core from *Xylopia vielana*: structural elucidation and biomimetic conversion. Org Lett. 2017;19:3013–6.

Chapter 3
Classification of Diverse Novel Diterpenoids

Diterpenoids serve as a vast source of terpenoidal compounds with a broad range of chemical and biological diversity and as well as providing a rich source of new drugs and therapeutic agents. About 208 new new skeletal diterpene compounds were isolated from 26 families in the last 23 years (1999–2021) (**126–165** Fig. 3.1, **166–174** Fig. 3.2, **175–185** Fig. 3.3, **186–205** Fig. 3.4, **206–223** Fig. 3.5, **224–242** Fig. 3.6, **243–256** Fig. 3.7, **257–269** Fig. 3.8, **270–296** Fig. 3.9, **297–333** Fig. 3.10; Table A2). Euphorbiaceae and Labiaceae accounted for a large proportion of new skeleton diterpene natural products, especially, the novel diterpenoids from *Isodon* species have been deeply investigated by Chinese scholar Sun Handong and his research group, their researchs have set a good example for those who study natural products. Here we focuse on their chemistry, distribution, and biological activities, as well as their structure characteristics.

3.1 Euphorbiaceae

3.1.1 Euphorbia

So far, the proportion of skeletal diterpenoids isolated from *Euphorbia* is the largest, which means that, on the one hand, this genus is rich of diterpenoids, on the other hand, the plants in this genus have attracted considerable attention. Interestingly, lagaspholones A and B (**126** and **127**), possessing a rare jatropholane-type skeleton and characteristic of a 5/6/7/3 fused ring system, were isolated from *Euphorbia lagascae* [1]. Heliojatrones A and B (**128** and **129**) are two jatrophane derived diterpenoids with an unprecedented *trans*-bicyclo[8.3.0]tridecane

Y. Cheng and D. Qin, *Novel Plant Natural Product Skeletons*,
https://doi.org/10.1007/978-981-99-7329-3_3

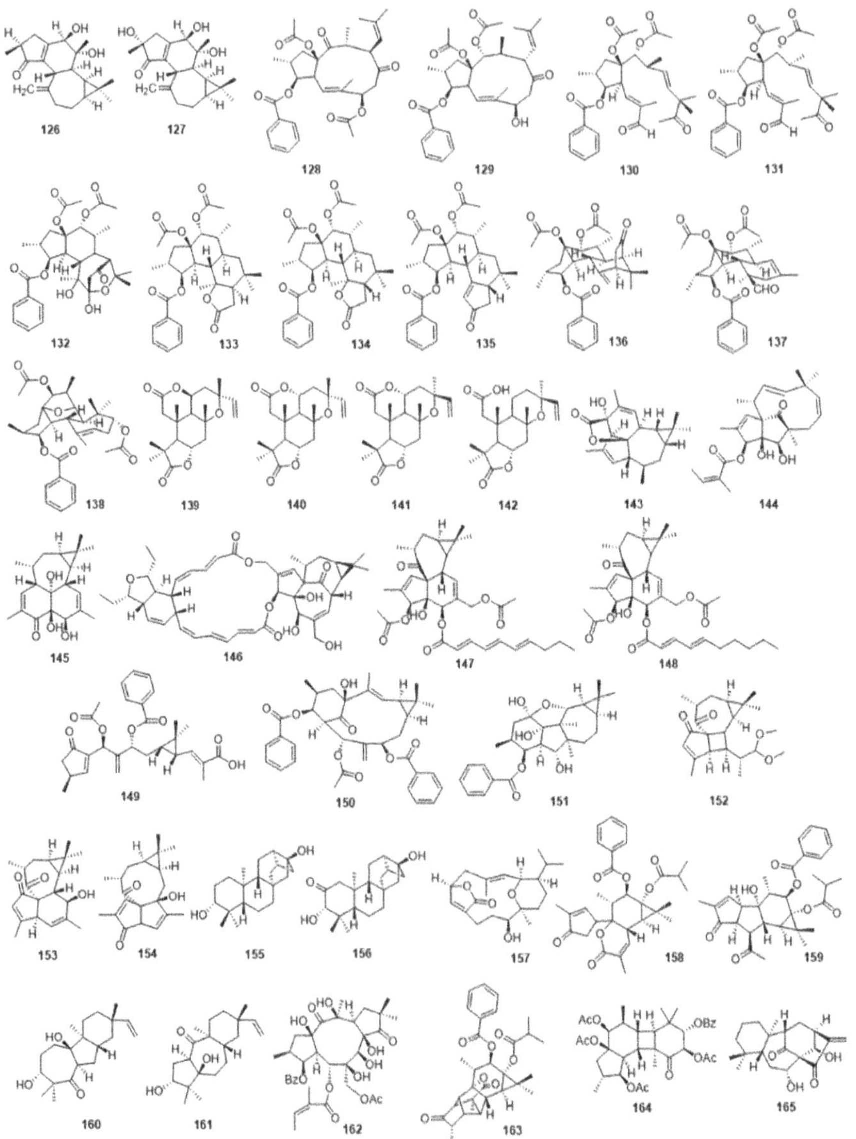

Fig. 3.1 Structures of new skeleton diterpenoids **126–165**

core, they were isolated from *Hedychium forrestii* and their absolute configurations were determined by electronic circular dichroism calculations and X-ray diffraction analyses [2]. Secoheliosphanes A (**130**) and B (**131**) and secoheliospholane A (**132**) are three diterpenoids from *Euphorbia helioscopia* possessing

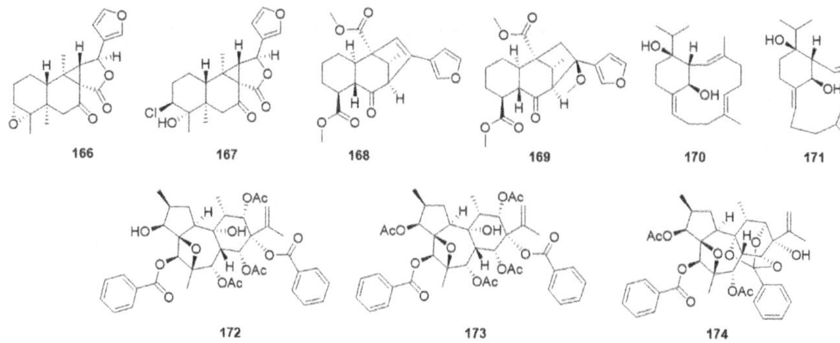

Fig. 3.2 Structures of new skeleton diterpenoids **166–174**

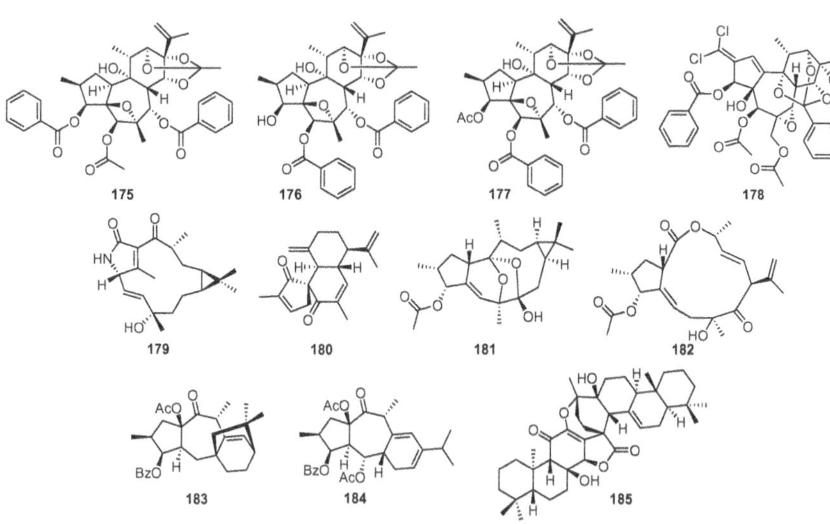

Fig. 3.3 Structures of new skeleton diterpenoids **175–185**

an unusual 7,8-*seco*-jatrophane skeleton and an unprecedented 9,10-*seco*-7,10-epoxyjatropholane skeleton, respectively [3]. Euphopias A−C (**133**−**135**), isolated from *Euphorbia helioscopia*, are another three rearranged jatrophane-type diterpenoids containing a ricyclo[8.3.0.$0^{2,7}$]tridecane core (**133** and **134**) and a tetracyclo-[11.3.0.$0^{2,10}$.$0^{3,7}$]hexadecane motif (**135**) [4]. Euphopia D−F (**136**−**138**), three novel diterpenoids were isolated from Euphorbia helioscopia L. **136** was defined by a unique caged tetracyclic[10.2.1.$0^{2,10}$.$0^{5,9}$]pentadecane core skeleton, **137** was the first example of an 8,9-norditerpenoid, possessing a tricyclic [8.3.0.$0^{2,8}$]tridecane core skeleton, and **138** had a tricyclic[10.3.0.$0^{2,8}$] pentadecane core skeleton [5]. Excolide A (**139**), 11-*epi*-excolide A (**140**), and 11,13-di-*epi*-excolide A (**141**), three tetracyclic 2,3-*seco*-labdanoids, and excolide B (**142**), a tricyclic 2,3-*seco*-labdanoid, are a new class of skeletal secolabdanoids, they were isolated from the

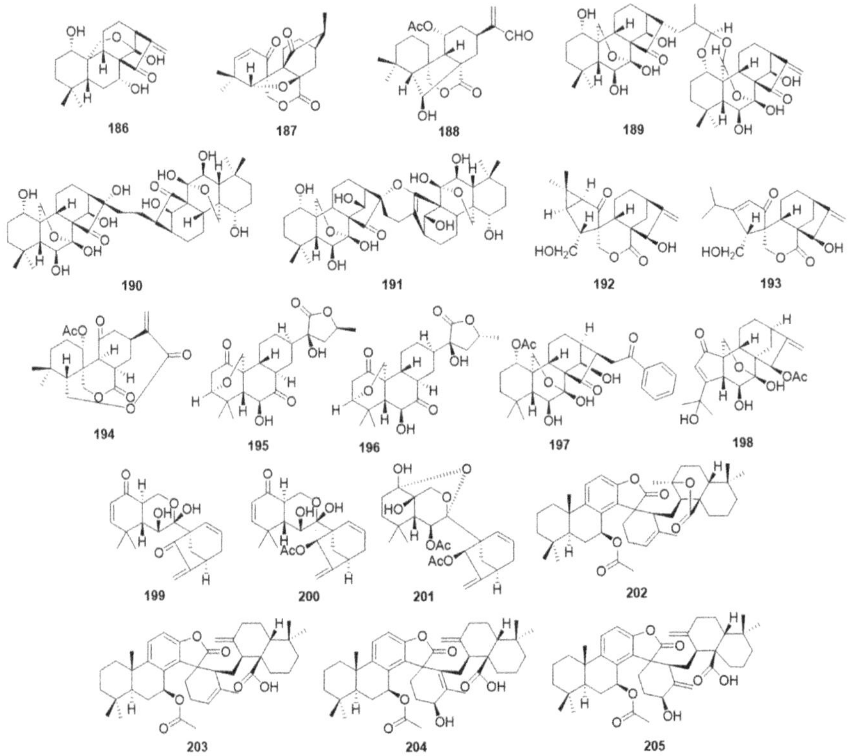

Fig. 3.4 Structures of new skeleton diterpenoids **186–205**

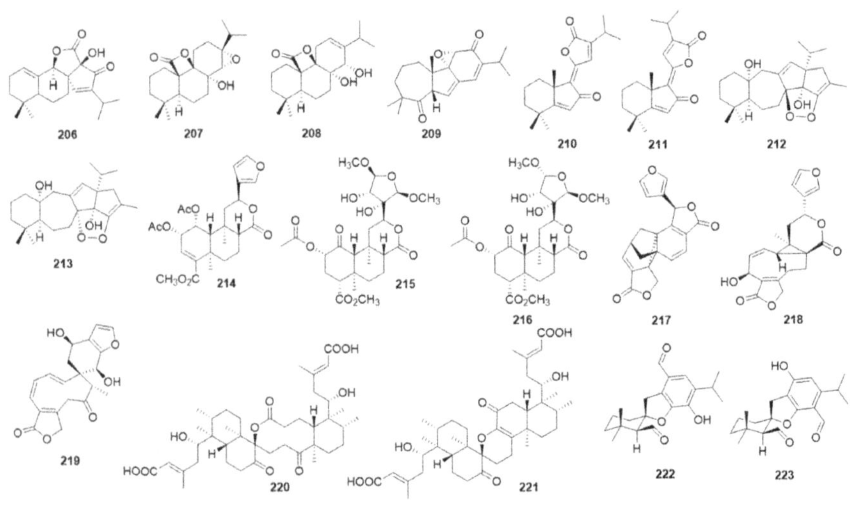

Fig. 3.5 Structures of new skeleton diterpenoids **206–223**

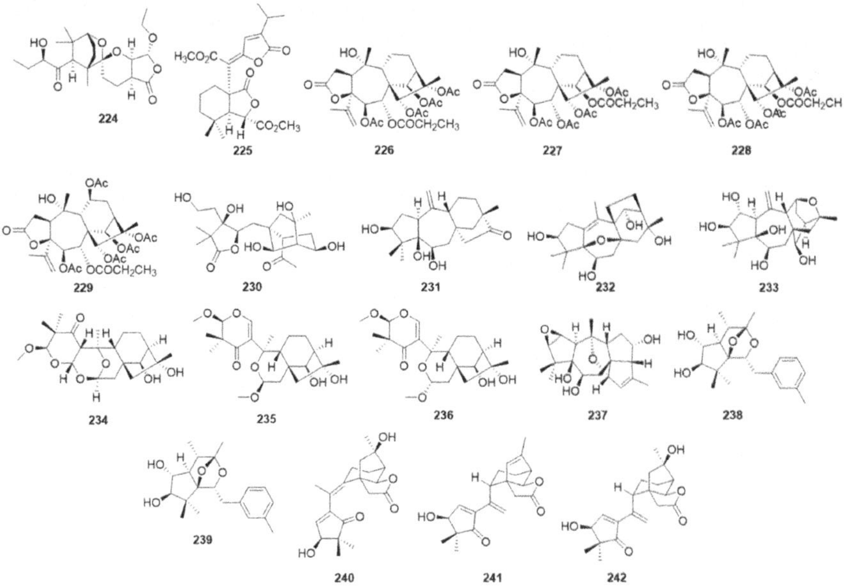

Fig. 3.6 Structures of new skeleton diterpenoids **224–242**

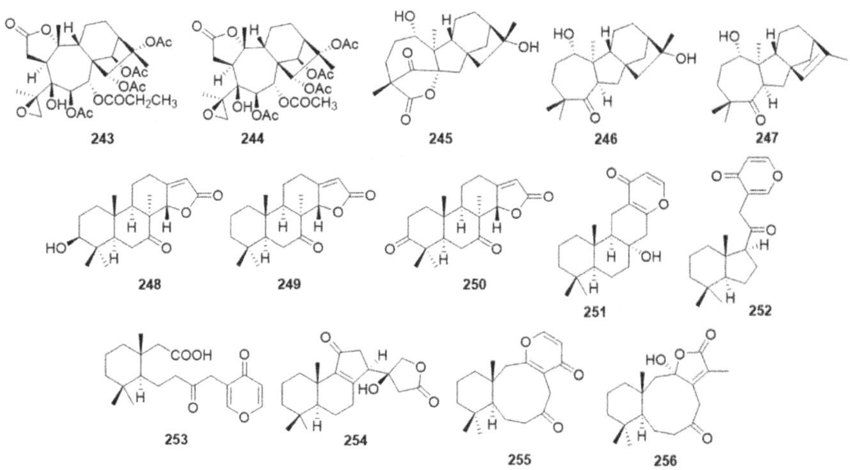

Fig. 3.7 Structures of new skeleton diterpenoids **243–256**

stems of *Excoecaria agallocha* [6]. Euphorikanin A (**143**) represents a special diterpenoid lactone having a fused 5/6/7/3 ring system from *Euphorbia kansui*, single-crystal X-ray diffraction analysis was used to assign its absolute stereochemistry [7]. Pepluanols C and D (**144** and **145**) are two different skeletal diterpenoids from *Euphorbia peplus* [8]. **144** features the presence of an unprecedented 5/5/10

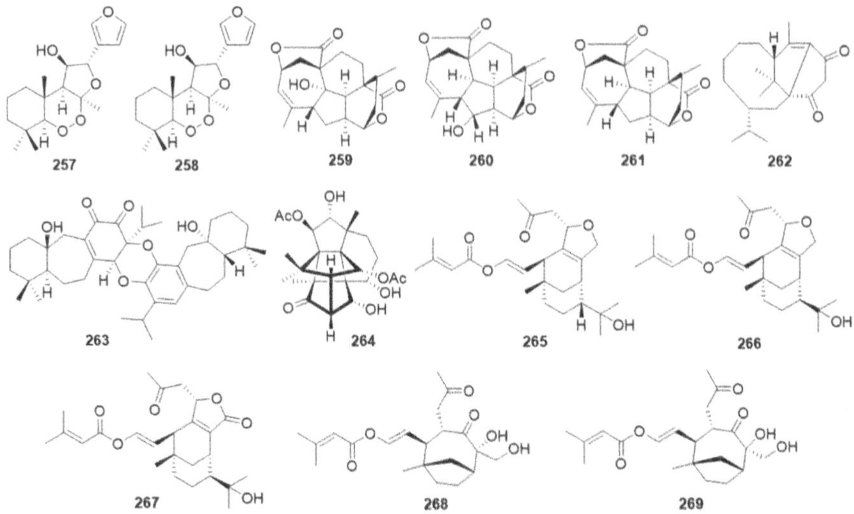

Fig. 3.8 Structures of new skeleton diterpenoids **257–269**

with *out,out*-[7.2.1]bicylcododecane core, whereas **145** possesses a 6/6/7/3 fused-ring skeleton. Euphorkanlide A (**146**), a highly modified ingenane diterpenoid, was isolated from the roots of *Euphorbia kansuensis* characteristic of a C_{24} appendage forming an additional hexahydroisobenzofuran-fused 19-membered macrocyclic *bis*-lactone ring system [9]. Euphorstranoids A (**147**) and B (**148**), two highly rearranged ingenane diterpenoids with an unusual 5/6/7/3 carbon ring system, were isolated from *Euphorbia stracheyi* [10]. Lathyranoic acid A (**149**) is the first seco-lathyrane diterpenoid with an unprecedented skeleton from *Euphorbia lathyris*, and Baeyer–Villiger oxidation was considered to be a key step in its biogenic pathway [11]. Lathyranone A (**150**) is a novel diterpenoid from *Euphorbia lathyris* with a rearrangement skeleton characteristic of the presence of an unprecedented cyclohexanone moiety in the structure [12]. Another novel diterpenoid, euphorbactin (**151**) with an unprecedented 6/5/7/3 fused-ring skeleton, was isolated from the roots of *Euphorbia micractina* [13]. Pepluacetal (**152**) and pepluanols A and B (**153** and **154**), three highly modified diterpenoids respectively exhibiting unprecedented 5/4/7/3, 5/6/7/3, and 5/5/8/3 ring systems, were isolated from *Euphorbia peplus* [14]. Another two interesting novel rearranged trachylobane diterpenoids, wallichanol A (**155**) and wallichanol B (**156**), were isolated from the roots of *Euphorbia wallichii* [15]. They are characteristic of an unprecedented pentacyclic skeleton named wallichane with a cyclobutane ring in the molecule. Quorumolide A (**157**), isolated from *Euphorbia antiquorum*, is so far the first example of a cembranoid embedding an α,β-unsaturated-γ-lactone and a tetrahydro-2*H*-pyran motif within the 14-membered ring [16]. It is particularly noteworthy that the stereochemistry at C-2 and C-12 in the pyran ring is opposite to those of marine-derived cembranoids. Euphordraculoate A (**158**), an unprecedented diterpenoid lactone, which contains a

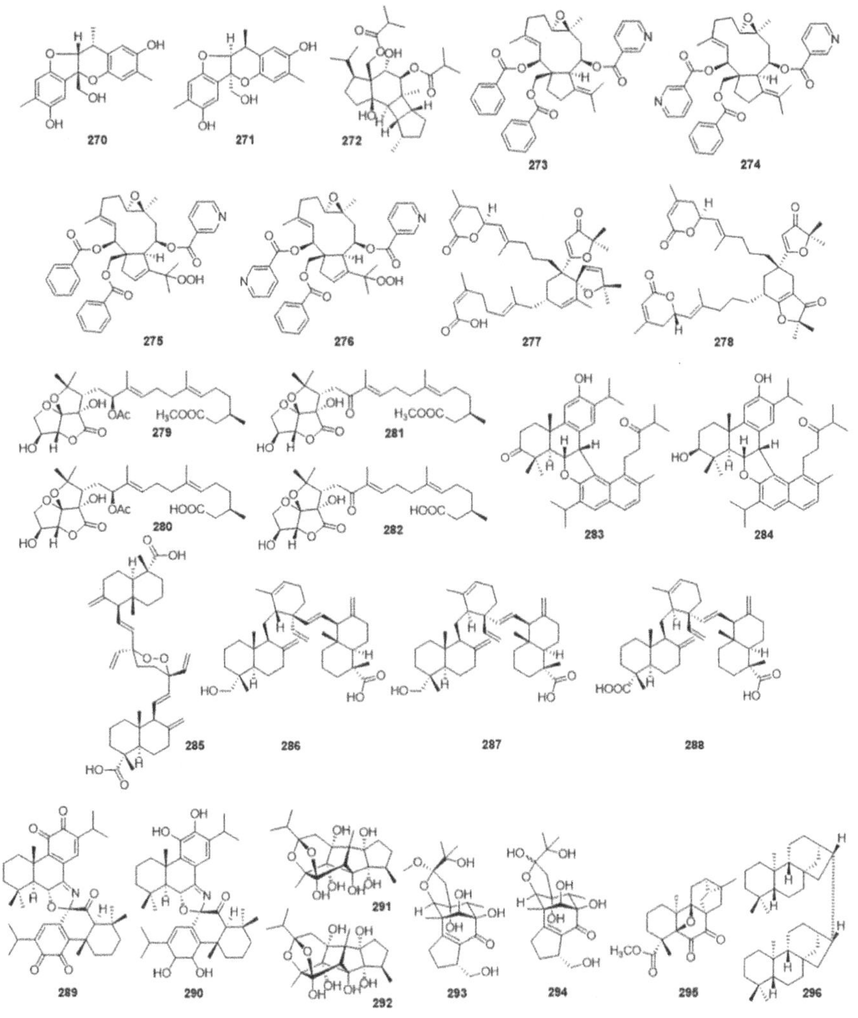

Fig. 3.9 Structures of new skeleton diterpenoids **270–296**

novel 6/6/3-fused ring system with a 2-methyl-2-cyclopentenone moiety, along with euphordraculoate B (**159**) with an unprecedented 5/5/6/3-fused tetracyclic ring core, were characterized as the secondary metabolites of *Euphorbia dracunculoides* [17]. Pedrolide (**160**), a diterpenoid with an unprecedented carbon skeleton, pedrolane, containing a bicycle[2.2.1]heptane system, was isolated from *Euphorbia pedroi* [18]. Two *ent*-rosane diterpenoids euphomilones A (**161**) and B (**162**) were discovered from *Euphorbia milii.* Of note, **161** is the first diterpenoid possessing a 7/5/6 tricyclic system and **162** represents the first example of a rosane-type diterpenoid featuring a 5/7/6 fused ring system [19]. Pre-segetanin (**163**), isolated from the sea spurge *Euphorbia paralias*, has an unprecedented carbon skeleton which provides

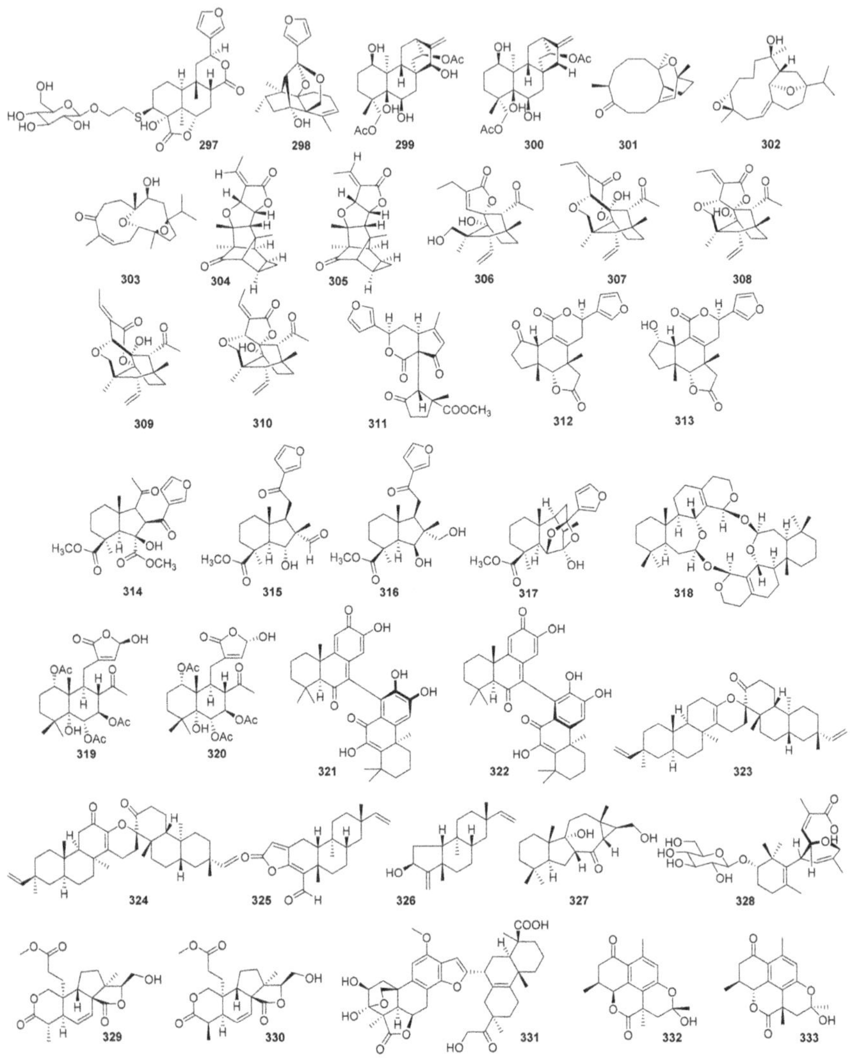

Fig. 3.10 Structures of new skeleton diterpenoids **297**–**333**

a first insight into the biosynthesis of diterpenoids with a segetane skeleton [20]. Gaditanone (**164**), possessing an unprecedented 5/6/4/6-fused gaditanane tetracyclic ring diterpenoidal skeleton was isolated from *Euphorbia gaditana* [21]. Interestingly, terpenoids with a four-membered ring are so far very scarce in the literature. Crokonoid A (**165**) is another highly rearranged diterpenoid featuring a dual-bridged tricyclo[4.4.1.$1^{1,4}$]dodecane-2,11-dione ring system, which was characterized from *Croton kongensis* [22].

3.1.2 *Croton*

Laevinoids A (**166**) and B (**167**) are the first examples of a new rearranged *ent*-clerodane skeleton with a unique 3/5 bicyclic ring system in the clerodane family [23]. It is noteworthy that **166** bearing the new skeleton was produced by *Croton laevigatus* as the major compound with a high yield (0.012% of dried plant material). **167** represents the first chlorinated member of the clerodane family. Crotonpenoids A (**168**) and B (**169**) are highly modified clerodane diterpenoids from *Croton yanhuii* with a new 10-(butan-2-yl)-1,6,12-trimethyltricyclo[7.2.1.$0^{2,7}$]dodecane backbone [24]. Single-crystal X-ray diffraction was used to clarify their absolute configurations. (+)- and (−)-Mangelonoids A and B (**170** and **171**), two pairs of novel macrocyclic diterpenoid enantiomers featuring an unprecedented bicyclo[9.3.1]-pentadecane core and a rare bridgehead double bond, was isolated from *Croton mangelong*, and their absolute configurations were unambiguously established using X-ray crystallography and electronic circular dichroism analysis [25]. Trigochinins A–C (**172–174**) are three highly oxygenated diterpenoids. The most interesting structural feature of **172–174** is the oxetane ring that formed between two oxygenated quaternary carbons (C-4 and C-6), which is unprecedented in the family of daphnane diterpenoids until quite recently a similar 4,6-oxetane ring with a different stereochemistry at C-6 of trigonothyrins A–C was reported [26].

3.1.3 *Trigonostemon*

Three highly functionalized daphnane diterpenoids, trigonothyrins A–C (**175–177**), were isolated from *Trigonostemon thyrsoideum*, which represent the first examples of daphnanes bearing an oxygen-bridged four-membered-ring system and a linkage mode of 12,13,14-orthoester in the backbone [27]. Trigocherrin A (**178**) is characteristic of an α,β-unsaturated dichlorovinyl moiety from *Trigonostemon cherrieri*, representing the first member of a new class of chlorinated diterpenoids [28].

3.1.4 *Jatropha*

Jatrophalactam (**179**), a novel diterpenoid lactam possessing an unprecedented 5/13/3 tricyclic skeleton, was isolated from the roots of *Jatropha curcas* [29]. Spirocurcasone (**180**) was a diterpenoid isolated from *Jatropha curcas*, a plant extensively cultivated throughout the world, which possesses an unprecedented spirorhamnofolane skeleton [30]. Jatrofolianes A (**181**) and B (**182**), are two highly modified lathyrane diterpenoids from *Jatropha gossypiifolia* [31]. Of which **181** incorporates an unusual transannular 1,3-dioxolane moiety to form a unique 5/6/5/8/3 ring system, while **182** possesses a new 10,11:13,14-diseco-lathyrane skeleton with a 12-membered macrocyclic lactone ring in the structure. Euphohyrisnoids A (**183**) and B (**184**), two highly rearranged lathyrane diterpenoids featuring a unique tetracyclo[10.2.2.0^{1,10}.0^{3,7}]cetane and tricyclo[8.4.1.0^{3,7}]pentadecane skeleton, respectively, were isolated from the seeds of *Euphorbia lathyris* [32]. Fischdiabietane A (**185**) was isolated from the roots of *Euphorbia fischeriana*, which is a novel asymmetric diterpenoid dimer with a unique nonacyclic 6/6/6/5/7/6/6/6/6 ring system possessing unprecedented 2-oxaspiro[4.5]decane-1-one and 2-oxabicyclo[3.2.2]nonane frameworks in D/E/F rings [33].

3.2 Lamiaceae

3.2.1 Isodon

Excisanin H (**186**) is a first skeletal *ent*-kaurene diterpenoid with an epoxy linkage between C-14 and C-20 isolated from *Rabdosia excise* [34]. Another two new skeletal kaurane diterpemoids: maoecrystal V (**187**), a novel C_{19} diterpenoid possessing a unique 6,7-*seco*-6-*nor*-15(8 → 9)-*abeo*-5,8-epoxy-*ent*-kaurane skeleton, [35] and maoecrystal Z (**188**), an unprecedented skeletal diterpenoid with a tetracyclic 6,7:8,15-di-*seco*-7,20-olide-6,8-cyclo-*ent*-kaurane motif in the structure, [36] were isolated from *Isodon eriocalyx* (Labiatae) leaves, a Chinese medicinal herb. Three new dimeric *ent*-kauranoids were isolated from *Isodon rubescens*: bisrubescensin A (**189**) contains an unprecedented C_{23} ent-kaurane unit, whereas bisrubescensins B and C (**190** and **191**) possesses a unique linkage of single carbon–carbon bond between two subunits [37]. Neolaxiflorin A (**192**), a skeletal *ent*-kaurane diterpenoid with a bicyclo[3.1.0]hexane unit, and its *seco*-derivative, neolaxiflorin B (**193**), were described from *Isodon eriocalyx* leaves [38]. It is worth noting that compounds hybridizing a bicyclo-[3.1.0]hexane unit are very rare in nature. **192** is the first example of *ent*-kaurane diterpenoid containing a rare 3/5/6/6/5 ring system, and **193** represents a new group of *ent*-kaurane diterpenoid having an α,β-unsaturated ketone unit in its five-membered ring and featuring a 5/6/6/5 ring system. Both **192** and **193** were isolated from *Isodon eriocalyx*. Ternifolide A (**194**) is a new diterpenoid featuring a unique 10-membered lactone ring formed between C-6 and C-15, its absolute configurations were confirmed by X-ray diffraction study [39]. Investigation of *Isodon eriocalyx* led to the isolation of two new diterpenoids, laxiflorolides A (**195**) and B (**196**), they are the first examples of *ent*-kauranoids bearing a unique C_{22} carbon framework [40]. Pharicusin A (**197**), composed of a benzoyl group coupled to a 7α,20-epoxy-*ent*-kauranoid, represents a skeletal C_{27} meroditerpenoid [41]. Maoeriocalysin A (**198**), a novel rearranged *ent*-kaurane diterpenoid with an unprecedented 4,5-*seco*-3,5-cyclo-7,20-epoxy-ent-kaurane scaffold, and maoeriocalysins B–D (**199–201**), another three rare 9,10-*seco*-7,20-epoxy-*ent*-kaurane diterpenoids, were also isolated from *Isodon eriocalyx* [42]. Hispidanins A–D (**202–205**), isolated from the rhizomes of *Isodon hispida*, are four unprecedented asymmetric dimeric diterpenoids [43].

3.2.2 *Salvia*

Przewalskin B (**206**), a skeletal and unique diterpenoid from *Salvia przewalskii*, is characteristic of a five-membered spiro ring and α-hydroxy-*β*-ketone lactone moieties [44]. Rubesanolides A (**207**) and B (**208**) were described from *Isodon rubescens*. They both possess an unprecedented *β*-lactone group formed between C-9 and C-20. In addition, the relationships between rings A/B and B/C are both *trans*-fused, representing the first examples of diterpenoids having such a conformation in the skeleton [45]. Salviyunnanone A (**209**), characterized from *Salvia yunnanensis*, features a diterpenoid possessing an unprecedented 7/5/6/3 ring system derived from a normal abietane skeleton [46]. Salpratlactones A (**210**) and B (**211**), a pair of abietane *cis–trans* tautomers from *Salvia prattii*, features a unique 6/5 carbocyclic rings bearing a *γ*-lactone ring via an exocyclic double bond [47]. Officinalins A (**212**) and B (**213**) are a pair of C_{23} terpenoidal epimers possessing a tetracyclic 6/7/5/5 system from *Salvia officinalis* [48]. They are characteristic of an unprecedented carbon skeleton with a tetracycline-[$9.6.0.0^{3,8}.0^{12,16}$]-heptadecane core and a peroxide bridge in the molecule. Salvinorin C (**214**) is a new *trans*-neoclerodane diterpenoid isolated from the leaves of the Mexican mint *Salvia divinorum* [49]. Salvinicins A (**215**) and B (**216**) were isolated from *Salvia divinorum* leaves and are structurally unique neoclerodanes possessing a 3,4-dihydroxy-2,5-dimethoxytetrahydrofuran ring [50]. Salvileucalin B (**217**), an unprecedented rearranged neoclerodane skeleton isolated from the aerial parts of *Salvia leucantha*, is characterized by a tricyclo[$3.2.1.0^{2,7}$]octane substructure [51]. Microphyllandiolide (**218**), isolated from *Salvia microphylla*, is an unprecedented rearranged clerodane-type diterpenoid featuring a 9/3 bicyclic ring system [52]. The absolute structure of **218** was confirmed by single crystal X-ray diffraction analysis. Teotihuacanin (**219**) was isolated from the leaves and flowers of *Salvia amarissima*, which possesses an unusual rearranged clerodane diterpene skeleton containing a spiro-10/6 bicyclic system [53]. Scospirosins A (**220**) and B (**221**) are two types of unprecedented dimeric *ent*-clerodanoids containing polycyclic 6/6/10/6 and 6/6/6/6/6ring systems, respectively, they were isolated from the aerial parts of *Isodon scoparius* [54]. Spirodesertols A (**222**) and B (**223**) are another highly modified spirocyclic diterpenoids with an unprecedented 6-isopropyl-3*H*-spiro[benzofuran2,1'-cyclohexane] motif [55].

3.2.3 *Leonurus*

Leonuketal (**224**) was isolated from the aerial parts of *Leonurus japonicus*, which possesses an unprecedented tetracyclic skeleton that comprised a bridged spiroketal moiety fused with a ketal-γ-lactone unit. The absolute configuration of **224** was determined by the modified Mosher's method and electronic circular dichroism calculations [56].

3.2.4 *Hyptis*

Hyptisolide A (**225**) was isolated from *Hyptis crenata*. It represents the first naturally occurring diterpenoid having the 7,8;11,12-bis-*seco*-abietane skeleton [57].

3.3 Ericaceae

3.3.1 *Rhododendron*

Secorhodomollolides A–D (**226–229**) are four highly acylated diterpenoids with a new 3,4-secograyanane skeleton [58]. They were isolated from *Rhododendron molle* and their structures were assigned by spectroscopic data interpretation and single-crystal X-ray crystallography. Mollolide A (**230**) was isolated from the roots of *Rhododendron molle* and features a new 1,10:2,3-disecograyanane skeleton [59]. Micranthanone A (**231**) represents a new tetracyclic diterpene carbon skeleton from *Rhododendron micranthum*, and its name "micranthane" was suggested for such a new skeleton type [60]. Two new grayanoids, mollanol A (**232**) possessing a new C-nor-D-homograyanane carbon skeleton and rhodomollein XXV (**233**) representing the first example of an 11,16-epoxygrayanane and featuring a caged oxatricyclo[3.3.1.$0^{3,7}$]nonane ring system, were described from *Rhododendron molle* [61]. Rhodomollacetal A (**234**) possesses a novel *cis*/*cis*/*cis*/*cis*-fused 6/6/6/6/5 pentacyclic ring system, featuring an unprecedented 11,13,18-trioxa-pentacyclo [8.7.1.$1^{5,8}$.$0^{2,8}$.$0^{12,17}$]nonadecane scaffold. Rhodomollacetals B and C (**235** and **236**) bear a rare 4-oxatricyclo[7.2.1.$0^{1,6}$]dodecane moiety and a 2,3-dihydro-4*H*-pyran-4-one unit. Diterpenoids **234–236** were all isolated from *Rhododendron molle* leaves [62]. Rhodomollanol A (**237**), was also described from the leaves of *Rhododendron molle*, which is found to possess a unique *cis*/*trans*/*trans*/*cis*/*cis*-fused 3/5/7/5/5/5 hexacyclic ring system and feature a rare 7-oxabicyclo[4.2.1]nonane core decorated with three cyclopentane units [63]. Mollebenzylanols A (**238**) and B (**239**) are highly modified and functionalized diterpenoids from *Rhododendron molle*, and they possess an unprecedented diterpene carbon skeleton featuring a unique 9-benzyl-8,10-dioxatricyclo[5.2.1.$0^{1,5}$]decane core [64]. From *Rhododendron molle*, Mollactones A–C (**240–242**) were also afforded [65]. They are naturally occurring highly functionalized 5,6-*seco*-grayanane diterpenoids characteristic of the presence of a unique 3-oxa-tricyclo[4,3,2,$0^{2,6}$]undecane motif.

3.3.2 *Pieris*

Pieris formosa is a poisonous plant to livestock and is used as an insecticide in rural areas of China. Pierisoids A and B (**243** and **244**), two novel polyesterified 3,4-*seco*-grayanane diterpenoids, were isolated from its flowers and their structures were identified by spectroscopic analysis and X-ray diffraction analysis [66]. Pierisketolide A (**245**) and pierisketones B and C (**246** and **247**) are diterpenoids with an unusual A-homo-B-nor-*ent*-kaurane carbon skeleton from *Pieris formosa* [67]. Structurally, **245** has a pentacyclic 7/5/5/6/5 ring system, and **246** and **247** bear a tetracyclic 7/5/6/5 ring system.

3.4 Zingiberaceae

3.4.1 *Amomum*

Kravanhins A–C (**248**–**250**) are isomerized isospongian diterpenoids from *Amomum kravanh*, possessing a *trans-anti-cis* fused tricyclic ring system [68]. Maximumins A–D (**251**–**254**) are four highly rearranged labdane-type diterpenoids and possess different new carbon skeletons [69]. Of which, **251** has a unique 6/6/6/6-fused ring system, and **252** and **253** are B-ring highly modified labdane diterpenoids. **254** represents the first example of 13(12 → 17)-*abeo*-12,17-cyclolabdane diterpenoid featuring a unique subunit of rings C and D. **251**–**254** were all described from *Amomum maximum*. Amomaxins A (**255**) and B (**256**), two skeletal diterpenoids from *Amomum maximum*, feature an unprecedented rearranged labdane norditerpene scarfold with a nine-membered ring [70]. Their absolute configurations were determined by CD data and single-crystal X-ray diffraction analysis.

3.4.2 *Hedychium*

Hedychins A (**257**) and B (**258**) were obtained from the rhizomes of *Hedychium forrestii*, representing unprecedented 6,7-dinorlabdane ditepenoids with a peroxide bridge [71]. Despite the fact that labdane-type norditerpenoids with the loss of the carbons from the side chain at C-9 are quite common, these are the first examples of the degradation of their C-6 and C-7 and new construction of a peroxide bridge between C-5 and C-8 in both **257** and **258**.

3.5 Taxaceae

3.5.1 *Cephalotaxus*

Mannolides A–C (**259–261**) are secondary metabolites from *Cephalotaxus mannii*, featuring a new intact C_{20} diterpenoid skeleton [72]. The discovery of **259–261** will shed new light on the biogenesis of *Cephalotaxus* troponoids, a rare class of antitumor C_{19} norditerpenoids.

3.5.2 *Amentotaxus*

Up to now, there is only one new skeleton deterpenoid being found from *Amentotaxus*. Amentoditaxone (**262**), possessing an unprecedented 5/5/9 fused ring in its structure, was isolated from *Amentotaxus formosana* [73].

3.5.3 *Torreya*

Grandione (**263**) is a heptacyclic rearranged abietane-type dimeric diterpenoid from *Torreya grandis*. **263** appears to be particularly interesting because the two monomers are linked through two ether linkages [74].

3.5.4 *Taxus*

Canataxpropellane (**264**) was isolated from the needles of the Canadian yew, *Taxus canadensis*. It is a novel taxane with an unprecedented 5/5/4/6/6/6-membered hexacyclic skeleton containing [3.3.2]propellane [75]. Biosynthesis of **264** was proposed to be derived from a normal 6/8/6-taxane via the intramolecular aldol reaction and [2 + 2] cycloaddition.

3.6 *Viburnum* (Caprifoliaceae)

Neovibsanin F (**265**), 14-*epi*-neovibsanin F (**266**), and 14-*epi*-18-oxoneovibsanin F (**267**), three new rearranged vibsane-type diterpenoids featuring a bicyclo[3.3.1]nonane ring have been isolated from *Viburnum suspensum* leaves [76]. Vibsatins A (**268**) and B (**269**), a pair of 14,15,16,17-tetranorvibsane-type diterpenoids that feature a bicyclo[4.2.1]nonane moiety formed by a new C-13/C-2 bond, was described from *Viburnum tinus* [77]. Their structures were elucidated by a combination of NMR spectra, optical rotation, and X-ray diffraction experiments.

3.7 Ranunculaceae

3.7.1 Paeonia

(+)- and (−)-Paeoveitol (**270** and **271**) from *Paeonia veitchii* are a pair of new norditerpene paeoveitol and feature an unexpected 6/5/6/6 fused tetracyclic ring system [78]. Another new diterpenoidal carbon skeleton from *Prunella vulgaris*, named vulgarisin A (**272**), possesses a rare 5/6/4/5 fused tetracyclic ring architecture, its absolute configuration was secured by single crystal X-ray diffraction analysis [79].

3.7.2 Nigella

In this genus, nigellamines A1 (**273**), A2 (**274**), B1 (**275**), and B2 (**276**) were isolated from the seeds of *Nigella sativa*. Structurally, they are novel dolabellane-type diterpenoidal alkaloids [80].

3.8 Meliaceae

3.8.1 Aphanamixis

Aphanamenes A (**277**) and B (**278**) are unprecedented acyclic diterpenoid dimers from *Aphanamixis grandifolia*, they are formed via a [4 + 2]-cycloaddition playing a pivotal role in the dimerization of the carbon skeleton of **277** and **278** [81].

3.8.2 Dysoxylum

Hongkonoids A–D (**279–282**) were isolated from *Dysoxylum hongkongense*, they are the first examples of ascorbylated terpenoids featuring a unique 5,5,5-fused tricyclic spiroketal butyrolactone moiety and diterpenoid-derived long chain [82].

3.9 Taxodiaceae

3.9.1 *Cunninghamia*

Bicunningines A (**283**) and B (**284**) are two unprecedented dimeric diterpenoids featuring a 2,3-dihydrofuran ring fusing an abietane and a 4,5-*seco*-abietane diterpenoid [83]. Their absolute configurations were determined by quantum chemical TDDFT ECD calculations, chemical transformations, and Mosher's method. Cunlanceloic acid A (**285**) was a symmetric terpenoid peroxide bearing an unprecedented dimeric labdane backbone linked through C-16/C-16' bonds, and cunlanceloic acids B–D (**286–288**) were three asymmetric diterpenoid dimers with a new carbon skeleton linked through C12/C-13' and C-15/C-16' bonds between two labdane units, they were isolated from the cones of *Cunninghamia lanceolata* [84].

3.9.2 *Taxodium*

Taxodikaloids A (**289**) and B (**290**) are two dimeric diterpenoids identified from the seeds of *Taxodium ascendens* [85]. Their structures feature a novel 2,3-dihydrofuranyl linkage between two abietane halves.

3.10 *Cinnamomum* (Lauraceae)

Cinnamomols A (**291**) and B (**292**) are two diterpenoids isolated from *Cinnamomum cassia*, representing an unprecedented diterpenoid carbon skeleton and featuring a cage-like, rigid, 5/5/5/5/5/6-fused hexacyclic ring system comprising a 9,10-dioxatricyclo[5.2.1.$0^{4,8}$]decane and a tricyclo[5.2.1.$0^{2,6}$]decane moieties [86]. Cassiabudanols A (**293**) and B (**294**) were isolated from cassia buds, they possess an unprecedented 11,14-*cyclo*-8,14:12,13-di-*seco*-isoryanodane (cassiabudane) carbon skeleton and feature a unique 3-oxatetracyclo-[6.6.1.$0^{2,6}$.$0^{10,14}$]pentadecane bridged system [87].

3.11 Annonaceae

3.11.1 Mitrephora

Mitrephorone A (**295**) was isolated and structurally identified from *Mitrephora glabra.* It possesses a hexacyclic ring system with adjacent ketone moieties and an oxetane ring, both of which are unprecedented among trachylobanes [88].

3.11.2 Annona

Annoglabayin (**296**) is a novel dimeric kaurane diterpenoid isolated from *Annona glabra* [89]. Interestingly, it contains a unique carbon bridge between two *nor-ent-*kaurane monomeric units.

3.12 Others

There are additional 15 plant families, and each of them contains a small number of new skeletal diterpenoids. Cordifolide A (**297**) is a novel unprecedented sulfur-containing clerodane diterpenoidal glycoside from *Tinospora cordifolia* (Menispermaceae) [90]. Scaparvin A (**298**) is a novel caged *cis*-clerodane diterpenoid from *Scapania parva* (Scapaniaceae) possessing an unprecedented C-6/C-11 bond and a ketal ring [91]. Atisanes 1 and 2 (**299** and **300**) is the first report of the atisane skeleton with an unprecedented level of oxygenation from a liverwort (Lepidolaenaceae).[92] populusone (**301**) was isolated from the exudates of *Populus euphratica*, which has an unprecedented, snail-shaped trinorditerpenoid skeleton [93]. Populusene A (**302**), an unprecedented carbon skeleton featuring a bicyclo[8.4.1]-pentadecane nucleus and a bridgehead double bond (anti-Bredt system), and populusin A (**303**), a cembrane-type diterpenoid possessing an uncommon dioxatricyclo[6.6.1.$1^{2,5}$]hexadecane scaffold, were isolated from the exudates of *Populus euphratica* [94]. Pallambins A (**304**) and B (**305**) are novel 19-*nor*-7,8-secolabdane diterpenoids with unprecedented tetracyclo[4.4.$0^{3,5}$.$0^{2,8}$]decane skeletons [95]. They were isolated from *Pallavicinia ambigua* (Pallaviciniaceae). Pallamolides A–E (**306–310**) are unprecedented labdane lactone from *pallavicinia ambigua* (*Pallaviciniaceae*), they possess a bicyclo[2.2.2]octane moiety in the molecule [96]. Hapmnioides A–C (**311–313**), isolated from *Haplomitrium mnioides*, are labdane-type diterpenoids with unprecedented scaffolds formed through cascade rearrangement [97]. Hypophyllins A–C (**314–316**) are also skeletal rearranged labdane-type diterpenoids. They were described from *Hypoestes phyllostachya* (Acanthaceae), hypophyllin D (**317**) is a caged labdane diterpenoid from *Hypoestes phyllostachya* (Acanthaceae) possessing a 8,9-dioxatricyclic[4.2.1.$1^{3,7}$]decane moiety [98]. Schaffnerine (**318**) has been isolated from *Acacia schaffneri* (Mimosaceae), which is an unprecedented macrocyclic dimeric diterpene containing a *C*2 symmetry axis [99]. Caesalminaxin A1 and A2 (**319** and **320**), isolated from *Caesalpinia minax* (Leguminosae), have an unprecedented carbon skeleton and were presumably derived from their analogue caesalminaxin B by cleavage of the C-13–C-14 bond [100]. (*M*)-Bicelaphanol A (**321**) and (*P*)-bicelaphanol A (**322**), two unprecedented dimeric trinorditerpenes existing as atropisomers representing the first examples of dimeric podocarpane-type trinorditerpenes, were isolated from *Celastrus orbiculatus* (Celastraceae) [101]. Tagalsins I (**323**) and J (**324**) are tetraditerpenoids with a novel bisdolabrane backbone from *Ceriops tagal* (Rhizophoraceae) [102]. Tagalide A (**325**) from *Ceriops tagal* (Rhizophoraceae) is the first C_{22} skeletal dolabrane with an unprecedented 5/6/6/6-fused tetracyclic core, whereas tagalol A (**326**) from *Ceriops tagal* (Rhizophoraceae) is the first 3-nordolabrane with a novel 5/6/6-fused tricarbocyclic scaffold [103]. Premnafulvol A (**327**) possesses an unprecedented 14,15-*cyclo*-C-homo-B-norisopimarane carbon core with a tetracyclic 6/5/7/3 ring system, which has been isolated from *Premna fulva* (Verbenaceae) [104]. From the stem bark of *Fraxinus sieboldiana* (Oleaceae), fraxinuacidoside (**328**) was isolated and it is a norditerpene glucopyranoside with a novel carbon skeleton [105]. Oliviformislactones A (**329**) and B (**330**) were isolated from the tuber of *Icacina oliviformis* (Icacinaceae) [106]. They are the first examples of rearranged 3,4-*seco*-pimarane possessing a 6/6/5/5 tetracyclic ring system featuring an unprecedented 4,12-dioxatetracyclo[8.6.0.$0^{2,7}$.$0^{10,14}$]hexadecane core. Trichanthol A (**331**) is the first example of a pimarane-derived diterpenoid dimer furnished by forming an undescribed C-16–C-7' linkage, which was obtained from the tuber of *Icacina trichantha* (Icacinaceae) [107]. Commiphoranes A and B (**332** and **333**) were isolated from *Resina Commiphora* (Burseraceae). They are aromatic dinorditerpenoids characteristic of a 6/6/6/6 ring system [108].

References

1. Duarte N, Ferreira MJU. Lagaspholones A and B: two new jatropholane-type diterpenes from *Euphorbia lagascae*. Org Lett. 2007;9:489–92.
2. Mai ZP, Ni G, Liu YF, Li L, Li JY, Yu DQ. Heliojatrones A and B, two jatrophane-derived diterpenoids with a 5/10 fused-ring skeleton from *Euphorbia helioscopia*: structural elucidation and biomimetic conversion. Org Lett. 2018;20:3124–7.
3. Mai ZP, Ni G, Liu YF, Li YH, Li L, Li JY, Yu DQ. Secoheliosphanes A and B and secoheliospholane A, three diterpenoids with unusual *seco* -jatrophane and *seco* -jatropholane skeletons from *Euphorbia helioscopia*. J Org Chem. 2018;83:167–73.
4. Shi QQ, Zhang XJ, Wang TT, Wang Q, Sun TT, Amin M, Zhang RH, Li XL, Xiao WL. Euphopias A–C: three rearranged jatrophane diterpenoids with tricyclo[8.3.0.$0^{2,7}$]tridecane and tetracyclo[11.3.0.$0^{2,10}$.0 ^{3,7}]hexadecane cores from *Euphorbia Helioscopia*. Org Lett. 2020;22:7820–7824.
5. Shi QQ, Zhang Y, Wang TT, Xiong F, Zhang RH, Li XL, Ji X, Zhang XJ, Wang WG, Xiao WL. Euphopias D–F from *Euphorbia L.*: quantum chemical calculation-based structure elucidation and their bioactivity of inhibiting NLRP3 inflammasome activation. Org Chem Front. 2021;8:3041–3046.

6. Annam SCHVAR, Ankireddy M, Sura MB, Ponnapalli MG, Sarma AVS,S,J.B. Epimeric excolides from the stems of *Excoecaria agallocha* and structural revision of rhizophorin A. Org Lett. 2015;17:2840–2843.
7. Fei DQ, Dong LL, Qi FM, Fan GX, Li HH, Li ZY, Zhang ZX. Euphorikanin A, a diterpenoid lactone with a fused 5/6/7/3 ring system from *Euphorbia kansui*. Org Lett. 2016;18:2844–7.
8. Wan LS, Nian Y, Peng XR, Shao LD, Li XN, Yang J, Zhou M, Qiu MH. Pepluanols C-D, two diterpenoids with two skeletons from *Euphorbia peplus*. Org Lett. 2018;20:3074–8.
9. Yan XL, Sang J, Chen SX, Li W, Tang GH, Gan LS, Yin S. Euphorkanlide A, a highly modified ingenane diterpenoid with a c_{24} appendage from *Euphorbia kansuensis*. Org Lett. 2019;21:4128–31.
10. Yuan FY, Pan YH, Yin AP, Li W, Huang D, Yan XL, Wu SQ, Tang GH, Pu R, Yin S. Euphorstranoids A and B, two highly rearranged ingenane diterpenoids from *Euphorbia stracheyi*: structural elucidation, chemical transformation, and lipid-lowering activity. Org Chem Front. 2022;9:775–80.
11. Liao SG, Zhan ZJ, Yang SP, Yue JM. Lathyranoic Acid A: first secolathyrane diterpenoid in nature from *Euphorbia lathyris*. Org Lett. 2005;7:1379–82.
12. Gao S, Liu HY, Wang YH, He HP, Wang JS, Di YT, Li CS, Fang X, Hao XJ. Lathyranone A: a diterpenoid possessing an unprecedented skeleton from *Euphorbia lathyris*. Org Lett. 2007;9:3453–5.
13. Tian Y, Guo Q, Xu W, Zhu C, Yang Y, Shi J. A minor diterpenoid with a new 6/5/7/3 fused-ring skeleton from *Euphorbia micractina*. Org Lett. 2014;16:3950–3.
14. Wan LS, Nian Y, Ye CJ, Shao LD, Peng XR, Geng CA, Zuo ZL, Li XN, Yang J, Zhou M, Qiu MH. Three minor diterpenoids with three carbon skeletons from *Euphorbia peplus*. Org Lett. 2016;18:2166–9.
15. Pan L, Zhou P, Zhang X, Peng S, Ding L, Qiu SX. Skeleton-rearranged pentacyclic diterpenoids possessing a cyclobutane ring from *Euphorbia wallichii*. Org Lett. 2006;8:2775–8.
16. Qi WY, Zhao JX, Wei WJ, Gao K, Yue JM. Quorumolides A-C, three cembranoids from *Euphorbia antiquorum*. J Org Chem. 2018;83:1041–5.
17. Wang L, Yang J, Kong LM, Deng J, Xiong Z, Huang J, Luo J, Yan Y, Hu Y, Li XN, Li Y, Zhao Y, Huang SX. Natural and semisynthetic tigliane diterpenoids with new carbon skeletons from *Euphorbia dracunculoides* as a wnt signaling pathway inhibitor. Org Lett. 2017;19:3911–4.
18. Ferreira RJ, Spengler G, Orthaber A, dos Santos DJVA, Ferreira MJU. Pedrolane, a polycyclic diterpene scaffold containing a bicyclo[2.2.1]heptane system, from *Euphorbia pedroi*. Org Lett. 2021;23:274–278.
19. Liu SN, Huang D, Morris Natschke S. L, Ma H, Liu Z, Seeram NP, Xu J, Lee KH, Gu Q. Euphomilones A and B, ent-rosane diterpenoids with 7/5/6 and 5/7/6 skeletons from *Euphorbia milii*. Org Lett. 2016;18:6132–6135.
20. Barile E, Lanzotti V. Biogenetical related highly oxygenated macrocyclic diterpenes from sea spurge *Euphorbia paralias*. Org Lett. 2007;9:3603–6.
21. Flores Giubi ME, Durán Peña MJ, Botubol Ares JM, Montaño FE, Zorrilla D, Macías Sánchez AJ, Hernández GR. Gaditanone, a diterpenoid based on an unprecedented carbon skeleton isolated from *Euphorbia gaditana*. J Nat Prod. 2017;80:2161–5.
22. Fan YY, Shi SQ, Deng GZ, Liu HC, Xu CH, Ding J, Wang GW, Yue JM. Crokonoids A-C, a highly rearranged and dual-bridged spiro diterpenoid and two other diterpenoids from *Croton kongensis*. Org Lett. 2020;22:929–33.
23. Wang GC, Zhang H, Liu HB, Yue JM. Laevinoids A and B: two diterpenoids with an unprecedented backbone from *Croton laevigatus*. Org Lett. 2013;15:4880–3.
24. Li W, Wang RM, Pan YH, Zhao YY, Yuan FY, Huang D, Tang GH, Bi HC, Yin S. Crotonpenoids A and B, two highly modified clerodane diterpenoids with a tricyclo[7.2.1.$0^{2,7}$]dodecane core from *Croton yanhuii*: isolation, structural elucidation, and biomimetic semisynthesis. Org Lett. 2020;2:4435–4439.
25. Zhang WY, Zhao JX, Sheng L, Fan YY, Li JY, Gao K, Yue JM. Mangelonoids A and B, two pairs of macrocyclic diterpenoid enantiomers from *Croton mangelong*. Org Lett. 2018;20:4040–3.

26. Chen HD, Yang SP, He XF, Ai J, Liu ZK, Liu HB, Geng MY, Yue JM. Trigochinins A–C: three new daphnane-type diterpenes from *Trigonostemon chinensis*. Org Lett. 2010;12:1168–71.
27. Zhang L, Luo RH, Wang F, Jiang MY, Dong ZJ, Yang LM, Zheng YT, Liu JK. Highly functionalized daphnane diterpenoids from *Trigonostemon thyrsoideum*. Org Lett. 2010;12:152–5.
28. Allard PM, Martin MT, Tran Huu Dau ME, Leyssen P, Guéritte F, Litaudon M. Trigocherrin A, the first natural chlorinated daphnane diterpene orthoester from *Trigonostemon cherrieri*. Org Lett. 2012;14:342–345.
29. Wang XC, Zheng ZP, Gan XW, Hu LH. Jatrophalactam, a novel diterpenoid lactam isolated from *Jatropha curcas*. Org Lett. 2009;11:5522–4.
30. Chianese G, Fattorusso E, Aiyelaagbe OO, Luciano P, Schröder HC, Müller WEG, Taglialatela SO. Spirocurcasone, a diterpenoid with a novel carbon skeleton from *Jatropha curcas*. Org Lett. 2011;13:316–9.
31. Li W, Tang YQ, Sang J, Fan RZ, Tang GH, Yin S. Jatrofolianes A and B: two highly modified lathyrane diterpenoids from *Jatropha gossypiifolia*. Org Lett. 2020;22:106–9.
32. Huang D, Pan YH, Yuan FY, Li W, Yan XL, Zou MF, Tang GH, Yin S. Euphohyrisnoids A and B, two highly rearranged lathyrane diterpenoids from *Euphorbia lathyris*. Org Lett. 2021;23:9602–5.
33. He Jun, Xu JK, Guo LB, Xia CY, Lian WW, Tian HY, Zhang J, Shi YX, and Zhang WK. Fischdiabietane A, an antitumoral diterpenoid dimer featuring an unprecedented carbon skeleton from *Euphorbia fischeriana*. J Org Chem. 2021;86:5894–5900.
34. Gui MY, Aoyagi Y, Jin YR, Li XW, Hasuda T, Takeya K. Excisanin H, a novel cytotoxic 14,20-epoxy-*ent*-kaurene diterpenoid, and three new *ent*-kaurene diterpenoids from *Rabdosia excisa*. J Nat Prod. 2004;67:373–6.
35. Li SH, Wang J, Niu XM, Shen YH, Zhang HJ, Sun HD, Li ML, Tian QE, Lu Y, Cao P, Zheng QT. Maoecrystal V, cytotoxic diterpenoidwith a novel C_{19} skeleton from *Isodon eriocalyx* (Dunn.) Hara. Org Lett. 2004;6:4327–4330.
36. Han QB, Cheung S, Tai J, Qiao CF, Song JZ, Tso TF, Sun HD, Xu HX. Maoecrystal Z, a cytotoxic diterpene from *Isodon eriocalyx* with a unique skeleton. Org Lett. 2006;8:4727–30.
37. Huang SX, Xiao WL, Li LM, Li SH, Zhou Y, Ding LS, Lou LG, Sun HD. Bisrubescensins A–C: three new dimeric *e nt* -kauranoids isolated from *Isodon rubescens*. Org Lett. 2006;8:1157–60.
38. Wang WG, Du X, Li XN, Wu HY, Liu X, Shang SZ, Zhan R, Liang CQ, Kong LM, Li Y, Pu JX, Sun HD. New Bicyclo[3.1.0]hexane unit *ent* -kaurane diterpene and its *seco* -derivative from *Isodon eriocalyx* Var. *laxiflora*. Org Lett. 2012;14:302–305.
39. Zou J, Du X, Pang G, Shi YM, Wang WG, Zhan R, Kong LM, Li XN, Li Y, Pu JX, Sun HD. Ternifolide A, a New diterpenoid possessing a rare macrolide motif from *Isodon ternifolius*. Org Lett 2012;14:3210–3213.
40. Wang WG, Li XN, Du X, Wu HY, Liu X, Su J, Li Y, Pu JX, Sun HD. Laxiflorolides A and B, epimeric bishomoditerpene lactones from *Isodon eriocalyx*. J Nat Prod. 2012;75:1102–7.
41. Hu ZX, Xu HC, Hu K, Liu M, Li XN, Li XR, Du X, Zhang YH, Puno PT, Sun HD. Structurally diverse diterpenoids from *Isodon pharicus*. Org Chem Front. 2018;5:2379–89.
42. Yang Q, Hu K, Yan BC, Liu M, Li XN, Sun HD, Puno PT. Maoeriocalysins A-D, four novel *ent*-kaurane diterpenoids from *Isodon eriocalyx* and their structure determination utilizing quantum chemical calculation in conjunction with quantitative interproton distance analysis. Org Chem Front. 2019;6:45–53.
43. Huang B, Xiao CJ, Huang ZY, Tian XY, Cheng X, Dong X, Jiang B. Hispidanins A-D: four new asymmetric dimeric diterpenoids from the rhizomes of *Isodon hispida*. Org Lett. 2014;16:3552–5.
44. Xu G, Hou AJ, Zheng YT, Zhao Y, Li XL, Peng LY, Zhao QS. Przewalskin B, a novel diterpenoid with an unprecedented skeleton from *Salvia przewalskii* Maxim. Org Lett. 2006;9:291–3.
45. Zou J, Pan L, Li Q, Zhao J, Pu J, Yao P, Gong N, Lu Y, Kondratyuk TP, Pezzuto JM, Fong HHS, Zhang H, Sun H. Rubesanolides A and B: diterpenoids from *Isodon rubescens*. Org Lett. 2011;13:1406–9.

46. Xia F, Zhang DW, Wu CY, Geng HC, Xu WD, Zhang Y, Yang XW, Qin HB, Xu G. Isolation, structural elucidation, and synthetic study of salviyunnanone a, an abietane derived diterpenoid with a 7/5/6/3 ring system from *Salvia yunnanensis*. Org Chem Front. 2018;5:1262–6.
47. Xia F, Li WY, Yang XW, Yang J, Li X, Nian Y, Xu G. Salpratlactones A and B: a pair of *cis–trans* tautomeric abietanes as $Ca_v3.1$ T-type calcium channel agonists from *Salvia prattii*. Org Lett. 2019;21:5670–5674.
48. Li LN, Liu XQ, Zhu DR, Chen C, Lin YL, Wang WL, Zhu L, Luo JG, Kong LY. Officinalins A and B, a pair of C_{23} terpenoid epimers with a tetracyclic 6/7/5/5 system from *Salvia officinalis*. Org Chem Front. 2019;6:3369–73.
49. Valdés LJ, Chang HM, Visger DC, Koreeda M. Salvinorin C, a new neoclerodane diterpene from a bioactive fraction of the Hallucinogenic Mexican Mint *Salvia divinorum*. Org Lett. 2001;3:3935–7.
50. Harding WW, Tidgewell K, Schmidt M, Shah K, Dersch CM, Snyder J, Parrish D, Deschamps JR, Rothman RB, Prisinzano TE. Salvinicins A and B, new neoclerodane diterpenes from *Salvia divinorum*. Org Lett. 2005;7:3017–20.
51. Aoyagi Y, Yamazaki A, Nakatsugawa C, Fukaya H, Takeya K, Kawauchi S, Izumi H, Salvileucalin B. A novel diterpenoid with an unprecedented rearranged neoclerodane skeleton from *Salvia Leucantha* Cav. Org Lett. 2008;10:4429–32.
52. Bautista E, Toscano RA, Ortega A. Microphyllandiolide, a new diterpene with an unprecedented skeleton from *Salvia microphylla*. Org Lett. 2013;15:3210–3.
53. Bautista E, Fragoso-Serrano M, Toscano RA, García-Peña MdelR, Ortega A. Teotihuacanin, a diterpene with an unusual spiro-10/6 system from *Salvia amarissima* with potent modulatory activity of multidrug resistance in cancer cells. Org Lett. 2015;17:3280–3282.
54. Li XR, Yan BC, Hu K, He SJ, Sun HD, Zuo JP, Puno PT. Spiro *ent*-clerodane dimers: discovery and green approaches for a scalable biomimetic synthesis. Org Lett. 2021;23:5647–51.
55. Zheng G, Kadir A, Zheng X, Jin P, Liu J, Maiwulanjiang M, Yao G, Aisa HA. Spirodesertols A and B, two highly modified spirocyclic diterpenoids with an unprecedented 6-isopropyl-3*H*-spiro[benzofuran-2,1′-cyclohexane] motif from *Salvia deserta*. Org Chem Front. 2020;7:3137–45.
56. Xiong L, Zhou QM, Zou Y, Chen MH, Guo L, Hu GY, Liu ZH, Peng C. Leonuketal, a spiroketal diterpenoid from *Leonurus japonicus*. Org Lett. 2015;17:6238–41.
57. Yun YS, Fukaya H, Nakane T, Takano A, Takahashi S, Takahashi Y, Inoue H. A new bis-*seco*-abietane diterpenoid from *Hyptis crenata* Pohlex Benth. Org Lett. 2014;16:6188–91.
58. Wang S, Lin S, Zhu C, Yang Y, Li S, Zhang J, Chen X, Shi J. Highly acylated diterpenoids with a new 3,4-secograyanane skeleton from the flower buds of *Rhododendron molle*. Org Lett. 2010;12:1560–3.
59. Li Y, Liu YB, Zhang JJ, Li YH, Jiang JD, Yu SS, Ma SG, Qu J, L HN. Mollolide A, a diterpenoid with a new 1,10:2,3-disecograyanane skeleton from the roots of *Rhododendron molle*. Org Lett. 2013;15:3074–3077.
60. Zhang M, Zhu Y, Zhan G, Shu P, Sa R, Lei L, Xiang M, Xue Y, Luo Z, Wan Q, Yao G, Zhang Y, Micranthanone A. A new diterpene with an unprecedented carbon skeleton from *Rhododendron micranthum*. Org Lett. 2013;15:3094–7.
61. Li Y, Liu YB, Liu YL, Wang C, Wu LQ, Li L, Ma SG, Qu J, Yu SS. Mollanol A, a diterpenoid with a new c-nor-d-homograyanane skeleton from the fruits of *Rhododendron Molle*. Org Lett. 2014;16:4320–3.
62. Zhou J, Sun N, Zhang H, Zheng G, Liu J, Yao G. Rhodomollacetals A-C, PTP1B Inhibitory diterpenoids with a 2,3:5,6-di-*seco*-grayanane skeleton from the leaves of *Rhododendron molle*. Org Lett. 2017;19:5352–5.
63. Zhou J, Zhan G, Zhang H, Zhang Q, Li Y, Xue Y, Yao G, Rhodomollanol A. A highly oxygenated diterpenoid with a 5/7/5/5 tetracyclic carbon skeleton from the leaves of *Rhododendron molle*. Org Lett. 2017;19:3935–8.
64. Zhou J, Liu J, Dang T, Zhou H, Zhang H, Yao G. Mollebenzylanols A and B, highly modified and functionalized diterpenoids with a 9-benzyl-8,10-dioxatricyclo[5.2.1.$0^{1,5}$]decane core from *Rhododendron molle*. Org Lett. 2018;20:2063–2066.

65. Zhou J, Zuo Z, Liu J, Zhang H, Zheng G, Yao G. Discovery of highly functionalized 5,6-*seco*-grayanane diterpenoids as potent competitive PTP1B inhibitors. Org Chem Front. 2020;7:820–8.
66. Li CH, Niu XM, Luo Q, Xie MJ, Luo SH, Zhou YY, Li SH. Novel polyesterified 3,4-*seco*-grayanane diterpenoids as antifeedants from *Pieris formosa*. Org Lett. 2010;12:2426–9.
67. Niu CS, Li Y, Liu YB, Ma SG, Liu F, Li L, Xu S, Wang XJ, Wang RB, Qu J, Yu SS. Pierisketolide A and pierisketones B and C, three diterpenes with an unusual carbon skeleton from the roots of *Pieris formosa*. Org Lett. 2017;19:906–9.
68. Yin H, Luo JG, Kong LY. Tetracyclic diterpenoids with isomerized isospongian skeleton and labdane diterpenoids from the fruits of *Amomum kravanh*. J Nat Prod. 2013;76:237–42.
69. Ji KL, Fan YY, Ge ZP, Sheng L, Xu YK, Gan LS, Li JY, Yue JM. Maximumins A-D, rearranged labdane-type diterpenoids with four different carbon skeletons from *Amomum maximum*. J Org Chem. 2019;84:282–8.
70. Yin H, Luo JG, Shan SM, Wang XB, Luo J, Yang MH, Kong LY. Amomaxins A and B, two unprecedented rearranged labdane norditerpenoids with a nine-membered ring from *Amomum maximum*. Org Lett. 2013;15:1572–5.
71. Zhao Q, Gao JJ, Qin XJ, Hao XJ, He HP, Liu HY. Hedychins A and B, 6,7-dinorlabdane diterpenoids with a peroxide bridge from *Hedychium Forrestii*. Org Lett. 2018;20:704–7.
72. Ni G, Zhang H, Fan YY, Liu HC, Ding J, Yue JM. Mannolides A-C with an intact diterpenoid skeleton providing insights on the biosynthesis of antitumor *Cephalotaxus* troponoids. Org Lett. 2016;18:1880–3.
73. Chen HL, Wang LW, Su HJ, Wei BL, Yang SZ, Lin CN. New terpenoids from *Amentotaxus Formosana*. Org Lett. 2006;8:753–6.
74. Galli B, Gasparrini F, Lanzotti V, Misiti D, Riccio R, Villani C, Guan FH, Zhong WM, Wan FY. Grandione, a new heptacyclic dimeric diterpene from *Torreya grandis* Fort. Tetrahedron. 1999;55:11385–94.
75. Huo CH, Su XH, Wang YF, Zhang XP, Shi QW, Kiyota H. Canataxpropellane, a novel taxane with a unique polycyclic carbon skeleton (tricyclotaxane) from the needles of *Taxus Canadensis*. Tetrahedron Lett. 2007;48:2721–4.
76. Fukuyama Y, Fujii H, Minami H, Takahashi H, Kubo M. Neovibsanin F and its congeners, rearranged vibsane-type diterpenes from *Viburnum suspensum*. J Nat Prod. 2006;69:1098–100.
77. Gao X, Shao LD, Dong LB, Cheng X, Wu XD, Liu F, Jiang WW, Peng LY, He J, Zhao QS. Vibsatins A and B, two new tetranorvibsane-type diterpenoids from *Viburnum tinus* Cv. Variegatus. Org Lett. 2014;16:980–3.
78. Liang WJ, Geng CA, Zhang XM, Chen H, Yang CY, Rong GQ, Zhao Y, Xu HB, Wang H, Zhou NJ, Ma YB, Huang XY, Chen JJ. (±)-Paeoveitol, a pair of new norditerpene enantiomers from *Paeonia veitchii*. Org Lett. 2014;16:424–7.
79. Lou H, Zheng S, Li T, Zhang J, Fei Y, Hao X, Liang G, Pan W, Vulgarisin A. A new diterpenoid with a rare 5/6/4/5 ring skeleton from the chinese medicinal plant *Prunella vulgaris*. Org Lett. 2014;16:2696–9.
80. Morikawa T, Xu F, Kashima Y, Matsuda H, Ninomiya K, Yoshikawa M. Novel dolabellane-type diterpene alkaloids with lipid metabolism promoting activities from the seeds of *Nigella s ativa*. Org Lett. 2004;6:869–72.
81. Zhang HJ, Luo J, Shan SM, Wang XB, Luo JG, Yang MH, Kong LY. Aphanamenes A and B, two new acyclic diterpene [4 + 2]-cycloaddition adducts from *Aphanamixis grandifolia*. Org Lett. 2013;15:5512–5.
82. Zhao JX, Yu YY, Wang SS, Huang SL, Shen Y, Gao XH, Sheng L, Li JY, Leng Y, Li J, Yue JM. Structural elucidation and bioinspired total syntheses of ascorbylated diterpenoid hongkonoids A-D. J Am Chem Soc. 2018;140:2485–92.
83. Hou XF, Yao S, Mándi A, Kurtán T, Tang CP, Ke CQ, Li XQ, Ye Y. Bicunningines A and B, two new dimeric diterpenes from *Cunninghamia lanceolata*. Org Lett. 2012;14:460–3.
84. Wu XD, Ding LF, Chen B, Li XN, Peng LY, Zhao QS. Cunlanceloic acids A-D: unprecedented labdane diterpenoid dimers with AChE inhibitory and cytotoxic activities from Cunninghamia lanceolata. Org Chem Front. 2021;8:5777–84.

85. Huang XH, Tao LX, Ke CQ, Tang C, Zhang HY, Ye Y, Lin LG, Yao S. Taxodikaloids A and B, two dimeric abietane-type diterpenoids from *Taxodium ascendens* possessing an oxazoline ring linkage. Org Lett. 2017;19:556–9.
86. Zhou L, Tuo Y, Hao Y, Guo X, Tang W, Xue Y, Zeng J, Zhou Y, Xiang M, Zuo J, Yao G, Zhang Y. Cinnamomols A and B, immunostimulative diterpenoids with a new carbon skeleton from the leaves of *Cinnamomum cassia*. Org Lett. 2017;19:3029–32.
87. Zhou HF, Guoruoluo Y DZ; Tuo Y, Zhou J, Zhang H, Wang W, Xiang M, Aisa HA, Yao G. Cassiabudanols A and B, immunostimulative diterpenoids with a cassiabudane carbon skeleton featuring a 3-oxatetracyclo[6.6.1.$0^{2,6}$.$0^{10,14}$]pentadecane scaffold from cassia buds. Org Lett. 2019;21:549–553.
88. Li C, Lee D, Graf TN, Phifer SS, Nakanishi Y, Burgess JP, Riswan S, Setyowati FM, Saribi AM, Soejarto DD, Farnsworth NR, Falkinham JO, Kroll DJ, Kinghorn AD, Wani MC, Oberlies NH. A hexacyclic *ent*-trachylobane diterpenoid possessing an oxetane ring from *Mitrephora glabra*. Org Lett. 2005;7:5709–12.
89. Chen CH, Hsieh TJ, Liu TZ, Chern CL, Hsieh PY, Chen CY. Annoglabayin, a novel dimeric kaurane diterpenoid, and apoptosis in hep g2 cells of annomontacin from the fruits of *Annona g labra*. J Nat Prod. 2004;67:1942–6.
90. Pan L, Terrazas C, Lezama-Davila CM, Rege N, Gallucci JC, Satoskar AR, Kinghorn AD. Cordifolide A, a sulfur-containing clerodane diterpene glycoside from *Tinospora cordifolia*. Org Lett. 2012;14:2118–21.
91. Guo DX, Zhu RX, Wang XN, Wang LN, Wang SQ, Lin ZM, Lou HX. Scaparvin A, a novel caged *cis*-clerodane with an unprecedented c-6/c-11 bond, and related diterpenoids from the liverwort *Scapania parva*. Org Lett. 2010;12:4404–7.
92. Perry NB, Burgess EJ, Baek SH, Weavers RT. The first atisane diterpenoids from a liverwort: polyols from *Lepidolaena clavigera*. Org Lett. 2001;3:4243–5.
93. Liu KX, Zhu YX, Yan YM, Zeng Y, Jiao YB, Qin FY, Liu JW, Zhang YY, Cheng YX. Discovery of populusone, a skeletal stimulator of umbilical cord mesenchymal stem cells from *Populus euphratica* exudates. Org Lett. 2019;21:1837–40.
94. Liu YY, Yan YM, Wang DW, Cheng YX. Populusene A, an anti-inflammatory diterpenoid with a bicyclo[8,4,1]pentadecane scaffold from *Populus euphratica* Resins. Org Lett. 2021;23:8657–61.
95. Wang LN, Zhang JZ, Li X, Wang XN, Xie CF, Zhou JC, Lou HX. Pallambins A and B, unprecedented hexacyclic 19- *nor* -secolabdane diterpenoids from the chinese liverwort *Pallavicinia ambigua*. Org Lett. 2012;14:1102–5.
96. Li Y, Xu Z, Zhu R, Zhou J, Zong Y, Zhang J, Zhu M, Jin X, Qiao Y, Zheng H, Lou H. Probing the interconversion of labdane lactones from the chinese liverwort *Pallavicinia ambigua*. Org Lett. 2020;22:510–4.
97. Zhou J, Zhang J, Li R, Liu J, Fan P, Li Y, Ji M, Dong Y, Yuan H, Lou H. Hapmnioides A-C, rearranged labdane-type diterpenoids from the chinese liverwort *Haplomitrium mnioides*. Org Lett. 2016;18:4274–6.
98. Wu XD, Luo D, Tu WC, Deng ZT, Chen XJ, Su J, Ji X, Zhao QS. Hypophyllins A-D, labdane-type diterpenoids with vasorelaxant activity from *Hypoestes phyllostachya* "Rosea." Org Lett. 2016;18:6484–7.
99. Manríquez-Torres JJ, Torres-Valencia JM, Velázquez-Jiménez R, Valdez-Calderón A, Alvarado-Rodríguez JG. Cerda García Rojas CM, Joseph-Nathan PA Macrocyclic dimeric diterpene with a c_2 symmetry axis. Org Lett. 2013;15:4658–61.
100. Zheng Y, Zhang SW, Cong HJ, Huang YJ, Xuan LJ. Caesalminaxins A-L, cassane diterpenoids from the seeds of *Caesalpinia minax*. J Nat Prod. 2013;76:2210–8.
101. Wang LY, Wu J, Yang Z, Wang XJ, Fu Y, Liu SZ, Wang HM, Zhu WL, Zhang HY, Zhao WM. (*m*)- and (*p*)-Bicelaphanol A, dimeric trinorditerpenes with promising neuroprotective activity from *Celastrus orbiculatus*. J Nat Prod. 2013;76:745–9.
102. Zhang Y, Lu Y, Mao L, Proksch P, Lin W. Tagalsins I and J, two novel tetraterpenoids from the Mangrove plant *Ceriops tagal*. Org Lett. 2005;7:3037–40.

103. Zhang XH, Yang Y, Liu JJ, Shen L, Shi Z, Wu J. Tagalide A and tagalol A, naturally occurring 5/6/6/6- and 5/6/6-fused cyclic dolabrane-type diterpenes: a new insight into the anti-breast cancer activity of the dolabrane scaffold. Org Chem Front. 2018;5:1176–83.
104. Pu DB, Du BW, Chen W, Gao JB, Hu K, Shi N, Li YM, Zhang XJ, Zhang RH, Li XN, Zhang HB, Wang F, Xiao WL. Premnafulvol A: a diterpenoid with a 6/5/7/3-fused tetracyclic core and its biosynthetically related analogues from *Premna fulva*. Org Lett. 2018;20:6314–7.
105. Xu MM, Zhou JF, Zeng LP, Xu JC, Onakpa MM, Duan JA, Che CT, Bi HK, Zhao M. Pimarane-derived diterpenoids with anti-*Helicobacter pylori* activity from the tuber of *Icacina trichantha*. Org Chem Front. 2021;8:3014–22.
106. Lin S, Wang S, Liu M, Gan M, Li S, Yang Y, Wang Y, He W, Shi J. Glycosides from the stem bark of *Fraxinus sieboldiana*. J Nat Prod. 2007;70:817–23.
107. Zhou J, Wu Z, Guo B, Sun M, Onakpa MM, Yao G, Zhao M, Che CT. Modified diterpenoids from the tuber of *Icacina oliviformis* as protein tyrosine phosphatase 1B inhibitors. Org Chem Front. 2020;7:355–67.
108. Dong L, Cheng LZ, Yan YM, Wang SM, Cheng YX. Commiphoranes A-D, carbon skeletal terpenoids from *Resina Commiphora*. Org Lett. 2017;19:286–9.

Chapter 4
Diverse Novel Sesterterpenoids

New skeleton sesterterpenoids are relatively rare in plants. Since 1999, only 20 new skeleton sesterterpenoids, assigned to 5 families 6 gena, have been isolated and identified (Fig. 4.1; Table A3), Labiaceae is the main source of novel sesterterpenoids. The names, structure classes, structure characteristics, biological activities, and plant sources of these new skeleton sesterterpenoids will go into depth in this chapter.

Bolivianine (**334**) is a novel sesterterpenoid with an unprecedented skeleton possessing a polycyclic structure. It was isolated from the trunk bark of *Hedyosmum angustifolium* (Chloranthaceae) [1]. Genepolide (**335**) was isolated from *Artemisia umbelliformis* (Compositae/Asteraceae), which is a sesterterpene γ-lactone with a novel carbon skeleton [2]. Leucosceptrine (**336**) represents a novel sesterterpenoid isolated from the medicinal plant of *Leucosceptrum canum* (Lamiaceae), its structure was determined by single-crystal X-ray diffraction [3]. Colquhounoids A–C (**337–339**) were identified from the peltate glandular trichomes of *Colquhounia coccinea*, they are clearly distinct from leucosceptroids by their adverse stereochemistries at C-6, C-7, and C-14. Moreover, compared to the leucosceptroids whose C-8 is never oxygenated and C-4 side chain is exclusively an isobutenyl residue [4]. Norleucosceptroids A–C (**340–342**) are novel C_{20} terpenoids, they were isolated from the leaves and flowers of *Leucosceptrum canum* (Lamiaceae) [5]. A pair of new C-14 epimeric sesterterpenoids, colquhounoid D (**343**) and 14-epi-colquhounoid D (**344**), and five degradation products featuring new C_{20} and C_{21} frameworks, norcolquhounoids A−E (**345−349**), were isolated from *Colquhounia coccinea* var. *mollis*. Their structures were elucidated by comprehensive spectroscopic analysis and single-crystal X-ray diffraction [6]. Gentianelloids A and B (**350** and **351**) are two sesterterpenoids possessing an unusual 10,11-*seco*-gentianellane skeleton, they were isolated from a traditional Uighur medicine *Gentianella turkestanorum* (Gemianaceae) [7]. Eurysoloids A (**352**) and B (**353**), two novel diastereomeric sesterterpenoids possessing a pentacyclic 5/6/5/10/5 framework with an unusual macrocyclic ether system, were isolated from *Eurysolen gracilis* Prain. Their structures were unambiguously determined by spectroscopic, single-crystal X-ray diffraction and DP4 + analyses [8].

Y. Cheng and D. Qin, *Novel Plant Natural Product Skeletons*,
https://doi.org/10.1007/978-981-99-7329-3_4

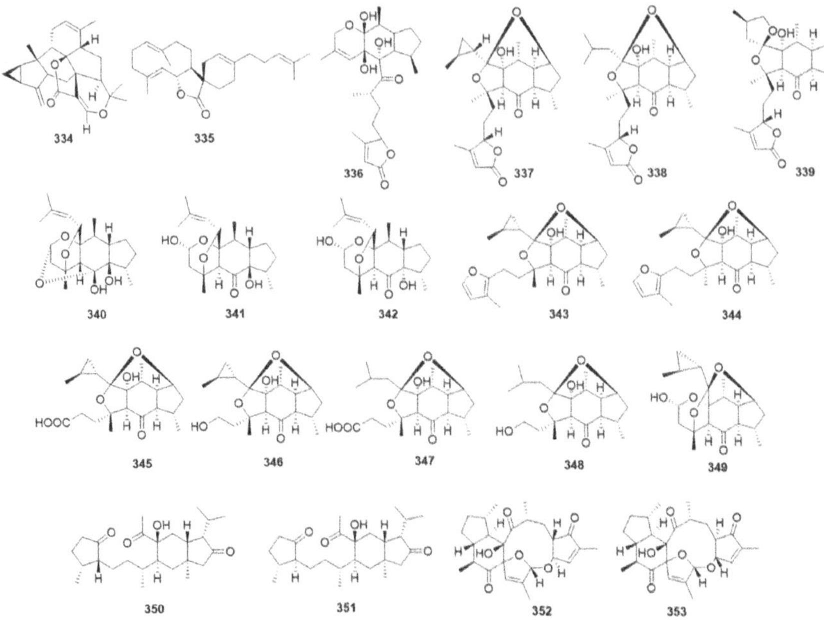

Fig. 4.1 Structures of new skeleton sesterterpenoids **334–353**

References

1. Acebey L, Sauvain M, Beck S, Moulis C, Gimenez A, Jullian V. Bolivianine, a new sesterpene with an unusual skeleton from *Hedyosmum angustifolium,* and its isomer, isobolivianine. Org Lett. 2007;9:4693–6.
2. Appendino G, Scafati OT, Romano A, Pollastro F, Avonto C, Rubiolo P. Genepolide, a sesterpene γ-lactone with a novel carbon skeleton from mountain wormwood (*Artemisia umbelliformis*). J Nat Prod. 2009;72:340–4.
3. Choudhary MI, Ranjit R, Atta-ur-Rahman, Shrestha TM, Yasin A, Parvez M. Leucosceptrine–A novel sesterterpene with prolylendopeptidase inhibitory activity from *Leucosceptrum canum.* J Org Chem. 2004;69:2906–2909.
4. Li CH, Jing SX, Luo SH, Shi W, Hua J, Liu Y, Li XN, Schneider B, Gershenzon J, Li SH. Peltate glandular trichomes of *Colquhounia coccinea* var. *mollis* harbor a new class of defensive sesterterpenoids. Org Lett. 2013;15:1694–1697.
5. Luo SH, Hua J, Li CH, Jing SX, Liu Y, Li XN, Zhao X, Li SH. New antifeedant C_{20} terpenoids from *Leucosceptrum canum.* Org Lett. 2012;14:5768–71.
6. Jing SX, Fu R, Li CH, Zhou TT, Liu YC, Liu Y, Luo SH, Li XN, Zeng F, Li S. Immunosuppresive sesterterpenoids and norsesterterpenoids from *Colquhounia coccinea* var. *mollis.* J Org Chem. 2021;86:11169−11176.
7. Guo K, Liu X, Zhou TT, Liu YC, Liu Y, Shi QM, Li XN, Li SH. Gentianelloids A and B: immunosuppressive 10,11-*seco*-gentianellane sesterterpenoids from the traditional uighur medicine *Gentianella turkestanorum.* J Org Chem. 2020;85;5511–5515.
8. Teng LL, Mu RF, Liu YC, Xiao CJ, Li DS, Gao JX, Guo K, Li XN, Liu Y, Zeng F, Li SH. Immunosuppressive and adipogenesis inhibitory sesterterpenoids with a macrocyclic ether system from *Eurysolen gracilis.* Org Lett. 2021;23:2232−2237.

Chapter 5
Classification of Diverse Triterpenoids

Triterpenoids contain many structurally diverse natural secondary metabolites, ring cracking and carbon skeleton rearrangement are the main factors for the formation of new skeleton triterpenoids. In particular, Schisandraceae and Iridaceae have yielded a wealth of novel triterpenoids. Schinortriterpenoids are the most structural diversity triterpenoids, the China scientist Handong Sun's research group have made great contributions in this field. About 134 new new skeletal triterpene compounds were isolated from 27 families in the last 23 years (1999–2021) (**354–391** Fig. 5.1, **392–415** Fig. 5.2, **416–456** Fig. 5.3, **457–487** Fig. 5.4; Table A4).

5.1 Schisandraceae

5.1.1 Schisandra

Micrandilactone A (**354**) possesses a highly oxidized, rearranged cycloartane skeleton, which was obtained from *Schisandra micrantha* [1]. Lancifodilactone F (**355**) is a nortriterpenoid possessing a unique skeleton, whereas lancifodilactone G (**356**) is a highly oxygenated nortriterpenoid featuring a partial enol structure [2, 3]. Both **355** and **356** were isolated from the medicinal plant *Schisandra lancifolia.* Rubriflordilactones A (**357**) and B (**358**) are two novel highly unsaturated rearranged bisnortriterpenoids possessing a biosynthetically modified aromatic D-ring from *Schisandra rubriflora* [4]. Their absolute structures were confirmed by X-ray crystallographic analysis. Schinalactone A (**359**) possesses a five-membered carbon ring featuring C-30 connected to C-1, henrischinins A–C (**360–362**), are another three novel triterpenoids featuring the unique motif of a 3-one-2-oxabicyclo[3.2.1]-octane, they were all isolated from the leaves and stems of *Schisandra henryi* [5, 6]. Sphenadilactones A (**363**) and B (**364**) are nortriterpenoids with diversity of highly oxygenated structures, they were isolated from the leaves and stems of *Schisandra*

Y. Cheng and D. Qin, *Novel Plant Natural Product Skeletons*,
https://doi.org/10.1007/978-981-99-7329-3_5

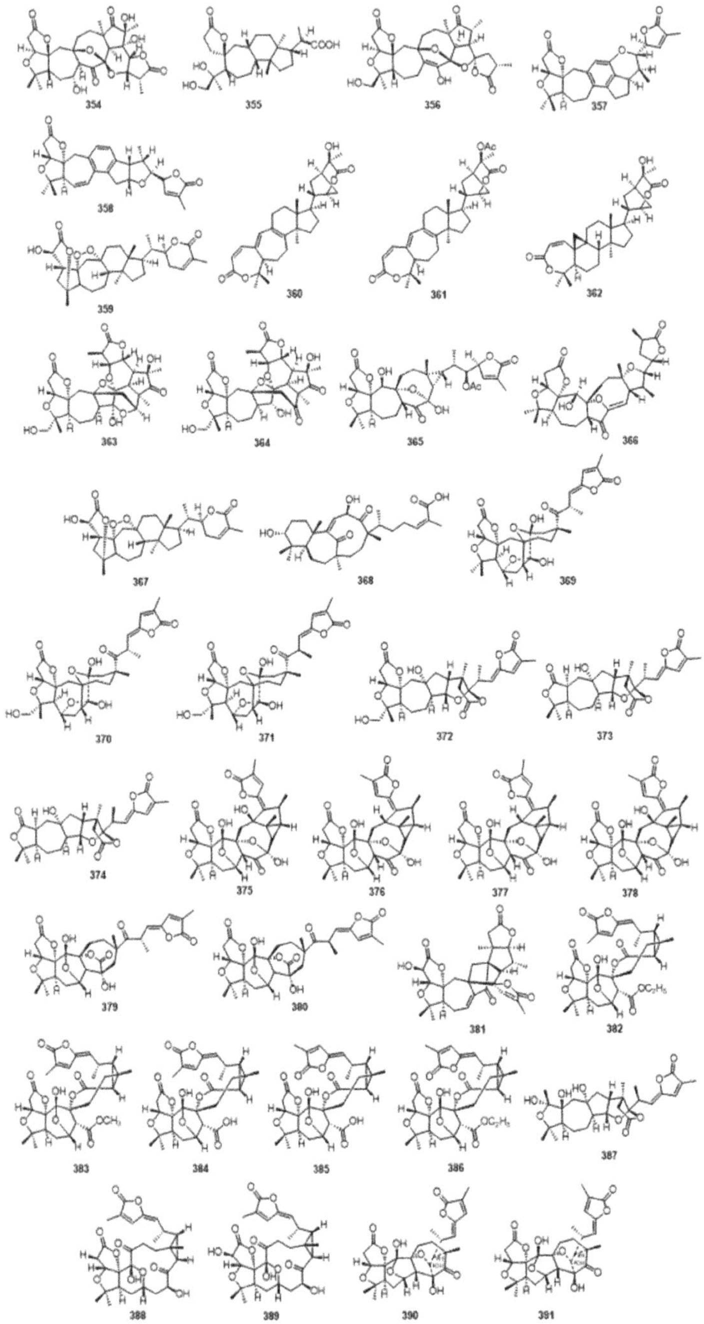

Fig. 5.1 Structures of new skeleton triterterpenoids **354–391**

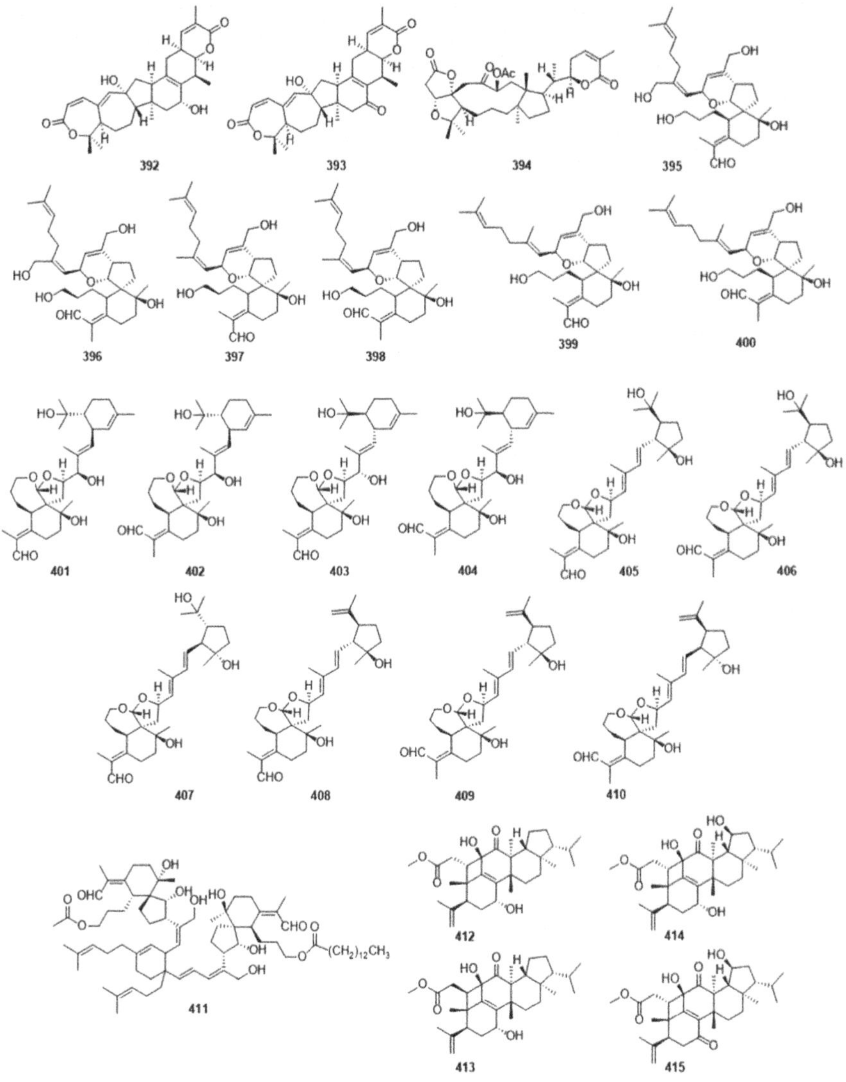

Fig. 5.2 Structures of new skeleton triterterpenoids **392–415**

sphenanthera [7]. Pre-schisanartanin (**365**), isolated from *Schisandra chinensis*, is a nortriterpene skeleton including a unique 7/8/3 consecutive carbocycle [8]. Arisandilactone A (**366**) was isolated from the fruits of *Schisandra arisanensis*, which has an unprecedented skeleton with a 5/5/7/5/8/5-fused hexacyclic ring system [9]. Schinarisanlactone A (**367**) has been isolated from *Schisandra arisanensis*, which is found to possess an unprecedented skeleton having a 5/7/7/5/7/5/6/5-fused octacyclic ring system [10]. Schiglautone A (**368**) is a unique 6/7/9-fused tricyclic carbon backbone triterpenoid, which was isolated from the stems of *Schisandra glaucescens* [11].

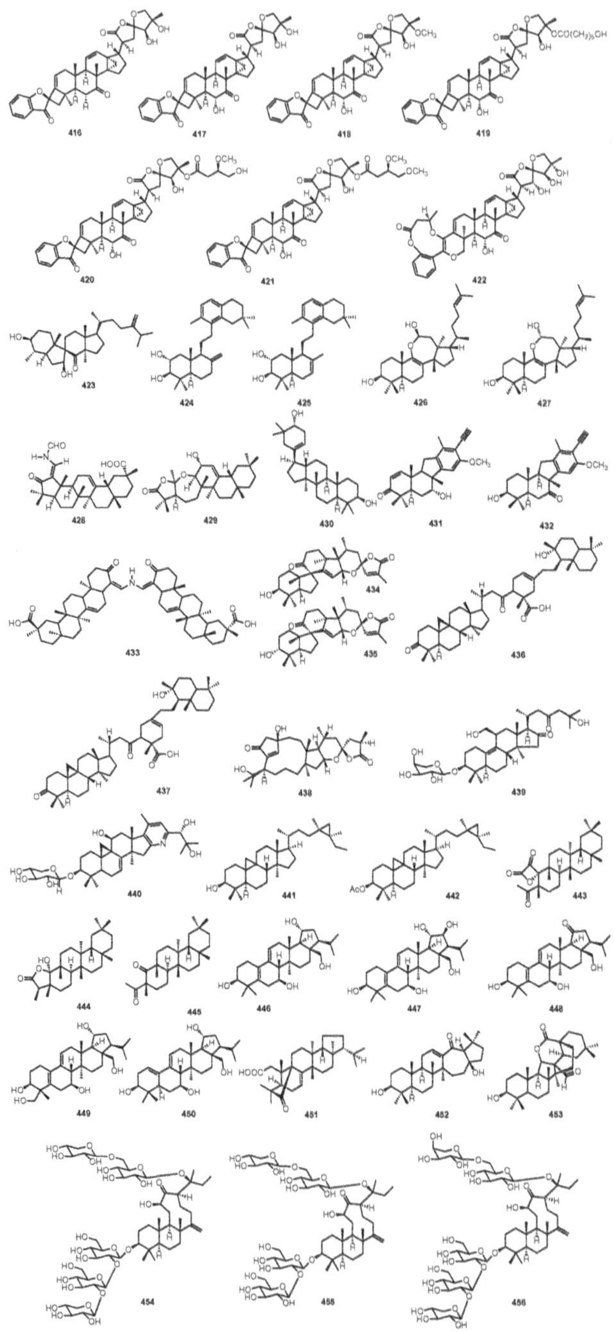

Fig. 5.3 Structures of new skeleton triterterpenoids **416–456**

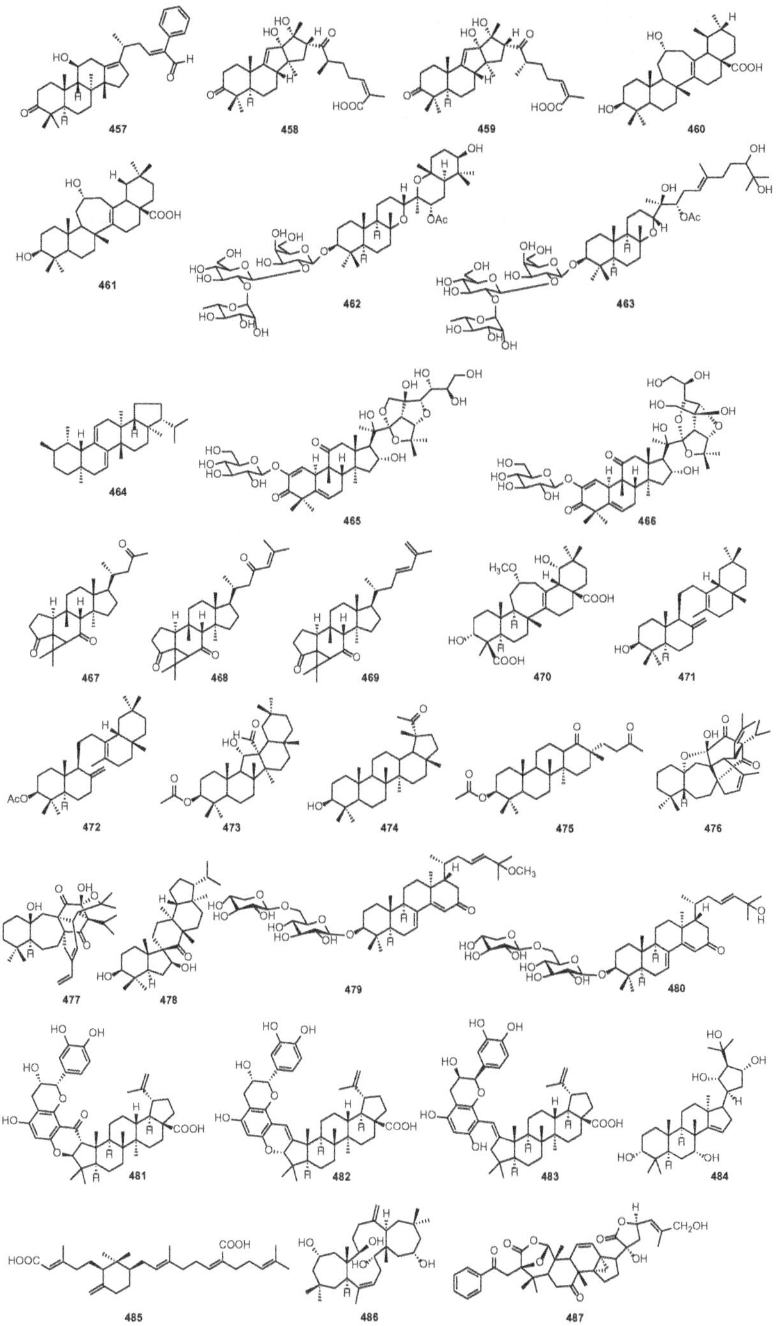

Fig. 5.4 Structures of new skeleton triterterpenoids **457–487**

Schicagenins A–C (**369–371**), isolated from *Schisandra chinensis*, are another three unprecedented nortriterpenoids characterized of a tetracyclic oxa-cage motif and a C_9 side chain [12]. Schilancitrilactones A–C (**372–374**) were isolated from the stems of *Schisandra lancifolia*. Of which, **372** possesses a 5/5/7/5/5/5-fused hexacyclic ring system with a C_{29} backbone, while **373** and **374** feature a C_{27} skeleton with a 5/7/5/5/5-fused pentacyclic ring system [13]. Lancolides A–D (**375–378**), described from *Schisandra lancifolia*, represents the first examples of natural products possessing a tricyclo[6.3.0.0^{2,11}]undecane-bridged system [14]. Lancifonins E and F (**379** and **380**) have been identified from *Schisandra lancifolia* and are found to possess a unique 7/7 fused carbocyclic core with an internal ester bridge between C-9 and C-14 [15]. Schincalide A (**381**) was isolated from the stems and leaves of *Schisandra incarnate*, which possesses a tricyclo[5.2.1.0^{1,6}]decane-bridged system in the structure [16]. Spiroschincarins A–E (**382–386**) are five novel spirocyclic schinortriterpenoids from *Schisandra incarnata*, featuring a unique 1-oxaspiro[6.6]tridecane motif, their absolute configurations were determined by single-crystal X-ray diffraction and computational methods [17]. Schilancidilactone C (**387**) is the first 3-norlancischiartane from *Schisandra lancifolia* with unusual configuration inversions occurring at C-1 and C-10 [18]. Schincalactones A (**388**) and B (**389**), featuring a unique 5/5/6/11/3 ring system, were isolated from *Schisandra incarnata* [19]. Incarnolides A (**390**) and B (**391**) are another two schinortriterpenoids from *Schisandra incarnata* featuring a tricyclo[9.2.1.0^{2,8}]tetradecane-bridged system [20].

5.1.2 *Kadsura*

Kadlongilactones A (**392**) and B (**393**) were isolated from the leaves and stems of *Kadsura longipedunculata*, they are two novel triterpene dilactones with an unprecedented rearranged hexacyclic skeleton [21]. Kadsuphilactone A (**394**) features an eleven-membered system, it was isolated from *Kadsura philippinensis* and its absolute configurations were confirmed by X-ray crystallographic analysis [22].

5.2 Indaceae

5.2.1 *Iris*

Spirioiridotectals A–F (**395–400**) are six novel iridal-type triterpenoids with a previously unreported 3,6-dihydro-2*H*-pyran moiety [23]. Polycycloiridals A–D (**401–404**) are four novel iridals with an unprecedented α-terpineol moiety resulting from cyclization of the homofarnesylside chain, while polycycloiridals E–J (**405–410**) contain an unprecedented cyclopentane ring [24, 25]. **395–410** were all isolated from the rhizomes of *Iris tectorum*.

5.2.2 *Belamcanda*

Dibelamcandal A (**411**), isolated from *Belamcanda chinensis*, is an unprecedented dimeric triterpenoid featuring a six-membered ring links two iridal type triterpenoid nuclei [26]. Belamchinanes A–D (**412–415**) feature a 4/6/6/6/5 polycyclic system were isolated from the seeds of *Belamcanda chinensis*, a four-membered carbocyclic ring bridging the C-1 and C-11 positions of a classical triterpenoid framework makes the structure of **412–415** unusual [27].

5.3 Euphorbiaceae

5.3.1 *Phyllanthus*

Phainanoids A–F (**416–421**) have been isolated from *Phyllanthus hainanensis*. They are six highly modified triterpenoids with a new carbon skeleton by incorporating two unique motifs of a 4,5- and a 5,5-spirocyclic systems [28]. Phainanolide A (**422**) is a highly modified triterpenoid from *Phyllanthus hainanensis* resulting from incorporating an unprecedented 6/9/6 heterotricyclic system and a highly oxygenated 5,5-spirocyclic ketal lactone. Its structure was completely elucidated by a combination of diverse methods including 2D NMR, quantum chemical NMR and ECD calculations, and NMR data analogy with model compounds [29].

5.3.2 *Euphorbia*

Spiropedroxodiol (**423**) is a new tetracyclic triterpenoid with an unusual spiro scaffold, which was obtained from *Euphorbia pedroi* [30]. Ebracpenes A (**424**) and B (**425**) are another two unusual ring C-*seco* and ring D-aromatic nortriterpenoids isolated from *Euphorbia ebracteolata* [31]. Euphorol J (**426**) and euphorstranol A (**427**) are the rare examples of 9,11-*seco* euphane or lanostane triterpenoids featuring an enol-hemiacetal functionality, they were isolated from *Euphorbia stracheyi* [32].

5.4 Meliaceae

5.4.1 Dysoxylum

Dysoxyhainanins A (**428**) and B (**429**) are two novel rearranged oleanane-type triterpenes from *Dysoxylum hainanense*. Among them, **428** possesses a unique 1,3-*cyclo*-2,3-*seco* A ring with a formamido-containing appendage and **429** features an unprecedented 1,2-dinor-3,10:9,10-*bisseco* skeleton [33].

5.4.2 Aglaia

Cyclodammarane (**430**) is a new type of pentacyclic triterpenoid, which has been isolated from *Aglaia odorata* [34].

5.4.3 Walsura

Walsucochins A (**431**) and B (**432**) were isolated from the leaves and twigs of *Walsura cochinchinensis*, they feature a phenylacetylene moiety fused into a contracted five-membered C-ring [35].

5.4.4 Turraea

Turraenine (**433**) is a nitrogen containing dimeric nor-multiflorane triterpenoid, which has been identified from a species in *Turraea* [36].

5.5 Pinaceae

5.5.1 Abies

Spirochensilides A (**434**) and B (**435**) are the first examples of triterpenoids possessing a unique 8,10-cyclo-9,10-*seco* and methyl-rearranged carbon skeleton [37] They were isolated from *Abies chensiensis* and their absolute configurations were determined by single crystal X-ray diffraction analysis and computational methods.

5.5.2 *Pseudolarix*

Pseudolaridimers A (**436**) and B (**437**), isolated from *Pseudolarix amabilis*, are two unprecedented heterodimers formed via a [4 + 2] Diels–Alder cycloaddition between a cycloartane triterpenoid unit and a labdane diterpenoid unit [38] Pseudolarenone (**438**), an unusual nortriterpenoid lactone bearing a fused 5/11/5/6/5 ring system featuring an unprecedented bicyclo[8.2.1]tridecane core, has been described from *Pseudolarix amabilis* [39].

5.6 Ranunculacea

5.6.1 *Actaea*

Podocarpaside (**439**) is a novel arabinoside possessing a unique triterpene skeleton and was isolated from *Actaea podocarpa* [40].

5.6.2 *Cimicifuga*

Cimicifugadine (**440**) is an unprecedented triterpene alkaloid glycoside and features a pyridine ring incorporated to a cycloartane triterpenoid nucleus. **382** was isolated from *Cimicifuga foetida* [41].

5.7 *Lygodium* (Lygodiaceae)

Lygodipenoids A (**441**) and B (**442**) are two novel C_{33} tetracyclic triterpenoids sharing a new 9,19:24,32-dicyclopropane skeleton. They were isolated from the whole grass of *Lygodium japonicum* [42].

5.8 Malpighia (Malpighiaceae)

Norfriedelanes A–C (**443–445**) were isolated from the branches and roots of *Malpighia emarginata.* Among them, **443** is a 3-norfriedelane possessing an α-oxo-β-lactone group, **444** is a 1,2-dinorfriedelane bearing a keto-lactone group, while **445** is a 1,2,3-trinorfriedelane [43].

5.9 *Sinocalamus* (Poaceae)

446–450, isolated from *Sinocalamus affinis*, are five triterpenoids with a new 25-norfern carbon skeleton, the absolute configuration of **446** was confirmed by single-crystal X-ray crystallographic analysis [44].

5.10 *Alstonia* (Apocynaceae)

Alstonic acids A (**451**) has been described from *Alstonia scholaris*, which is a new 2,3-secofernane-type triterpenoid. The cyclization pattern between C-3 and C-9 within **393** is uncommon in triterpenoids [45] Alstoscholarinoid A (**452**) and B (**453**) are two rearranged triterpenoids, representing new subtypes of pentacyclic triterpenoids, **452** possessed an unprecedented rearranged 6/6/6/7/5 fused ring system with degraded C-28, and **453** possessed a rare cleavage and reconstructed 6/6/5/6/6/6 six-ring system with an additional unique C28 → C11-olide ring-F, they were isolated from *Alstonia scholaris* [46].

5.11 *Panax* (Araliaceae)

Nototronesides A–C (**454–456**), possessing an unprecedented 6/6/9 fused tricyclic tetranordammarane core, have been isolated from the leaves of *Panax notoginseng* [47].

5.12 Others

There are some skeletal triterpenoids which are sporadically distributed in various plant families. Alismanin A (**457**) is a novel aromatic triterpenoid from *Alisma orientale* (Alismaceae) with a C_{34} skeleton [48]. Kadcoccitones A (**458**) and B (**459**) are a pair of new triterpenoid epimers isolated from *Kadsura coccinea* (Magnoliaceae), they feature an unprecedented carbon skeleton with a 6/6/5/5-fused tetracyclic ring system unit and a C_9 side chain [49]. Ilelic acids A and B (**460** and **461**) represent a new type of triterpenoids with a seven-membered ring in the molecule. They have been isolated from *Ilex latifolia* (Aquifoliaceae) [50]. Phyteumosides A (**462**) and B (**463**) isolated from *Phyteuma orbiculare* (Campanulaceae) are two saponins with unprecedented triterpenoid aglycones [51]. **462** features two additional tetrahydropyran rings, while **463** possesses a 17-polypodene aglycone. Canarene (**464**) was isolated from *Canarium schweinfurthii* (Burseraceae) and is the first member of a new class of triterpenoids, its structure was unambiguously deduced by single-crystal X-ray diffraction [52]. Machilusides A (**465**) and B (**466**) were isolated from *Machilus yaoshansis* (Lauraceae). They are novel homocucurbitane triterpenoid glycosides possessing an unprecedented C_{36} skeleton with a D-fructose moiety incorporated into a cucurbitane nucleus to form unique cage-like tricyclicring moieties [53]. Cucurbalsaminones A–C (**467–469**) are characteristic of a unique 5/6/3/6/5-fused pentacyclic carbon skeleton and they were isolated from *Momordica balsamina* (Cucurbitaceae) [54]. Duboscic acid (**470**) was isolated from *Duboscia macrocarpa* (Tiliaceae) and is a triterpenoid with a unique carbon backbone [55]. **471** and **472** are two *seco*-ring C oleanane triterpenoids isolated from *Stevia eupatoria* (Compositae/Asteraceae) [56]. **473–475** were isolated from *Ficus microcarpa* (Moraceae) and **473** is characteristic of the presence of a special five-membered C ring, the unique skeletons of **474** and **475** are proposed to arise from the same biogenetic pathway [57]. Salvadione-A (**476**) and salvadione-B (**477**) were isolated from the hexane soluble fraction of *Salvia bucharica* (Labiatae) and have novel carbon skeletons as shown [58]. Teuviscins A (**478**) was isolated from *Teucrium viscidum* (Labiatae), which possesses a rare 7(8 → 9)abeo-9*R*-D:C-friedo-B':A'-*neo*-gammacerane skeleton [59]. Two novel triterpenoid saponins, colqueleganoids A (**479**) and B (**480**), with the first methyl-30 incorporated 6/6/6/6-cyclized carbon skeleton (named colquelegane), were isolated from the root of *Colquhounia elegans*. Their structures including absolute configuration were determined by spectroscopic methods and X-ray crystallographic analyses [60]. *ent*-Epicatechinoceanothic acids A (**481**) and B (**482**) and epicatechino-3-deoxyceanothetric acid A (**483**) are three unprecedented ceanothane-type triterpenoids from *Zizyphus jujuba* (Rhamnaceae), they feature C–C bond linkages with catechin moieties [61]. Picraquassin A (**484**), possessing an unprecedented 21,24-cycloapotirucallane skeleton, was isolated from *Picrasma quassioides* (Simarubaceae) [62]. Mispyric acid (**485**) is a monocyclic triterpenoid with a novel skeleton from *Mischocarpus pyriformis* (Sapindaceae). [63]. Volvalerenol A (**486**) features an unprecedented type of triterpenoid with a 7/12/7 tricyclic ring system in the molecule, which has been described from the roots of *Valeriana hardwickii* (Valerianaceae). [64]. Longipetalol A (**487**) is an unprecedented highly modified triterpenoid with a unique 1,2-*seco*-3-(2-oxo-phenylethyl)-17α-13,30-cyclodammarane skeleton, featuring an acetallactone fragment. It was isolated from Dichapetalum longipetalum [65].

References

1. Li RT, Zhao QS, Li SH, Han QB, Sun HD, Lu Y, Zhang LL, Zheng QT. Micrandilactone A: a novel triterpene from *Schisandra micrantha*. Org Lett. 2003;5:1023–6.
2. Xiao WL, Li RT, Li SH, Li XL, Sun HD, Zheng YT, Wang RR, Lu Y, Wang C, Zheng QT. Lancifodilactone F: a novel nortriterpenoid possessing a unique skeleton from *Schisandra lancifolia* and its anti-hiv activity. Org Lett. 2005;7:1263–6.

3. Xiao WL, Zhu HJ, Shen YH, Li RT, Li SH, Sun HD, Zheng YT, Wang RR, Lu Y, Wang C, Zheng QT. Lancifodilactone G: a unique nortriterpenoid isolated from *Schisandra lancifolia* and its anti-hiv activity. Org Lett. 2005;7:2145–8.
4. Xiao WL, Yang LM, Gong NB, Wu L, Wang RR, Pu JX, Li XL, Huang SX, Zheng YT, Li RT, Lu Y, Zheng QT, Sun HD. Rubriflordilactones A and B, two novel bisnortriterpenoids from *Schisandra rubriflora* and their biological activities. Org Lett. 2006;8:991–4.
5. He F, Pu JX, Huang SX, Wang YY, Xiao WL, Li LM, Liu JP, Zhang HB, Li Y, Sun HD. Schinalactone A, a new cytotoxic triterpenoid from *Schisandra sphenanthera*. Org Lett. 2010;12:1208–11.
6. Xue YB, Yang JH, Li XN, Du X, Pu JX, Xiao WL, Su J, Zhao W, Li Y, Sun HD. Henrischinins A-C: three new triterpenoids from *Schisandra henryi*. Org Lett. 2011;13:1564–7.
7. Xiao WL, Pu JX, Chang Y, Li XL, Huang SX, Yang LM, Li LM, Lu Y, Zheng YT, Li RT, Zheng QT, Sun HD. Sphenadilactones A and B, two novel nortriterpenoids from *Schisandra sphenanthera*. Org Lett. 2006;8:1475–8.
8. Huang SX, Li RT, Liu JP, Lu Y, Chang Y, Lei C, Xiao WL, Yang LB, Zheng QT, Sun HD. Isolation and characterization of biogenetically related highly oxygenated nortriterpenoids from *Schisandra chinensis*. Org Lett. 2007;9:2079–82.
9. Cheng YB, Liao TC, Lo IW, Chen YC, Kuo YC, Chen SY, Chien CT, Shen YC. Arisandilactone A, a new triterpenoid from the fruits of *Schisandra arisanensis*. Org Lett. 2010;12:1016–9.
10. Lin YC, Lo IW, Chen SY, Lin PH, Chien CT, Chang S, Shen YC. Schinarisanlactone A, a new bisnortriterpenoid from *Schisandra arisanensis*. Org Lett. 2011;13:446–9.
11. Meng FY, Sun JX, Li X, Yu HY, Li SM, Ruan HL. Schiglautone A, a new tricyclic triterpenoid with a unique 6/7/9-fused skeleton from the stems of *Schisandra glaucescens*. Org Lett. 2011;13:1502–5.
12. Shi YM, Li XY, Li XN, Luo X, Xue YB, Liang CQ, Zou J, Kong LM, Li Y, Pu JX, Xiao WL, Sun HD. Schicagenins A-C: three cagelike nortriterpenoids from leaves and stems of *Schisandra chinensis*. Org Lett. 2011;13:3848–51.
13. Xiao WL, Huang SX, Zhang L, Tian RR, Wu L, Li L, Pu JX, Zheng YT, Lu Y, Li RT, Zheng QT, Sun HD. Nortriterpenoids from *Schisandra lancifolia*. J Nat Prod. 2006;69:650–3.
14. Shi YM, Wang XB, Li XN, Luo X, Shen ZY, Wang YP, Xiao WL, Sun HD. Lancolides, antiplatelet aggregation nortriterpenoids with tricyclo[6.3.0.0 ^{2,11}]undecane-bridged system from *Schisandra lancifolia*. Org Lett. 2013;15:5068–5071.
15. Shi YM, Yang J, Xu L, Li XN, Shang SZ, Cao P, Xiao WL, Sun HD. Structural characterization and antioxidative activity of lancifonins: unique nortriterpenoids from *Schisandra lancifolia*. Org Lett. 2014;16:1370–3.
16. Zhou M, Liu Y, Song J, Peng XG, Cheng Q, Cao H, Xiang M, Ruan HL. Schincalide A, a schinortriterpenoid with a tricyclo[5.2.1.0 ^{1,6}]decane-bridged system from the stems and leaves of *Schisandra incarnate*. Org Lett. 2016;18:4558–4561.
17. Song J, Liu Y, Zhou M, Cao H, Peng XG, Liang JJ, Zhao XY, Xiang M, Ruan HL. Spiroschincarins A-E: five spirocyclic nortriterpenoids from the fruit of *Schisandra incarnata*. Org Lett. 2017;19:1196–9.
18. Shi YM, Hu K, Pescitelli G, Liu M, Li XN, Du X, Xiao WL, Sun HD, Puno PT. Schinortriterpenoids with identical configuration but distinct ECD spectra generated by nondegenerate exciton coupling. Org Lett. 2018;20:1500–4.
19. Song J, Zhou M, Zhou J, Liang JJ, Peng XG, Liu J, Ruan HL. Schincalactones A and B, two 5/5/6/11/3 fused schinortriterpenoids with a 13-membered carbon ring system from *Schisandra incarnata*. Org Lett. 2018;20:2499–502.
20. Chen J, Peng XG, Liu J, Gao Y, Zhou M, Zhou J, Song J, Ruan HL. Incarnolides A and B, two schinortriterpenoids with a tricyclo[9.2.1.0^{2,8}]tetradecane-bridged system from *Schisandra incarnat*a. Org Lett. 2020;22:1071–1075.
21. Pu JX, Xiao WL, Lu Y, Li RT, Li HM, Zhang L, Huang SX, Li X, Zhao QS, Zheng QT, Sun HD. Kadlongilactones A and B, two novel triterpene dilactones from *Kadsura longipedunculata* possessing a unique skeleton. Org Lett. 2005;7:5079–82.

22. Shen YC, Lin YC, Chiang MY, Yeh SF, Cheng YB, Liao CC. Kadsuphilactones A and B, two new triterpene dilactones from *Kadsura philippinensis*. Org Lett. 2005;7:3307–10.
23. Zhang CL, Wang Y, Liu YF, Ni G, Liang D, Luo H, Song XY, Zhang WQ, Chen RY, Chen NH, Yu DQ. Iridal-type triterpenoids with neuroprotective activities from *Iris tectorum*. J Nat Prod. 2014;77:411–5.
24. Zhang CL, Liu YF, Wang Y, Liang D, Jiang ZB, Li L, Hao ZY, Luo H, Shi GR, Chen RY, Cao ZY, Yu DQ. Polycycloiridals A-D, four iridal-type triterpenoids with an α-terpineol moiety from *Iris tectorum*. Org Lett. 2015;17:5686–9.
25. Zhang CL, Hao ZY, Liu YF, Wang Y, Shi GR, Jiang ZB, Chen RY, Cao ZY, Yu DQ. Polycycloiridals with a cyclopentane ring from *Iris tectorum*. J Nat Prod. 2017;80:156–61.
26. Song ZJ, Xu XM, Deng WL, Peng SL, Ding LS, Xu HH. A new dimeric iridal triterpenoid from *Belamcanda chinensis* with significant molluscicide activity. Org Lett. 2011;13:462–5.
27. Song YY, Miao JH, Qin FY, Yan YM, Yang J, Qin DP, Hou FF, Zhou LL, Cheng YX. Belamchinanes A-D from *Belamcanda chinensis*: triterpenoids with an unprecedented carbon skeleton and their activity against age-related renal fibrosis. Org Lett. 2018;20:5506–9.
28. Fan YY, Zhang H, Zhou Y, Liu HB, Tang W, Zhou B, Zuo JP, Yue JM. Phainanoids A-F, a new class of potent immunosuppressive triterpenoids with an unprecedented carbon skeleton from *Phyllanthus hainanensis*. J Am Chem Soc. 2015;137:138–41.
29. Fan YY, Gan LS, Liu HC, Li H, Xu CH, Zuo JP, Ding J, Yue JM. Phainanolide A, highly modified and oxygenated triterpenoid from *Phyllanthus hainanensis*. Org Lett. 2017;19:4580–3.
30. Ferreira, RJ, Kincses A, Gajdács M, Spengler G, dos Santos DJVA, Molnár J, Ferreira MJU. Terpenoids from *Euphorbia pedroi* as multidrug-resistance reversers. J Nat Prod. 2018;81:2032–2040.
31. Wang C, Yan QS, Wang YF, Huang SS, Ning J, Feng L, Sun CP, Zhang BJ, Li DW, Ma XC. Ebracpenes A and B, Unusual Ring C-*seco* and Ring D-Aromatic Nor-Triterpenoids, from *Euphorbia ebracteolata* and lipase inhibitory evaluation. J Org Chem. 2019;84:1624–1629.
32. Yuan FY, Xu F, Fan RZ, Li W, Huang D, Tang GH, Yuan T, Gan LS, Yin S. Structural elucidation of three 9,11-*seco*tetracyclic triterpenoids enables the structural revision of euphorol J. J Org Chem. 2021;86:7588−7593.
33. He XF, Wang XN, Gan LS, Yin S, Dong L, Yue JM. Two novel triterpenoids from *Dysoxylum hainanense*. Org Lett. 2008;10:4327–30.
34. Cai XH, Luo XD, Zhou J, Hao XJ. Compound representatives of a new type of triterpenoid from *Aglaia odorata*. Org Lett. 2005;7:2877–9.
35. Zhou ZW, Yin S, Zhang HY, Fu Y, Yang SP, Wang XN, Wu Y, Tang XC, Yue JM. Walsucochins A and B with an unprecedented skeleton isolated from *Walsura cochinchinensis*. Org Lett. 2008;10:465–8.
36. Rasamison VE, Rakotondraibe LH, Slebodnick C, Brodie PJ, Ratsimbason M, TenDyke K, Shen Y, Randrianjanaka LM, Kingston DGI. Nitrogen-containing dimeric *nor*-multiflorane triterpene from a *Turraea* sp. Org Lett. 2014;16:2626–9.
37. Zhao QQ, Song QY, Jiang K, Li GD, Wei WJ, Li Y, Gao K. Spirochensilides A and B, two new rearranged triterpenoids from *Abies chensiensis*. Org Lett. 2015;17:2760–3.
38. Li B, Kong DY, Shen YH, Yuan H, Yue RC, He YR, Lu L, Shan L, Li HL, Ye J, Yang XW, Su J, Liu RH, Zhang WD. Pseudolaridimers A and B, hetero-cycloartane–labdane diels-alder adducts from the cone of *Pseudolarix amabilis*. Org Lett. 2012;14:5432–5.
39. Li B, Kong DY, Shen YH, Fu KL, Yue RC, Han ZZ, Yuan H, Liu QX, Shan L, Li HL, Yang XW, Zhang WD. Pseudolarenone, an unusual nortriterpenoid lactone with a fused 5/11/5/6/5 ring system featuring an unprecedented bicyclo[8.2.1]tridecane core from *Pseudolarix amabilis*. Chem Commun. 2013;49:1187.
40. Ali Z, Khan SI, Ferreira D, Khan IA. Podocarpaside, a triterpenoid possessing a new backbone from *Actaea podocarpa*. Org Lett. 2006;8:5529–32.
41. Dan C, Zhou Y, Ye D, Peng S, Ding L, Gross ML, Qiu SX. Cimicifugadine from *Cimicifuga foetida*, a new class of triterpene alklaoids with novel reactivity. Org Lett. 2007;9:1813–6.
42. Han Q, Liu X, Yao W, Cheng Z, Lin T, Song C, Yin S. Unusual 9,19: 24,32-dicyclotetracyclic triterpenoids from Lygodium japonicum. Planta Med. 2012;78:1971–5.

43. Liu JQ, Peng XR, Li XY, Li TZ, Zhang WM, Shi L, Han J, Qiu MH. Norfriedelins A-C with acetylcholinesterase inhibitory activity from acerola tree (*Malpighia emarginata*). Org Lett. 2013;15:1580–3.
44. Xiong L, Zhu M, Zhu CG, Lin S, Yang YC, Shi JG. Structure and bioassay of triterpenoids and steroids isolated from *Sinocalamus affinis*. J Nat Prod. 2012;75:1160–1166.
45. Wang F, Ren FC, Liu JK. Alstonic acids A and B, unusual 2,3-secofernane triterpenoids from *Alstonia scholaris*. Phytochemistry. 2009;70:650–654.
46. Hu BY, Zhao YL, Xiong DS, He YJ, Zhou ZS, Zhu PF, Wang ZJ, Wang YL, Zhao LX, Luo XD. Potent antihyperuricemic triterpenoids based on two unprecedented scaffolds from the leaves of *Alstonia scholaris*. Org Lett. 2021;23:4158−4162.
47. Liu XY, Li CJ, Chen FY, Ma J, Wang S, Yuan YH, Li L, Zhang DM. Nototronesides A-C, three triterpene saponins with a 6/6/9 fused tricyclic tetranordammarane carbon skeleton from the leaves of *Panax notoginseng*. Org Lett. 2018;20:4549–53.
48. Wang C, Huo XK, Luan ZL, Cao F, Tian XG, Zhao XY, Sun CP, Feng L, Ning J, Zhang BJ, Ma XC. Alismanin A, a triterpenoid with a c_{34} skeleton from *Alisma orientale* as a natural agonist of human pregnane X receptor. Org Lett. 2017;19:5645–8.
49. Liang CQ, Shi YM, Luo RH, Li XY, Gao ZH, Li XN, Yang LM, Shang SZ, Li Y, Zheng YT, Zhang HB, Xiao WL, Sun HD. Kadcoccitones A and B, two new 6/6/5/5-fused tetracyclic triterpenoids from *Kadsura coccinea*. Org Lett. 2012;14:6362–5.
50. Wang CQ, Wang L, Fan CL, Zhang DM, Huang XJ, Jiang RW, Bai LL, Shi JM, Wang Y, Ye WC. Ilelic acids a and b, two unusual triterpenes with a seven-membered ring from *Ilex latifolia*. Org Lett. 2012;14:4102–5.
51. Abbet C, Neuburger M, Wagner T, Quitschau M, Hamburger M, Potterat O. Phyteumosides A and B: new saponins with unique triterpenoid aglycons from *Phyteuma orbiculare* L. Org Lett. 2011;13:1354–7.
52. Kamdem RST, Wafo P, Yousuf S, Ali Z, Adhikari A, Rasheed S, Khan IA, Ngadjui BT, Fun HK, Choudhary MI. Canarene: A triterpenoid with a unique carbon skeleton from *Canarium schweinfurthii*. Org Lett. 2011;13:5492–5.
53. Liu M, Gan M, Lin S, Zhang Y, Zi J, Song W, Fan X, Liu Y, Yang Y, Shi J. Machilusides A and B: structurally unprecedented homocucurbitane glycosides from the stem bark of *Machilus Yaoshansis*. Org Lett. 2011;13:2856–9.
54. Mónico A, Ramalhete C, André V, Spengler G, Mulhovo S, Duarte MT, Ferreira MJU. Cucurbalsaminones A-C, rearranged triterpenoids with a 5/6/3/6/5-fused pentacyclic carbon skeleton from *Momordica balsamina*, as multidrug resistance reversers. J Nat Prod. 2019;82:2138–43.
55. Wafo P, Kamdem RST, Ali Z, Anjum S, Khan SN, Begum A, Krohn K, Abegaz BM, Ngadjui BT, Choudhary MI. Duboscic Acid: a potent α-glucosidase inhibitor with an unprecedented triterpenoidal carbon skeleton from *Duboscia macrocarpa*. Org Lett. 2010;12:5760–3.
56. Román LU, Guerra Ramírez D, Morán G, Martínez I, Hernández JD, Cerda García Rojas CM, Torres Valencia JM, Joseph Nathan P. First *seco*-C oleananes from nature. Org Lett. 2004;6:173–176.
57. Chiang YM, Kuo YH. Novel triterpenoids from the aerial roots of *Ficus microcarpa*. J Org Chem. 2002;67:7656–61.
58. Ahmad VU, Zahid M, Ali MS, Ali Z, Jassbi AR, Abbas M, Clardy J, Lobkovsky E, Tareen RB, Iqbal MZ. Salvadiones-A and -B: two terpenoids having novel carbon skeleta from *Salvia bucharica*. J Org Chem. 1999;64:8465–7.
59. Li ZY, Qi FM, Zhi DJ, Hu QL, Liu YH, Zhang ZX, Fei DQ. A novel spirocyclic triterpenoid and a new taraxerane triterpenoid from *Teucrium viscidum*. Org Chem Front. 2017;4:42–6.
60. Hua J, Liu YC, Luo SH, Liu Y, Xiao CJ, Li XN, Li SH. Immunostimulatory 6/6/6/6 tetracyclic triterpenoid saponins with the methyl-30 incorporated cyclization from the root of *Colquhounia elegans*. Org Lett. 2021;23:7462–6.
61. Kang KB, Kim HW, Kim JW, Oh WK, Kim J, Sung SH. Catechin-bound ceanothane-type triterpenoid derivatives from the roots of *Zizyphus Jujuba*. J Nat Prod. 2017;80:1048–54.
62. Xu J, Xiao D, Lin QH, He JF, Liu WY, Xie N, Feng F, Qu W. Cytotoxic tirucallane and apotirucallane triterpenoids from the stems of *Picrasma quassioides*. J Nat Prod. 2016;79:1899–910.

63. Sun DA, Deng JZ, Starck SR, Hecht SM. Mispyric Acid, a new monocyclic triterpenoid with a novel skeleton from *Mischocarpus pyriformis* that inhibits dna polymerase β. J Am Chem Soc. 1999;121:6120–4.
64. Wang PC, Ran XH, Luo HR, Ma QY, Liu YQ, Dai HF, Zhou J, Zhao YX. Volvalerenol A, a new triterpenoid with a 12-membered ring from *Valeriana hardwickii*. Org Lett. 2013;15:2898–901.
65. Zhang DL, Li M, Han GF, Li SY, Jin DJ, Tang SA. Longipetalol A: a highly modified triterpenoid from *Dichapetalum longipetalum*. J Nat Prod. 2021;84:1556–62.

Chapter 6
Classification of Diverse Novel Limonoids

Limonoids in the plant kingdom occur mainly in the Meliaceae and Rutaceae families and less frequently in the Cneoraceae. Structurally, limonoids are formed by loss of four terminal carbons of the side chain in the apotirucallane or apoeuphane skeleton and then cyclized to form the 17β-furan ring. [1] There are 66 new skeleton limonoids have been excavated, they mainly experienced the breaking of the ring carbon–carbon bond and the subsequent rearrangement of the carbon skeleton (**488–502** Fig. 6.1, **503–538** Fig. 6.2, **539–553** Fig. 6.3; Table A5). In this chapter, we focus on the unique structures in limonoids and present their chemistry, distribution and structure characteristics.

6.1 Demolition of a Single Ring

*6.1.1 Ring A-*Seco *Group*

This class is characteristic of the cleavage of C-3/4 followed by 3,19-linked formation of a spirio-pentacyclic ring occurring in the ring A-*seco* group. Walsuronoid A (**488**) is the first limonoid peroxide isolated from the Meliaceae family, which features an unprecedented A-*seco* limonoid skeleton incorporating a 3,4-peroxide bridge [2]. Walrobsins A (**489**) and B (**490**), featuring an unprecedented 5-oxatricyclo[5.4.1^{1,4}]hendecane ring system and possessing a stable hemiketal structure formed between the OH-11 and 3-carbonyl group in the hexatomic oxoheterocyclic ring, have been isolated from the root barks of *Walsura robusta* [3]. Aphananoid A (**491**), isolated from *Aphanamixis polystachya*, is a limonoid which features a rare C_{24} appendage and new 5/6/5 fused-ring framework [4]. **491** was proposed to arise from the triterpene by the 3,4-*seco*-7,8-*seco*-6,8 *cyclo*-7,30-*decarbon* key pattern.

Y. Cheng and D. Qin, *Novel Plant Natural Product Skeletons*,
https://doi.org/10.1007/978-981-99-7329-3_6

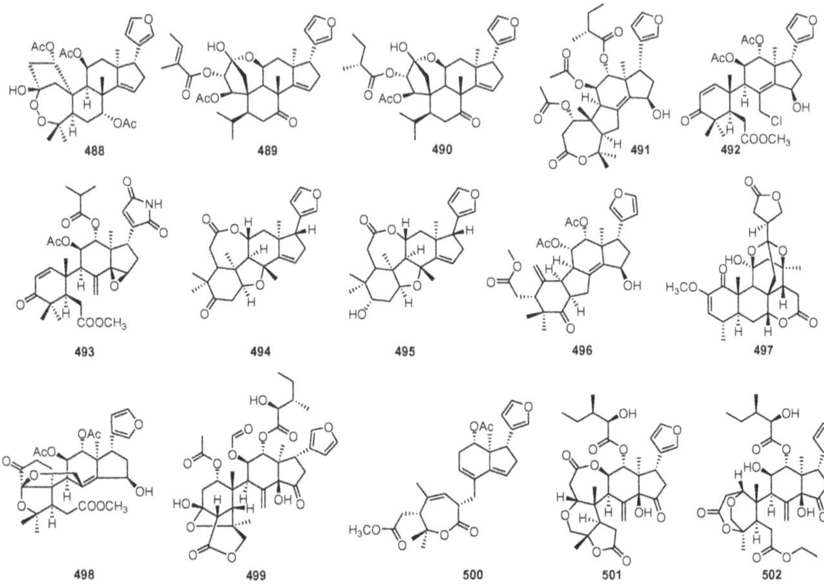

Fig. 6.1 Structures of new skeleton limonoids **488–502**

6.1.2 *Ring B*-Seco *Group*

Until now, the C-7/8 bond breakaged novel skeleton limonoids from Meliaceae were found only in the *Turraea* and *Toona* genera. Turrapubesins A (**492**) and B (**493**), representing the first examples of halogenated and maleimide-bearing limonoids, were isolated from the twigs and leaves of *Turraea pubescens*, their absolute configurations were determined by X-ray crystallography of **492** and by CD analysis of a dihydrogenated derivative of **493** [5]. Ciliatonoids A (**494**) and B (**495**) obtained from *Toona ciliate* feature an unprecedented limonoid architecture by furnishing a very unique *cis*-fused central motif of methylhexahydro-3*H*,6*H*-furo[3,4-c]oxepin-6-one [6]. Tooniliatone A (**496**), a novel limonoid with an unprecedented 6/5/6/5 (ABCD) tetracarbocyclic framework, was isolated and identified [7]. The divergent biomimetic sequence based on a carbocation from an epoxide moiety provides an inspiration for constructing complex polycyclic carbon skeletons.

6.1.3 *Ring D*-Seco *Group*

Baeyer–Villiger oxidation breaks the C13/C17 bond to produce structural diversity of ring D-*seco* group. Perforalactone A (**497**), a new 20*S* quassinoid with aunique cagelike 2,4-dioxaadamantane ring system and amigrated side chain, was isolated

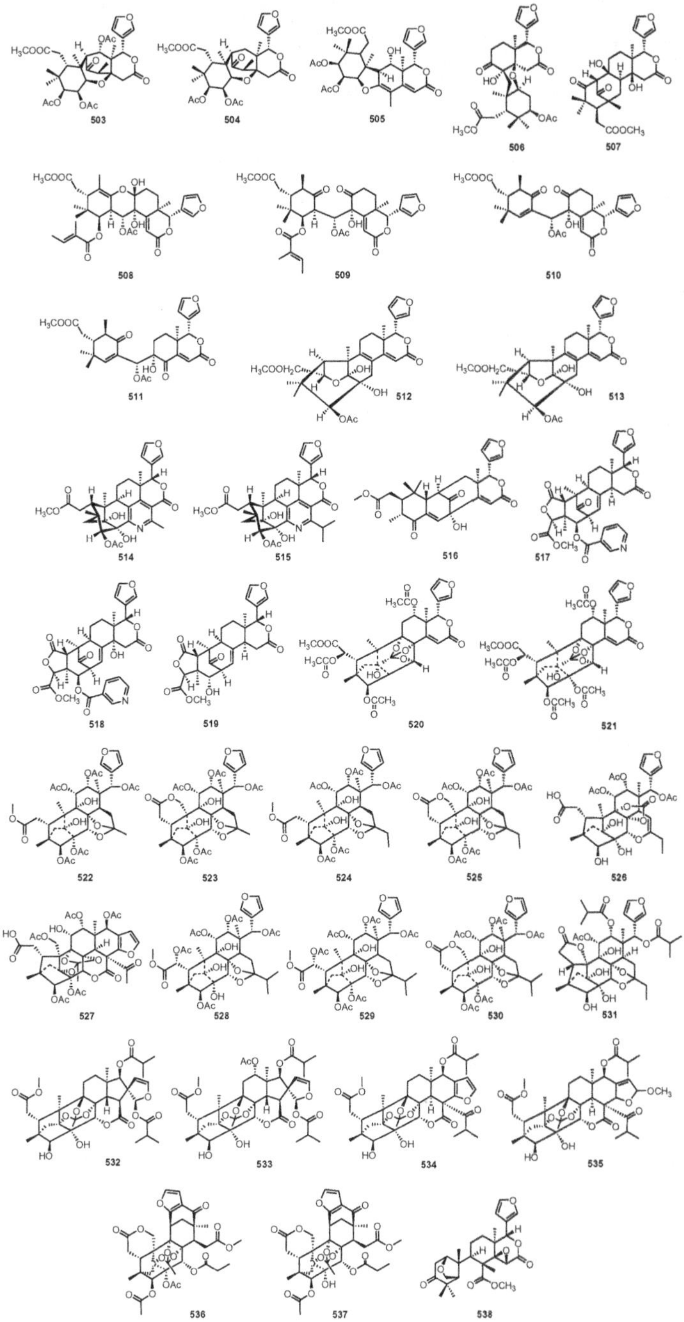

Fig. 6.2 Structures of new skeleton limonoids **503–538**

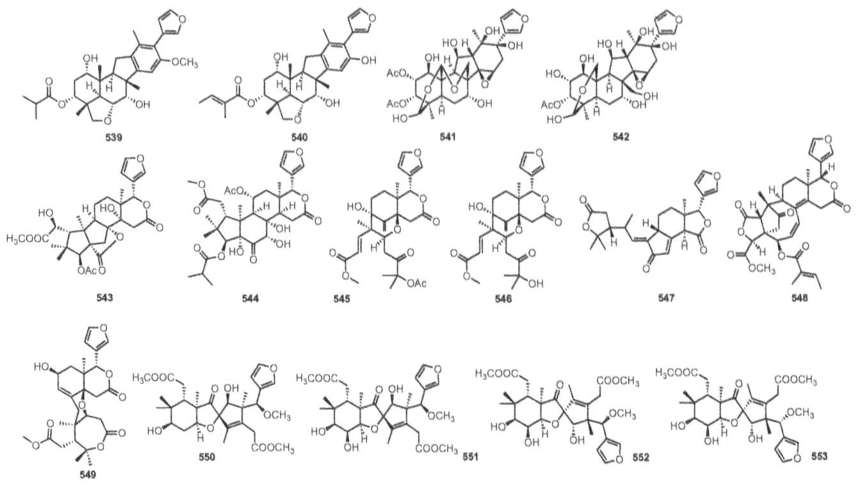

Fig. 6.3 Structures of new skeleton limonoids **539–553**

from *Harrisonia perforata* (Simaroubaceae), providing valuable inspiration for the development of new pesticides [8].

6.2 Demolition of Two Rings

6.2.1 Rings A,B-Seco Group

The rings A,B-*seco* group were depicted as arising from cleavages of C-3/4 (ring A) and C-7/8 (ring B) and formation of the complex limonoids. Musidunin (**498**), containing a unique acetal annulation, was isolated from *Croton jatrophoides* by bioassay-guided fractionation [9]. Aphanamolide A (**499**), featuring an unprecedented carbon skeleton via the formation of a C-3–C-6 bond, was isolated from the seeds of *Aphanamixis polystachya* [10]. Aphanamixoid A (**500**), was isolated from the leaves and twigs of *Aphanamixis polystachya*, which is formed via the unique cleavage of a C-9–C-10 bond as well as the formation of a C-2–C-30 bond by means of 3,3-rearrangement [11]. The absolute stereochemistry of **500** was determined by spectroscopic analysis and X-ray crystallography. Another two limonoids, aphanaonoid A (**501**) possessing an unprecedented fused 7/6/5 tricyclic skeleton and aphanaonoid B (**502**) featuring a 2,6-dioxabicyclo[3.2.2]nonan-3-one caged ring A system, were also identified [12].

6.2.2 *Rings B,D-Seco Group*

I. The cleavages of C-7/8 and C-16/17 followed by the formation of δ-lactonic D ring: Cipadesins A–C (**503–505**), possess a novel carbon skeleton, in which rings A and C are joined via C-10 and C-11 [13].

II. The cleavages of C-7/8 and C-16/17 and the formation of δ-lactonic D ring and an additional C–C bond between C-11 and C-30: Cipadonoid A (**506**), representing a new type of limonoid, was characterized by a rearranged tetrahydropyranyl ring B incorporating usually exocyclic C-30 and was isolated from *Cipadessa cinerasecns* [14].

III. The cleavages of C-7/8 and C-16/17 and the formation of δ-lactonic D ring and an additional C–C bond between C-10 and C-30: Trichiconin A (**507**), obtained as a colorless crystal from *Trichilia connaroides*, possessed a new carbon skeleton of a rearranged A,B-ring system [15].

IV. The cleavages of C-7/8 and C-16/17 and the formation of δ-lactonic D ring and an additional C–C bond between C-2 and C-30: Xylogranatin A (**508**) featuring by a unique 1,9-oxygen bridge, was isolated from the seeds of a Chinese marine mangrove *Xylocarpus granatum* and structurally secured by single-crystal X-ray diffraction, xylogranatins B (**509**) and C (**510**), possessing an unusual 9,10-*seco* skeleton, and xylogranatin D (**511**) featuring an unprecedented skeleton of C-30–C-9 linkage, have been described from *Xylocarpus granatum* [16]. Khayalenoids A (**512**) and B (**513**) with an unprecedented 8-oxatricyclo[4.3.2.0^{2,7}]undecane [] motif in the nortriterpenoid core, were isolated from the stems of *Khaya senegalensis* [17]. Thaixylomolins B (**514**) and C (**515**), limonoids containing a unique pentasubstituted pyridine scaffold, were obtained from the seeds of a Thai mangrove, *Xylocarpus moluccensis* [18]. Xylomexicanin I (**516**) represents an unprecedented limonoid with a bridged skeleton between the B- and C-rings, which has been isolated from *Xylocarpus granatum* [19]. Triconoids A–C (**517–519**), possessing a new rearranged mexicanolide skeleton and sharing an F ring of methyl 5-oxotetrahydrofuran-2-carboxylate postulated to be formed biosynthetically via a very unique chemical cascade, were isolated from the Nepalese plant *Trichilia connaroides* [20].

V. The cleavages of C-7/8 and C-16/17 and the formation of δ-lactonic D ring and a bridge ring between C-1 and C-29: Two highly oxidized octacyclic B and D-*seco*-limonoids sxyloccensins O (**520**) and P (**521**), belonging to a completely new type of phragmalins and naming 8,9,30-phragmalin ortho esters, were isolated from the mangrove plant *Xylocarpus granatum* [21]. Chuktabularins A–D (**522–525**), four novel 16-norphragmalin-type limonoids, were isolated from the stem bark of *Chukrasia tabularis*, whose structures are characteristic of unprecedented skeletons with a biosynthetically extended C2 or C3 unit at C-15 forming a unique 2,7-dioxabicyclo[2.2.1]heptane system [22]. Chuktabrin A (**526**), featuring an unprecedented 1,3-dioxolan-2-one and a 3,4-dihydro-2*H*-pyran formed via an ether bond between C-30 and C-1′ biosynthetically extended C3 unit at C-15, in contrast to **526**, chuktabrin B. (**527**) possesses an unprecedented polycyclic skeleton with a biosynthetically extended C2 unit (acetyl) at C-15. **526** and **527** were both isolated from *Chukrasia tabularis* [23]. Chukvelutins A–C (**528–530**) were isolated from the stem bark of *Chukrasia tabularis* var. Velutina collected in Xishuangbanna, China, which possess unprecedented 16-norphragmalin limonoid skeletons featuring with a characteristic that ketal moiety between the phragmalin skeleton and a biosynthetically extended isobutyryl group at C-15 to form a characteristic 2,7-dioxabicyclo[2.2.1]heptane moiety [24]. Chukrasone B (**531**) is the first 16,19-dinor limonoid backbone featuring a biosynthetically extended unusual 2,7-dioxabicyclo[2.2.1]heptane motif [25]. The proposed biosynthetic routes for **531** suggest new avenues of chemical transformation for this versatile class. Chukfuransins A (**532**) and B (**533**) characterized by a C-15/C-20 linkage and possessing an unique 2-oxaspiro[4.4]non-3-ene fragment along with chukfuransins C (**534**) and D (**535**) featuring C-15/C-21 bonding but bearing C-15 extended C4 unit, were attained from the twigs and leaves of *Chukrasia tabularis* [26]. Another two new interesting carbon skeletal limonoids, guianolides A (**536**) and B (**537**), were isolated from the seeds of *Carapa guianensis*, they feature an unprecedented carbon skeleton via the formation of C-11–C-21 bond in phragmalin-1,8,9-orthoacetate [27].

VI. The cleavages of C-6/7 and C-16/17 and the formation of δ-lactonic D ring: thaixylomolin A (**538**) is a secomahoganin-type limonoid with a novel 6-oxabicyclo[3.2.1]octan-3-one motif, its absolute configurations were determined by single-crystal X-ray diffraction analysis and circular-dichroism spectroscopy in combination with quantum-chemical calculations [18].

*6.2.3 Ring C,D-*Seco *Group*

The cleavages of C-13/14 and the formation of carbon–carbon bond C12/14: Walsucochinoids A (**539**) and B (**540**), isolated from *Walsura cochinchinensis*, are noteworthy in that they feature a rearranged motif of C/D rings, with a five-membered C ring fused with a six-membered aromatic D ring, adding to their structural complexity, an extra F ring of tetrahydrofuran is also formed connecting C-6 and C-28 [28]. Phyllanthoids A (**541**) and B (**542**), featuring a rare 6/6/5/6 ring system with a six-membered aliphatic D ring, were isolated from *Phyllanthus cochinchinensis* [29]. Furthermore, **541** has an oxide bridge between C-19 and C-30 to furnish an extra tetrahydropyran ring, in addition to the C-19/C-29 lactol bridge.

6.3 Demolition of Three Rings

*6.3.1 Ring A,B,D-*Seco *Group*

The cleavages of C-7/8 in ring B and C-16/17 in ring D and other cleavages of carbon–carbon bonds in ring A give rise to the structural diversity of this kind of new skeleton compounds. Grandifotane A (**543**), having a complex hexacyclic carbon skeleton without precedent among the known limonoid families, the enzymatic Baeyer–Villiger oxidation may be the key step in its biogenetic route [30]. Chukrasone A (**544**), isolated from *Chukrasia tabularis* incorporating a highly rearranged A/B ring system, represents a unique rearranged scaffold which could be originated from the mexicanolide-type limonoid [31]. Trichiconins B (**545**) and C (**546**) obtained from *Trichilia connaroides*, features an unprecedented A,B,D-seco skeleton [15]. Perforanoid A (**547**) has an all-carbon quaternary stereocenter at C13 and a novel BCD tricyclic ring system, it was isolated from the leaves of *Harrisonia perforata* and its structure was confirmed via total synthesis in 10 steps [32]. Another limonoids triconoid D (**548**) isolated from *Trichilia connaroides*, is found to furnish a new rearranged 1,2-*seco*-phragmalin skeleton, and its structure was fully accomplished by spectroscopic data and electrostatic circular dichroism analysis [20]. Harpertrioate A (**549**) was furnished from the twigs of *Harrisonia perforata*, which possesses an A,B,D-*seco*-limonoid skeleton with a highly rearranged ring B incorporating unusually exocyclic C-30. Its absolute configurations were established by single-crystal X-ray diffraction [33].

*6.3.2 Ring B,C,D-*Seco *Group*

Cipacinoids A–D (**550–553**) feature unprecedented spirocyclic skeletons which have been isolated from *Cipadessa cinerascens* [34]. For **551–553**, the 17*S* configuration structurally determined by the solid evidence of X-ray crystallography, is the first examples identified in the limonoid family. This finding suggests that the formally identified D-ring demolished limonoids bearing either a 17*R*-OMe or a 17*R*-OAc, whose absolute configurations were tentatively assigned by comparing NMR data and biosynthetic consideration, should be reconsidered.

References

1. Tan QG. Luo XD. Meliaceous limonoids: chemistry and biological activities. Chem Rev 2011;111:7437–7522.
2. Yin S, Wang XN, Fan CQ, Liao SG, Yue JM. The first limonoid peroxide in the meliaceae family: walsuronoid A from *Walsura robusta*. Org Lett. 2007;9:2353–6.

3. An FL, Sun DM, Li RJ, Zhou MM, Yang MH, Yin Y, Kong LY, Luo J. Walrobsins A and B, two anti-inflammatory limonoids from root barks of *Walsura robusta*. Org Lett. 2017;19:4568–71.
4. Yang BJ, Fan SR, Cai JY, Wang YT, Jing C, Guo JJ, Chen DZ, Hao XJ. Aphananoid A is an anti-inflammatory limonoid with a new 5/6/5 fused ring featuring a C_{24} carbon skeleton from *Aphanamixis polystachya*. J Org Chem. 2020;85:8597–602.
5. Wang XN, Yin S, Fan CQ, Wang FD, Lin LP, Ding J, Yue JM. Turrapubesins A and B, first examples of halogenated and maleimide-bearing limonoids in nature from *Turraea pubescens*. Org Lett. 2006;8:3845–8.
6. Liu CP, Wang GC, Gan LS, Xu CH, Liu QF, Ding J, Yue JM. Ciliatonoids A and B, two limonoids from *Toona ciliata*. Org Lett. 2016;18:2894–7.
7. Luo J, Huang WS, Hu SM, Zhang PP, Zhou XW, Wang XB, Yang MH, Luo JG, Wang C, Liu C, Yao HQ, Zhang C, Sun HB, Chen YJ, Kong LY. Rearranged limonoids with unique 6/5/6/5 tetracarbocyclic skeletons from *Toona ciliata* and biomimetic structure divergence. Org Chem Front. 2017;4:2417–21.
8. Fang X, Di YT, Zhang Y, Xu ZP, Lu Y, Chen QQ, Zheng QT, Hao XJ. Unprecedented quassinoids with promising biological activity from *Harrisonia perforata*. Angew Chem Int Ed. 2015;54:5592–5.
9. Nihei K, Asaka Y, Mine Y, Yamada Y, Iigo M, Yanagisawa T, Kubo I. Musidunin and musiduol, insect antifeedants from *Croton jatrophoides*. J Nat Prod. 2006;69:975–7.
10. Yang SP, Chen HD, Liao SG, Xie BJ, Miao ZH, Yue JM. Aphanamolide A, a new limonoid from *Aphanamixis polystachya*. Org Lett. 2011;13:150–3.
11. Cai JY, Zhang Y, Luo SH, Chen DZ, Tang GH, Yuan CM, Di YT, Li SH, Hao XJ, He HP. Aphanamixoid A, a potent defensive limonoid, with a new carbon skeleton from *Aphanamixis polystachya*. Org Lett. 2012;14:2524–7.
12. Zhang P, Xue S, Huang W, Wang C, Cui Z, Luo J, Kong L. Diverse prieurianin-type limonoids with oxygen-bridged caged skeletons from two *Aphanamixis* Species: discovery and biomimetic conversion. Org Chem Front. 2021;8:566–71.
13. Yuan XH, Li BG, Zhou M, Qi HY, Zhang GL. Cipadesins A−C: novel tetranortriterpenoids from *Cipadessa cinerascens*. Org Lett. 2005;7:5051–3.
14. Fang X, Di YT, He HP, Liu HY, Zhang Z, Ren YL, Gao ZL, Gao S, Hao XJ. Cipadonoid A, a novel limonoid with an unprecedented skeleton, from *Cipadessa cinerasecns*. Org Lett. 2008;10:1905–8.
15. Liu CP, Xu JB, Han YS, Wainberg MA, Yue JM. Trichiconins A-C, limonoids with new carbon skeletons from *Trichilia connaroides*. Org Lett. 2014;16:5478–81.
16. Yin S, Fan CQ, Wang XN, Lin LP, Ding J, Yue JM. Xylogranatins A−D: novel tetranortriterpenoids with an unusual 9,10-*seco* scaffold from marine mangrove *Xylocarpus granatum*. Org Lett. 2006;8:4935–8.
17. Yuan T, Yang SP, Zhang CR, Zhang S, Yue JM. Two limonoids, khayalenoids a and b with an unprecedented 8-oxa-tricyclo[4.3.2.$0^{2,7}$]undecane motif, from *Khaya senegalensis*. Org Lett. 2009;11:617–20.
18. Li J, Li MY, Bruhn T, Katele FZ, Xiao Q, Pedpradab P, Wu J, Bringmann G. Thaixylomolins A-C: limonoids featuring two new motifs from the Thai Xylocarpus moluccensis. Org Lett. 2013;15:3682–5.
19. Wu YB, Wang YZ, Ni ZY, Qing X, Shi QW, Sauriol F, Vavricka CJ, Gu YC, Kiyota H. Xylomexicanins I and J: limonoids with unusual B/C rings from *Xylocarpus granatum*. J Nat Prod. 2017;80:2547–50.
20. Wang GC, Fan YY, Shyaula SL, Yue JM. Triconoids A-D, four limonoids possess two rearranged carbon skeletons from Trichilia connaroides. Org Lett. 2017;19:2182–5.
21. Wu J, Xiao Q, Huang J, Xiao Z, Qi S, Li Q, Zhang S. Xyloccensins O and P, unique 8,9,30-phragmalin ortho esters from *Xylocarpus granatum*. Org Lett. 2004;1841(6):1844.
22. Zhang CR, Yang SP, Liao SG, Fan CQ, Wu Y, Yue JM. Chuktabularins A−D, four new limonoids with unprecedented carbon skeletons from the stem bark of *Chukrasia tabularis*. Org Lett. 2007;9:3383–6.

23. Zhang CR, Fan CQ, Zhang L, Yang SP, Wu Y, Lu Y, Yue JM. Chuktabrins A and B, two novel limonoids from the twigs and leaves of *Chukrasia tabularis*. Org Lett. 2008;10(3183):3186.
24. Luo J, Wang JS, Luo JG, Wang XB, Kong LY. Chukvelutins A–C, 16-norphragmalin limonoids with unprecedented skeletons from Chukrasia tabularis var Velutina. Org Lett. 2009;11:2281–4.
25. Liu HB, Zhang H, Li P, Gao ZB, Yue JM. Chukrasones A and B: potential Kv1.2 potassium channel blockers with new skeletons from *Chukrasia tabularis*. Org Lett. 2012;14:4438–441.
26. Hu K, Liu JQ, Li X, Chen JC, Zhang WM, Li Y, Li L, Guo L, Ma W, Qiu MH. Chukfuransins A-D, four new phragmalin limonoids with β-furan ring involved in skeleton reconstruction from Chukrasia tabularis. Org Lett. 2013;15(3902):3905.
27. Inoue T, Matsui Y, Kikuchi T, In Y, Yamada T, Muraoka O, Matsunaga S, Tanaka R. Guianolides A and B, new carbon skeletal limonoids from the seeds of *Carapa guianensis*. Org Lett. 2013;15:3018–21.
28. Han ML, Zhang H, Yang SP, Yue JM. Walsucochinoids A and B: new rearranged limonoids from *Walsura cochinchinensis*. Org Lett. 2012;14(486):489.
29. Zhao JQ, Wang YM, He HP, Li SH, Li XN, Yang CR, Wang D, Zhu HT, Xu M, Zhang YJ. Two new highly oxygenated and rearranged limonoids from *Phyllanthus cochinchinensis*. Org Lett. 2013;15:2414–7.
30. Yuan T, Zhu RX, Zhang H, Odeku OA, Yang SP, Liao SG, Yue JM. Structure determination of grandifotane A from *Khaya grandifoliola* by NMR, X-ray diffraction, and ECD calculation. Org Lett. 2010;12:252–5.
31. Liu HB, Zhang H, Li P, Gao ZB, Yue JM. Chukrasones A and B: potential Kv1.2 potassium channel blockers with new skeletons from *Chukrasia tabularis*. Org Lett 2012;14:4438–441.
32. Lv C, Yan X, Tu Q, Di Y, Yuan C, Fang X, Ben David Y, Xia L, Gong J, Shen Y, Yang Z, Hao X. Isolation and asymmetric total synthesis of perforanoid A. Angew Chem Int Ed. 2016;55:7539–43.
33. Tang XH, Luo RC, Ye MS, Tang HY, Ma YL, Chen YN, Wang XM, Lu QY, Liu S, Li XN, Yan Y, Yang J, Ran XQ, Fang X, Zhou Y, Yao YG, Di YT, Hao XJ. Harpertrioate A, an A, B, D-seco-limonoid with promising biological activity against alzheimer's disease from Twigs of Harrisonia perforata (Blanco) Merr. Org Lett. 2021;23:262–7.
34. Yu JH, Liu QF, Sheng L, Wang GC, Li J, Yue JM. Cipacinoids A-D, four limonoids with spirocyclic skeletons from Cipadessa cinerascens. Org Lett. 2016;18:444–7.

Chapter 7
Classification of Diverse Novel Phloroglucinols

Phloroglucinol derivatives are a major class of secondary metabolites with fascinating chemical structures and intriguing biological activities. A large number of differently substituted and structurally diverse phloroglucinol derivatives have been isolated from the land plants, meanwhile, a few phloroglucinols with unique skeletons are also reported. The polycyclic polyprenylated acylphloroglucinols (PPAPs) feature a highly oxygenated and densely substituted bicyclo[3.3.1]nonane-2,4,9-trione or bicyclo[3.2.1]octane-2,4,8-trione core decorated with C_5H_9 or $C_{10}H_{17}$ (prenyl, geranyl, etc.) side chains (the latter in several isomeric forms). The variability of phloroglucinols usually resulted in the construction of highly functionalized polycyclic scaffolds with complex structures. 195 new skeleton phloroglucinols were unearthed and sorted out (**554–595** Fig. 7.1, **596–613** Fig. 7.2, **614–638** Fig. 7.3, **639–652** Fig. 7.4, **653–679** Fig. 7.5, **680–718** Fig. 7.6, **719–748** Fig. 7.7; Table A6), which represent only a small portion of phloroglucinols that have been discovered from 1999 to 2021. Most of the novel fascinating architectures are derived from the two families Hypericaceae and Myrtaceae.

7.1 Bicyclic Polyprenylated Acylphloroglucinols (BPAPs)

7.1.1 Type A BPAPs

This subclass includes 15 BPAPs in which the acyl group is located at the C-1 position. Interestingly, garsubelone A (**554**) is the first dimeric PPAP-type metabolite having an unprecedented 6/6/6/6/6/6/6 heptacyclic system with 10 chiral centers. **554** was isolated from *Garcinia subelliptica* and its absolute configurations were determined by comprehensive spectroscopic and X-ray diffraction analyses [1]. Hyperforones A–C (**555–557**), isolated from *Hypericum forrestii*, are

Y. Cheng and D. Qin, *Novel Plant Natural Product Skeletons*,
https://doi.org/10.1007/978-981-99-7329-3_7

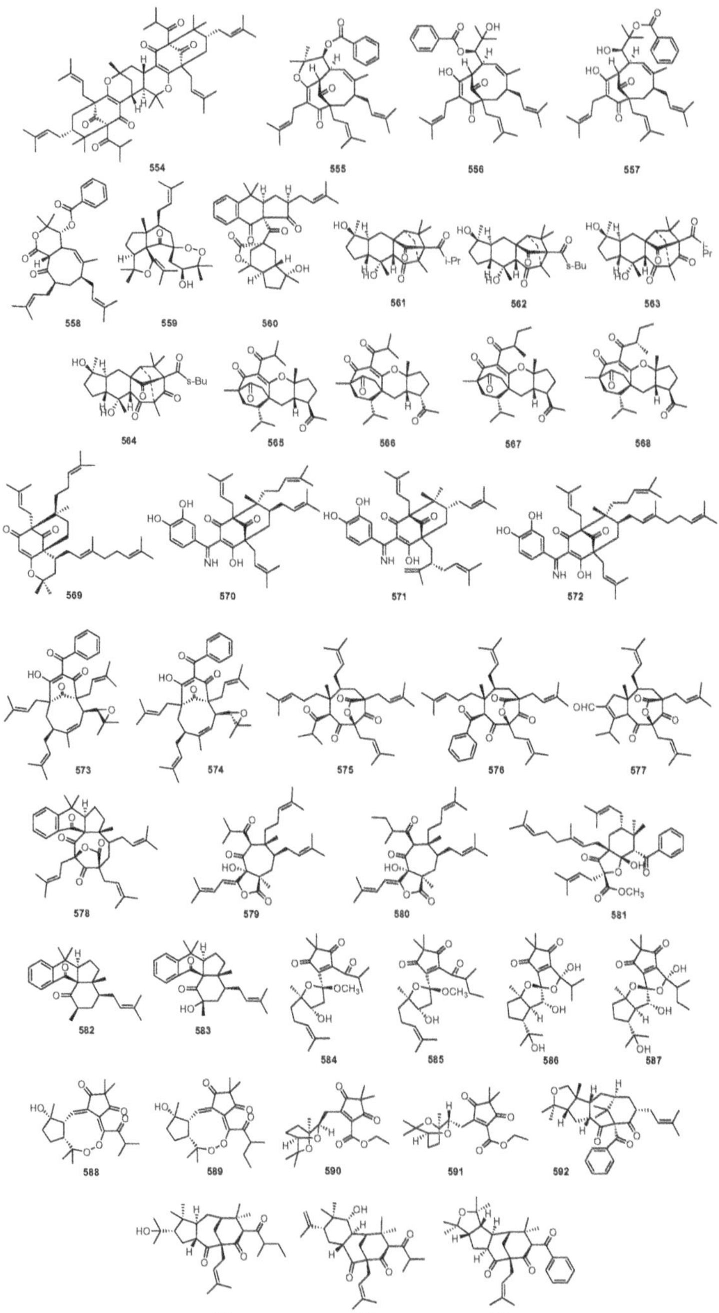

Fig. 7.1 Structures of new skeleton phloroglucinols **553–595**

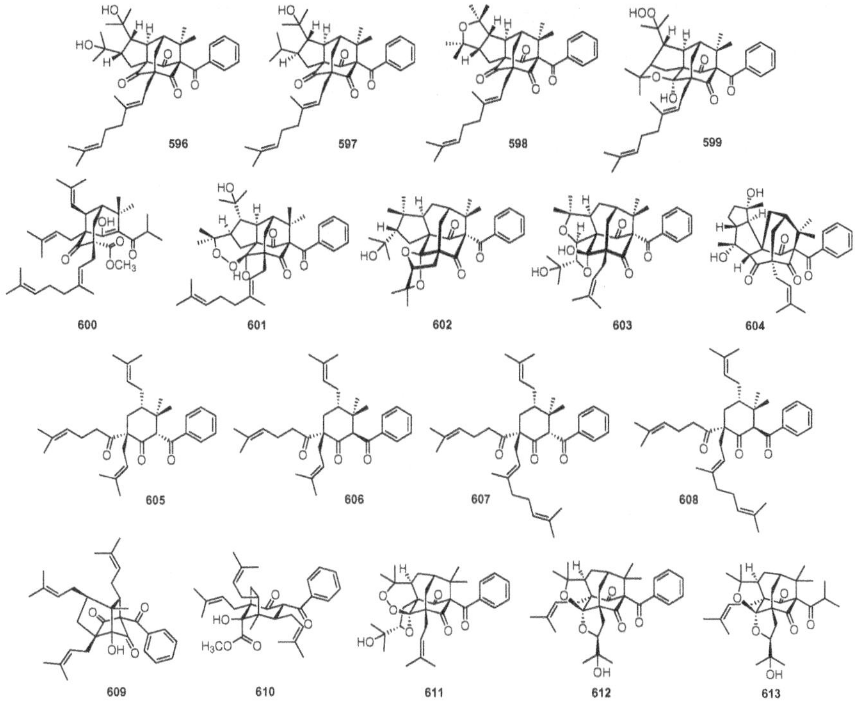

Fig. 7.2 Structures of new skeleton phloroglucinols **596–613**

the first examples of naturally occurring benzoyl-PPAPs with a unique C-1 H-substituted bicyclo[5.3.1]hendecane scaffold. Of note, the formation of the benzoyl-migrated C-1 H-substituted bicyclo[5.3.1]hendecane framework in **555–557** may undergo a sterically favored hemiketalization/Retro-Claisen cascade [2]. Hypaluton A (**558**), an unprecedented nor-polycyclic polyprenylated acylphloroglucinol (PPAP) bearing a new 8/6 bicyclic architecture, was obtained from *Hypericum patulum* [3]. Norwilsonnol A (**559**), a structurally complex polycyclic polyprenylated acylphloroglucinol (PPAP) bearing an unprecedented scaffold, was isolated from *Hypericum wilsonii* [4]. Hyperbeanone A (**560**), a novel 5,6-*seco*-polycyclic polyprenylated acylphloroglucinol (PPAP) derivative characterized by an undescribed benz[f]indene-1,9(4*H*)-dione ring system fused to a tricyclic γ-lactone unit via a ketone carbonyl, were isolated from the aerial parts of *Hypericum beanii* [5]. Hymoins A–D (**561–564**), two pairs of light-induced transformative polyprenylated acylphloroglucinols with an unprecedented pentacyclic skeleton, were isolated from the flowers of *Hypericum monogynum*. This is the first report on PPAPs with rare caged pentacyclic 5/6/6/5/5 architectures (**561** and **562**) and pentacyclic 5/6/6/6/5

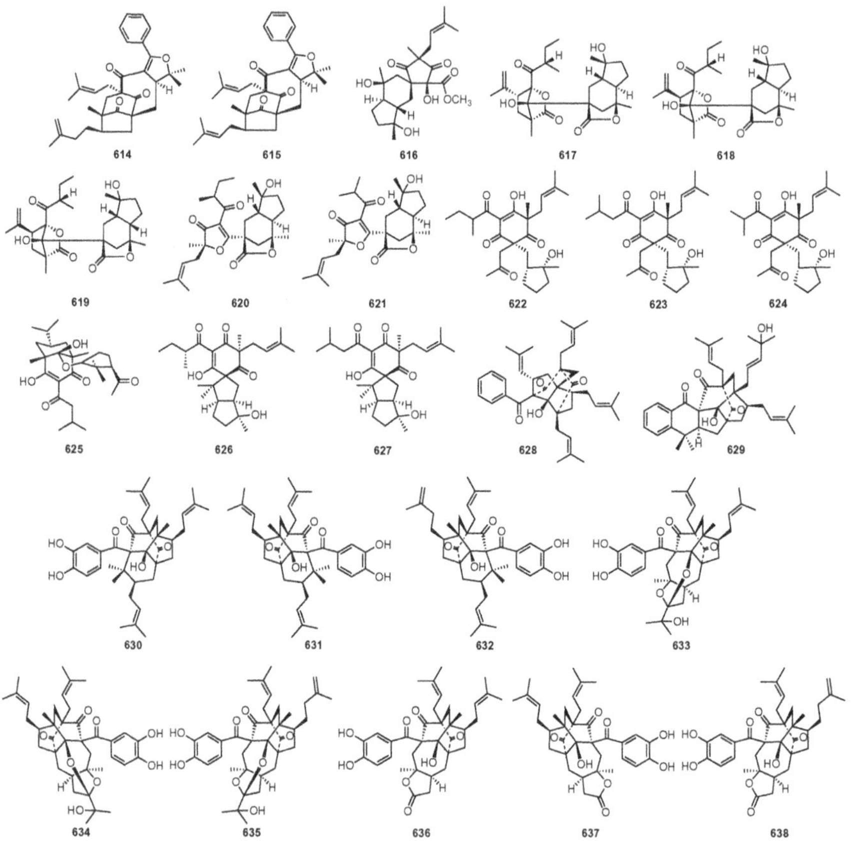

Fig. 7.3 Structures of new skeleton phloroglucinols **614–638**

(**563** and **564**) ring systems [6]. Hypermonins A–D (**565–568**), four rearranged nor-polycyclic polyprenylated acylphloroglucinols (PPAPs) with unprecedented skeletons, were isolated and identified from the flowers of *Hypericum monogynum*. **565–568** represented the first examples of highly modified norPPAPs characterized by a rare 7/6/6/5-tetracyclic system [7].

7.1.2 *Type B BPAPs*

This subclass, whose acyl groups are located at the C-3 position. Garcicowin A (**569**) is the first compound of this type without any substitution at C-2 and C-6, which was isolated from *Garcinia cowa* [8]. Garciyunnanimines A–C (**570–572**) from *Garcinia yunnanensis* is found to contain unprecedented imine functionality at C-10 [9]. Hyperberins A (**573**) and B (**574**) are the first examples of type B

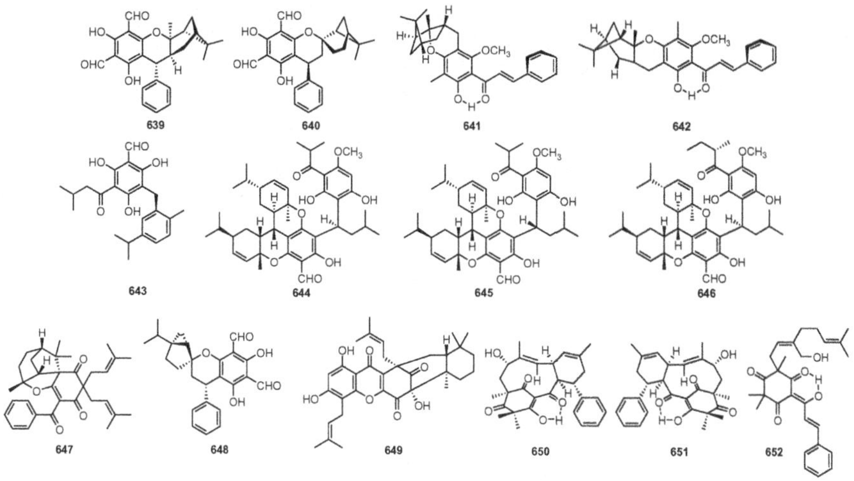

Fig. 7.4 Structures of new skeleton phloroglucinols **639–652**

polycyclic polyprenylated acylphloroglucinols with a bicyclo[5.3.1]hendecane core in the structure, which was isolated from *Hypericum beanii* and divergent cationic cyclization is the key steps in the biosynthesis [10].

7.1.3 *Seco*-BPAPs

Compounds **575–578** belong to type A derivatives and are suggested to be 1,9-*seco*-BPAPs. Hyphenrones A (**575**) and B (**576**) bear an unprecedented *seco*-PPAP skeleton formed by the cleavage of the C-1/C-9 bond of the normal *endo*-BPAPs, which has been described from *Hypericum henryi*, hyphenrones C (**577**) and D (**578**) were also isolated from *Hypericum henryi*, which feature fascinating 5/8/5 and 6/6/5/8/5 fused ring systems and were presumably synthetically derived from **575** and **576** via key aldol condensation and Diels–Alder cycloaddition reactions [11]. Ascyronone A (**579**) and B (**580**), another two *seco*-BPAPs, sharing an unusual seven-membered carbon core fused with a c-lactone ring, were isolated from *Hypericum ascyron* [12]. Hyperhexanone A (**581**) was characterized from *Hypericum sampsonii* and is the first PPAP with a 1,2-*seco*-bicyclo[3.3.1]-PPAP core arising from the cleavage of C-1/C-2 bond via a retro-Claisen condensation reaction [13]. Norascyronones A and B (**582** and **583**), isolated from *Hypericum ascyron*, are secondary metabolites resulting from bicyclic polyprenylated acylphloroglucinols via losing eight carbons [14]. It is of particular interest that **582** and **583** share an intriguing 6/6/5/6 ring system which could be generated via [4+2] intramolecular cyclization. Elodeoidins A–H (**584–591**) are eight unprecedented rearranged acylphloroglucinol meroterpenoids from *Hypericum elodeoides* [15]. Structurally, the six-membered acylphloroglucinol core

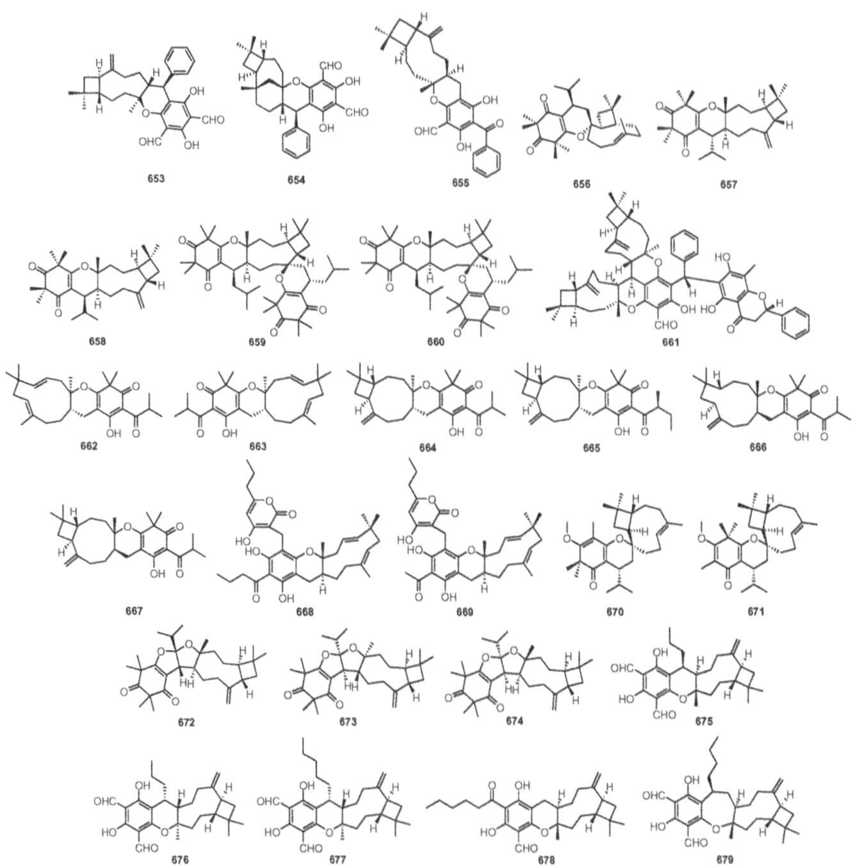

Fig. 7.5 Structures of new skeleton phloroglucinols **653**–**679**

rearranging to a five-membered β-diketone unit with an exocyclic carbonyl group (C-1 or C-3) through an α-ketol rearrangement makes them unusual. In detail, **584** and **585** are characteristic of a 2-cyclopentyltetrahydrofuran moiety via key cyclizations with C-1 or C-3. **586** and **587** feature a [5,5]-spiroketal-fused 5/5/5/5/5 skeleton. **588** and **589** bear a 1,2-dioxonane-bridged 5/9/5 framework. Whereas **590** and **591** possess a 2,5-dioxabicyclo[2.2.2]octane caged structure. Soniiglucinols A–D (**592**–**595**) are another four new polycyclic polyprenylated acylphloroglucinols (PPAPs) with diverse architectures from *Hypericum wilsonii* [16]. **592** and **593** represent the first examples of PPAPs that possess a fascinating tricyclo-[7.3.1.0^{3,7}] tridecane core bearing an unusual 5/7/6 carbon skeleton. **594** and **595** are the first PPAPs sharing intriguing tricyclo-[7.3.1.0^{2,7}]tridecane and tricyclo-[6.3.1.0^{2,6}]dodecane moieties.

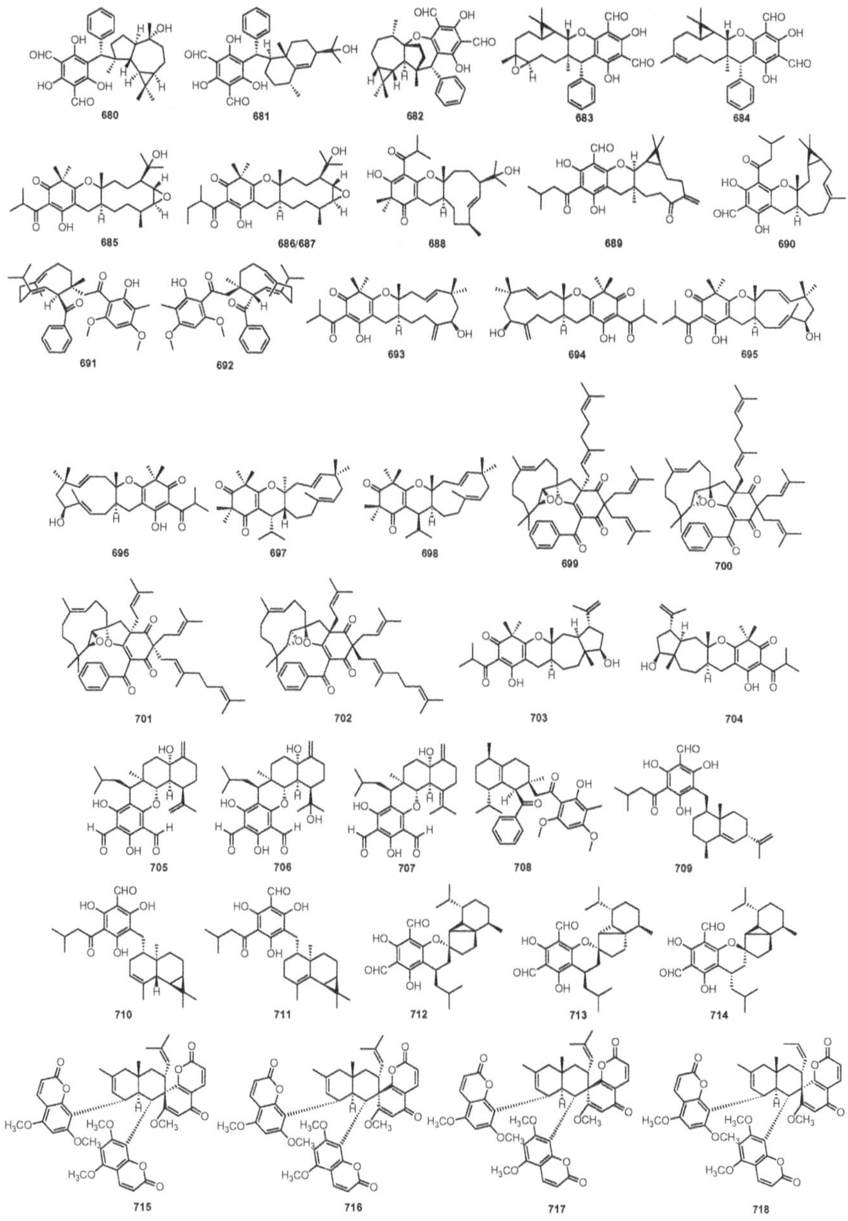

Fig. 7.6 Structures of new skeleton phloroglucinols **680–718**

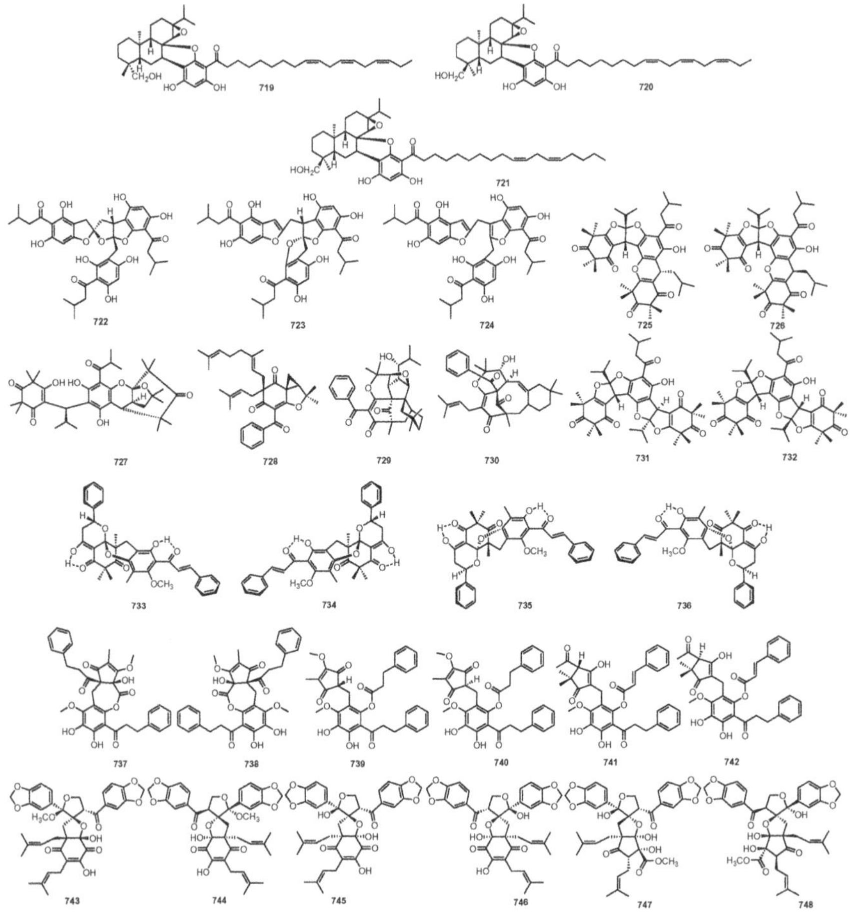

Fig. 7.7 Structures of new skeleton phloroglucinols **719–748**

7.2 Caged PPAPs with Adamantane and Homoadamantane Skeletons

7.2.1 Adamantane-Type PPAPs

This subclass includes 6 PPAPs shaped as a "diamond" caged core (**596** – **599**, **601**) or related *seco*-scaffold (**600**). They are all derived from corresponding type A BPAP precursors. Hyperisampsins A–D (**596–599**) represent the first examples of natural PPAPs with rigid caged tetracyclo[6.3.1.1^{3,10}.0^{3,7}]tridecane-2,11,13-trione skeletons, which were described from *Hypericum sampsonii* [17]. In addition, **598** featuring the presence of an additional furan ring on the western hemisphere, and **599** is characteristic of an additional pyran ring incorporated into the tetracyclo[6.3.1.1^{3,10}.0^{3,7}]tridecane ring systems and represents the first example of adamantyl PPAPs with an unusual hemiketal carbon in the adamantyl skeleton. Hypersubone A (**600**) from *Hypericum subsessile* is the first secoadamantane PPAP, which is proposed to be biosynthetically formed by the cleavage of the C-1/C-9 bond of normal adamantane, hypersubone B (**601**) was elucidated to possess a tetracyclo[6.3.1.1^{3,10}.0^{4,8}]-tridecane carbon skeleton with an unusual peroxide ring [18].

7.2.2 Homoadamantane-Type PPAPs

As shown in Table A6, 7 members of the homoadamantane-type PPAPs, which share a "diamond-like" caged core (**602–604**) or related *seco*-scaffold (**605–608**), all have type A structures [19–21]. Two novel PPAPs, dioxasampsones A and B (**602** and **603**), with unusual epoxy-ring-fused skeletons by new cyclization pattern, were isolated from the aerial parts of *Hypericum sampsonii* [19]. **602** possesses an unexpected hexacyclic skeleton with a rare 2,7-dioxabicyclo[2.2.1]heptane moiety, whereas **603** features the presence of a unique tetrahydrofuro[3,4-*b*]furan-fusedtricyclo[4.3.1.1^{5,7}]undecane undecane skeleton. Hypatulone A (**604**), an unprecedented homoadamantane architecture based on a tricyclo-[4.3.1.1^{3,8}]-undecane core and a unique 5/5/7/6/ 6 pentacyclic ring system, was isolated from *Hypericum patulum* [20]. Wagner–Meerwein rearrangement is considered to be the key step in its biosynthetic pathway. Norsampsones A–D (**605–608**) were isolated from *Hypericum sampsonii*. They feature an unprecedented carbon skeleton with the loss of C-2 carbonyl in the phloroglucinol ring. Retro-Claisen and decarboxylation reactions are involved in the formation of these norsampsones [21]. Hypatulins A (**609**) and B (**610**) are prenylated benzophenones with cage-like architectures from *Hypericum patulum* [22].

Of which, **609** has a unique densely substituted tricyclic octahydro-1,5-methanopentalene core, while **610** bears a bicyclo[3.2.1]octane moiety. Wilsonglucinols A–C (**611–613**) are novel homoadamantane architectures with unusual epoxy-ring-fused systems. Structurally, **611** features a new hexahydrofuro[2,3-*c*][1,2]dioxine motif, whereas **612** and **613** possess unusual decahydro-1,8:6a,10-dimethanocycloocta[1,2-*b*:1,8-*b*′]difuran fragments. **611–613** were all isolated from *Hypericum wilsonii* [23].

7.2.3 Other Caged PPAPs

Garcibractinones A (**614**) and B (**615**) bears an unprecedented caged tricyclo-[4.4.1.1^{1,4}] dodecane skeleton were described from *Garcinia bracteata*, whose isolation will enrich the metabolic diversity of *Garcinia* species [24].

7.3 Other PPAPs

7.3.1 *Spirocyclic PPAPs with Octahydrospiro[Cyclohexan-1,5′-Indene] Core*

This structure family comprises 12 PPAPs (Table A6). In general, they are considered to be the further intramolecular cyclization products derived from the less complicated MPAPs exemplified as hypercalin C. Biyouyanagiol (**616**), likely derived from spirophloroglucinols, possesses a spiro-skeleton with a monoterpene moiety as well as a unique cyclopenta-1,3-dione moiety. **616** was isolated from *Hypericum chinense* [25]. Biyoulactones A–C (**617–619**), three novel pentacyclic meroterpenoids with a unique dilactone structure containing C–C bonded bi- and tricyclic γ-lactone moieties, were formed and isolated from the roots of *Hypericum chinense* [26]. They might be biosynthesized undergoing intramolecular cyclization, dehydration, cyclization, Baeyer–Villiger oxidation, and lactonization. Furanmonogones A and B (**620** and **621**) from *Hypericum monogynum* possessing a novel 4,5-*seco*-3(2*H*)-furanone skeleton, is presumably biosynthesized from chinesins I and II via a key Baeyer–Villiger oxidation [27]. Interestingly, three unprecedented 12,13-*seco*-spirocyclic PPAPs named hyperilongenols A–C (**622–624**) with an enolizable β,β'-tricarbonyl system, were isolated from the stems and leaves of *Hypericum longistylum.* Structural determination of PPAPs with enolizable β-dicarbonyl and β,β'-tricarbonyl systems was suggested to use ^{1}H NMR spectrum recorded in chloroform-*d* in the range of 0–20 ppm [28]. Longisglucinol A–C (**625–627**) are another PPAPs with new skeletons from *Hypericum longistylum* [29]. Among them, **625** was characterized to possess a 8-oxatetracyclo-[8.3.3.0^{1,9}.0^{3,7}]cetane fragment bearing a new 6/6/6/5 fused architecture. Whereas **626** and **627** feature a 6/5/5 spirocyclic skeleton system, which were defined by the presence of a rare spiro[4.5]decane moiety.

7.3.2 *Complex PPAPs via Intramolecular [4+2] Cycloadditions from MPAPs*

The complicated PPAPs are derived from intramolecular [4+2] cycloadditions of MPAPs rather than via formation of the BPAPs. Hyperuralone A (**628**), a polycyclic polyprenylated acylphloroglucinol possessing an unprecedented tetracyclo[5.3.1.1^{4,9}.0^{f}]-dodecane core, and hyperuralone B (**629**), a congener with another complex caged skeleton formed via intramolecular Diels-Alder reactions, were characterized from *Hypericum uralum* [30]. (+)- and (−)-Garcimulins A (**630** and **631**) as well as garcimulin B (**632**), featuring the unique caged tetracyclo[5.4.1.1^{1,5}.0^{9,13}] tridecane skeleton, have been described from *Garcinia multiflora* [31]. It is interesting that the first examples of PPAPs characterized by the coupling of two novel cages, 2,11-dioxatricyclo[4.4.1.$0^{3,9}$]undecane and tricyclo[4.3.1.$0^{3,7}$]decane, (+)- and (−)-garmultins A, and (−)-garmultin B (**633–635**), along with three biogenetically related analogues, (+)- and (−)-garmultins F, and (−)-garmultin G (**636–638**) possessing an unusual 31,35-γ-lactone ring, were isolated from *Garcinia multiflora* [32].

7.4 Phloroglucinol–Monoterpenoid Meroterpenoids (PMMs)

7.4.1 PMMs Containing Sabinene Moiety

Four phloroglucinol-sabinene adducts were obtained from Myrtaceae. Guadials B (**639**) and C (**640**) are two new monoterpene-based meroterpenoids with unprecedented skeletons from the Leaves of *Psidium guajava* [33], while operculatols A (**641**) and B (**642**) represent the first examples of phloroglucinol-terpene adducts (PTAs) bearing a 2,4-dimethyl-cinnamyl-phloroglucinol moiety isolated from *Ophiorrhiza japonica* (Myrtaceae) [34].

7.4.2 PMMs Containing Pinene Moiety

Eucalrobusone F (**643**) is a novel adduct formed between aformyl-derived carbon-atom on the phloroglucinol ring and monoterpene [35]. Eucalyptusdimers A–C (**644–646**), three dimeric phellandrene-derived meroterpenoids, possess an unprecedented fused skeleton between two phellandrene and two acylphloroglucinol subunits, were isolated from *Eucalyptus robusta* [36].

7.4.3 Other PMMs

Hypatone A (**647**), a hybrid consisting of dearomatized isoprenylated acylphloroglucinol (DIAP) and monoterpenoid, was characterized from *Hypericum patulum* [37]. Structurally, **647** possesses an unprecedented spiro[bicyclo[3.2.1]octane-6,1′-cyclohexan]-2′,4′,6′-trione core as elucidated by extensive spectroscopic and X-ray crystallographic analyses. Guadial A (**648**) is the first monoterpene-based meroterpenoid with unprecedented skeleton isolated from the leaves of *Psidium guajava* [38]. Oliganthin M (**649**), a unique xanthone derivative from the leaves of *Garcinia oligantha*, is the first novel hybrid monoterpene-tetrahydroxanthone

[39]. Interestingly, another three novel PTAs (**650–652**) featuring a rare 2,2,4-trimethylcinnamyl-β-triketone motif were also isolated from the buds of *Cleistocalyx operculatus* [40]. Among them, (+)- and (−)-cleistocaltones A (**650** and **651**) are a pair of PTA enantiomers possessing an unprecedented, highly functionalized tricyclo[11.3.1.0^{3,8}]heptadecane bridged core skeleton with an unusual bridgehead enol. Moreover, the acylphloroglucinol and terpenoid motifs are connected through three C–C bonds (C-4–C-1′, C-8–C-5′, and C-9–C-8′) thereof, which is unprecedented in PTAs. Cleistocaltone B (**652**) is a novel PTA comprising a geranyl moiety and a 2,2,4-trimethylcinnamyl-β-triketone unit. Notably, **652** was isolated as a pair of inseparable tautomers due to the existence of an enolizable β,β'-tricarbonyl motif in the molecule.

7.5 Phloroglucinol–Sesquiterpenoid Meroterpenoids (PSMs)

7.5.1 PSMs Containing Caryophyllene Moiety

Among the caryophyllene-based meroterpenoids, guajadial (**653**), a novel caryophyllene-based meroterpenoid, was isolated from the leaves of *Psidium guajava*, which is formed via Diels-Alder reaction between the phloroglucinol moiety and caryophyllene moiety [41]. Psiguadials A (**654**) is a caryolane-based meroterpenoid from the leaves of *Psidium guajava*, which features a sesquiterpenoid-diphenylmethane meroterpenoid with unusual skeleton possessing an unusual coupling pattern [42]. Guapsidial A (**655**), isolated from leaves of *Psidium guajava*, is a novel sesquiterpene-based *Psidium* meroterpenoid with an unusual coupling pattern, in which C-14′ and C-5′ rather than C-1′ and C-3′ of the acylphloroglucinol moiety are involved in the construction of the dihydropyran ring [33].

It is noteworthy that because of this unusual coupling pattern, the overall structure of **655** behaves helical-like shape rather than propeller-like, as most common *Psidium* meroterpenoids, based on the lowest-energy conformer obtained from DFT calculation. **656–658** are three new carbon skeleton sesquiterpene-based meroterpenoids isolated from *Myrtus communis* [43]. Especially, **656** is characteristic of an unprecedented octahydrospiro[bicyclo[7.2.0]undecane-2,2′-chromene] tetracyclic ring system. Rhodomyrtials A and B (**659** and **660**), likely arising from the hetero-Diels-Alder (HDA) reaction, feature a triketone-sesquiterpene-triketone framework constructed from the conjugation of two-1,1,3,3-tetramethyl-isopentyl cyclohexatrione and caryophyllene moieties [44]. **659** and **660** were isoalted from *Rhodomyrtus tomentosa*. It's remarkable that guajavadimier A (**661**), isolated from the leaves of *Psidium guajava*, represents the most complex sesquiterpene-based meroterpenoid dimer identified so far, which possesses an unprecedented two caryophyllenes, a benzylphlorogulcinol, and a flavonone-fused complicated stereochemical skeleton [45]. (+)-Hyperjapone A (**662**) and (−)-hyperjapone A

(**663**), possessing a 11/6/6 fused ring system, was obtained as a racemic mixture from *Hypericum japonicum* and was separated by a column coated with cellulose tris(4methylbenzoate) after attempts with various chiral materials, hyperjapones B–E (**664–667**), containing a caryophyllane-type sesquiterpenoid moiety in their molecules, are two pairs of diastereoisomers from *Hypericum japonicum* [46]. They are the first examples of the hybridization between a sesquiterpenoid unit and a trimethylated acylphloroglucinol in *Hypericum* species, expanding plant resources for diverse meroterpenoids. Drychampones A and C (**668** and **669**) feature a new carbon skeleton with the incorporation of a sesquiterpenoid moiety to an unusual phloroglucinol derivative via a hetero-Diels-Alder cycloaddition to form the unexpected 11/6/6 ring system and they were isolated from *Dryopteris championii* [47]. Frutescones A and C (**670** and **671**) from the aerial parts of *Baeckea frutescens* possess a rare carbon skeleton with an unprecedented oxa-spiro[5.8] tetradecadiene ring system, existing as two favored equilibrating conformers in $CDCl_3$ solution, identified by variable-temperature NMR [48]. Rhodomyrtusials A–C (**672–674**), are the first examples of triketone-sesquiterpene meroterpenoids from *Rhodomyrtus tomentosa*, which feature a unique 6/5/5/9/4 fused pentacyclic ring system [49]. More excitingly, bioinspired total syntheses of **672–674** were achieved in six steps utilizing areactive enetrione intermediate generated in situ from alreadily available hydroxy-endoperoxide precursor. The latest new skeletal phloroglucinol–caryophyllene adducts are littordials A–E (**675–679**) until 2019, they feature unusual acyl phloroglucinol units and were isolated from the leaves of *Psidium littorale* [50]. Particularly, **679** possesses a rare 6/7/9/4-fused tetracyclic ring system.

7.5.2 PSMs Containing Germacrene Moiety

Psidials B (**680**) and C (**681**) represent the new skeleton of the 3,5-diformylbenzyl phloroglucinol-coupled sesquiterpenoid from the leaves of *Psidium guajava*, with 3,5-dimethyl-2,4,6-trihydroxybenzophenon as the key biogenetic precursor [51]. Another three isolates psiguadials A, C and D (**682–684**), obtained from the leaves of *Psidium guajava*, represent the novel sesquiterpenoid-diphenylmethane meroterpenoids with unprecedented skeletons [38, 42]. Hyperjaponols D–G (**685–688**), possessing hybrid structures of acylfilicinic acid moieties and germacrene moiety bearing unusual 6/6/10 ring system, were isolated from *Hypericum japonicum*, Hitherto, antipodes of **685–688** have not been found in the present study [52]. Eucalrobusones D and E (**689** and **690**) are formed by connecting a chroman ring to a bicyclogermacrane nucleus at the C-3/C-4 position, and they were characterized from the leaves of *Eucalyptus robusta* [35]. (−)- and (+)-Leptosperols A (**691** and **692**) feature an unprecedented 1-benzyl-2-(2-phenylethyl) cyclodecane backbone, they were isolated from *Leptospermum scoparium* and represent the first example of phloroglucinol derivatives biogenetically constructed by a De Mayo reaction [53].

7.5.3 PSMs Containing Humulene Moiety

Fantastically, (±)-hyperjaponol A and (±)-hyperjaponol B (**693–696**), possessing 6/6/11 ring systems, belong to the range of filicinic acid based meroterpenoids [52]. They were described from *Hypericum japonicum.* Another two humulene-based meroterpenoids (**697** and **698**) with new skeletons were isolated from the leaves of *Myrtus communis.* Interestingly, they are a pair of enantiomers [43]. (−)-Hypulatone A, (+)-hypulatone A, (−)-hypulatone B and (+)-hypulatone B (**699–702**) are two racemic meroterpenoids from *Hypericum patulum*, which possess an unprecedented spiro[benzofuran2,1′-cycloundecan]-4′-ene-4,6(5*H*)-dione core, Importantly, they might be derived from an interesting radical cyclization [54].

7.5.4 PSMs Containing Daucene Moiety

Among the investigated references, there are only two phloroglucinol–daucene meroterpenoids, (±)-hyperjaponol C (**703/704**), were obtained from *Hypericum japonicum* [52].

7.5.5 PSMs Containing Cadinane Moiety

Particularly, cadinane-type sesquiterpenoids have never been found in the genus of *Eucalyptus*, while eucalyptals A–C (**705–707**) with a new skeleton of 3,5-diformyl-isopentyl phloroglucinol-coupled cadinane were isolated from the fruits of *Eucalyptus globulus*, the biogenetic precursor of the key cadinane-type intermediate was thus proposed to be litseagermacrane, a coexisting major compound isolated from the fruits of *Eucalyptus globulus* [55] Leptosperol B (**708**) is another cinnamoylphloroglucinol-sesquiterpenoid hybrid from *Leptospermum scoparium* featuring an unprecedented 2-benzyl-3-phenylethyl decahydronaphthalene backbone [53].

7.5.6 PSMs Containing Eudesmene Moiety

Assembled via jensenone (4,6-diformyl-2-isopentanoylphloroglucinol) and daeudesmane-type sesquiterpene moieties, eucalrobusone C (**709**), a novel formyl-phloroglucinol meroterpenoid were acquired from the leaves of *Eucalyptus robusta* [35].

7.5.7 *PSMs Containing Maaliene Moiety*

The only two maaliene-based meroterpenoids, eucalrobusones A and B (**710** and **711**), have an unprecedented skeleton in which amaaliane sesquiterpene moiety is attached at a formyl-derived position the phloroglucinol ring, and they were characterized from the leaves of *Eucalyptus robusta* [35].

7.5.8 *PSMs Containing Cubebene Moiety*

Until to now, eucalrobusones G–I (**712–714**) are the first examples of cubebane-based FPMs connected by an unusual 1-oxaspiro[5.5]undecane core, which were isolated from the leaves of *Eucalyptus robusta* [35].

7.5.9 *Other PSMs*

(±)-Spirotriscoumarin A and (±)-spirotriscoumarin B (**715–718**), two pairs of oligomeric coumarin enantiomers with a spirodienone-sesquiterpene skeleton from *Toddalia asiatica*, not only feature the first spirodienone-sesquiterpene fused skeleton but also the most complex oligomeric coumarins. Intriguingly, the cyclic sesquiterpene moiety is normally derived from the fundamental precursor farnesyl diphosphate (FPP), which is generally synthesized via the addition of a C5 IPP unit to GPP in an extension of the GPP synthase reaction [56].

7.6 Phloroglucinol–Diterpenoid Meroterpenoids (PDMs)

Chlorabietols A–C (**719–721**), isolated from the roots of the rare Chloranthaceae plant *Chloranthus oldhamii*, contains an unprecedented skeleton featuring an *ent*-abietane-type diterpenoid coupled with an alkenyl phloroglucinol moiety by forming an unexpected 2,3-dihydrofuran ring. Up to now the occurrence of phloroglucinol-diterpene adducts from natural sources has never been reported [57].

7.7 Prenylated Phloroglucinols (PPs)

Three novel phloroglucinol derivatives of lysidicins A–C (**722–724**) have been isolated from the roots of *Lysidice rhodostegia* [58]. **722** and **724** possess spirocyclic benzodihydrofuran skeleton. Tomentosones A and B (**725** and **726**), possessing a ring system new to science, feature constructing a novel hexacyclic ring system [59]. Another PPs with an unprecedented carbon skeleton, myrtucommuacetalone (**727**), were afforded from *Myrtus communis* [60]. Hypercohin K (**728**) from *Hypericum cohaerens* is the first PPAP that with a cyclopropane ring fused to the phloroglucinol core [61]. Hyperprins A (**729**) and B (**730**) are two polyprenylated acylphloroglucinol related meroterpenoids with undescribed carbon skeletons from *Hypericum przewalskii* [62]. **729** possesses a new 6/6/6/6/5/5 hexacyclic system with an unprecedented tetracyclo[10.3.1.$0^{3,8}$.$0^{8,12}$] hexadecane motif, while **730** features a unique 6/8/6/6 tetracyclic scaffold. Rhodomentosones A and B (**731** and **732**), two pairs of novel enantiomeric phloroglucinol trimers featuring a unique 6/5/5/6/5/5/6-fused ring system were isolated from *Rhodomyrtus tomentosa* [63].

7.8 Phloroglucinol–Phenylpropanoids

In this unit, 16 unusual skeletons phloroglucinol–phenylpropanoids (**733–748**) were reported. (±)-Cleistoperlone A and (±)-cleistoperlone B (**733–736**) are two pairs of new enantiomeric phloroglucinol dimers possessing an unprecedented polycyclic skeleton with a highly functionalized dihydropyrano[3,2-d]xanthene tetracyclic core [34]. They were described from *Cleistocalyx operculatus*. Structurally, (±)-xanthchrysone A (**737** and **738**), isolated from *Xanthostemon chrysanthus*, are the first examples of phloroglucinol dimers linked by a hexanolactone unit, in which an unusual *bis*-phenylpropanoyl-benzo[*b*]cyclopent[*e*]oxepine backbone is present, (±)-xanthchrysone B and (±)-xanthchrysone C (**739–742**) are the first examples of phloroglucinols from *Xanthostemon chrysanthus*, featuring an unprecedented 1-(cyclopentylmethyl)-3-(3-phenylpropanoyl)benzene scaffold [64]. (±)-Melipatulinones A–C (**743–748**) are three unique enantiomeric pairs of lignan–phloroglucinol hybrids from *Melicope patulinervia*. Among them, melipatulinones A and B share a novel spiro[hydrobenzofuran-2,3′-furan] 5/5/6 tricyclic ring system, while melipatulinone C features an unprecedented spiro[cyclopenta[b]hydrofuran-2,3′-furan] 5/5/5 tricyclic framework [65].

References

1. Wang YL, Ye YS, Fu WW, Wu R, Xiang Q, Lao YZ, Yang JL, Tan HS, Yang XW, Yang BC, Xu HX, Xu G. Garsubelone A, the first dimeric polycyclic polyprenylated acylphloroglucinols with complicated heptacyclic architecture from *Garcinia subelliptica*. Org Lett. 2019;21:1534–7.
2. Lu WJ, Xu WJ, Zhang YQ, Li YR, Zhou X, Li QJ, Zhang H, Luo J, Kong LY. Hyperforones A–C, benzoyl-migrated [5.3.1]-type polycyclic polyprenylated acylphloroglucinols from *Hypericum forrestii*. Org Chem Front. 2020;7:1070–76.
3. Duan YL, Xie SS, Bu PF, Guo Y, Shi ZY, Guo Y, Cao YF, Sun WG, Qi CX, Zhang YH. Hypaluton A, an immunosuppressive 3,4-nor-polycyclic polyprenylated acylphloroglucinol from *Hypericum patulum*. J Org Chem. 2021;86:6478−85.

4. Xie SS, Zhou Y, Tan XS, Sun WG, Duan YL, Feng H, Sun LJ, Guo Y, Shi Z, Hao XC, Chen G, Qi CX, Zhang YH. Norwilsonnol A, an immunosuppressive polycyclic polyprenylated acylphloroglucinol with a spiro[5-oxatricyclo[6.4.0.03,7]dodecane-6′,1-1′,2′-dioxane] system from *Hypericum wilsonii*. Org Chem Front. 2021;8:2280–86.
5. Yang BY, Huang JZ, Lin S, Tong QY, Yao ZY, Li FL, Ye Y, Hu ZX, Zhang YH. Hyperbeanone A, a 5,6-seco-spirocyclic polycyclic polyprenylated acylphloroglucinol derivative with an unprecedented skeleton from *Hypericum beanii*. Org Chem Front. 2021;8:6411–8.
6. Zeng YR, Yuan CM, Li YN, Huang LJ, Hu ZX, Gu W, Hao XJ. Hymoins A–D: two pairs of polyprenylated acylphloroglucinols from *Hypericum monogynum* and their light-induced transformation. Org Lett. 2021;23:3125–9.
7. Zeng YR, Li YN, Zhang ZZ, Hu ZX, Gu W, Huang LJ, Li YM, Yuan CM, Hao XJ. Hypermoins A–D: rearranged nor-polyprenylated acylphloroglucinols from the flowers of *Hypericum monogynum*. J Org Chem. 2021;86:7021–7.
8. Xu G, Kan WLT, Zhou Y, Song JZ, Han QB, Qiao CF, Cho CH, Rudd JA, Lin G, Xu HX. Cytotoxic acylphloroglucinol derivatives from the twigs of *Garcinia cowa*. J Nat Prod. 2010;73:104–8.
9. Zheng D, Zhang H, Zheng CW, Lao YZ, Xu DQ, Xiao LB, Xu HX. Garciyunnanimines A–C, Novel cytotoxic polycyclic polyprenylated acylphloroglucinol imines from *Garcinia yunnanensis*. Org Chem Front. 2017;4:2102–8.
10. Xu WJ, Tang PF, Lu WJ, Zhang YQ, Wang XB, Zhang H, Luo J, Kong LY. Hyperberins A and B, type B polycyclic polyprenylated acylphloroglucinols with bicyclo[5.3.1]hendecane core from *Hypericum beanii*. Org Lett. 2019;21:8558–62.
11. Yang XW, Ding Y, Zhang JJ, Liu X, Yang LX, Li XN, Ferreira D, Walker LA, Xu G. New acylphloroglucinol derivatives with diverse architectures from *Hypericum henryi*. Org Lett. 2014;16:2434–7.
12. Kong LM, Long XW, Yang XW, Xia F, Khan A, Yan H, Deng J, Li X, Xu G. *Seco*-Polycyclic polyprenylated acylphloroglucinols with unusual carbon skeletons from *Hypericum ascyron*. Tetrahedron Lett. 2017;58:2113–7.
13. Zhu H, Chen C, Yang J, Li D, Zhang J, Guo Y, Wang J, Luo Z, Xue Y, Zhang Y. Hyperhexanone A, a crucial intermediate from bicyclo[3.3.1]- to cyclohexanone monocyclic-polycyclic polyprenylated acylphloroglucinols. *Tetrahedron* 2016;72:4655–59.
14. Hu YL, Hu K, Kong LM, Xia F, Yang XW, Xu G. Norascyronones A and B, 2,3,4-*nor*-polycyclic polyprenylated acylphloroglucinols from *Hypericum ascyron*. Org Lett. 2019;21:1007–10.
15. Li QJ, Tang PF, Zhou X, Lu WJ, Xu WJ, Luo J, Kong LY. Elodeoidins A-H, acylphloroglucinol meroterpenoids possessing diverse rearranged skeletons from *Hypericum elodeoides*. Org Chem Front. 2021;8:1409–14.
16. Xie S, Qi C, Duan Y, Xu Q, Liu Y, Huang Y, Yin X, Sun W, Zhou Y, Zhang Y. Discovery of new polycyclic polyprenylated acylphloroglucinols with diverse architectures as potent cyclooxygenase-2 inhibitors. Org Chem Front. 2020;7:1349–57.
17. Zhu H, Chen C, Yang J, Li XN, Liu J, Sun B, Huang S,X, Li D, Yao G, Luo Z, Li Y, Zhang J, Xue Y, Zhang Y. Bioactive acylphloroglucinols with adamantyl skeleton from *Hypericum Sampsonii*. Org Lett. 2014;16:6322–25.
18. Liao Y, Liu X, Yang J, Lao YZ, Yang XW, Li XN, Zhang JJ, Ding Z, Xu HX, Xu G. Hypersubones A and B, new polycyclic acylphloroglucinols with intriguing adamantane type cores from *Hypericum subsessile*. Org Lett. 2015;17:1172–5.
19. Tian WJ, Qiu YQ, Yao XJ, Chen HF, Dai Y, Zhang XK, Yao XS. Dioxasampsones A and B, two polycyclic polyprenylated acylphloroglucinols with unusual epoxy-ring-fused skeleton from *Hypericum sampsonii*. Org Lett. 2014;16:6346–9.
20. Liu YY, Ao Z, Xue GM, Wang XB, Luo JG, Kong LY. Hypatulone A, a homoadamantane-type acylphloroglucinol with an intricately caged core from *Hypericum patulum*. Org Lett. 2018;20:7953–6.
21. Tian WJ, Yu Y, Yao XJ, Chen HF, Dai Y, Zhang XK, Yao XS. Norsampsones A–D, four new decarbonyl polycyclic polyprenylated acylphloroglucinols from *Hypericum sampsonii*. Org Lett. 2014;16:3448–51.

22. Tanaka N, Yano Y, Tatano Y, Kashiwada Y. Hypatulins A and B, meroterpenes from *Hypericum patulum*. Org Lett. 2016;18:5360–3.
23. Xie S, Qi C, Duan Y, Hao X, Guo Y, Deng M, Qiao Y, Shi Z, Tao L, Cao Y, Gu L, Zhou Y, Zhang Y. Wilsonglucinols A–C, homoadamantane-type polycyclic polyprenylated acylphloroglucinols with unusual fused epoxy ring skeletons from *Hypericum wilsonii*. Org Chem Front. 2020;7:464–71.
24. Chen Y, Xue Q, Teng H, Qin R, Liu H, Xu J, Mei Z, Yang G. Acylphloroglucinol derivatives with a tricyclo-[4.4.1.1^{1,4}] dodecane skeleton from *Garcinia bracteata* fruits. J Org Chem. 2020;85:6620–25.
25. Tanaka N, Kashiwada Y, Kim SY, Hashida W, Sekiya M, Ikeshiro Y, Takaishi Y. Acylphloroglucinol, biyouyanagiol, biyouyanagin b, and related spiro-lactones from *Hypericum chinense*. J Nat Prod. 2009;72:1447–52.
26. Tanaka N, Abe S, Hasegawa K, Shiro M, Kobayashi J. Biyoulactones A-C, new pentacyclic meroterpenoids from *Hypericum chinense*. Org Lett. 2011;13:5488–91.
27. Xu WJ, Luo J, Li RJ, Yang MH, Kong LY. Furanmonogones A and B: two rearranged acylphloroglucinols with a 4,5-*seco*-3(2*h*)-furanone core from the flowers of *Hypericum monogynum*. Org Chem Front. 2017;4:313–7.
28. Zhang N, Shi Z, Guo Y, Xie S, Qiao Y, Li XN, Xue Y, Luo Z, Zhu H, Chen C, Hu L, Zhang Y. The absolute configurations of hyperilongenols A–C: rare 12,13-*seco*-spirocyclic polycyclic polyprenylated acylphloroglucinols with enolizable β, β′-tricarbonyl systems from *Hypericum longistylum* oliv. Org Chem Front. 2019;6:1491–502.
29. Zhang N, Shi Z, Xu Q, Sun W, Gu L, Xie S, Guo Y, Duan Y, Zhang K, Qi C, Zhang Y. Longisglucinols A–C, structurally intriguing polycyclic polyprenylated acylphloroglucinols with anti-inflammatory activity from *Hypericum longistylum*. Org Lett. 2020;22:7926–9.
30. Zhang JJ, Yang J, Liao Y, Yang XW, Ma JZ, Xiao QL, Yang LX, Xu G. Hyperuralones A and B, new acylphloroglucinol derivatives with intricately caged cores from *Hypericum uralum*. Org Lett. 2014;16:4912–5.
31. Fan YM, Yi P, Li Y, Yan C, Huang T, Gu W, Ma Y, Huang LJ, Zhang JX, Yang CL, Li Y, Yuan CM, Hao XJ. Two unusual polycyclic polyprenylated acylphloroglucinols, including a pair of enantiomers from *Garcinia multiflora*. Org Lett. 2015;17:2066–9.
32. Tian DS, Yi P, Xia L, Xiao X, Fan YM, Gu W, Huang LJ, Ben David Y, Di YT, Yuan CM, Hao XJ. Garmultins A–G, biogenetically related polycyclic acylphloroglucinols from *Garcinia multiflora*. Org Lett. 2016;18:5904–7.
33. Jian YQ, Huang XJ, Zhang DM, Jiang RW, Chen MF, Zhao BX, Wang Y, Ye WC. Guapsidial A and guadials B and C: three new meroterpenoids with unusual skeletons from the leaves of *Psidium guajava*. Chem Eur J. 2015;21:9022–7.
34. Su JC, Wang S, Cheng W, Huang XJ, Li MM, Jiang RW, Li YL, Wang L, Ye WC, Wang Y. Phloroglucinol derivatives with unusual skeletons from *Cleistocalyx operculatus* and their in vitro antiviral activity. J Org Chem. 2018;83:8522–32.
35. Shang ZC, Yang MH, Jian KL, Wang XB, Kong LY. ^{1}H NMR-guided isolation of formyl-phloroglucinol meroterpenoids from the leaves of *Eucalyptus robusta*. Chem Eur J. 2016;22:11778–84.
36. Qin XJ, Feng MY, Liu H, Ni W, Rauwolf T, Porco JA, Yan H, He L, Liu HY. Eucalyptusdimers A–C, dimeric phloroglucinol–phellandrene meroterpenoids from *Eucalyptus robusta*. Org Lett. 2018;20:5066–70.
37. Ye YS, Li WY, Du SZ, Yang J, Nian Y, Xu G. Congenetic hybrids derived from dearomatized isoprenylated acylphloroglucinol with opposite effects on $Ca_v3.1$ low voltage-gated Ca^{2+} channel. J Med Chem. 2020;63:1709–16.
38. Shao M, Wang Y, Jian YQ, Huang XJ, Zhang DM, Tang QF, Jiang RW, Sun XG, Lv ZP, Zhang XQ, Ye WC. Guadial A and psiguadials C and D, three unusual meroterpenoids from *Psidium guajava*. Org Lett. 2012;14:5262–5.
39. Tang YX, Fu WW, Xi ZC, Yang JL, Zheng CW, Lu Y, Shen ZW, Xu HX. Xanthone derivatives from the leaves of *Garcinia oligantha*. Eur J Med Chem. 2019;181:1–13.

40. Song JG, Su JC, Song QY, Huang RL, Tang W, Hu LJ, Huang XJ, Jiang RW, Li YL, Ye WC, Wang Y. Cleistocaltones A and B, antiviral phloroglucinol–terpenoid adducts from *Cleistocalyx operculatus*. Org Lett. 2019;21:9579–83.
41. Yang XL, Hsieh KL, Liu JK. Guajadial: an unusual meroterpenoid from guava leaves *Psidium guajava*. Org Lett. 2007;9:5135–8.
42. Shao M, Wang Y, Liu Z, Zhang DM, Cao HH, Jiang RW, Fan CL, Zhang XQ, Chen HR, Yao XS, Ye WC. Psiguadials A and B, two novel meroterpenoids with unusual skeletons from the Leaves of *Psidium guajava*. Org Lett. 2010;12:5040–3.
43. Liu C, Ang S, Huang XJ, Tian HY, Deng YY, Zhang DM, Wang Y, Ye WC, Wang L. Meroterpenoids with new skeletons from *Myrtus communis* and structure revision of myrtucommulone K. Org Lett. 2016;18:4004–7.
44. Zhang YL, Chen C, Wang XB, Wu L, Yang MH, Luo J, Zhang C, Sun HB, Luo JG, Kong LY. Rhodomyrtials A and B, two meroterpenoids with a triketone-sesquiterpene-triketone skeleton from *Rhodomyrtus tomentosa*: structural elucidation and biomimetic synthesis. Org Lett. 2016;18:4068–71.
45. Li CJ, Ma J, Sun H, Zhang D, Zhang DM. Guajavadimer A, a dimeric caryophyllene-derived meroterpenoid with a new carbon skeleton from the leaves of *Psidium guajava*. Org Lett. 2016;18:168–71.
46. Yang XW, Li YP, Su J, Ma WG, Xu G. Hyperjapones A–E, terpenoid polymethylated acylphloroglucinols from *Hypericum japonicum*. Org Lett. 2016;18:1876–9.
47. Chen NH, Zhang YB, Huang XJ, Jiang L, Jiang SQ, Li GQ, Li YL, Wang GC. Drychampones A–C: three meroterpenoids from *Dryopteris championii*. J Org Chem. 2016;81:9443–8.
48. Hou JQ, Guo C, Zhao JJ, He QW, Zhang BB, Wang H. Frutescone A-G, tasmanone-based meroterpenoids from the aerial parts of *Baeckea frutescens*. J Org Chem. 2017;82:1448–57.
49. Qin X, Rauwolf TJ, Li P, Liu H, McNeely J, Hua Y, Liu H, Porco JA. Isolation and synthesis of novel meroterpenoids from *Rhodomyrtus tomentos* a: investigation of a reactive enetrione intermediate. Angew Chem Int Ed. 2019;58:4291–6.
50. Xu J, Zhu HL, Zhang J, Liu WY, Luo JG, Pan K, Cao WY, Bi QR, Feng F, Qu W. Littordials A–E, novel formyl-phloroglucinol-β-caryophyllene meroterpenoids from the leaves of *Psidium littorale*. Org Chem Front. 2019;6:1667–73.
51. Fu HZ, Luo YM, Li CJ, Yang JZ, Zhang DM. Psidials A–C, three unusual meroterpenoids from the leaves of *Psidium guajava* L. Org Lett. 2010;12:656–9.
52. Hu L, Zhang Y, Zhu H, Liu J, Li H, Li XN, Sun W, Zeng J, Xue Y, Zhang Y. Filicinic acid based meroterpenoids with anti-epstein–barr virus activities from *Hypericum japonicum*. Org Lett. 2016;18:2272–5.
53. Gu JH, Wang WJ, Chen JZ, Liu JS, Li NP, Cheng MJ, Hu LJ, Li CC, Ye WC, Wang L. Leptosperols A and B, two cinnamoylphloroglucinol–sesquiterpenoid hybrids from *Leptospermum scoparium*: structural elucidation and biomimetic synthesis. Org Lett. 2020;22:1796–800.
54. Ye YS, Du SZ, Jiang NN, Xu HX, Yang J, Fu WW, Nian Y, Xu G. Novel meroterpenoids from *Hypericum patulum*: highly potent late $Na_V1.5$ sodium current inhibitors. Org Lett. 2020;22:6339–43.
55. Yin S, Xue JJ, Fan CQ, Miao ZH, Ding J, Yue JM. Eucalyptals A−C with a new skeleton isolated from *Eucalyptus globulus*. Org Lett. 2007;9:5549–52.
56. Tang ZH, Liu YB, Ma SG, Li L, Li Y, Jiang JD, Qu J, Yu SS. Antiviral spirotriscoumarins A and B: two pairs of oligomeric coumarin enantiomers with a spirodienone–sesquiterpene skeleton from *Toddalia asiatica*. Org Lett. 2016;18:5146–9.
57. Xiong J, Hong ZL, Gao LX, Shen J, Liu ST, Yang GX, Li J, Zeng H, Hu JF. Chlorabietols A–C, phloroglucinol-diterpene adducts from the chloranthaceae plant *Chloranthus oldhamii*. J Org Chem. 2015;80:11080–5.
58. Hu Y, Wu X, Gao S, Yu S, Liu Y, Qu J, Liu J, Liu Y. Novel phloroglucinol derivatives from the roots of *Lysidice rhodostegia*. Org Lett. 2006;8:2269–72.
59. Hiranrat A, Mahabusarakam W, Carroll AR, Duffy S, Avery VM. Tomentosones A and B, hexacyclic phloroglucinol derivatives from the Thai shrub *Rhodomyrtus tomentosa*. J Org Chem. 2012;77:680–3.

60. Choudhary MI, Khan N, Ahmad M, Yousuf S, Fun HK, Soomro S, Asif M, Mesaik MA, Shaheen F. New inhibitors of ROS generation and t-cell proliferation from *Myrtus communis*. Org Lett. 2013;15:1862–5.
61. Yang XW, Yang J, Liao Y, Ye Y, Li YP, Yang SY, Xia F, Xu G. Hypercohin K, a polycyclic polyprenylated acylphloroglucinol with an unusual spiro-fused cyclopropane ring from *Hypericum cohaerens*. Tetrahedron Lett. 2015;56:5537–40.
62. Zong JF, Hu Z, Shao YY, Shi Q, Zhang MM, Zhou YB, Li J, Hou AJ. Hyperprins A and B, two complex meroterpenoids from *Hypericum przewalskii*. Org Lett. 2020;22:2797–800.
63. Deng LM, Hu LJ, Bai Yang TZ, Wang J, Qin GQ, Song QY, Su JC, Huang XJ, Jiang RW, Tang W, Li YL, Li CC, Ye WC, Wang Y. Rhodomentosones A and B: two pairs of enantiomeric phloroglucinol trimers from *Rhodomyrtus tomentosa* and their asymmetric biomimetic synthesis. Org Lett. 2021;23:4499–504.
64. Liu F, Tian HY, Huang XL, Wang WJ, Li NP, He J, Ye WC, Wang L. Xanthchrysones A–C: rearranged phenylpropanoyl–phloroglucinol dimers with unusual skeletons from *Xanthostemon chrysanthus*. J Org Chem. 2019;84:15355–61.
65. Vu VT, Chen XL, Kong LY, Luo JG. Melipatulinones A–C, three lignan–phloroglucinol hybrids from *Melicope patulinervia*. Org Lett. 2020;22:1380–4.

Chapter 8
Classification of Diverse Novel Meroterpenoids

Meroterpenoids are derived from the mixed biogenic pathway of terpenoids and other natural products. New skeleton meroterpenoids are more structural complexity and diversity. To date, over 72 new skeleton meroterpenoids have been isolated and characterized from 18 families (**749–761** Fig. 8.1, **762–795** Fig. 8.2, **796–820** Fig. 8.3; Table A7). In this chapter, the chemistry, distribution, classification and structure characteristics of novel meroterpenoids will be displayed.

8.1 Monoterpenoid Meroterpenoids

8.1.1 Verbenaceae

Littoralisone (**749**) is a novel neuritogenic iridolactone having an unprecedented heptacyclic skeleton including four- and nine-membered rings consisting of glucose, which was isolated from *Verbena littoralis* [1].

8.1.2 Ericaceae

(−)-Rhodonoid A (**750**) and (+)-rhodonoid A (**751**) are the first examples of meromonoterpenes featuring a unique 6/6/6/4 ring system, which were isolated from *Rhododendron capitatum*. Their absolute configurations were determined by X-ray crystallography and electronic circular dichroism analysis [2]. (+)-/(−)-Rhodonoids C (**752** and **753**) are the first pair of meromonoterpenes incorporating an unprecedented 6/6/6/5 ring system, and (+)-/(−)-rhodonoids D (**754** and **755**) are the first examples of meromonoterpenes featuring a unique 6/6/5/5 ring system, they were all obtained from the aerial parts of *Rhododendron capitatum* [3].

Y. Cheng and D. Qin, *Novel Plant Natural Product Skeletons*,
https://doi.org/10.1007/978-981-99-7329-3_8

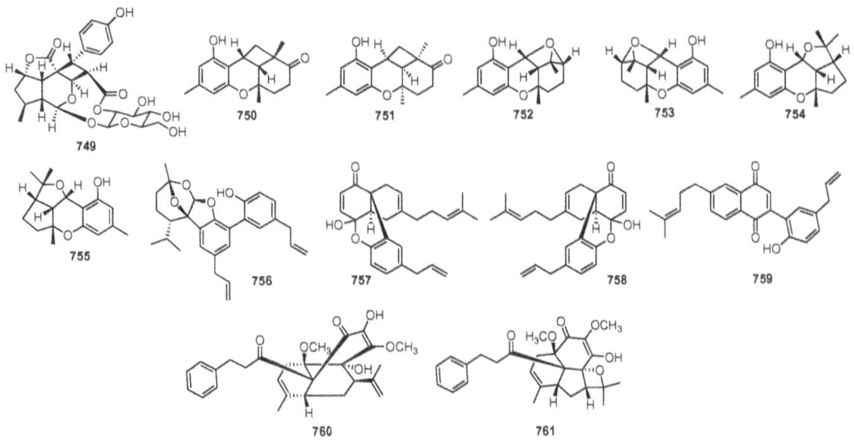

Fig. 8.1 Structures of new skeleton meroterpenoids **749–761**

8.1.3 *Magnoliaceae*

Magterpenoid A (**756**), possessing a rare 4,6,11-trioxatricyclo[5.3.1.$0^{1,5}$]undecane framework with an irregular monoterpenoid moiety, (+)-magterpenoid B (**757**) and (−)-magterpenoid B (**758**) featuring an unprecedented 6/6/6/6 polycyclic skeleton, and magterpenoid C (**759**), a novel terpenoid quinone with a C6–C3 unit, they were all isolated from the bark of *Magnolia officinalis* [4]. (±)-fissisternoids A (**760**) and B (**761**) were isolated from the branches and leaves of *Fissistigma bracteolatum*. **760** represents an unprecedented meroterpenoid featuring a unique tricyclo [3,3,1,$0^{1',5'}$] decane central framework, and **761** possesses a rare 6/6/5/4 tetracyclic carbon skeleton, both of which were derived from quinodihydrochalcone and monoterpenoid via a key [4+2] Diels–Alder cyclization and Prins reaction [5].

8.2 Sesquiterpenoid Meroterpenoids

8.2.1 *Annonaceae*

Fissistigmatins A–D (**762–765**) are the first examples of a group of natural products consisting of a flavonoid and a sesquiterpene moiety, which were isolated from *Fissistigma bracteolatum* [6]. The absolute configuration of **762** was accomplished by molecular modelling calculations and X-ray crystallographic analysis.

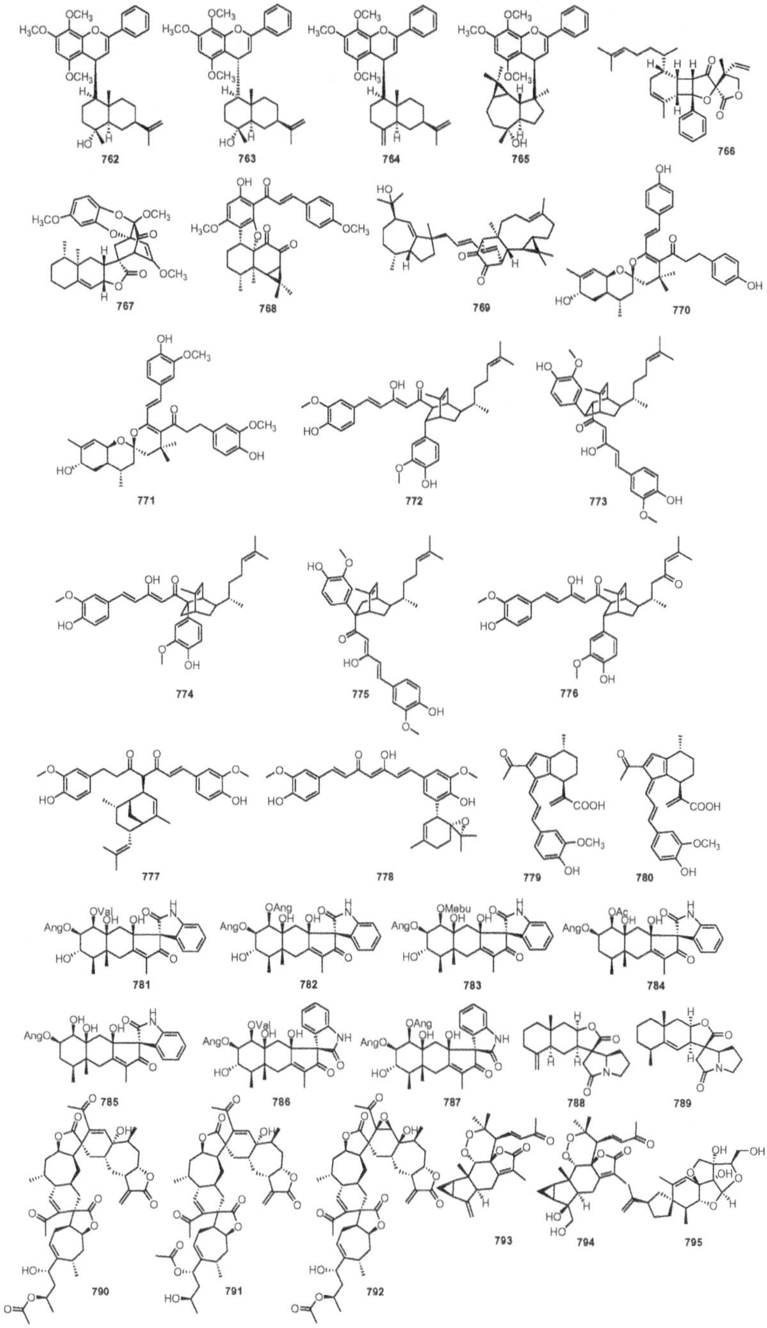

Fig. 8.2 Structures of new skeleton meroterpenoids **762–795**

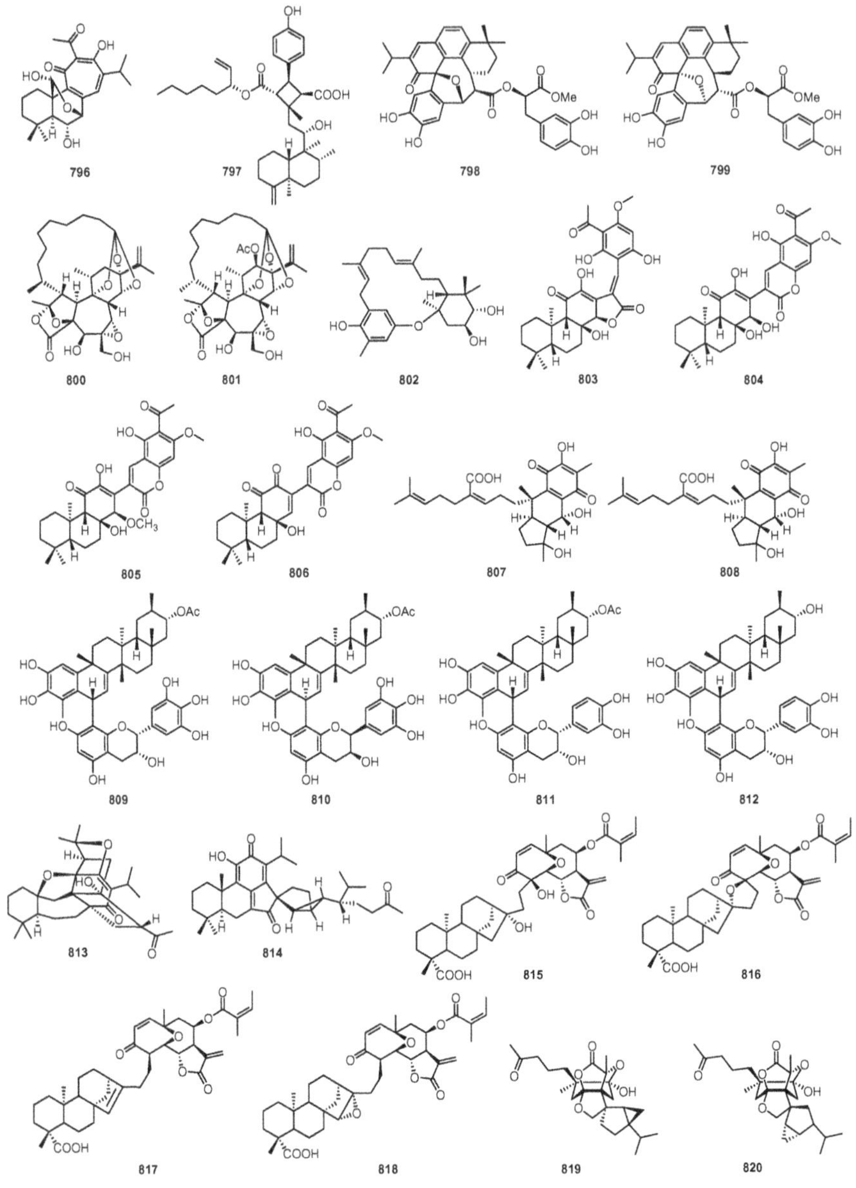

Fig. 8.3 Structures of new skeleton meroterpenoids **796–820**

8.2.2 *Hypericaceae*

Biyouyanagin A (**766**) is a structurally unique hydrophobic compound containing sesquiterpene, cyclobutane, and spirolactone moieties, which has been described from *Hypericum chinense* [7]. Hyperdioxane A (**767**) is a conjugate of dibenzo-1,4-dioxane and sesquiterpene with an unprecedented heptacyclic ring system isolated from *Hypericum ascyron*, its absolute configuration was confirmed by a modified Mosher's method [8].

8.2.3 *Valerianaceae*

Nardoaristolone A (**768**) represents the first reported aristolane-chalcone derivative, which was isolated from the underground parts of *Nardostachys chinensis* and its absolute configuration was established by single-crystal X-ray diffraction analysis [9].

8.2.4 *Piperaceae*

Nudibaccatumone (**769**) is a trimer comprising a phenylpropanoid and two sesquiterpene moieties, which was obtained from *Piper nudibaccatum* [10].

8.2.5 *Zingiberaceae*

Terpecurcumins J–Q (**770–777**) and V (**778**) represent four unprecedented skeletons featuring an unusual core of hydrobenzannulated[6,6]-spiroketal (**770** and **771**), bicyclo[2.2.2]octene (**772–776**), bicyclo[3.1.3]octene (**777**), and spiroepoxide (**778**), respectively [11]. They were all isolated from the rhizomes of *Curcuma longa*. The configuration of **770** was confirmed by single-crystal X-ray diffraction (Cu Kα).

8.2.6 *Compositae/Asteraceae*

Arteannoides B and C (**779** and **780**) are two novel heterodimers incorporating a rearranged cadinene sesquiterpenoid and a phenylpropanoid. They were obtained from the famous traditional Chinese medicine *Artemisia annua* L. (Qinghao), and their unique constructions increase the structural diversity of sesquiterpenoids in nature [12]. Parasubindoles A–G (**781–787**) are seven eremophilanyl indoles with an unprecedented 12*H*-cyclopentane[*b*]naphthalenespiro-1,3′-indole skeleton, they were isolated from the whole plant of *Parasenecio albus* [13]. Finally, spiroalanpyrroids A (**788**) and B (**789**) are two sesquiterpene alkaloids with an unprecedented eudesmanolide–pyrrolizidine spiro[5.5] framework, which were isolated from *Inula helenium* [14]. Xanthanoltrimer A–C (**790–792**), three xanthanolide sesquiterpene trimers, were first isolated from the fruits of *Xanthium italicum* Moretti [15].

8.2.7 Chloranthaceae

Sarglaperoxides A (**793**) and B (**794**) represent a pair of unusual sesquiterpene-normonoterpene conjugates with a peroxide bridge, they were isolated from the seeds of *Sarcandra glabra* [16].

8.2.8 Solanaceae

Nicotabin A (**795**) is a sesquiterpenoid derivative possessing a fused 5/6/5/5/5 ring system, which has been described from *Nicotiana tabacum* and its absolute configurations were confirmed by single crystal X-ray diffraction [17].

8.3 Diterpenoid Meroterpenoids

8.3.1 Lamiaceae

Przewalskin A (**796**) is a novel C_{23} terpenoid with a 6/6/7 carbon ring skeleton, which was isolated from *Salvia przewalskii* [18]. Scopariusic acid (**797**) is a new *ent*-clerodane-based meroditerpenoid from *Isodon scoparius* with a unique cyclobutane ring and an unusual 1-octen-3-ol substituent, whose absolute configuration was determined by single-crystal X-ray diffraction analysis [19]. Perovsfolins A (**798**) and B (**799**) are two structurally unique C_{28} terpenoids. They were isolated from the aerial parts of *Perovskia scrophulariifolia* and their chemical structures feature an unprecedented 6/8/6/6/6 pentacyclic carbon skeleton with a C_6–C_3 ester moiety [20].

8.3.2 *Thymelaeaceae*

Pimelotides A and B (**800** and **801**) possess an unusual carbon skeleton which represents the first known example of daphnane ketallactone-type diterpenoid orthoesters from a natural source (*Perovskia scrophulariifolia*) [21].

8.3.3 *Meliaceae*

Dysohonin A (**802**) is a merditerpenoid incorporating an unprecedented 6,15,6-fused heterotricyclic ring system, which has been isolated from *Dysoxylum hongkongense* [22].

8.3.4 *Euphorbiaceae*

Fischernolides A–D (**803–806**) are four meroterpenoids based on diterpene and acylphloroglucinol possessing an unprecedented 28-carbon skeleton with a novel scaffold, they were isolated from the roots of *Euphorbia fischeriana* [23].

8.3.5 *Myristicaceae*

Pycnanthuquinones A (**807**) and B (**808**) were isolated from leaves and stems of the African plant *Pycnanthus angolensis* [24]. Their structures are characteristic of terpenoid-quinone structures containing a fused 6,6,5-ring skeleton.

8.4 Triterpenoid Meroterpenoids

8.4.1 *Celastraceae*

Celamonols A–D (**809–812**) feature an unusual pattern of conjunction between a 24,29-dinorfriedelanetype triterpenoid and a catechin, they were isolated from the stems of *Celastrus monospermu* [25].

8.5 Others

Hydrangenone (**813**) is a new heptacyclic isoprenoid with a 6/7/6/5/5 membered carbon ring skeleton, which has been isolated from the aerial parts of *Salvia hydrangea* [26]. Cryptotrione (**814**) is a novel C_{35}-terpene, which possesses an unprecedented skeleton having an abietane diterpene with a unique bicyclic sesquiterpene. It was characterized from the bark of *Cryptomeria japonica* [27]. Four novel compounds helikaurolides A–D (**815–818**), representing unprecedented skeletons combining a sesquiterpene lactone and a kaurane diterpene acid, were isolated from *Helianthus annuus* [28]. Hitorins A (**819**) and B (**820**) are two novel C_{25} terpenoids with a 6/5/5/5/5/3 hexacyclic skeleton including one γ-lactone ring and two tetrahydrofuran rings, they were isolated from the aerial parts of *Chloranthus japonicas* [29].

References

1. Li YS, Matsunaga K, Ishibashi M, Ohizumi Y. Littoralisone, a novel neuritogenic iridolactone having an unprecedented heptacyclic skeleton including four- and nine-membered rings consisting of glucose from *Verbena littoralis*. J Org Chem. 2001;66:2165–7.
2. Liao HB, Lei C, Gao LX, Li JY, Li J, Hou AJ. Two enantiomeric pairs of meroterpenoids from *Rhododendron capitatum*. Org Lett. 2015;17:5040–3.
3. Liao HB, Huang GH, Yu MH, Lei C, Hou AJ. Five pairs of meroterpenoid enantiomers from *Rhododendron capitatum*. J Org Chem. 2017;82:1632–7.
4. Li C, Li CJ, Ma J, Chen FY, Li L, Wang XL, Ye F, Zhang DM. Magterpenoids A–C, three polycyclic meroterpenoids with ptp1b inhibitory activity from the bark of *Magnolia officinalis* var. *biloba*. Org Lett 2018;20:3682–86.
5. Xue GM, Zhao CG, Xue JF, Chen H, Zhao ZZ, Si YY, Du K, Zhi YL, Feng WS. Fissisternoids A and B, two 2′,5′-quinodihydrochalcone-based meroterpenoid enantiomers with unusual carbon skeletons from *Fissistigma bracteolatum*. Org Chem Front 2022;9:190–6.
6. Porzel A, Phuong Lien T, Schmidt J, Drosihn S, Wagner C, Merzweiler K, Van Sung T, Adam G. Fissistigmatins A-D: novel type natural products with flavonoid–sesquiterpene hybrid structure from *Fissistigma bracteolatum*. Tetrahedron. 2000;56:865–72.
7. Tanaka N, Okasaka M, Ishimaru Y, Takaishi Y, Sato M, Okamoto M, Oshikawa T, Ahmed SU, Consentino LM, Lee KH. Biyouyanagin A, an anti-HIV agent from *Hypericum chinense* L. var. *salicifolium*. Org Lett 2005;7:2997–9.
8. Niwa K, Tanaka N, Kim SY, Kojoma M, Kashiwada Y. Hyperdioxane A, a conjugate of dibenzo-1,4-dioxane and sesquiterpene from *Hypericum ascyron*. Org Lett. 2018;20:5977–80.
9. Liu ML, Duan YH, Hou YL, Li C, Gao H, Dai Y, Yao XS. Nardoaristolones A and B, two terpenoids with unusual skeletons from *Nardostachys chinensis* batal. Org Lett. 2013;15:1000–3.
10. Liu HX, Chen K, Sun QY, Yang FM, Hu GW, Wang YH, Long CL. Nudibaccatumone, a trimer comprising a phenylpropanoid and two sesquiterpene moieties from *Piper nudibaccatum*. J Nat Prod. 2013;76:732–6.
11. Lin X, Ji S, Qiao X, Hu H, Chen N, Dong Y, Huang Y, Guo D, Tu P, Ye M. Density functional theory calculations in stereochemical determination of terpecurcumins J-W, cytotoxic terpene-conjugated curcuminoids from *Curcuma longa* L. J Org Chem. 2013;78:11835–48.
12. Qin DP, Pan DB, Xiao W, Li HB, Yang B, Yao XJ, Dai Y, Yu Y, Yao XS. Dimeric cadinane sesquiterpenoid derivatives from *Artemisia annua*. Org Lett. 2018;20:453–6.

13. Zhou M, Zhou J, Liu J, Liang JJ, Peng XG, Duan FF, Ruan HL. Parasubindoles A-G, seven eremophilanyl indoles from the whole plant of *Parasenecio albus*. J Org Chem. 2018;83:12122–8.
14. Cai YS, Wu Z, Zheng XQ, Wang C, Wang JR, Zhang XX, Qiu G, Zhu K, Cao S, Yu J. Spiroalanpyrroids A and B, sesquiterpene alkaloids with a unique spiro-eudesmanolide–pyrrolizidine skeleton from *Inula helenium*. Org Chem Front. 2020;7:303–9.
15. Fu J, Wang YN, Ma SG, Li L, Wang XJ, Li Y, Liu YB, Qu J, Yu SS. Xanthanoltrimer A-C: three xanthanolide sesquiterpene trimers from the fruits of *Xanthium italicum* Moretti isolated by HPLC-MS-SPE-NMR. Org Chem Front. 2021;8:1288–93.
16. Wang P, Li RJ, Liu RH, Jian KL, Yang MH, Yang L, Kong LY, Luo J. Sarglaperoxides A and B, sesquiterpene–normonoterpene conjugates with a peroxide bridge from the seeds of *Sarcandra glabra*. Org Lett. 2016;18:832–5.
17. Feng T, Li XM, He J, Ai HL, Chen HP, Li XN, Li ZH, Liu JK. Nicotabin A, a sesquiterpenoid derivative from *Nicotiana tabacum*. Org Lett. 2017;19:5201–3.
18. Xu G, Hou AJ, Wang RR, Liang GY, Zheng YT, Liu ZY, Li XL, Zhao Y, Huang SX, Peng LY, Zhao QS. Przewalskin A: a new c_{23} terpenoid with a 6/6/7 carbon ring skeleton from *Salvia przewalskii* Maxim. Org Lett. 2006;8:4453–6.
19. Zhou M, Zhang HB, Wang WG, Gong NB, Zhan R, Li XN, Du X, Li LM, Li Y, Lu Y, Pu JX, Sun HD. Scopariusic acid, a new meroditerpenoid with a unique cyclobutane ring isolated from *Isodon scoparius*. Org Lett. 2013;15:4446–9.
20. Tanaka N, Niwa K, Kajihara S, Tsuji D, Itoh K, Mamadalieva NZ, Kashiwada Y. C_{28} terpenoids from lamiaceous plant *Perovskia scrophulariifolia*: their structures and anti-neuroinflammatory activity. Org Lett. 2020;22:7667–70.
21. Hayes PY, Chow S, Somerville MJ, De Voss JJ, Fletcher MT. Pimelotides A and B, diterpenoid ketal-lactone orthoesters with an unprecedented skeleton from *Pimelea elongata*. J Nat Prod. 2009;72:2081–3.
22. Zhao JX, Liu CP, Zhang MM, Li J, Yue JM. Dysohonin A, a meroditerpenoid incorporating a 6,15,6-fused heterotricyclic ring system from *Dysoxylum hongkongense*. Org Chem Front. 2018;5:2202–7.
23. Zhang J, He J, Cheng YC, Zhang PC, Yan Y, Zhang HJ, Zhang WK, Xu JK. Fischernolides A-D, four novel diterpene-based meroterpenoid scaffolds with antitumor activities from *Euphorbia fischeriana*. Org Chem Front. 2019;6:2312–8.
24. Fort DM, Ubillas RP, Mendez CD, Jolad SD, Inman WD, Carney JR, Chen JL, Ianiro TT, Hasbun C, Bruening RC, Luo J, Reed MJ, Iwu M, Carlson TJ, King SR, Bierer DE, Cooper R. Novel antihyperglycemic terpenoid-quinones from *Pycnanthus angolensis*. J Org Chem. 2000;65:6534–9.
25. Qin JJ, Fu YF, Chen X, Liu YT, Zhao JH, Zuo JP, Li YM, Zhu WL, Zhao WM. Celamonols A-D, four triterpenoid and catechin conjugates with immunosuppressive activities from the stems of *Celastrus monospermus*. Org Chem Front. 2019;6:3786–92.
26. Farimani MM, Taheri S, Ebrahimi SN, Bahadori MB, Khavasi HR, Zimmermann S, Brun R, Hamburger M. Hydrangenone, a new isoprenoid with an unprecedented skeleton from *Salvia hydrangea*. Org Lett. 2012;14:166–9.
27. Chen CC, Wu JH, Yang NS, Chang JY, Kuo CC, Wang SY, Kuo YH. Cytotoxic C_{35} terpenoid cryptotrione from the bark of *Cryptomeria japonica*. Org Lett. 2010;12:2786–9.
28. Torres A, Molinillo JMG, Varela RM, Casas L, Mantell C, Martínez de la Ossa EJ, Macías FA. Helikaurolides A–D with a diterpene-sesquiterpene skeleton from supercritical fluid extracts of *Helianthus annuus* L. var. Arianna. Org Lett 2015;17:4730–3.
29. Kim SY, Nagashima H, Tanaka N, Kashiwada Y, Kobayashi J, Kojoma M. Hitorins A and B, hexacyclic C_{25} terpenoids from *Chloranthus japonicus*. Org Lett. 2016;18:5420–3.

Chapter 9
Classification of Diverse Novel Flavonoid Hybrids

New skeleton flavonoid hybrids are a diverse class of compounds having complex structural features with many hybrids. The important pharmacological activities and structural complexity of the new skeleton flavonoid hybrids have long interested scientists due to their medicinal uses. Since 1999, 51 new skeleton flavonoid hybrids, assigned to 12 families 16 genuses, have been isolated and identified from plants (**821–848** Fig. 9.1, **849–871** Fig. 9.2; Table A8). Leguminosae have the largest number of novel flavonoid hybrids in all families. The names, classes, structure characteristics, activities, the types of compounds, and plant sources of these new skeleton flavonoid hybrids are collated here. This review will be a detailed update of the naturally occurring new skeleton flavonoid hybrids reported from the plant kingdom from 1999 to 2021, providing a multi-perspective understanding of these compounds.

9.1 Leguminosae

9.1.1 Millettia

Caeruleanone A (**821**) is a novel rotenoid with an unprecedented arrangement of the D ring, which was isolated from the fruits of *Millettia caerulea* [1]. (−)-Millpuline A (**822**) and (+)-millpuline A (**823**) were isolated from *Millettia pulchra.* Their structures are interesting with a biflavone skeleton constructed by a four-membered carbocyclic ring [2]. They were characterized by spectral analyses and single-crystal X-ray diffraction.

Y. Cheng and D. Qin, *Novel Plant Natural Product Skeletons*,
https://doi.org/10.1007/978-981-99-7329-3_9

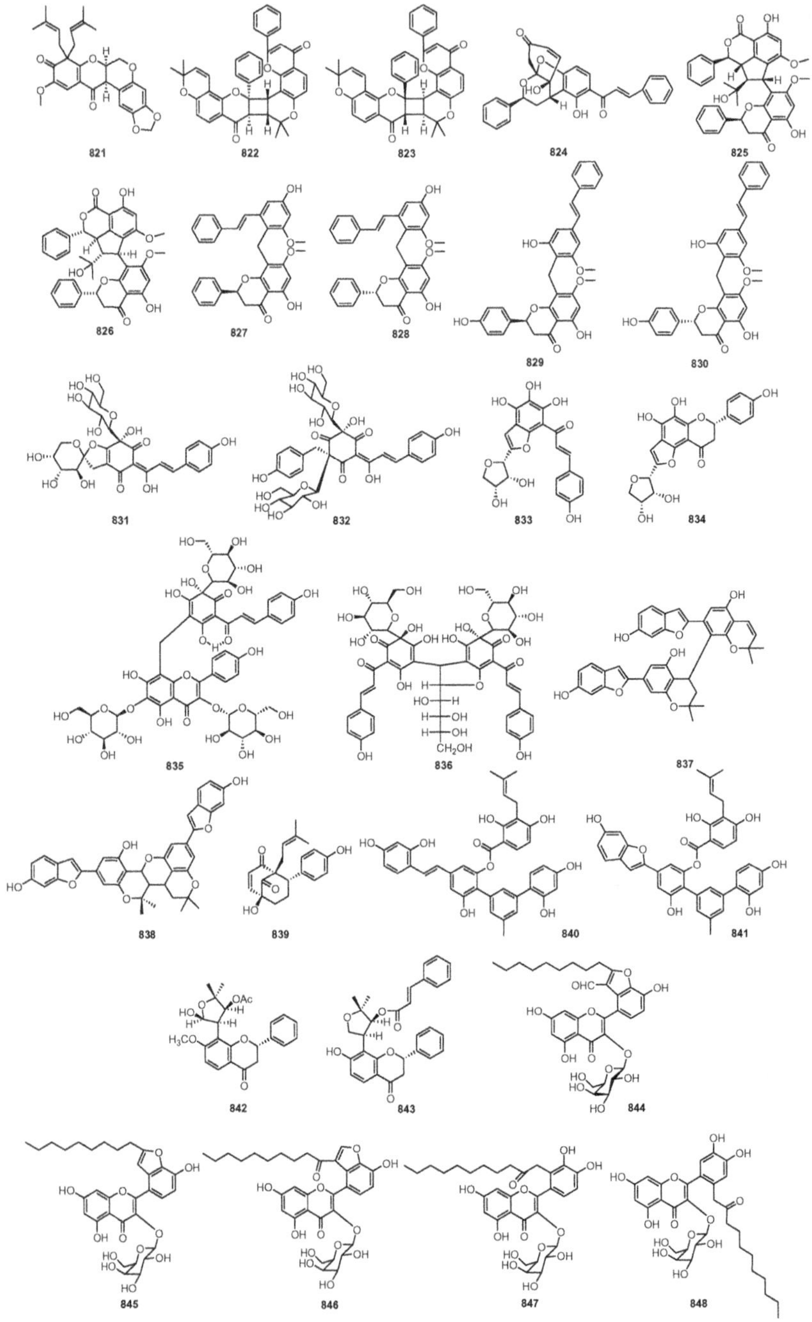

Fig. 9.1 Structures of new skeleton flavonoid hybrids **821–848**

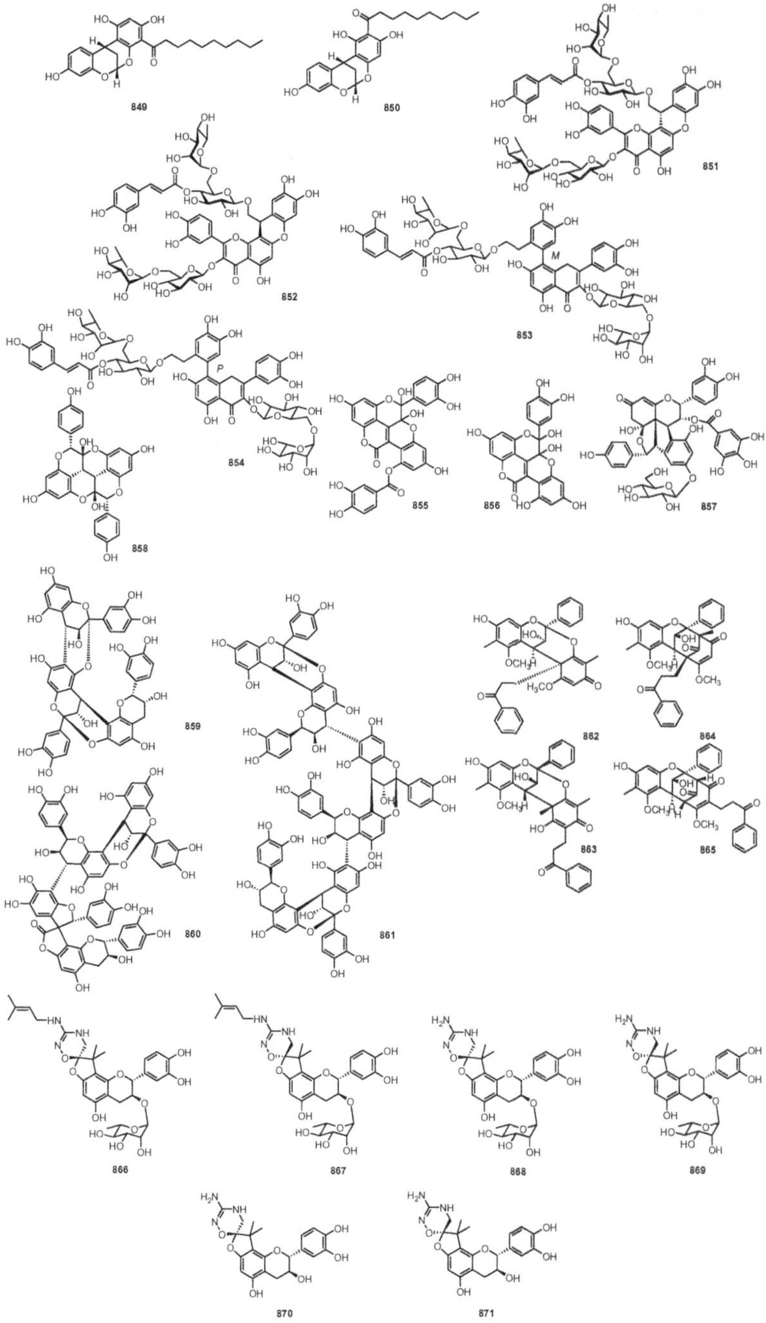

Fig. 9.2 Structures of new skeleton flavonoid hybrids **849–871**

9.1.2 *Caesalpinia*

Caesalpinnone A (**824**) is an unprecedented hybrid of flavan and chalcone, possessing a 10,11-dioxatricyclic[5.3.3.$0^{1,6}$]tridecane-bridged system, which has been isolated from the twigs and leaves of *Caesalpinia enneaphylla* [3].

9.1.3 *Cajanus*

(+)-Cajanusflavanol A (**825**) and (−)-cajanusflavanol A (**826**) possess an unprecedented carbon skeleton featuring a unique highly functionalized cyclopenta[1,2,3-*de*]isobenzopyran-1-one tricyclic core, while (+)-cajanusflavanol B (**827**), (−)-cajanusflavanol B (**828**), (+)-cajanusflavanol C (**829**), and (−)-cajanusflavanol C (**830**) are the first examples of methylene unit linked flavonostilbenes. **825–830** were all isolated from *Cajanus cajan* [4].

9.2 *Carthamus* (Compositae/Asteraceae)

Saffloquinosides A (**831**) and B (**832**) were isolated from the florets of *Carthamus tinctorius* [5]. Of which **831** has an uncommon six-five member dioxaspirocycle and **832** has a cyclohexatrione skeleton with a benzyl group and two C-glycosyl units.

Safflofiavonesides A (**833**) and B (**834**) are two new rearranged derivatives of flavonoid C-glycosides isolated from *Carthamus tinctorius*, they are unprecedented chalcone and flavanone derivatives possessing a furan conjoining tetrahydrofuran ring [6]. Carthorquinoside A (**835**) is an unprecedented quinochalcone-flavonol structure linked via a methylene bridge, and carthorquinoside B (**836**) comprises two glucopyranosylquinochalcone moieties linked via the formyl carbon of an acyclic glucosyl unit [7]. Both **835** and **836** were isolated from *Carthamus tinctorius* and their structures were established by spectroscopic data and chemical degradation.

9.3 Moraceae

9.3.1 *Morus*

Morusyunnansins A and B (**837** and **838**) are novel 2-arylbenzofuran dimers isolated from *Morus yunnanensis* [8]. This is the first time that the two monomers in **837** and **838** are linked through the isoprenoid groups. As for the reported bioflavonoids consisting of a flavan and a deoxotetrahydrochalcone moiety, the linkage position of the two monomers is usually at the A ring of flavan.

9.3.2 Brosimum

Acutifolin A (**839**), isolated from the bark of *Brosimum acutifolium*, is so far the only compound with a bicyclo[3.3.1]non-3-ene-2,9-dione ring in the structure [9]. Morunigrines A (**840**) and B (**841**) were isolated from the twigs of *Morus nigra*, morunigrines A (**840**) and B (**841**) are a novel class of Diels-Alder adducts with unprecedented carbon skeletons featuring a rearranged chalcone-stilbene/2-arylbenzofuran core decorated with a unique methylbiphenyl moiety [10].

9.4 *Tephrosia* (Papilionaceae)

(+)-Tephrorins A (**842**) and B (**843**) are flavanones containing an unusual tetrahydrofuran moiety. They were isolated from *Tephrosia purpurea* and their absolute configurations were determined by Mosher ester method [11].

9.5 *Houttuynia* (Saururaceae)

Houttuynoids A–E (**844–848**) are a new type of flavonoids with houttuynin tethered to hyperoside, they were isolated from the whole plant of *Houttuynia cordata* [12].

9.6 *Horsfieldia* (Myristicaceae)

Myristicyclins A (**849**) and B (**850**) are procyanidin-like congeners of myristinins lacking a pendant aromatic ring, they were obtained from *Horsfieldia spicata* [13].

9.7 *Forsythia* (Oleaceae)

Forsythoneosides A–D (**851–854**) are four unusual adducts consisting of a flavonoid unit fused to a phenylethanoid glycoside through a pyran ring or carbon–carbon bond. All these hybrids were isolated from the fruits of *Forsythia suspensa* [14].

9.8 *Aronia* (Rosaceae)

Melanodiol 4″-*O*-protocatechuate (**855**) and melanodiol (**856**) represent novel flavonoid derivatives isolated from the dried black chokeberry (*Aronia melanocarpa*) fruit juice [15]. They possess an unprecedented fused pentacyclic core with two contiguous hemiketals.

9.9 *Polygonum* (Polygonaceae)

Polyflavanostilbene A (**857**) is a new flavanol-fused stilbene glycoside, which has been isolated from the rhizome of *Polygonum cuspidatum* and possesses an unprecedented rearranged flavanol skeleton fused to stilbene via a hexahydrocyclopenta[*c*]furan moiety [16].

9.10 Pinaceae

9.10.1 *Abies*

Abiesanol A (**858**) is a novel biflavanol with unique six connective hexacyclic rings, it was isolated from *Abies georgei* [17].

9.10.2 *Pinus*

Three novel flavonoid hybrids (**859**–**861**) with rare A-type, spiro-type, and highly oligomeric proanthocyanidins were characterized from *Pinus massoniana* [18]. Pinutwindoublin (**859**) is the first reported trimer with double A-type interflavanyl linkages ($2\alpha \rightarrow O \rightarrow 5, 4\alpha \rightarrow 6$ and $2\alpha \rightarrow O \rightarrow 7, 4\alpha \rightarrow 8$). Pinuspirotetrin (**860**) represents the first reported proanthocyanidin tetramer with a heterodimeric framework consisting of one spiro-type and one A-type dimer. Whereas, pinumassohexin (**861**) was identified as a mixed A + B-type hexamer consisting of a peanut derived tetramer, peanut procyanidin E, and an A-type dimer.

9.11 *Daemonorops* (Palmaceae)

Dragonins A–D (**862–865**) were characterized from the traditional Chinese medicine Sanguis Draconis, which are novel A-type flavan-3-ol-dihydroretrochalcone dimers [19].

9.12 *Alchornea* (Euphorbiaceae)

Rugonidines A–F (**866–871**), three pairs of novel configurationally semistable diastereomers featuring an unprecedented 1,6-dioxa-7,9-diazaspiro[4.5]dec-7-en-8-amine scaffold, were isolated from *Alchornea rugosa* [20].

References

1. Bueno Pérez L, Pan L, Muñoz Acuña U, Li J, Chai HB, Gallucci JC, Ninh TN.; Carcache de Blanco EJ, Soejarto DD, Kinghorn AD. Caeruleanone A, a rotenoid with a new arrangement of the d-ring from the fruits of *Millettia caerulea.* Org Lett 2014;16:1462–5.
2. Wang W, Tang Y, Liu Y, Yuan L, Wang J, Lin B, Zhou D, Sun L, Huang R, Chen G, Li N. Isolation, structure elucidation, and synthesis of (±)-millpuline a with a suppressive effect in mir-144 expression. Org Chem Front. 2019;6:2850–9.
3. Zhang LJ, Bi DW, Hu J, Mu WH, Li YP, Xia GH, Yang L, Li XN, Liang XS, Wang LQ. Four hybrid flavan–chalcones, caesalpinnone a possessing a 10,11-dioxatricyclic [5.3.3.$0^{1,6}$]tridecane-bridged system and caesalpinflavans A–C from *Caesalpinia enneaphylla.* Org Lett. 2017;19:4315–8.
4. He QF, Wu ZL, Huang XJ, Zhong YL, Li MM, Jiang RW, Li YL, Ye WC, Wang Y. Cajanusflavanols A-C, three pairs of flavonostilbene enantiomers from *Cajanus cajan.* Org Lett. 2018;20:876–9.
5. Jiang JS, He J, Feng ZM, Zhang PC. Two new quinochalcones from the florets of *Carthamus tinctorius.* Org Lett. 2010;12:1196–9.
6. He J, Yang YN, Jiang J, Feng ZM, Zhang PC. Saffloflavonesides A and B, two rearranged derivatives of flavonoid *C*-glycosides with a furan–tetrahydrofuran ring from *Carthamus tnctorius.* Org Lett. 2014;16:5714–7.
7. Yue SJ, Qu C, Zhang PX, Tang YP, Jin Y, Jiang JS, Yang YN, Zhang PC, Duan JA. Carthorquinosides A and B, quinochalcone *C*-glycosides with diverse dimeric skeletons from *Carthamus tinctorius.* J Nat Prod. 2016;79:2644–51.
8. Hu X, Wu JW, Wang M, Yu MH, Zhao QS, Wang HY, Hou AJ. 2-Arylbenzofuran, flavonoid, and tyrosinase inhibitory constituents of *Morus yunnanensis.* J Nat Prod. 2012;75:82–7.
9. Takashima J, Ohsaki A. Acutifolins A−F, a new flavan-derived constituent and five new flavans from *Brosimum acutifolium.* J Nat Prod. 2001;64:1493–6.
10. Qu KJ, Wang B, Jiang CS, Xie BG, Liu AH, Li SW, Guo YW, Li J, Mao SC. Rearranged Diels−Alder adducts and prenylated flavonoids as potential ptp1b inhibitors from *Morus nigra.* J Nat Prod. 2021;84:2303–11.
11. Chang LC, Chávez D, Song LL, Farnsworth NR, Pezzuto JM, Kinghorn AD. Absolute configuration of novel bioactive flavonoids from *Tephrosia purpurea.* Org Lett. 2000;2:515–8.
12. Chen SD, Gao H, Zhu QC, Wang YQ, Li T, Mu ZQ, Wu HL, Peng T, Yao XS. Houttuynoids A-E, anti-herpes simplex virus active flavonoids with novel skeletons from *Houttuynia cordata.* Org Lett. 2012;14:1772–5.

13. Lu Z, Van Wagoner RM, Pond CD, Pole AR, Jensen JB, Blankenship D, Grimberg BT, Kiapranis R, Matainaho TK, Barrows LR, Ireland CM. Myristicyclins A and B: antimalarial procyanidins from *Horsfieldia spicata* from Papua New Guinea. Org Lett. 2014;16:346–9.
14. Zhang F, Yang YN, Song XY, Shao SY, Feng ZM, Jiang JS, Li L, Chen NH, Zhang PC. Forsythoneosides A-D, neuroprotective phenethanoid and flavone glycoside heterodimers from the fruits of *Forsythia suspensa*. J Nat Prod. 2015;78:2390–7.
15. Naman CB, Li J, Moser A, Hendrycks JM, Benatrehina PA, Chai H, Yuan C, Keller WJ, Kinghorn AD. Computer-assisted structure elucidation of black chokeberry (*Aronia melanocarpa*) fruit juice isolates with a new fused pentacyclic flavonoid skeleton. Org Lett. 2015;17:2988–91.
16. Li F, Zhan Z, Liu F, Yang Y, Li L, Feng Z, Jiang J, Zhang P. Polyflavanostilbene A, a new flavanol-fused stilbene glycoside from *Polygonum cuspidatum*. Org Lett. 2013;15:674–7.
17. Yang XW, Li SM, Feng L, Shen YH, Tian JM, Zeng HW, Liu XH, Shan L, Su J, Zhang C, Zhang WD. Abiesanol A, a novel biflavanol with unique six connective hexacyclic rings isolated from *Abies georgei*. Tetrahedron Lett. 2008;49:3042–4.
18. Zhou B, Alania Y, Reis MC, McAlpine JB, Bedran-Russo AK, Pauli GF, Chen SN. Rare A-type, spiro-type, and highly oligomeric proanthocyanidins from *Pinus massoniana*. Org Lett. 2020;22:5304–8.
19. Kuo PC, Hung HY, Hwang TL, Du WK, Ku HC, Lee EJ, Tai SH, Chen FA, Wu TS. Anti-inflammatory flavan-3-ol-dihydroretrochalcones from *Daemonorops draco*. J Nat Prod. 2017;80:783–9.
20. Doan TP, Park EJ, Cho HM, Ryu B, Lee BW, Thuong PT, Oh WK. Rugonidines A−F, diastereomeric 1,6-dioxa-7,9-diazaspiro[4.5]dec-7-en-8-amines from the leaves of *Alchornea rugosa*. J Nat Prod 2021;84:3055−63.

Chapter 10
Diverse Novel Lignins

New skeleton lignins are also few in nature plants, since 1999, 33 new skeleton lignins, assigned to 9 families 9 genuses, have been isolated and identified from plants (**872–904**, Fig. 10.1; Table A9). Schisandraeaceae contributed more disesquiterpenes than other families. This review will focus on the names, classes, structure characteristics, activities, the types of compounds, and plant sources of these new skeleton lignins.

Taiwanschirins A–C (**872–874**) are three novel C_{19} homolignans with a 3,4-pentano-2,3-dihydrobenzo[*b*]furan skeleton, they were isolated from *Schizandra arisanensis* (Schisandraceae) [1]. Taiwankadsurins A–C (**875–877**) are three novel C_{19} homolignans isolated from the aerial parts of *Kadsura philippinensis* (Schisandraceae) [2]. The structures of **875–877** share a 3,4-{1′-[(*Z*)-2″-methoxy-2″-oxo-ethylidene]}-pentano(2,3-dihydro-benzo[b]-furano)-3-(2‴-methoxycarbonyl-2‴-hydroxy-2‴,3′-epoxide) skeleton. **878** and **879** are two rare 7,8-*seco*-lignans from *Schisandra glaucescens* (Schisandraceae), their absolute configurations were mainly determined by comparing their experimental CD data with those of calculated electronic circular dichroism [3]. Kadsuraols A–C (**880–882**) are tetrahydrocyclobutaphenanthrofuranone-type lignans with a new carbon skeleton comprising a four-membered ring across C-1′–C-8, they were isolated from the roots of *Kadsura longipedunculata* (Schisandraceae) [4]. (±)-Torreyunlignans A–D (**883–886**) are four pairs of 8–9′ linked neolignan enantiomers featuring a rare (*E*)-2-styryl-1,3-dioxane moiety, they were isolated from the trunk of *Torreya yunnanensis* (Taxaceae) [5]. (±)-Subaveniumins A (**887**) and B (**888**), isolated from *Cinnamomum subavenium* (Lauraceae), are two pairs of racemic spirodienone neolignans with a rare 2-oxaspiro[4.5]deca-6,9-dien-8-one motif, their absolute configurations were determined by comparing the experimental CD data with those of and calculated electronic circular dichroism [6]. Aiphanol (**889**) is a novel stilbenolignan representing a novel carbon skeleton having a stilbene-phenylpropane unit with a dioxane moiety, it was isolated by bioassay-guided fractionation from the seeds of *Aiphanes aculeata* (Palmaceae) [7]. Chimarrhinin (**890**) is a new C6-C3

Y. Cheng and D. Qin, *Novel Plant Natural Product Skeletons*,
https://doi.org/10.1007/978-981-99-7329-3_10

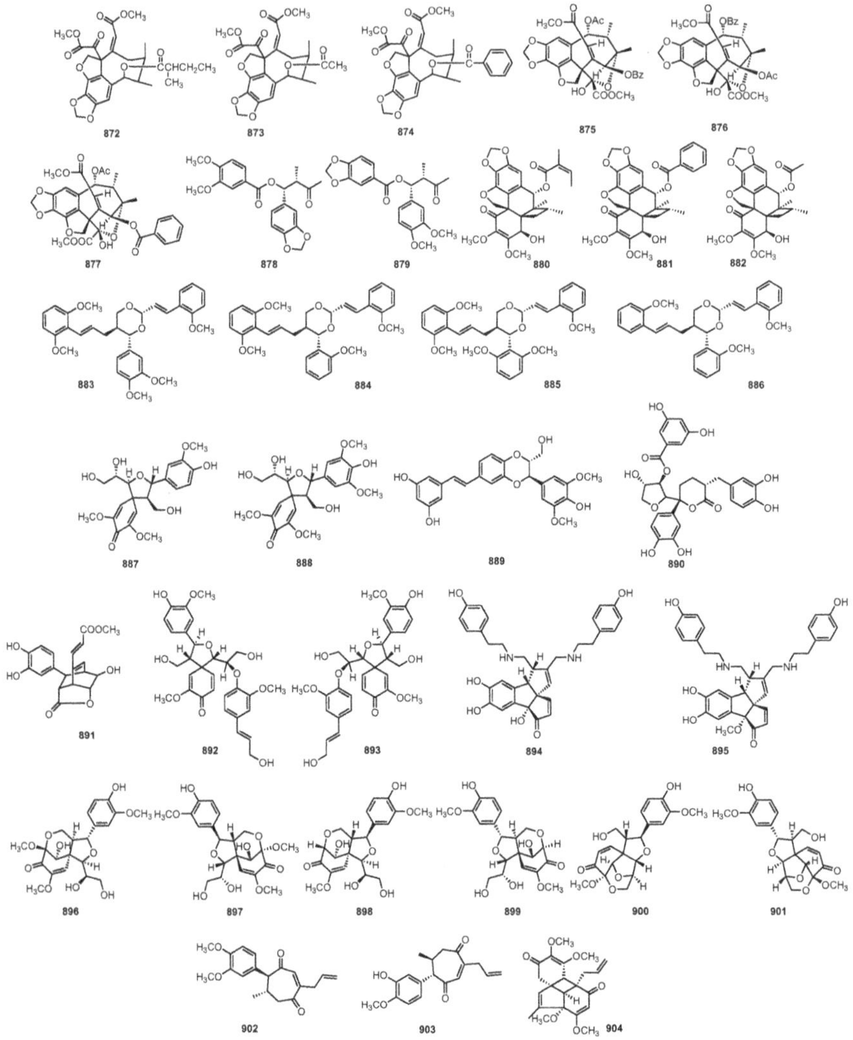

Fig. 10.1 Structures of new skeleton lignins **872–904**

lignan skeleton type, which was isolated from an extract of the leaves of *Chimarrhis turbinata* (Rubiaceae) [8]. Rufescenolide (**891**) has been isolated from the stems of *Cordia rufescens* (Boraginaceae), which is an unusual lignan with a bicyclic [2.2.2] octene skeleton [9]. (+)- and (−)-Sibiricumins A (**892** and **893**) are a pair of enantiomeric spirodienone neolignans. They were isolated from the fruits of *Xanthium sibiricum* (Compositae/Asteraceae), and their absolute configurations were determined by a combination of X-ray diffraction and ECD calculations. This is the first time that the relative and absolute configurations of this type of spirodienone neolignans were unambiguously determined [10]. (±)-Sativamides A

(**894**) and B (**895**) are two pairs of norlignanamide enantiomers isolated from the fruits of *Cannabis sativa* (Moraceae) and featured a unique benzo-angular triquinane skeleton [11]. (−)- and (+)-Pinnatifidaones A, (−)- and (+)-pinnatifidaones B, (−)- and (+)-pinnatifidaones C (**896–901**) are three pairs of highly modified spirocyclohexenone neolignan enantiomers, which feature a 5/6/6 tricyclic or a 5/5/6/6 tetracyclic system with a 2-oxaspiro[4.5]deca-6-en-8-one motif. All these isomers were isolated from *Crataegus pinnatifida* (Rosaceae) [12]. Pibeneolignans A and B (**902** and **903**) are new naturally occurring neolignan skeletons, based on the cyclohept-2-ene-1,4-dione framework, they were obtained an extract of the leaves of *Piper betle* [13]. (±)-piperhancins A (**904**) is an unprecedented 1′,2:1,2′-dicyclo-8,3′-neolignane and isolated from the stems of *Piper hancei*. Its structures and absolute configurations were elucidated by X-ray diffraction [14].

References

1. Kuo YH, Huang HC, Kuo LMY, Chen CF. Novel C_{19} homolignans, taiwanschirin A, B, and cytotoxic taiwanschirin C, and a new C_{18} lignan, schizanrin A, from *Schizandra arisanensis*. J Org Chem. 1999;64:7023–7.
2. Shen YC, Lin YC, Cheng YB, Kuo YH, Liaw CC. Taiwankadsurins A, B, and C, three new C19 homolignans from *Kadsura philippinensis*. Org Lett. 2005;7:5297–300.
3. Yu HY, Chen ZY, Sun B, Liu J, Meng FY, Liu Y, Tian T, Jin A, Ruan HL. Lignans from the fruit of *Schisandra glaucescens* with antioxidant and neuroprotective properties. J Nat Prod. 2014;77:1311–20.
4. Wang X, Liu J, Pandey P, Fronczek FR, Doerksen RJ, Chen J, Qi X, Zhang P, Ferreira D, Valeriote FA, Sun H, Li S, Hamann MT. Computationally assisted assignment of the kadsuraols, a class of chemopreventive agents for the control of liver cancer. Org Lett. 2018;20:5559–63.
5. Cheng ZB, Lu X, Bao JM, Han QH, Dong Z, Tang GH, Gan LS, Luo HB, Yin S. (±)-Torreyunlignans A-D, rare 8–9′ linked neolignan enantiomers as phosphodiesterase-9a inhibitors from *Torreya yunnanensis*. J Nat Prod. 2014;77:2651–7.
6. Lai Y, Liu T, Sa R, Wei X, Xue Y, Wu Z, Luo Z, Xiang M, Zhang Y, Yao G. Neolignans with a rare 2-oxaspiro[4.5]deca-6,9-dien-8-one motif from the stem bark of *Cinnamomum subavenium*. J Nat Prod. 2015;78:1740–4.
7. Lee D, Cuendet M, Vigo JS, Graham JG, Cabieses F, Fong HHS, Pezzuto JM, Kinghorn AD. A novel cyclooxygenase-inhibitory stilbenolignan from the seeds of *Aiphanes aculeata*. Org Lett. 2001;3:2169–71.
8. Cardoso CL, Castro Gamboa I, Bergamini GM, Cavalheiro AJ, Silva DHS, Lopes MN, Araújo AR, Furlan M, Verli H, Bolzani VdaS. An unprecedented neolignan skeleton from *Chimarrhis turbinata*. J Nat Prod. 2011;74:487–91.
9. do Vale AE, David JM, dos Santos EO, David JP, e Silva LCRC, Bahia MV, Brandão HN. An unusual caffeic acid derived bicyclic [2.2.2]octane lignan and other constituents from *Cordia rufescens*. Phytochemistry. 2012;76:158–61.
10. Shi Y, Liu Y, Li Y, Li L, Qu J, Ma S, Yu S. Chiral resolution and absolute configuration of a pair of rare racemic spirodienone sesquineolignans from *Xanthium sibiricum*. Org Lett. 2014;16:5406–9.
11. Zhu GY, Yang J, Yao XJ, Yang X, Fu J, Liu X, Bai LP, Liu L, Jiang ZH. (±)-sativamides A and B, two pairs of racemic nor-lignanamide enantiomers from the fruits of *Cannabis sativa*. J Org Chem. 2018;83:2376–81.

12. Guo R, Zhao P, Yu X, Yao G, Lin B, Huang X, Song S. (±)-Pinnatifidaones A-D, four pairs of highly modified neolignan enantiomers with a rare spirocyclohexenone skeleton from *Crataegus pinnatifida*. Org Chem Front. 2021;8:953–60.
13. Xiao CY, Sun ZL, Huang J, Li RS, He JM, Gibbons S, Ju DW, Mu Q. Neolignans from *Piper betle* have synergistic activity against antibiotic-resistant *Staphylococcus aureus*. J Org Chem. 2021;86:11072–85.
14. Yang F, Su BJ, Hu YJ, Liu JL, Li H, Wang YQ, Liao HB, Liang D. Piperhancins A and B, two pairs of antineuroinflammatory cycloneolignane enantiomers from *Piper hancei*. J Org Chem. 2021;86:5284–91.

Chapter 11
Classification of Diverse Novel Alkaloids

Plants have proved to be a rich and important source of a huge number of new skeleton alkaloids with a large variety of chemical structures and diverse biological activities.

Importantly, alkaloids are the largest and most structurally diverse classes of new skeletons from plants. This review covers nearly 331 new skeleton alkaloids (**905–936** Fig. 11.1, **937–959** Fig. 11.2, **960–993** Fig. 11.3, **994–1017** Fig. 11.4, **1018–1039** Fig. 11.5, **1040–1065** Fig. 11.6, **1066–1088** Fig. 11.7, **1089–1108** Fig. 11.8, **1109–1145** Fig. 11.9, **1146–1162** Fig. 11.10, **1163–1200** Fig. 11.11, **1201–1235** Fig. 11.12; Table A10). Apocynaceae, Daphniphyllaceae, Lycopodiaceae, Leguminosae, Euphorbiaceae, Rutaceae are thought to be the most prolific of the novel alkaloids, especially, Daphniphyllum alkaloids and lycopodium alkaloids are typical alkaloid classes and carried on an extensive and thorough research. We wish to provide an overall and in-depth view of this type of new skeletons.

11.1 Euphorbiaceae

11.1.1 Flueggea

Flueggenines A (**905**) and B (**906**), are two unprecedented C,C-linked dimeric indolizidine alkaloids, they were isolated from the roots of *Flueggea virosa* [1]. Flueggines A (**907**) and B (**908**) are two unprecedented C,C-linked dimeric indolizidine alkaloids, they were isolated from the twigs and leaves of *Flueggea virosa* [2]. Suffrutines A (**909**) and B (**910**) are photochemical *Z*/*E* isomeric indolizidine alkaloids isolated from *Flueggea suffruticosa* with a unique and highly conjugated C_{20} skeleton [3]. Interestingly, **909** and **910** appear to be interconvertible induced by light. Fluvirosaones A (**911**) and B (**912**) were isolated from *Flueggea virosa* and represent the first examples of a pentacyclic *Securinega* alkaloids containing a pentacyclic system and an α,β-unsaturated ketone in the molecule [4]. Virosaines A (**913**) and B (**914**) are

Y. Cheng and D. Qin, *Novel Plant Natural Product Skeletons*,
https://doi.org/10.1007/978-981-99-7329-3_11

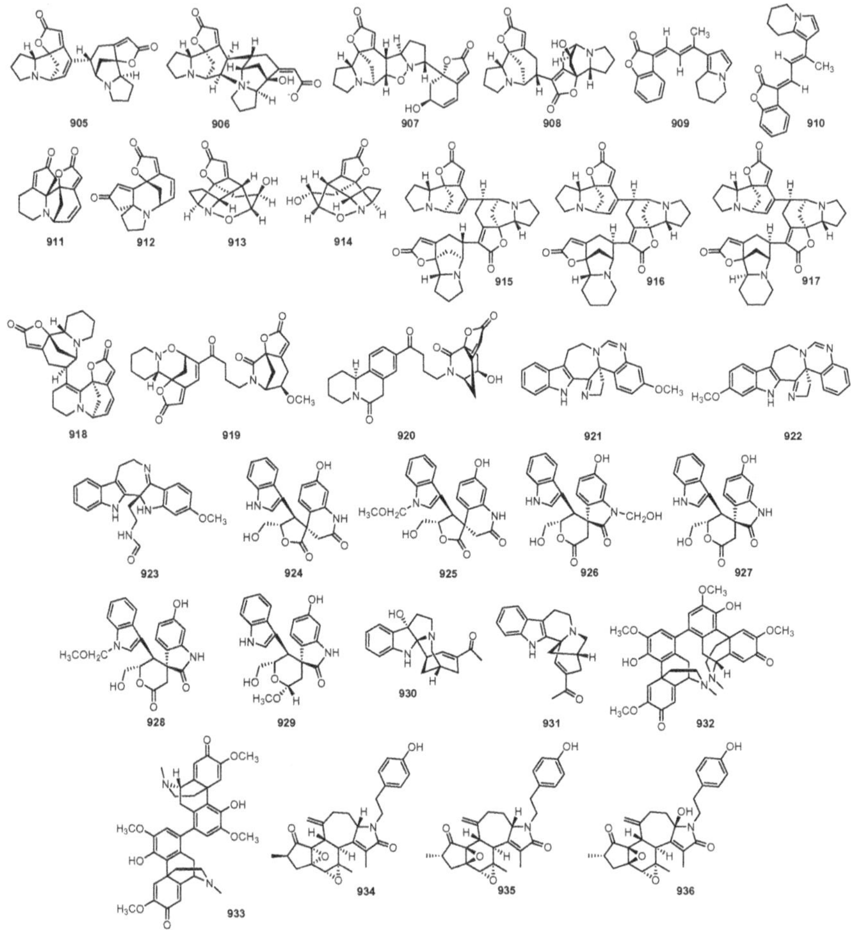

Fig. 11.1 Structures of new skeleton lignins **905–936**

the first examples of *Securinega* alkaloids bearing a 7-oxa-1-azabicyclo[3.2.1]octane ring system, they were isolated from the twigs and leaves of *Flueggea virosa* [5]. Fluevirosines A–C (**915–917**) are three unprecedented C,C-linked trimeric *Securinega* alkaloids. They were isolated from *Flueggea virosa* and their absolute structures were assigned by spectroscopic data and computational analysis [6]. Flueggeacosines A–C (**918–920**), described from *Flueggea suffruticosa*, are three dimeric securinine-type alkaloid analogues with unprecedented skeletons [7]. Among them, **918** and **919** are the first examples of C-3–C-15′ connected dimeric securininetype alkaloids, while **920** is an unprecedented heterodimer of securinine-type and benzoquinolizidine alkaloids.

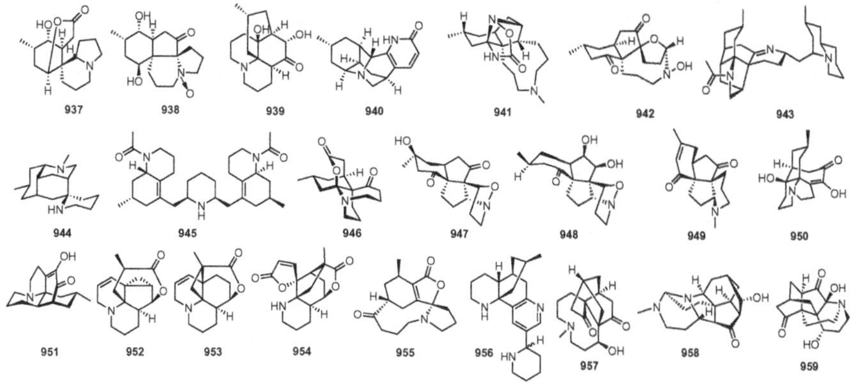

Fig. 11.2 Structures of new skeleton alkaloids **937–959**

11.1.2 *Trigonostemon*

Trigonoliimines A–C (**921–923**) are unprecedented indole alkaloids possessing a unique polycyclic system and were isolated from the leaves of *Trigonostemon lii* [8]. Trigolutesins A and B (**924** and **925**) and trigolutes A–D (**926–929**) are six unprecedented bisindole alkaloids. Of which, **924** and **925** have a unique polycyclic skeleton, while **926–929** feature another polycyclic skeleton. **924–929** were all isolated from the twigs of *Trigonostemon lutescens* [9]. Voatinggine (**930**) and tabertinggine (**931**) are characterized as new skeleton natural products featuring a pentacyclic system, and they were both isolated from *Tabernaemontana* species [10].

11.1.3 *Croton*

Saludimerines A (**932**) and B (**933**) are the first biarylic bis-morphinanedienone alkaloids, they were isolated from a tree of *Croton flavens* [11]. Cascarinoids A–C (**934–936**) are a new class of diterpenoid alkaloids with unpredicted conformations, they were isolated from *Croton cascarilloides* and represent the first examples of crotofolane diterpenoid alkaloids [12].

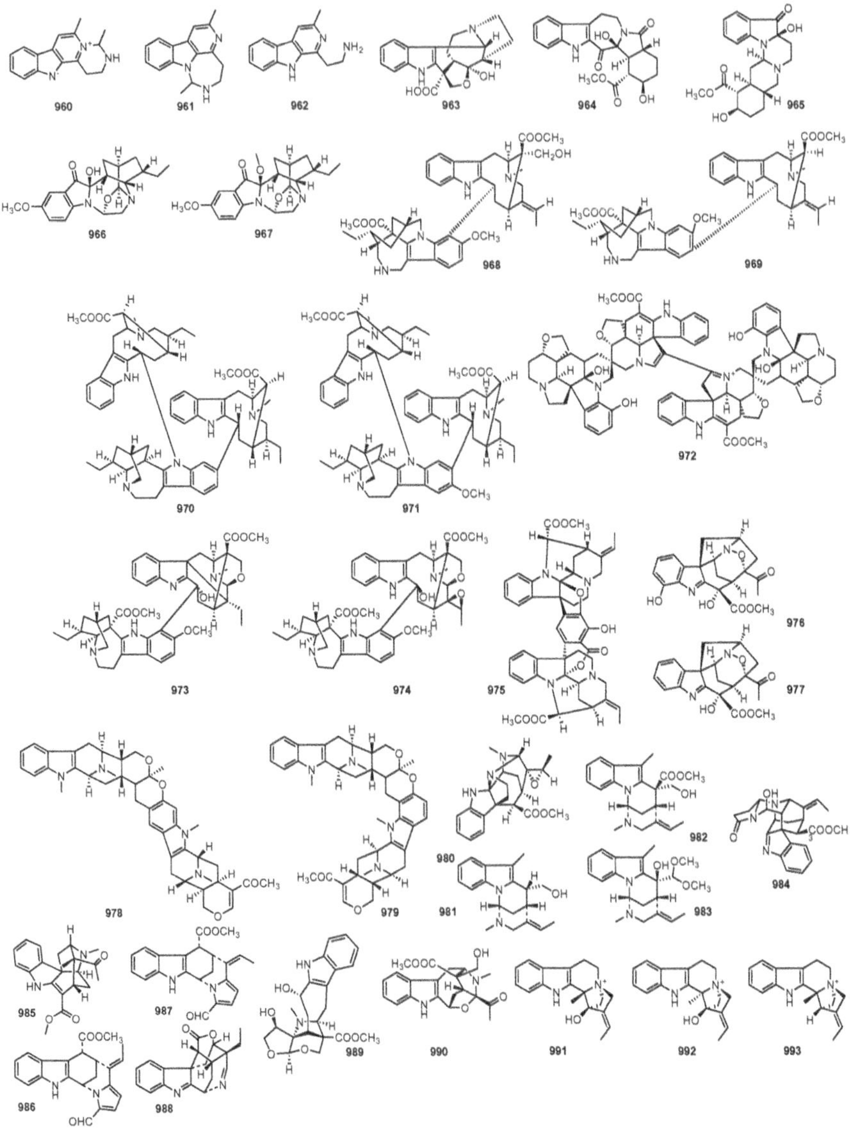

Fig. 11.3 Structures of new skeleton alkaloids **960–993**

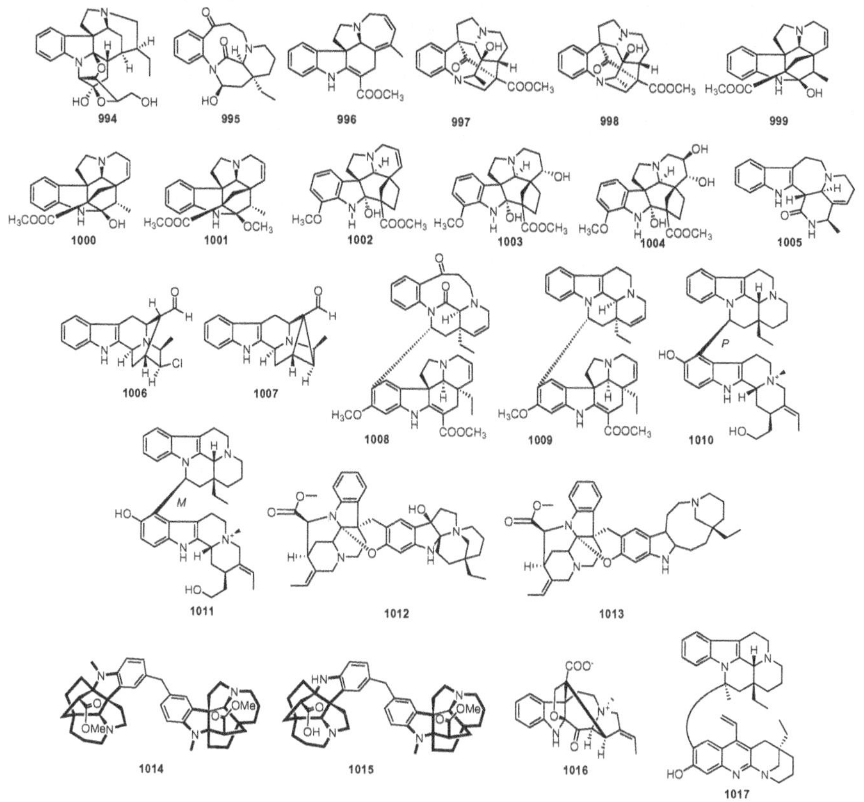

Fig. 11.4 Structures of new skeleton alkaloids **994–1017**

11.2 Lycopodiaceae

11.2.1 Lycopodium

Serratezomines A–C (**937–939**) were isolated from the club moss *Lycopodium serratum*. **937–939** are all new skeletons representing a *seco*-serratinine type, a serratinine type, and a lycodoline type, respectively [13]. Lyconadin A (**940**) is a novel alkaloid with an unprecedented skeleton consisting of three six-membered, one five-membered, and one α-pyridone rings. It was isolated from the club moss *Lycopodium complanatum* [14]. Lycoposerramine-A (**941**) has a 1,2,4-oxadiazolidin-5-one residue in the molecule, it was isolated from *Lycopodium serratum* and its absolute configuration was determined by X-ray analyses [15]. Sieboldine A (**942**) was isolated from the club moss *Lycopodium sieboldii*, which is a novel *Lycopodium* alkaloid with an unprecedented fused-tetracyclic ring system consisting of an azacyclononane ring having a *N*-hydroxy group, a cyclohexanone, a cyclopentanone,

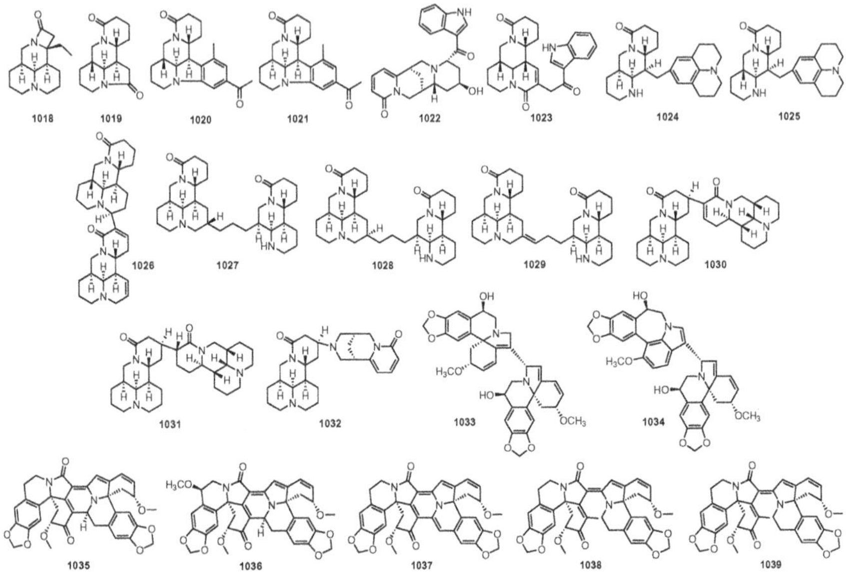

Fig. 11.5 Structures of new skeleton alkaloids **1018–1039**

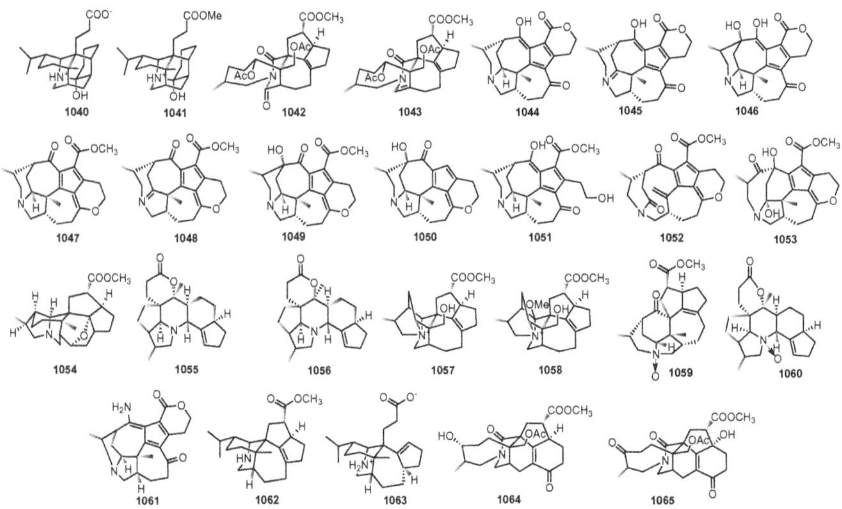

Fig. 11.6 Structures of new skeleton alkaloids **1040–1065**

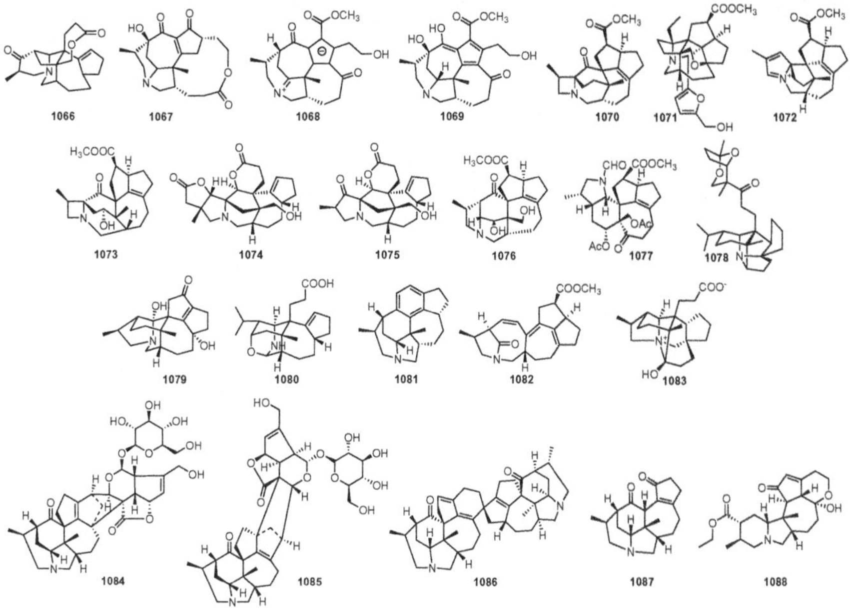

Fig. 11.7 Structures of new skeleton alkaloids **1066–1088**

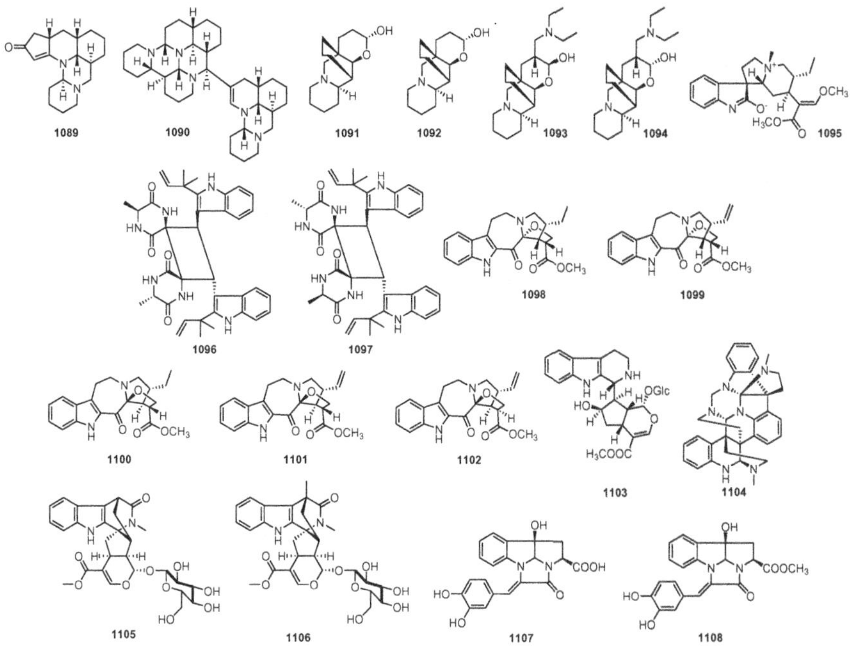

Fig. 11.8 Structures of new skeleton alkaloids **1089–1108**

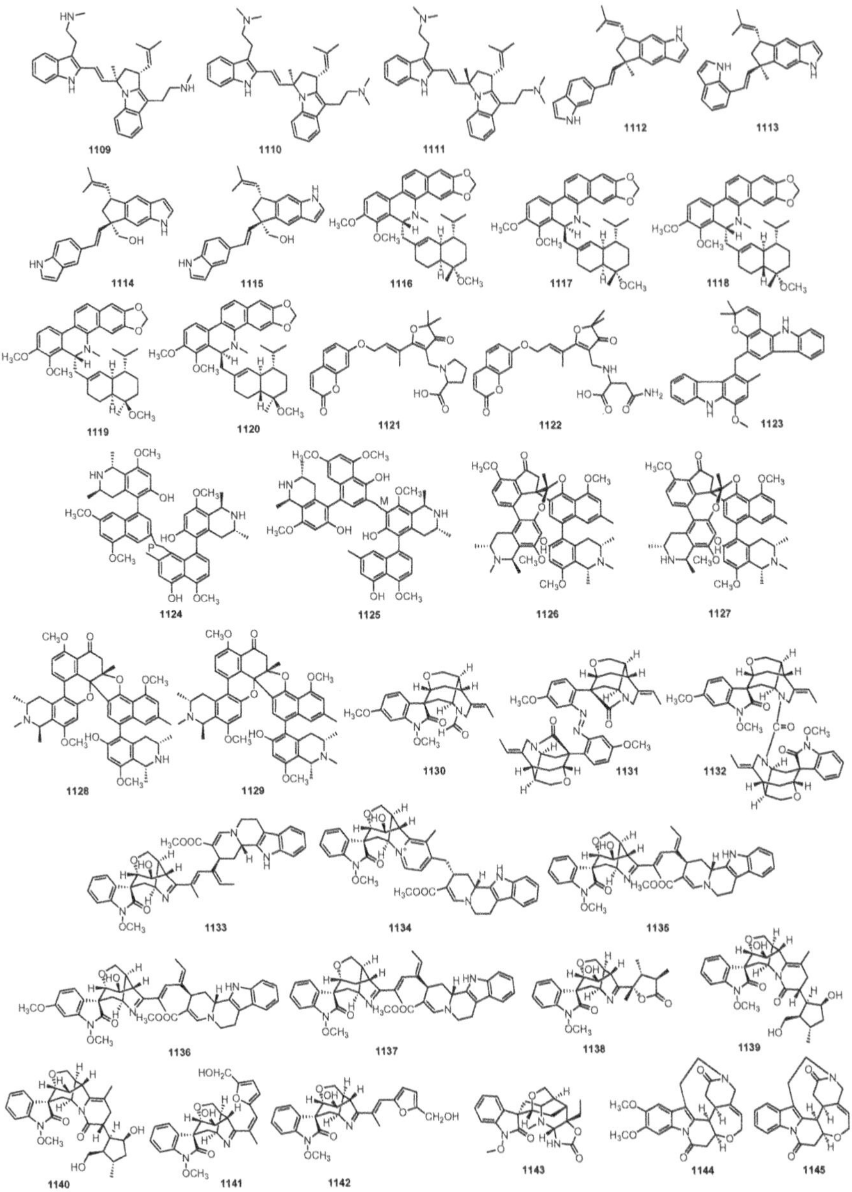

Fig. 11.9 Structures of new skeleton alkaloids **1109–1145**

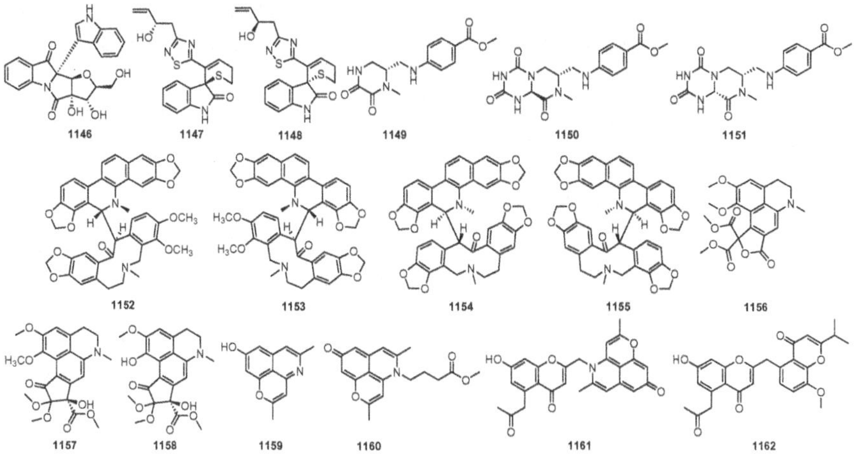

Fig. 11.10 Structures of new skeleton alkaloids **1146–1162**

and a tetrahydrofuran ring [16]. Himeradine A (**943**) is another novel $C_{27}N_3$-type *Lycopodium* alkaloid from *Lycopodium chinense*, which consists of a fastigiatine-type skeleton ($C_{16}N_2$) and a quinolizidine moiety ($C_{11}N$) [17]. Nankakurine A (**944**) is an unique fused-tetracyclic *Lycopodium* alkaloid consisting of a cyclohexane ring and a 3-aza-bicyclo[3.3.1]nonane ring connected to a piperidine ring through a spiro carbon [18]. From the club moss *Lycopodium hamiltonii*, lycoperine A (**945**) was also isolated which is a novel $C_{27}N_3$-type pentacyclic *Lycopodium* alkaloid and consists of two octahydroquinoline rings and a piperidine ring [19]. Lycojapodine A (**946**) represents a novel $C_{16}N$-type *Lycopodium* alkaloid with an unprecedented 6/6/6/7 tetracyclic ring system, which was isolated from the club moss *Lycopodium japonicum* [20]. Lycojaponicumins A–C (**947–949**) are trace alkaloids isolated from *Lycopodium japonicum*, they represent a unique heterocyclic skeleton formed by the new linkage C4–C9. In particular, lycojaponicumins A and B (**947** and **948**) are the first examples of natural products possessing a 5/5/5/5/6 pentacyclic ring system with a 1-*aza*-7-oxabicyclo[2.2.1]heptane moiety [21]. Lycojaponicumin D (**950**), described from *Lycopodium japonicum*, possesses an unprecedented 5/7/6/6 tetracyclic skeleton formed by an unusual C3–C13 linkage, which was first reported in *Lycopodium* alkaloids [22]. Lycospidine A (**951**) is the first example of a *Lycopodium* alkaloid which contains an unprecedented five-membered A ring. The unique five-membered A ring in **951** indicates that carbons 2–5 in **951** are presumably derived from proline instead of the lysine biosynthetically, which suggests that **951** represent a new class of *Lycopodium* alkaloid [23]. Annotinolides A–C (**952–954**) are three novel 7,8-*seco*-lycopodane-derived 8,5-lactones from *Lycopodium annotinum*. Among them, **952** possesses an unusual cyclopropane ring constructed through a hitherto unknown C-6/C-12 bond, **953** represents the first 7,8-*seco*lycopodane-derived alkaloid with a rare cyclobutane ring formed by a new C-12/C-15 linkage, while the C-8/C-15 bond remains, **954** contains an unprecedented 12-spiro-9,12-γ-lactone

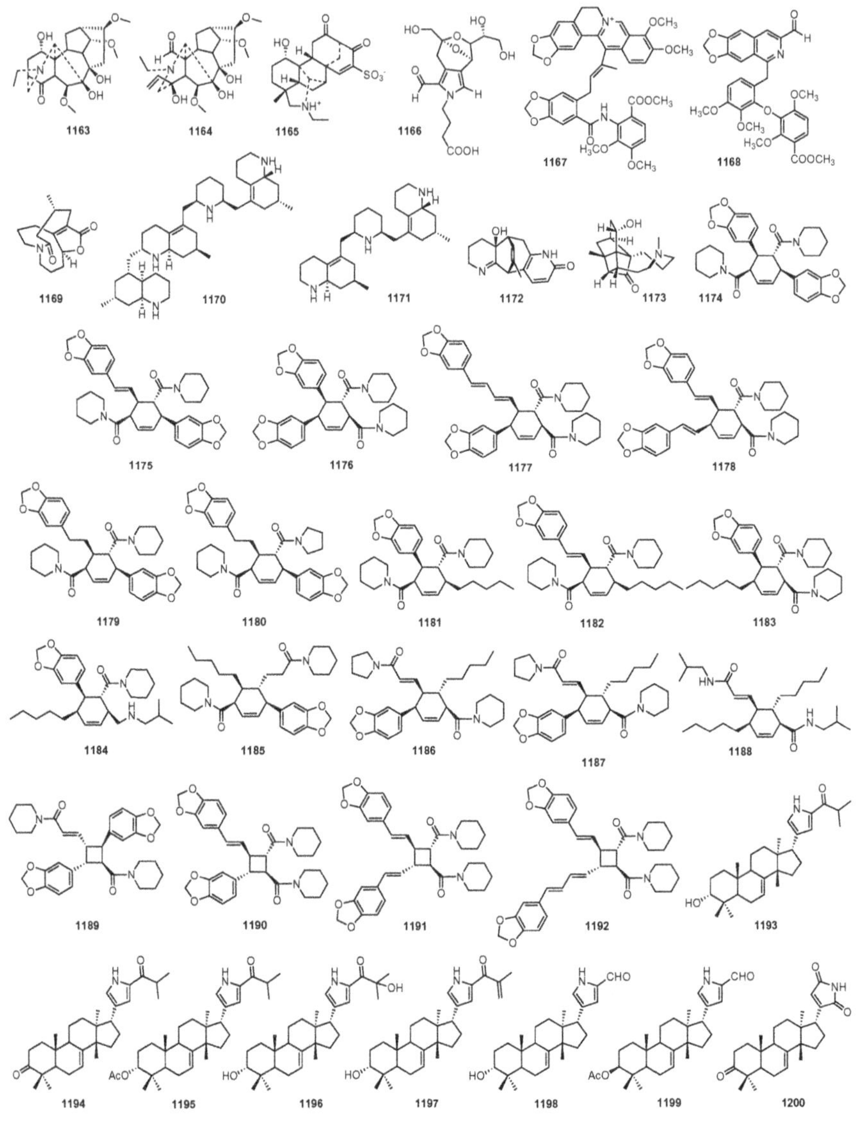

Fig. 11.11 Structures of new skeleton alkaloids **1163–1200**

moiety [24]. Lycoplanine A (**955**) was isolated from *Lycopodium complanatum* and is a *Lycopodium* alkaloid with a 6/9/5 tricyclic ring skeleton fused with the γ-lactone ring and featuring an unusual 1-oxa-6-azaspiro[4.4]nonane moiety and an unprecedented 3-azabicyclo[6.3.1]dodecane unit [25]. The absolute configurations of **955** were identified by single-crystal X-ray diffraction. Lycoplatyrine A (**956**) has been

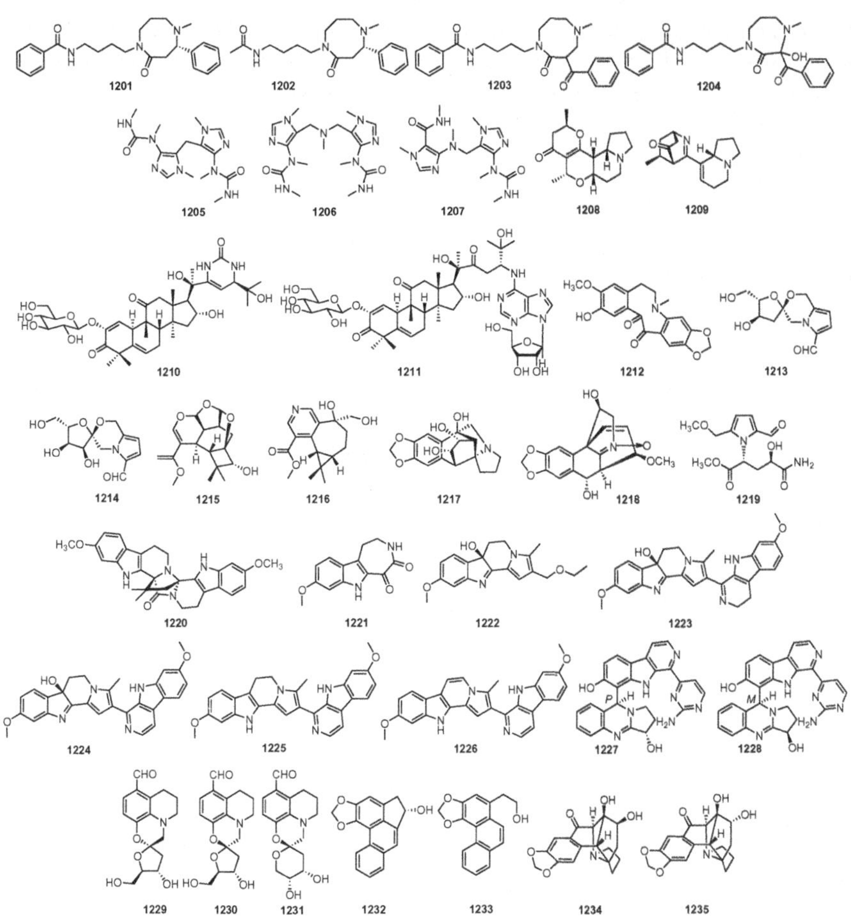

Fig. 11.12 Structures of new skeleton alkaloids **1201–1235**

isolated from *Lycopodium platyrhizoma* representing an unusual lycodine-piperidine adduct [26].

11.2.2 *Palhinhaea*

Palhinine A (**957**) was isolated from *Palhinhaea cernua* and is a novel C_{16}N-type *Lycopodium* alkaloid with a unique 5/6/6/9 tetracyclic ring system, it represents the first example of *Lycopodium* alkaloids with C-16 being fused to a new ring via a C-16–C-4 lingkage [27]. Lycopalhine A (**958**) possesses a fused hexacyclic (5/5/5/6/6/6) ring system comprising a 5,9-diaza-tricyclo[6.2.1.$0^{4,9}$]undecane moiety and a tricyclo[5.2.1.0^{4,8}]decane moiety, which was isolated from *Palhinhaea cernua* [28]. Isopalhinine A (**959**) is a new pentacyclic (5/6/6/6/7) *Lycopodium* alkaloid isolated from *Palhinhaea cernua*, which possesses a sterically congested architecture built with a tricyclo[4.3.1.03,7]decane (isotwistane) moiety and a 1-azabicyclo[4.3.1]decane moiety [29].

11.3 Apocynaceae

11.3.1 Tabernaemontana

Tabernines A–C (**960–962**) are three novel β-carboline indole alkaloids from *Tabernaemontana elegans* [30]. Of which, **960** and **961** contain a two-carbon unit attached to a structurally related β-carboline skeleton as a part of an additional six-membered ring in **960** and a seven-membered ring in **961**. This is the first report of β-carboline indole alkaloids from *Tabernaemontana* plants. Actinophyllic acid (**963**) was isolated from the Leaves of *Alstonia actinophylla* possessing a unique 2,3,6,7,9,13c-hexahydro-1*H*-1,7,8-(methanetriyloxymethano)-pyrrolo[1′,2′:1,2]azocino[4,3-*b*]indole-8(5*H*)-carboxylic acid skeleton [31]. Structurally, criofolinine (**964**) incorporate a pyrroloazepine motif within a pentacyclic ring system and vernavosine (**965**) incorporate a pyridopyrimidine moiety embedded within a pentacyclic carbon framework [32]. Erchinines A (**966**) and B (**967**), isolated from *Ervatamia chinensis*, are found to possess a unique 1,4-diazepine fused with oxazolidine architecture and three hemiaminals [33]. Bistabercarpamines A (**968**) and B (**969**) are two novel dimeric monoterpene indole alkaloids possessing unprecedented *bis*-vobasinyl-chippiine-type skeleton [34]. They were isolated from the leaves of *Tabernaemontana corymbosa* and their absolute configurations were determined by CD excition chirality method. Ervadivamines A (**970**) and B (**971**) are two unprecedented trimeric monoterpenoid indole alkaloids. They were isolated from *Ervatamia divaricata* and are the first examples of vobasine-iboga-vobasine-type alkaloids with both C–C and C–N linkage patterns [35]. Alasmontamine A (**972**) is a novel tetrakis monoterpene indole alkaloid, which consists of *bis*-vobtusine-type skeletons and was isolated from the leaves of *Tabernaemontana elegans* [36]. Tabercorymines A (**973**) and B (**974**) are two vobasinyl–ibogan-type bisindole alkaloids with an unprecedented skeleton [37]. Especially, **973** represents a novel bisindole alkaloid, characteristic of a caged heteropentacyclic ring system incorporating an unprecedented C-7/C-20 bond in the vobasinyl.

11.3.2 Alstonia

Bipleiophylline (**975**) is a bisindole alkaloid possessing an unprecedented structure in which two indole moieties are bridged by an aromatic spacer unit, **975** was isolated from *Alstonia angustifolia* [38]. Alsmaphorazines A and B (**976** and **977**) are novel indole alkaloids from *Alstonia pneumatophora* possessing a new skeleton consisting of an 1,2-oxazinane and an isoxazolidine chromophore [39]. Lumutinines A (**978**) and B (**979**) represent the first examples of linear, ring A/F-fused macroline–macroline-type bisindoles [40]. They were isolated from *Alstonia macrophylla.* Alistonitrine A (**980**) is a new monoterpene indole alkaloid from *Alstonia scholaris* incorporating a third nitrogen atom possessing an unprecedented caged skeleton with a unique 6/5/6/5/5/6 ring system [41]. Alstoscholarisines H–J (**981–983**) are monoterpenoid indole alkaloids from *Alstonia scholaris* with an unprecedented skeleton created via the formation of a C-3/N-1 bond [42]. Alstoscholarisine K (**984**), an indole alkaloid with eight chiral carbons and featuring a novel 6/5/6/6/6/6/6/5 octacyclic architecture, was found to be specific to the gall-infected leaves of Alstonia scholaris. Its structure was elucidated by spectroscopy, computational analysis, and single-crystal X-ray diffraction [43]. Alstonlarsine A (**985**) possesses a new carbon skeleton with a cage-shaped 9-azatricyclo[4.3.1.0^{3,8}]decane motif, it was isolated from *Alstonia scholaris* and its absolute configuration was determined by X-ray crystal diffraction [44]. (19,20) *E*-alstoscholarine (**986**) and (19,20) *Z*-alstoscholarine (**987**) are a pair of geometrically isomeric monoterpenoid indole alkaloids with a skeleton rearrangement and two additional carbons. They were obtained from *Alstonia scholaris* leaves [45]. Scholarisine A (**988**) is an unprecedented cage-like alkaloid from *Alstonia scholaris*, which was proposed to be derived from picrinine via oxygenation, rearrangement, and lactonization [46]. Alstolarines A and B (**989** and **990**) are novel monoterpenoid indole alkaloids from *Alstonia scholaris* [47]. Of which, **989** is the first vobasinyl-type alkaloid with a 6/5/8/5/6/5 hexacyclic architecture featuring an unprecedented 21-oxa-4-azaspiro[4,4]nonane unit. Whereas **990** is the first 21-*nor*-vobasinyl-type alkaloid with a unique 3-oxa-4-azatricyclo[5.1.1.1^{3,20}]undecane ring system.

11.3.3 Aspidosperma

Subincanadines A–C (**991–993**) are novel quaternary indole alkaloids with an unprecedented 1-azoniatricyclo[4.3.3.0^{1,5}]undecane moiety and they were isolated from *Aspidosperma subincanum* [48].

11.3.4 Melodinus

Melohenine A (**994**) is a monoterpenoid indole alkaloid with additional skeletal carbons arranged compactly in eight rings, and melohenine B (**995**) is an alkaloid with an unprecedented 6/9/6/6 tetracyclic ring system regarded as a key intermediate from indole to quinoline alkaloids [49]. Both **994** and **995** were isolated from *Melodinus henryi*. Melotenine A (**996**) is an unprecedented skeleton with a 6/5/5/6/7 pentacyclic rearranged ring system, it was isolated from *Melodinus tenuicaudatus* and its absolute configuration was confirmed by X-ray diffraction analysis [50]. Melocochines A (**997**) and B (**998**) are a pair of epimers with an unprecedented skeleton from *Melodinus cochinchinensis*, which represent a class of novel alkaloids characterized by a rare 1*H*-benzo[*b*]azepane ring system within monoterpenoid indole alkaloid categories [51]. Meloyunnanines A–C (**999–1001**) are unprecedented skeleton alkaloids featuring a caged-6/6/5/6/5/5 ring system. They were isolated from the fruits of *Melodinus yunnanensis* [52].

11.3.5 Kopsia

Kopsifolines A–C (**1002–1004**) are indole alkaloids from *Malayan kopsia* possessing a novel hexacyclic carbon skeleton [53]. Arboflorine (**1005**) is a new indole alkaloid from the Malayan *Kopsia arborea* possessing a novel pentacyclic carbon skeleton and incorporating a third nitrogen atom [54].

11.3.6 Rauvolfia

Rauvomine A (**1006**) and rauvomine B (**1007**) are two unusual normonoterpenoid indole alkaloids from *Rauvolfia vomitoria*. **1006** is a C_{18} sarpagine-type normonoterpenoid indole alkaloid with a chlorine atom in the structure. Interestingly, **1007** represents a novel 6/5/6/6/3/5-fused hexcyclos sarpagine-type normonoterpenoid indole alkaloid possessing a cyclopropane ring and C_{18} skeleton [55].

11.3.7 Bousigonia

Mekongenine A (**1008**) and mekongenine B (**1009**) were both isolated from *Bousigonia mekongensis*. Of which, **1008** possesses an unprecedented structure constituted from the union of a rare 2,7-*seco* eburnamine half and an aspidospermine alkaloid, whereas **1009** consists of an eburnamine-aspidospermine type skeleton [56].

11.3.8 *Hunteria*

Bisnicalaterines B and C (**1010** and **1011**) are bisindole alkaloids from the bark of *Hunteria zeylanica*, consisting of an eburnane and a corynanthe type of skeletons [57].

11.3.9 *Gonioma*

Goniomedines A (**1012**) and B (**1013**) have been isolated from the stem bark of *Gonioma malagasy*. They are two novel bisindole alkaloids possessing an unprecedented quebrachamine-pleioarpamine-type skeleton where indole moieties are fused via a dihydropyran unit [58].

11.3.10 *Pleiocarpa*

Pleiokomenines A and B (**1014** and **1015**) are the first examples of dimeric aspidofractinine alkaloids linked by a methylene bridge, they were isolated from the stem bark of *Pleiocarpa mutica* [59].

11.3.11 *Winchia*

Calophyline A (**1016**) is a novel unprecedented rearranged monoterpenoid indole alkaloid, it has been isolated from *Winchia calophylla* and its absolute configuration was confirmed by X-ray crystallographic analysis [60].

11.3.12 *Leuconotis*

Leucophyllidine (**1017**) is a bisindole alkaloid from *Leuconotis griffithii*, which possessing an unprecedented structure assembled from the union of an eburnan half and a novel vinylquinoline alkaloid [61].

11.4 Leguminosae

11.4.1 Sophora

Sophalines A–D (**1018**–**1021**) are four novel matrine-based alkaloids from the Seeds of *Sophora alopecuroides*. **1018** and **1019** possess unprecedented 6/6/6/4 and 6/5/6/6 ring systems, respectively, while **1020** and **1021** are a pair of stereoisomeric matrine-acetophenone alkaloids with an unusual skeleton [62]. Sophalines E–I (**1022**–**1026**) are five novel quinolizidine-based alkaloids from *Sophora alopecuroides*. **1022** and **1023** are the first examples of sparteineindolizine and matrine-indolizine alkaloids, respectively, whereas **1024** and **1025** are epimeric normatrine-julolidine alkaloids with unusual skeletons, and **1026** is a novel matrine-type alkaloid dimer with a C-14–C-10′ connection [63]. Flavesines A–F (**1027**–**1032**) are six unusual matrine-type alkaloid dimers from the Roots of *Sophora flavescens*, **1027**–**1031** represent the first natural matrine-type alkaloid dimers, and **1032** represents an unprecedented dimerization pattern constructed by matrine and (−)-cytisine [64].

11.4.2 Erythrina

Erythrivarines A (**1033**) and B (**1034**) are unprecedented dimeric *Erythrina* alkaloids. They feature spirocyclic (6/5/6/6) and spiro-fused (6/5/7/6) rings, respectively and were isolated from the cultivated plant *Erythrina variegata* [65]. Five dimeric Erythrina alkaloids, named erythrivarines J–N (**1035**–**1039**), were isolated from the barks of *Erythrina variegata* L. (Fabaceae). The erythrivarines J–L featured a 6/6/5/6/6/5/6/6/6 ring system and super conjugated double bond systems, causing intense color from blue to wine red, while erythrivarines M–N looked orange [66].

11.5 *Daphniphyllum* (Daphniphyllaceae)

Daphnezomines A (**1040**) and B (**1041**) were isolated from the leaves of *Daphniphyllum humile* and are novel alkaloids with a unique aza-adamantane core containing an amino ketal bridge in the molecule [67]. Daphnezomines F (**1042**) and G (**1043**) are novel alkaloids with a 1-azabicyclo[5.2.2]undecane ring system, which were isolated from *Daphniphyllum humile* [68]. Daphnicyclidins A–H (**1044**–**1051**) are highly modified *Daphniphyllum* alkaloids with unprecedented fused hexa- or pentacyclic skeletons, they were isolated from the stems of *Daphniphyllum humile* and *D. teijsmanni* [69]. Daphnicyclidins J (**1052**) and K (**1053**) are *Daphniphyllum* alkaloids with unprecedented polycyclic skeletons, they were isolated from the stems of *Daphniphyllum humile* [70]. Daphcalycine (**1054**) possesses an unusual framework: a central quinuclidine like tricycle produced by fusion of a piperidine, a tetrahydropyran, and an oxazine ring in turn condensed to surrounding three penta-, one hexa-, and one hepta-membered rings. It was isolated from the stem bark of *Daphniphyllum calycinum* [71]. Deoxycalyciphylline B (**1055**) and deoxyisocalyciphylline B (**1056**) are two novel major alkaloids isolated from the stem of *Daphniphyllum subverticillatum* which possess a unique fused hexacyclic skeleton in the backbone [72]. Daphniglaucins A (**1057**) and B (**1058**) are quaternary Daphniphyllum alkaloids with an unprecedented fused-polycyclic skeleton containing a 1-azoniatetracyclo[5.2.2.0.^{1,6}0.^{4,9}]-undecane ring system, they were isolated from *Daphniphyllum glaucescens* leaves [73]. Calyciphyllines A (**1059**) and B (**1060**) are two types of *Daphniphyllum* alkaloids with unprecedented fused-hexacyclic ring systems, they were isolated from *Daphniphyllum calycinum* leaves [74]. Daphnipaxinin (**1061**), isolated from the stem of *Daphniphyllum paxianum* is the first diamino *Daphniphyllum* alkaloid with an unprecedented hexacyclic fused skeleton [75]. Paxdaphnidine A and B (**1062** and **1063**) are two novel alkaloids with a unique pentacyclic skeleton and an uncommon tetracyclic skeleton, respectively, they were isolated from the stems and leaves of *Daphniphyllum paxianum* [76]. The structures of daphmanidins C (**1064**) and D (**1065**) are characteristic of alkaloids with an unprecedented fused-pentacyclic skeleton, they were isolated from the leaves of *Daphniphyllum teijsmanii* [77]. Calycilactone A (**1066**) is a novel *Daphniphyllum* alkaloid with a rearranged fused-hexacyclic ring system, it was isolated from the leaves of *Daphniphyllum calycillum* [78]. Macropodumines A–C (**1067–1069**) are three novel alkaloids from *Daphniphyllum macropodum*, interestingly, the structure of macropodumine A (**1067**) was characterized as having a fused pentacyclic system including an unusual eleven-membered macrolactone ring, whereas macropodumine B (**1068**) contains a rare cyclopentadienyl carbanion, which is stabilized as a zwitterion by an internal iminium cation [79]. Calyciphylline C (**1070**) is a novel *Daphniphyllum* alkaloid with unprecedented fused-hexacyclic ring system, it was isolated from the leaves of *Daphniphyllum calycinum* [80]. Daphlongeramine A (**1071**), isolated from the fruits of *Daphniphyllum longeracemosum*, is a novel *Daphyniphyllum* alkaloid with a unique fused octacyclic skeleton [81]. Calyciphylline G (**1072**) is a novel *Daphniphyllum* alkaloid isolated from the stem of *Daphniphyllum calycinum*, which represents an unprecedented fused-hexacyclic skeleton containing a 5-azatricyclo[6.2.1.0^{1,5}]undecane ring [82]. Calydaphninone (**1073**), containing a 4-azatricyclo[5.2.2.0^{1,4}]undecan ring system, was isolated from *Daphniphyllum calycillum*. The conformations of **1073** in solution are discussed on the basis of NMR spectral analysis and computational results [83]. Daphlongeranines A (**1074**) and B (**1075**) are two novel alkaloids with an unprecedented ring system, they were isolated from the fruits of *Daphniphyllum longeracemosum* [84]. Macropodumines D (**1076**) and E (**1077**) are two new complex polycyclic alkaloids, they were isolated from the leaves and barks of *Daphniphyllum macropodum* [85]. Calyciphylline D (**1078**), a novel Daphniphyllum alkaloid with an unprecedented fused-pentacyclic skeleton containing a 8-azatricyclo[4.2.1.0.^{4,8}]-nonane ring system, has been isolated from the leaves of Daphniphyllum calycinum (Daphniphyllaceae) [86]. Calycinumines A (**1079**) and B (**1080**) are two novel alkaloids isolated from the twigs of *Daphniphyllum calycinum*. **1079** is the first example of C-22-*nor* yuzurimine-type alkaloids, and **1080** features an unprecedented heteroatom-containing adamantane-like western hemisphere of the alkaloid [87]. Daphenylline (**1081**), possessing an unprecedented rearranged 22-*nor*-calyciphylline skeleton, was isolated from the fruits of *Daphniphyllum longeracemosum* [88]. Daphhimalenine A (**1082**) is a novel alkaloid with a rearrangement C-21 skeleton containing a unique 1-azabicyclo[5.2.1]decane ring system, it has been isolated from the leaves of *Daphniphyllum himalense* [89]. Angustimine (**1083**) was isolated from *Daphniphyllum angustifolium* twigs. **1083** has an unprecedented hexacyclic fused skeleton which was proposed to be derived from a bukittinggine-type alkaloid, caldaphnidine P, via the cleavage of a C-6–C-7 bond and the formation of a C-6–N bond as the key biogenic steps [90]. Hybridaphniphyllines A and B (**1084** and **1085**) are two *Daphniphyllum* alkaloid and iridoid hybrids with the hitherto unknown decacyclic fused skeletons, they were isolated from the stems and leaves of *Daphniphyllum longeracemosum* [91]. Logeracemin A (**1086**) is the first *Daphniphyllum* alkaloid dimer featuring an unprecedented carbon skeleton with a unique conjugated trispiro[4,5] decane backbone, it has been isolated from *Daphniphyllum longeracemosum* and its absolute configuration was established on the basis of X-ray crystallography [92]. Himalensine A (**1087**) features a 13,14,22-trinorcalyciphylline A type backbone, while himalensine B (**1088**) incorporates a 22-*nor*-1,13-secodaphnicyclidin framework with the ethyl group being likely introduced during the extraction process. Both **1087** and **1088** were isolated from *Daphniphyllum himalense* [93].

11.6 Rubiaceae

11.6.1 Myrioneuron

Myriberine A (**1089**) possesses an unprecedented heteropentacyclic carbon skeleton, it was isolated from *Myrioneuron faberi* [94]. Myrifabine (**1090**) is the first example of a dimer with 12 chiral centers embraced in a decacyclic novel skeleton, it was isolated from *Myrioneuron faberi* and its absolute configuration was determined by X-ray diffraction [95]. (±)-β-Myrifabral A (**1091**), (±)-α-myrifabral A (**1092**), (±)-β-myrifabral B (**1093**), and (±)-α-myrifabral B (**1904**), isolated from *Myrioneuron faberi*, all possess novel cyclohexane-fused octahydroquinolizine skeletons representing the first examples of quinolizidine alkaloids from the genus *Myrioneuron* [96].

11.6.2 Uncaria

Macrophyllionium (**1095**), an unusual oxindole alkaloid inner salt, has been isolated from *Uncaria macrophylla* [97]. (±)-Uncarilins A and B (**1096** and **1097**) are two pairs of unusual dimeric isoechinulin-type enantiomers with a symmetric four-membered core, they were isolated from *Uncaria rhynchophylla* [98]. Rhynchines A–E (**1098–1102**), five new indole alkaloids with an unprecedented skeleton, were isolated from *Uncaria rhynchophylla.* The new skeleton was characterized by an indole moiety and a 2-oxa-8-azatricyclo[6,5,$0^{1,5}$,$0^{1,8}$]tridecane core, forming a unique 6/5/7/5/5 ring system [99].

11.6.3 Psychotria

Brachycerine (**1103**) is an unusual monoterpenoid indole alkaloid from the leaves of *Psychotria brachyceras* [100]. Psychotripine (**1103**) is a trimeric pyrroloindoline derivative with an unprecedented hendecacyclic system bearing a hexahydro-1,3,5-triazine unit, it was isolated from the leaves of *Psychotria pilifera* [101].

11.6.4 Ophiorrhiza

Ophiorrhines A (**1105**) and B (**1106**) are monoterpenoid indole alkaloids possessing a novel spirocyclic ring system, they were obtained from *Ophiorrhiza japonica* [102].

11.6.5 Coffea

Robustanoids A (**1107**) and B (**1108**) are novel pyrrolo[2,3-b]indole alkaloids featuring an unprecedented 1,2,3,4,8b,8c-hexahydro-2a,4a-diazapentaleno[1,6-*ab*]indene moiety, they were isolated from *Coffea canephora* [103].

11.7 Rutaceae

11.7.1 Flindersia

Flinderole A–C (**1109–1111**) are three novel indole alkaloids containing an unprecedented rearranged skeleton, which were described from *Flindersia Species* [104].

11.7.2 Raputia

Raputindoles A–D (**1112–1115**) are found to have a cyclopentyl moiety fused on the benzene part of the indole ring, originating from the combination of prenylated indole monomers, they were isolated from the Amazonian plant *Raputia simulans* [105].

11.7.3 Zanthoxylum

Zanthomuurolanine (**1116**), *epi*-zanthomuurolanine (**1117**), zanthocadinanines A (**1118**) and B (**1119**), and epizanthocadinanine B (**1120**), are novel alkaloids composed of dihydrochelerythrine and a cadinane-type sesquiterpene linked by a methylene bridge. All these compounds were isolated from the stem bark of *Zanthoxylum nitidum* [106].

11.7.4 Geijera

Parvifloranines A and B (**1121** and **1122**) are novel alkaloids featuring an unusual 11-carbon skeleton linked with amino acids. They were isolated from *Geijera parviflora* [107]. Clausanisumine (**1123**) was isolated from the fruits of *Clausena anisum-olens*. **1123** was an uncommon prenylated bicarbazole alkaloid, possessing an unprecedented carbon skeleton, which was composed of a simple carbazole alkaloid and a prenylated carbazole alkaloid [108].

11.8 *Ancistrocladus* (Ancistrocladaceae)

Mbandakamines A (**1124**) and B (**1125**), isolated from a Congolese *Ancistrocladus* species, are the first dimeric naphthylisoquinoline alkaloids bearing an unsymmetrically coupled central biaryl axis. Their novel 6′,1″-coupling pattern implies a hitherto unprecedented periperi coupling in one of the naphthalene parts, leading to the as yet highest steric hindrance at the central axis and a total of seven elements of chirality [109]. Spirombandakamines A1 (**1126**) and A2 (**1127**) are two novel naphthylisoquinoline dimers from a Congolese *Ancistrocladus* plant. They feature a complex molecular architecture displaying an unprecedented cage-like molecular framework with a five membered carbon ring and five- and seven-membered oxygen heterocycles [110]. Cyclombandakamines A1 (**1128**) and A2 (**1129**) both possess an unprecedented pyrane-cyclohexenone-dihydrofuran sequence and six stereocenters and two chiral axes, they were isolated from a Congolese *Ancistrocladus* liana and represent the first oxygen-bridged dimeric naphthylisoquinoline alkaloids [111].

11.9 Loganiaceae

11.9.1 Gelsemium

Geleganidines A–C (**1130–1132**) are three unusual monoterpenoid indole alkaloids from *Gelsemium elegans.* Among them, **1130** is the sole monoterpenoid indole alkaloid existing as a pair of rotamers caused by restricted amide bond rotation, **1131** is the first naturally occurring aromatic azo-linked dimeric monoterpenoid indole alkaloid, and **1132** represents the first example of a urea-linked dimeric monoterpenoid indole alkaloid [112]. Gelsecorydines A–E (**1133–1137**) are five monoterpenoid bisindole alkaloids with new carbon skeletons isolated from the fruits of *Gelsemium elegans.* **1133–1137** represent the first examples of heterodimeric frameworks from the fruits of *Gelsemium elegans* composed of a gelsedine-type alkaloid and a modified corynanthe-type one [113]. Notably, **1134** features an unprecedented caged skeleton with a 6/5/7/6/5/6 heterohexacyclic ring system, which possesses a pyridine ring that linked the two monomers. Five monoterpenoid indole alkaloids (MIAs) with unusual skeletons, gelserancines A–E (**1138–1142**) were isolated from the roots of Gelsemium elegans. **1138** features a new skeleton with an unusual trimethyl-dihydrofuranone unit. **1139** and **1140** are two novel gelsedine-iridoid adducts constructed through unusual C_{19}–$C_{11'}$ and N_4–$C_{3'}$ linkages forming an additional pyridine ring. **1141** and **1142** are a pair of photochemical E/Z tautomeric MIAs with a highly conjugated skeleton [114]. Another novel triamino monoterpene indole alkaloid with an unprecedented skeleton, gelstriamine A (**1143**) was isolated from Gelsemium elegans [115].

11.9.2 Strychnos

Strynuxlines A (**1144**) and B (**1145**) are two novel indole alkaloids with an unprecedented 6/5/9/6/7/6 hexacyclic ring system, they were isolated from the seeds of *Strychnos nuxvomica* [116].

11.10 Cruciferae

11.10.1 Isatis

Isatisine A (**1146**) is a novel alkaloid possessing an unprecedented fused-pentacyclic skeleton (fragment C-9 to C-13). It was isolated from the leaves of *Isatis indigotica* [117]. **1147** and **1148** are a pair of enantiomers of an indole alkaloid containing dihydrothiopyran and 1,2,4-thiadiazole rings, they were isolated from the water soluble extract of *Isatis indigotica* roots [118].

11.10.2 Orychophragmus

Orychophragines A–C (**1149–1151**) are three new alkaloids with an novel 2-piperazinone-fused 2,4-dioxohexahydro-1,3,5-triazine skeleton, they were isolated from the seeds of *Orychophragmus violaceus* [119].

11.11 *Macleaya* (Papaveraceae)

(+)- and (−)-macleayins A and (+)- and (−)-macleayins B (**1152–1155**) represent a novel dimerization pattern of two different types of alkaloids via a C–C σ-bond, they were isolated from the aerial parts of *Macleaya cordata* [120]. Dactyllactone A (**1156**), isolated from *Dactylicapnos scandens*, is an isoquinoline alkaloid with a rearranged and reconstructed D ring likely derived from a common aporphine skeleton, which might undergo biogenetic rearrangement and cleavage of the aromatic ring [121]. Dactylicapnosines A (**1157**) and B (**1158**) are two reconstructed aporphines with unprecedented five-membered carbon ring D, they were isolated from *Dactylicapnos scandens* [122].

11.12 *Cassia* (Caesalpiniaceae)

Cassiarins A (**1159**) and B (**1160**) are two novel alkaloids with an unprecedented tricyclic skeleton, they were isolated from the leaves of *Cassia siamea* [123]. Fistulains A and B (**1161** and **1162**) are two novel bischromones with unique coupling patterns, they were isolated from the barks of *Cassia fistula* [124].

11.13 Ranunculaceae

11.13.1 Aconitum

Puberunine (**1163**) and puberudine (**1164**) are two C_{18}-diterpenoid alkaloids, they were isolated from *Aconitum barbatum* [125]. Aconicarmisulfonine A (**1165**) is a novel sulfonated C_{20}-diterpenoid alkaloid with an unprecedented carbon skeleton, its absolute configuration was confirmed by single-crystal X-ray diffraction [126].

11.13.2 Ranunculus

Ternatusine A (**1166**) is a novel alkaloid with an unprecedented epoxyoxepino[4,5-*c*] pyrrole ring, which has been isolated from the roots of *Ranunculus ternates* [127]. Baicalensines A (**1167**) and B (**1168**) are two isoquinoline alkaloids isolated from the roots of *Thalictrum baicalense*, while **1175** represents a new class of alkaloid dimer containing berberine conjugated to a ring-opened isoquinoline, and **1167** is the first reported natural benzylisoquinoline bearing a formyl group at C-3 [128].

11.14 Huperziaceae

11.14.1 Huperzia

Huperzine R (**1169**) is a novel 15-carbon *Lycopodium* alkaloid isolated from the whole plant of *Huperzia serrata* [129]. Hupercumine A (**1170**) is a novel class of $C_{38}N_4$ *Lycopodium* alkaloid consisting of two octahydroquinolines, a decahydroquinoline, and a piperidine, and hupercumine B (**1171**) is a new $C_{27}N_3$-type alkaloid. Both **1170** and **1171** were isolated from *Huperzia cunninghamioides* and their structures were elucidated on the basis of spectroscopic and chemical methods as well as a biogenetic point of view [130].

11.14.2 Phlegmariurus

Phlefargesiine A (**1172**) is a novel $C_{16}N_2$ *Lycopodium* alkaloid isolated from *Phlegmariurus fargesii*, which possesses an unprecedented [6/7/6/6]-tetracyclic framework where the seven-membered ring (ring B) is constructed through a hitherto unknown C-4–C-14 linkage [131]. Phlegmadine A (**1173**) is a *Lycopodium* alkaloid from *Phlegmariurus phlegmaria* with a unique cyclobutane ring and features a complex tetracyclo-[4.2.2.0^{3,8}.0^{3,10}]decane-bridged system, its absolute configurations were determined by X-ray analysis [132].

11.15 *Piper* (Piperaceae)

Nigramides A–O (**1174–1188**) are 15 novel dimeric amide alkaloids possessing a cyclohexene ring, while nigramides P–S (**1189–1192**) are four novel dimeric amide alkaloids possessing a cyclobutane ring [133]. **1174–1188** were all isolated from the roots of *Piper nigrum*. In addition, the biosynthetic pathway of nigramides A–O (**1174–1188**) was proposed indicative of an intermolecular Diels-Alder reaction from the corresponding monomeric amides.

11.16 *Dysoxylum* (Meliaceae)

Laxiracemosin A–H (**1193–1200**) are eight novel tirucallane-type alkaloids isolated from the barks of *Dysoxylum laxiracemosum*, which possess a pyrrole substituent in the side chain. Of note, this is the first isolation of tirucallane-type alkaloids [134].

11.17 Others

Dovyalicins A–D (**1201–1204**) possess a spermidine as part of a perhydro1,5-diazocine or a perhydro-1,4-diazepine moiety and constitute a new group of polyamine-derived alkaloids, they were isolated from *Dovyalis macrocalyx* (Flacourtiaceae) [135]. Camellimidazole A–C (**1205–1207**) are three new dimeric imidazole alkaloids with a methylene bridge, they were isolated from *Keemum black tea*(Theaceae) [136]. Grandisine A (**1208**) and B (**1209**) are two novel indolizidine alkaloids from *Elaeocarpus grandis* (Elaeocarpaceae) [137]. Of which, grandisine A (**1208**) is a structurally unique pyrano[2′,3′:4,5]pyrano[2,3-*g*]indolizin-4-one alkaloid, while grandisine B (**1209**) is the only alkaloid isolated to date containing both an indolizidine and an isoquinuclidinone moiety. Machilaminosides A (**1210**) and B (**1211**) are two unusual glycosidic triterpene alkaloids, they were isolated from the stem barks of *Machilus yaoshansis* (Lauraceae), representing the first isolation of glycosidic triterpene alkaloids derived from cucurbitane derivatives [138]. Oubatchensine (**1212**) is a novel *seco*-dibenzopyrrocoline alkaloid isolated from *Cryptocarya oubatchensis* (Lauraceae) [139]. Acortatarins A (**1213**) and B (**1214**) are two novel spiroalkaloids with a naturally unusual morpholine motif, they were isolated from the rhizomes of *Acorus tatarinowii* (Araceae) [140]. Their absolute configurations were determined by X-ray diffraction analysis and Mosher's method. Plagiochianin A (**1215**) and plagiochianin B (**1216**) are two novel *ent*-aromadendrane derivatives, while **1215** possesses an unprecedented 2,3:6,7-di-*seco*-6,8-cycloaromadendrane carbon scaffold conjugated with three cyclic acetals, and **1216** is an exceptional pyridine type aromadendrane alkaloid. Both **1215** and **1216** were isolated from the Chinese liverwort *Plagiochila duthiana* (Plagiochilaceae) [141]. Cephalocyclidin A (**1217**) is a novel alkaloid with an unprecedented fused-pentacyclic skeleton and six-consecutive asymmetric centers, it was isolated from the fruits of *Cephalotaxus harringtonia* (Cephalotaxaceae) [142]. Hostasinine A (**1218**) is a benzylphenethylamine alkaloid with an unprecedented skeleton featuring a C-4–C-6 linkage and a nitrone moiety. **1218** was isolated from *Hosta plantaginea* (Liliaceae) and its absolute configuration was confirmed by X-ray diffraction [143]. Hemerocallisamine I (**1219**) is a rare glutamine derivative from *Hemerocallis fulva* var. *kwanso* (Asphodelaceae). It is of interest that this is the first report of a glutamine derivative with a pyrrole ring from plants [144]. Peganumine A (**1220**) is a new dimeric β-carboline alkaloid characterized by a unique 3,9-diazatetracyclo-[6.5.2.$0^{1,9}$.$0^{3,8}$]pentadec-2-one scaffold, it was isolated from the seeds of *Peganum harmala* (Zygophyllaceae) [145]. Pegaharines A–F (**1221–1226**) are six novel β-carboline alkaloids representing three types of skeletons from *Peganum harmala* (Zygophyllaceae) where **1221** is a peculiar β-carboline alkaloid characterized by the unprecedented carbon skeleton of an azepine–indole system. **1222** is a rare tetracyclic β-carboline alkaloid featuring a classic tricyclic β-carboline fused with an additional pyrrole ring. **1223–1226** represent the first reported β-carboline heterodimers featuring an unusual 6/5/6/5–6/5/6 skeleton, in which a rare tetracyclic β-carboline is connected to a classic tricyclic β-carboline via a C-15–C-1′ bridge [146]. Pegaharmols A (**1227**) and B (**1228**) are two nonbiaryl axially chiral β-carboline-quinazoline dimers, they were isolated from the roots of *Peganum harmala* (Zygophyllaceae). This is the first report that the β-carboline at the C-8 position is bonded to the vasicine at the C-9 position [147]. Three tetrahydroquinoline alkaloids, lycibarbarines A–C (**1229–1231**), possessing a unique tetracyclic tetrahydroquinoline–oxazine–ketohexoside fused motif, were isolated from the fruits of *Lycium barbarum.* Their structures were determined by spectroscopic analysis and quantum-chemical calculations [148]. Two unusual phenanthrene derivatives related to aporphine alkaloids, artapilosines A (**1232**) and B (**1233**) were separated and purified from Artabotrys pilosus. **1232** is the first compound representative of a new class of phenanthrene derivatives having an unprecedented carbon skeleton, in which the six-membered nitrogen-containing heterocyclic structure in a typical aporphine alkaloid was substituted with a unique five-membered carbocyclic ring. This is the first report of the formation of a carbon–carbon bond between C-5 and C-6a in **1232** with the loss of the nitrogen atom N-6 in the classic aporphine alkaloid. **1233** is a novel phenanthrene derivative having a hydroxyethyl as a substituent on the phenanthrene ring [149]. Fortuneicyclidins A (**1234**) and B (**1235**), a pair of epimeric pyrrolizidine alkaloids containing an unprecedented 7-azatetracyclo-[5.4.3.0.$0^{2,8}$]tridecane core, were isolated from the seeds of *Cephalotaxus fortune* [150].

References

1. Gan LS, Fan CQ, Yang SP, Wu Y, Lin LP, Ding J, Yue JM. Flueggenines A and B, two novel c, c-linked dimeric indolizidine alkaloids from *Flueggea virosa*. Org Lett. 2006;8:2285–8.
2. Zhao BX, Wang Y, Zhang DM, Jiang RW, Wang GC, Shi JM, Huang XJ, Chen WM, Che CT, Ye WC. Flueggines A and B, two new dimeric indolizidine alkaloids from *Flueggea virosa*. Org Lett. 2011;13:3888–91.
3. Wu ZL, Zhao BX, Huang XJ, Tang GY, Shi L, Jiang RW, Liu X, Wang Y, Ye WC. Suffrutines A and B: a pair of *Z*/*E* isomeric indolizidine alkaloids from the roots of *Flueggea suffruticosa*. Angew Chem Int Ed. 2014;53:5796–9.
4. Luo XK, Cai J, Yin ZY, Luo P, Li CJ, Ma H, Seeram NP, Gu Q, Xu J. Fluvirosaones A and B, two indolizidine alkaloids with a pentacyclic skeleton from *Flueggea virosa*. Org Lett. 2018;20:991–4.
5. Zhao BX, Wang Y, Zhang DM, Huang XJ, Bai LL, Yan Y, Chen JM, Lu TB, Wang YT, Zhang QW, Ye WC. Virosaines A and B, two new birdcage-shaped *securinega* alkaloids with an unprecedented skeleton from *Flueggea virosa*. Org Lett. 2012;14:3096–9.
6. Zhang H, Zhang CR, Zhu KK, Gao AH, Luo C, Li J, Yue JM. Fluevirosines A-C: a biogenesis inspired example in the discovery of new bioactive scaffolds from *Flueggea virosa*. Org Lett. 2013;15:120–3.
7. Wu ZL, Huang XJ, Xu MT, Ma X, Li L, Shi L, Wang WJ, Jiang RW, Ye WC, Wang Y. Flueggeacosines A-C, dimeric securinine-type alkaloid analogues with neuronal differentiation activity from *Flueggea suffruticosa*. Org Lett. 2018;20:7703–7.
8. Tan CJ, Di YT, Wang YH, Zhang Y, Si YK, Zhang Q, Gao S, Hu XJ, Fang X, Li SF, Hao XJ. Three new indole alkaloids from *Trigonostemon lii*. Org Lett. 2010;12:2370–3.
9. Ma SS, Mei WL, Guo ZK, Liu SB, Zhao YX, Yang DL, Zeng YB, Jiang B, Dai HF. Two new types of bisindole alkaloid from *Trigonostemon lutescens*. Org Lett. 2013;15:1492–5.
10. Nge CE, Gan CY, Low YY, Thomas NF, Kam TS. Voatinggine and tabertinggine, pentacyclic indole alkaloids derived from an iboga precursor via a common cleavamine-type intermediate. Org Lett. 2013;15:4774–7.
11. Bracher F, Eisenreich WJ, Mühlbacher J, Dreyer M, Bringmann G. Saludimerines A and B, novel-type dimeric alkaloids with stereogenic centers and configurationally semistable biaryl axes. J Org Chem. 2004;69:8602–8.
12. Gao XH, Xu YS, Fan YY, Gan LS, Zuo JP, Yue JM. Cascarinoids A-C, a class of diterpenoid alkaloids with unpredicted conformations from *Croton cascarilloides*. Org Lett. 2018;20:228–31.
13. Morita H, Arisaka M, Yoshida N, Kobayashi J. Serratezomines A−C, new alkaloids from *Lycopodium serratum* Var. *serratum*. J Org Chem. 2000;65:6241–45.
14. Kobayashi J, Hirasawa Y, Yoshida N, Morita H, Lyconadin A. A novel alkaloid from *Lycopodium complanatum*. J Org Chem. 2001;66:5901–4.
15. Takayama H, Katakawa K, Kitajima M, Seki H, Yamaguchi K, Aimi N. A new type of lycopodium alkaloid, lycoposerramine-A, from *Lycopodium serratum* Thunb. Org Lett. 2001;3:4165–7.
16. Hirasawa Y, Morita H, Shiro M, Kobayashi J. Sieboldine A, a novel tetracyclic alkaloid from *Lycopodium sieboldii*, inhibiting acetylcholinesterase. Org Lett. 2003;5:3991–3.
17. Morita H, Hirasawa Y, Kobayashi J. Himeradine A, a novel $C_{27}N_3$-type alkaloid from *Lycopodium chinense*. J Org Chem. 2003;68:4563–6.
18. Hirasawa Y, Morita H, Kobayashi J. Nankakurine A, a novel $C_{16}N_2$-type alkaloid from *Lycopodium hamiltonii*. Org Lett. 2004;6:3389–91.
19. Hirasawa Y, Kobayashi J, Morita H. Lycoperine A, a novel $C_{27}N_3$-type pentacyclic alkaloid from *Lycopodium hamiltonii*, inhibiting acetylcholinesterase. Org Lett. 2006;8:123–6.
20. He J, Chen XQ, Li MM, Zhao Y, Xu G, Cheng X, Peng LY, Xie MJ, Zheng YT, Wang YP, Zhao QS. Lycojapodine A, a novel alkaloid from *Lycopodium japonicum*. Org Lett. 2009;11:1397–400.

21. Wang XJ, Zhang GJ, Zhuang PY, Zhang Y, Yu SS, Bao XQ, Zhang D, Yuan YH, Chen NH, Ma S, Qu J, Li Y. Lycojaponicumins A-C, three alkaloids with an unprecedented skeleton from *Lycopodium japonicum*. Org Lett. 2012;14:2614–7.
22. Wang XJ, Liu YB, Li L, Yu SS, Lv HN, Ma SG, Bao XQ, Zhang D, Qu J, Li Y. Lycojaponicumins D and E: two new alkaloids from *Lycopodium japonicum*. Org Lett. 2012;14:5688–91.
23. Cheng JT, Liu F, Li XN, Wu XD, Dong LB, Peng LY, Huang SX, He J, Zhao QS. Lycospidine A, a new type of *Lycopodium* alkaloid from *Lycopodium complanatum*. Org Lett. 2013;15:2438–41.
24. Tang Y, Xiong J, Zhang JJ, Wang W, Zhang HY, Hu JF. Annotinolides A-C, three lycopodane-derived 8,5-lactones with polycyclic skeletons from *Lycopodium annotinum*. Org Lett. 2016;18:4376–9.
25. Zhang ZJ, Nian Y, Zhu QF, Li XN, Su J, Wu XD, Yang J, Zhao QS. Lycoplanine A, a $C_{16}N$ *Lycopodium* alkaloid with a 6/9/5 tricyclic skeleton from *Lycopodium complanatum*. Org Lett. 2017;19:4668–71.
26. Yeap JSY, Lim KH, Yong KT, Lim SH, Kam TS, Low YY. *Lycopodium* alkaloids: lycoplatyrine a, an unusual lycodine–piperidine adduct from *Lycopodium platyrhizoma* and the absolute configurations of lycoplanine D and lycogladine H. J Nat Prod. 2019;82:324–9.
27. Zhao FW, Sun QY, Yang FM, Hu GW, Luo JF, Tang GH, Wang YH, Long CL. Palhinine A, a novel alkaloid from *Palhinhaea cernua*. Org Lett. 2010;12:3922–5.
28. Dong LB, Yang J, He J, Luo HR, Wu XD, Deng X, Peng LY, Cheng X, Zhao QS. Lycopalhine A, a novel sterically congested lycopodium alkaloid with an unprecedented skeleton from *Palhinhaea cernua*. Chem Commun. 2012;48:9038–40.
29. Dong LB, Gao X, Liu F, He J, Wu XD, Li Y, Zhao QS. Isopalhinine A, a unique pentacyclic *Lycopodium* alkaloid from *Palhinhaea cernua*. Org Lett. 2013;15:3570–3.
30. Mansoor TA, Ramalhete C, Molnár J, Mulhovo S, Ferreira MJU. Tabernines A−C, *β*-carbolines from the leaves of *Tabernaemontana elegans*. J Nat Prod. 2009;72:1147–50.
31. Carroll AR, Hyde E, Smith J, Quinn RJ, Guymer G, Forster PI. Actinophyllic acid, a potent indole alkaloid inhibitor of the coupled enzyme assay carboxypeptidase u/hippuricase from the leaves of *Alstonia a ctinophylla* (Apocynaceae). J Org Chem. 2005;70:1096–9.
32. Nge CE, Gan CY, Lim KH, Ting KN, Low YY, Kam TS. Criofolinine and vernavosine, new pentacyclic indole alkaloids incorporating pyrroloazepine and pyridopyrimidine moieties derived from a common yohimbine precursor. Org Lett. 2014;16:6330–3.
33. Yu HF, Qin XJ, Ding CF, Wei X, Yang J, Luo JR, Liu L, Khan A, Zhang LC, Xia CF, Luo XD. Nepenthe-like indole alkaloids with antimicrobial activity from *Ervatamia chinensis*. Org Lett. 2018;20:4116–20.
34. Ma K, Wang JS, Luo J, Yang MH, Yao HQ, Sun HB, Kong LY. Bistabercarpamines A and B, first vobasinyl-chippiine-type bisindole alkaloid from *Tabernaemontana corymbosa*. Tetrahedron Lett. 2014;55:101–4.
35. Liu ZW, Zhang J, Li ST, Liu MQ, Huang XJ, Ao YL, Fan CL, Zhang DM, Zhang QW, Ye WC, Zhang XQ. Ervadivamines A and B, two unusual trimeric monoterpenoid indole alkaloids from *Ervatamia divaricata*. J Org Chem. 2018;83:10613–8.
36. Hirasawa Y, Miyama S, Hosoya T, Koyama K, Rahman A, Kusumawati I, Zaini NC, Morita H. Alasmontamine A, a first tetrakis monoterpene indole alkaloid from *Tabernaemontana elegans*. Org Lett. 2009;11:5718–21.
37. Yuan YX, Zhang Y, Guo LL, Wang YH, Goto M, Morris Natschke SL, Lee KH, Hao XJ. Tabercorymines A and B, two vobasinyl–ibogan-type bisindole alkaloids from *Tabernaemontana corymbosa*. Org Lett. 2017;19:4964–7.
38. Kam TS, Tan SJ, Ng SW, Komiyama K. Bipleiophylline, an unprecedented cytotoxic bisindole alkaloid constituted from the bridging of two indole moieties by an aromatic spacer unit. Org Lett. 2008;10:3749–52.
39. Koyama K, Hirasawa Y, Nugroho AE, Hosoya T, Hoe TC, Chan KL, Morita H. Alsmaphorazines A and B, novel indole alkaloids from *Alstonia pneumatophora*. Org Lett. 2010;12:4188–91.

40. Lim SH, Tan SJ, Low YY, Kam TS. Lumutinines A-D, linearly fused macroline–macroline and macroline–sarpagine bisindoles from *Alstonia macrophylla*. J Nat Prod. 2011;74:2556–62.
41. Zhu GY, Yao XJ, Liu L, Bai LP, Jiang ZH. Alistonitrine A, a caged monoterpene indole alkaloid from *Alstonia scholaris*. Org Lett. 2014;16:1080–3.
42. Pan Z, Qin XJ, Liu YP, Wu T, Luo XD, Xia C. Alstoscholarisines H-J, indole alkaloids from *Alstonia scholaris*: structural evaluation and bioinspired synthesis of alstoscholarisine H. Org Lett. 2016;18:654–7.
43. Yu HF, Ding CF, Zhang LC, Wei X, Cheng GG, Liu YP, Zhang RP, Luo XD. Alstoscholarisine K, an antimicrobial indole from gall-induced leaves of *Alstonia scholaris*. Org Lett. 2021;23:5782–6.
44. Zhu XX, Fan YY, Xu L, Liu QF, Wu JP, Li JY, Li J, Gao K, Yue JM. Alstonlarsines A-D, four rearranged indole alkaloids from *Alstonia scholaris*. Org Lett. 2019;21:1471–4.
45. Cai XH, Du ZZ, Luo XD. Unique monoterpenoid indole alkaloids from *Alstonia scholaris*. Org Lett. 2007;9:1817–20.
46. Cai XH, Tan QG, Liu YP, Feng T, Du ZZ, Li WQ, Luo XD. A cage-monoterpene indole alkaloid from *Alstonia scholaris*. Org Lett. 2008;10:577–80.
47. Zhang J, Song M, Ao YL, Li Y, Zou XY, Xu J, Wang Y, Zhang DM, Zhang XQ, Ye WC. Alstolarines A and B, two unusual monoterpenoid indole alkaloids with an acetal moiety from *Alstonia scholaris*. Org Chem Front. 2020;7:3468–73.
48. Kobayashi J, Sekiguchi M, Shimamoto S, Shigemori H, Ishiyama H, Ohsaki A. Subincanadines A−C, novel quaternary indole alkaloids from *Aspidosperma subincanum*. J Org Chem. 2002;67:6449–55.
49. Feng T, Cai XH, Li Y, Wang YY, Liu YP, Xie MJ, Luo XD. Melohenines A and B, two unprecedented alkaloids from *Melodinus henryi*. Org Lett. 2009;11:4834–7.
50. Feng T, Li Y, Liu YP, Cai XH, Wang YY, Luo XD. Melotenine A, a cytotoxic monoterpenoid indole alkaloid from *Melodinus tenuicaudatus*. Org Lett. 2010;12:968–71.
51. Zhang Y, Ding X, Shao S, Guo LL, Zhao Q, Hao XJ, He HP. Melocochines A and B, two alkaloids from the fruits of *Melodinus cochinchinensis*. Org Lett. 2019;21:9272–5.
52. Wu J, Zhao SM, Shi BB, Bao MF, Schinnerl J, Cai XH. Cage-Monoterpenoid quinoline alkaloids with neurite growth promoting effects from the fruits of *Melodinus yunnanensis*. Org Lett. 2020;22:7676–80.
53. Kam TS, Choo YM. Kopsifolines A, B, and C, indole alkaloids with a novel hexacyclic carbon skeleton from *Kopsia*. Tetrahedron Lett. 2003;44:1317–9.
54. Lim KH, Kam TS. Arboflorine, an unusual pentacyclic monoterpenoid indole alkaloid incorporating a third nitrogen atom. Org Lett. 2006;8:1733–5.
55. Zeng J, Zhang DB, Zhou PP, Zhang QL, Zhao L, Chen JJ, Gao K. Rauvomines A and B, two monoterpenoid indole alkaloids from *Rauvolfia vomitoria*. Org Lett. 2017;19:3998–4001.
56. Fu Y, H H, Di Y, Li S, Zhang Y, Hao X. Mekongenines A and B, two new alkaloids from *Bousigonia mekongensis*. Tetrahedron Lett. 2012;53:3642–46.
57. Hirasawa Y, Hara M, Nugroho AE, Sugai M, Zaima K, Kawahara N, Goda Y, Awang K, Hadi AHA, Litaudon M, Morita H. Bisnicalaterines B and C, atropisomeric bisindole alkaloids from *Hunteria zeylanica*, showing vasorelaxant activity. J Org Chem. 2010;75:4218–23.
58. Beniddir MA, Martin MT, Tran Huu Dau ME, Grellier P, Rasoanaivo P, Guéritte F, Litaudon M. Goniomedines A and B: unprecedented bisindole alkaloids formed through fusion of two indole moieties via a dihydropyran unit. Org Lett. 2012;14:4162–5.
59. N'Nang Obiang EO, Genta Jouve G, Gallard JF, Kumulungui B, Mouray E, Grellier P, Evanno L, Poupon E, Champy P, Beniddir MA. Pleiokomenines A and B: dimeric aspidofractinine alkaloids tethered with a methylene group. Org Lett. 2017;19:6180–3.
60. Li LM, Yang T, Liu Y, Liu J, Li MH, Wang YT, Yang SX, Zou Q, Li GY. Calophyline A, a new rearranged monoterpenoid indole alkaloid from *Winchia calophylla*. Org Lett. 2012;14:3450–3.
61. Gan CY, Robinson WT, Etoh T, Hayashi M, Komiyama K, Kam TS. Leucophyllidine, a cytotoxic bisindole alkaloid constituted from the union of an eburnan and a new vinylquinoline alkaloid. Org Lett. 2009;11:3962–5.

62. Zhang YB, Zhang XL, Chen NH, Wu ZN, Ye WC, Li YL, Wang GC. Four matrine-based alkaloids with antiviral activities against HBV from the seeds of *Sophora alopecuroides*. Org Lett. 2017;19:424–7.
63. Zhang YB, Yang L, Luo D, Chen NH, Wu ZN, Ye WC, Li YL, Wang GC. Sophalines E-I, five quinolizidine-based alkaloids with antiviral activities against the hepatitis b virus from the seeds of *Sophora alopecuroides*. Org Lett. 2018;20:5942–6.
64. Zhang YB, Zhan LQ, Li GQ, Wang F, Wang Y, Li YL, Ye WC, Wang GC. Dimeric matrine-type alkaloids from the roots of *Sophora flavescens* and their anti-hepatitis B virus activities. J Org Chem. 2016;81:6273–80.
65. Zhang BJ, Bao MF, Zeng CX, Zhong XH, Ni L, Zeng Y, Cai XH. Dimeric *Erythrina* alkaloids from the flower of *Erythrina variegata*. Org Lett. 2014;16:6400–3.
66. Tang YT, Wu J, Yu Y, Bao MF, Tan QG, Johann S, Cai XH. Colored dimeric alkaloids from the barks of *Erythrina variegata* and their neuroprotective effects. J Org Chem. 2021;86:13381−7.
67. Morita H, Yoshida N, Kobayashi J. Daphnezomines A and B, novel alkaloids with an aza-adamantane core from *Daphniphyllum humile*. J Org Chem. 1999;64:7208–12.
68. Morita H, Yoshida N, Kobayashi J. Daphnezomines F and G: novel alkaloids with 1-azabicyclo[5.2.2]undecane moiety from *Daphniphyllum humile*. J Org Chem. 2000;65:3558–62.
69. Kobayashi J, Inaba Y, Shiro M, Yoshida N, Morita H. Daphnicyclidins A−H, novel hexa- or pentacyclic alkaloids from two species of *Daphniphyllum*. J Am Chem Soc. 2001;123:11402–8.
70. Morita H, Yoshida N, Kobayashi J. Daphnicyclidins J and K, unique polycyclic alkaloids from *Daphniphyllum humile*. J Org Chem. 2002;67:2278–82.
71. Jossang A, Bitar HE, Pham VC, Sévenet T. Daphcalycine, a novel heptacycle fused ring system alkaloid from *Daphniphyllum calycinum*. J Org Chem. 2003;68:300–4.
72. Yang SP, Yue JM. Two novel alkaloids with a unique fused hexacyclic skeleton from *Daphniphyllum subverticillatum*. J Org Chem. 2003;68:7961–6.
73. Kobayashi J, Takatsu H, Shen YC, Morita H. Daphniglaucins A and B, novel polycyclic quaternary alkaloids from *Daphniphyllum glaucescens*. Org Lett. 2003;5:1733–6.
74. Morita H, Kobayashi J. Calyciphyllines A and B, two novel hexacyclic alkaloids from *Daphniphyllum calycinum*. Org Lett. 2003;5:2895–8.
75. Yang SP, Yue JM. First diamino *Daphniphyllum* alkaloid, daphnipaxinin, with an unprecedented heterohexacyclic skeleton from *Daphniphyllum paxianum*. Org Lett. 2004;6:1401–4.
76. Zhan ZJ, Yang SP, Yue JM. Paxdaphnidines A and B, novel penta- and tetracyclic alkaloids from *Daphniphyllum paxianum*. J Org Chem. 2004;69:1726–9.
77. Morita H, Ishioka N, Takatsu H, Shinzato T, Obara Y, Nakahata N, Kobayashi J. Daphmanidins C and D, novel pentacyclic alkaloids from *Daphniphyllum teijsmanii*. Org Lett. 2005;7:459–62.
78. Di YT, He HP, Liu HY, Du ZZ, Tian JM, Yang XW, Wang YH, Hao XJ. Calycilactone A, a novel hexacyclic alkaloid from *Daphniphyllum calycillum*. Tetrahedron Lett. 2006;47:5329–31.
79. Zhang W, Guo YW, Krohn K. Macropodumines A-C: novel pentacyclic alkaloids with an unusual skeleton or zwitterion moiety from *Daphniphyllum macropodum* Miq. Chem Eur J. 2006;12:5122–7.
80. Saito S, Kubota T, Fukushi E, Kawabata J, Zhang H, Kobayashi J. Calyciphylline C, a novel daphniphyllum alkaloid from *Daphniphyllum calycinum*. Tetrahedron Lett. 2007;48:1587–9.
81. Li C, He H, Di Y, Wang Y, Mu S, Li S, Gao S, Gao Z, Hao X. Calyciphylline G, aovel daphniphyllum alkaloid from *Daphniphyllum longeracemosum*. Tetrahedron Lett. 2007;48:2737–40.
82. Saito S, Kubota T, Kobayashi J. Calyciphylline G, a novel alkaloid with an unprecedented fused-hexacyclic skeleton from *Daphniphyllum calycinum*. Tetrahedron Lett. 2007;48:5693–5.
83. Di YT, He HP, Wang YS, Li LB, Lu Y, Gong JB, Fang X, Kong NC, Li SL, Zhu HJ, Hao XJ. Isolation, X-ray crystallography, and computational studies of calydaphninone, a new alkaloid from *Daphniphyllum calycillum*. Org Lett. 2007;9:1355–8.

84. Li CS, Di YT, He HP, Gao S, Wang YH, Lu Y, Zhong JL, Hao XJ. Daphlongeranines A and B, two novel alkaloids possessing unprecedented skeletons from *Daphniphyllum longeracemosum*. Org Lett. 2007;9:2509–12.
85. Li ZY, Chen P, Xu HG, Yang YM, Peng SY, Zhao ZZ, Guo YW. Macropodumines D and E, two new alkaloids with unusual skeletons from *Daphniphyllum macropodum* Miq. Org Lett. 2007;9:477–80.
86. Saito S, Kubota T, Fukushi E, Kawabata J, Zhang H, Kobayashi J. Calyciphylline D, a novel alkaloid with an unprecedented fused-pentacyclic skeleton from *Daphniphyllum calycinum*. Org Lett. 2007;9:1207–9.
87. Zhang CR, Liu HB, Dong SH, Zhu JY, Wu Y, Yue JM. Calycinumines A and B, two novel alkaloids from *Daphniphyllum calycinum*. Org Lett. 2009;11:4692–5.
88. Zhang Q, Di YT, Li CS, Fang X, Tan CJ, Zhang Z, Zhang Y, He HP, Li SL, Hao XJ. Daphenylline, a new alkaloid with an unusual skeleton, from *Daphniphyllum longeracemosum*. Org Lett. 2009;11:2357–9.
89. Zhang Y, Di YT, Zhang Q, Mu SZ, Tan CJ, Fang X, He HP, Li SL, Hao XJ. Daphhimalenine A, a new alkaloid with an unprecedented skeleton, from *Daphniphyllum himalense*. Org Lett. 2009;11:5414–7.
90. Zhang CR, Fan CQ, Dong SH, Liu HB, Zhou WB, Wu Y, Yue JM. Angustimine and angustifolimine: two new alkaloids from *Daphniphyllum angustifolium*. Org Lett. 2011;13:2440–3.
91. Wang F, Mao MF, Wei GZ, Gao Y, Ren FC, Liu JK. Hybridaphniphyllines A and B, *Daphniphyllum* alkaloid and iridoid hybrids suggestive of diels–alder cycloaddition in *Daphniphyllum longeracemosum*. Phytochemistry. 2013;95:428–35.
92. Xu JB, Zhang H, Gan LS, Han YS, Wainberg MA, Yue JM. Logeracemin A, an anti-HIV *Daphniphyllum* alkaloid dimer with a new carbon skeleton from *Daphniphyllum longeracemosum*. J Am Chem Soc. 2014;136:7631–3.
93. Zhang H, Shyaula SL, Li JY, Li J, Yue JM. Himalensines A and B, alkaloids from *Daphniphyllum himalense*. Org Lett. 2016;18:1202–5.
94. Huang SD, Zhang Y, Cao MM, Di YT, Tang GH, Peng ZG, Jiang JD, He HP, Hao XJ. Myriberine A, a new alkaloid with an unprecedented heteropentacyclic skeleton from *Myrioneuron faberi*. Org Lett. 2013;15:590–3.
95. Cao MM, Huang SD, Di YT, Yuan CM, Zuo GY, Gu YC, Zhang Y, Hao XJ. Myrifabine, the first dimeric *Myrioneuron* alkaloid from *Myrioneuron faberi*. Org Lett. 2014;16:528–31.
96. Cao MM, Zhang Y, Li XH, Peng ZG, Jiang JD, Gu YC, Di YT, Li XN, Chen DZ, Xia CF, He HP, Li SL, Hao XJ. Cyclohexane-fused octahydroquinolizine alkaloids from *Myrioneuron faberi* with activity against hepatitis C virus. J Org Chem. 2014;79:7945–50.
97. Wang K, Zhou XY, Wang YY, Li MM, Li YS, Peng LY, Cheng X, Li Y, Wang YP, Zhao QS. Macrophyllionium and macrophyllines A and B, oxindole alkaloids from *Uncaria macrophylla*. J Nat Prod. 2011;74:12–5.
98. Geng CA, Huang XY, Ma YB, Hou B, Li TZ, Zhang XM, Chen JJ. (±)-Uncarilins A and B, dimeric isoechinulin-type alkaloids from *Uncaria rhynchophylla*. J Nat Prod. 2017;80:959–64.
99. Zhou HF, Li WY, Peng LY, Li XN, Zuo ZL, Zhao QS. Rhynchines A−E: Cav3.1 calcium channel blockers from *Uncaria rhynchophylla*. Org Lett. 2021;23:9463−67.
100. Kerber VA, Gregianini TS, Paranhos JT, Schwambach J, Farias F, Fett JP, Fett-Neto AG, Zuanazzi JÂS, Quirion JC, Elizabetsky E, Henriques AT. Brachycerine, a novel monoterpene indole alkaloid from *Psychotria brachyceras*. J Nat Prod. 2001;64:677–9.
101. Li XN, Zhang Y, Cai XH, Feng T, Liu YP, Li Y, Ren J, Zhu HJ, Luo XD. Psychotripine: a new trimeric pyrroloindoline derivative from *Psychotria pilifera*. Org Lett. 2011;13:5896–9.
102. Feng T, Duan KT, He SJ, Wu B, Zheng YS, Ai HL, Li ZH, He J, Zuo JP, Liu JK. Ophiorrhines A and B, two immunosuppressive monoterpenoid indole alkaloids from *Ophiorrhiza japonica*. Org Lett. 2018;20:7926–8.
103. Han J, Niu ST, Liu Y, Gan L, Wang T, Lu CD, Yuan T. Robustanoids A and B, two novel pyrrolo[2,3-*b*]indole alkaloids from *Coffea canephora*: isolation and total synthesis. Org Chem Front. 2018;5:586–9.

104. Fernandez LS, Buchanan MS, Carroll AR, Feng YJ, Quinn RJ, Avery VM. Flinderoles A–C: antimalarial bis-indole alkaloids from *Flindersia* species. Org Lett. 2009;11:329–32.
105. Vougogiannopoulou K, Fokialakis N, Aligiannis N, Cantrell C, Skaltsounis AL. The raputindoles: novel cyclopentyl bisindole alkaloids from *Raputia simulans*. Org Lett. 2010;12:1908–11.
106. Yang CH, Cheng MJ, Chiang MY, Kuo YH, Wang CJ, Chen IS. Dihydrobenzo[*c*]phenanthridine alkaloids from stem bark of *Zanthoxylum nitidum*. J Nat Prod. 2008;71:669–73.
107. Shou Q, Banbury LK, Renshaw DE, Smith JE, He X, Dowell A, Griesser HJ, Heinrich M, Wohlmuth H. Parvifloranines A and B, two 11-carbon alkaloids from *Geijera parviflora*. J Nat Prod. 2013;76:1384–7.
108. Liu YP, Guo JM, Xie Z, Suo XY, Liu ZY, Qiao ZH, Guan RQ, Bian Y, Qiang L, Fu YH. Clausanisumine, a prenylated bicarbazole alkaloid from the fruits of *Clausena anisumolens* and its potential anti-HIV activity. J Org Chem. 2021;86:17722–6.
109. Bringmann G, Lombe BK, Steinert C, Ioset KN, Brun R, Turini F, Heubl G, Mudogo V. Mbandakamines A and B, unsymmetrically coupled dimeric naphthylisoquinoline alkaloids, from a congolese *Ancistrocladus* species. Org Lett. 2013;15:2590–3.
110. Lombe BK, Bruhn T, Feineis D, Mudogo V, Brun R, Bringmann G. Antiprotozoal spirombandakamines A_1 and A_2, fused naphthylisoquinoline dimers from a congolese *Ancistrocladus* plant. Org Lett. 2017;19:6740–3.
111. Lombe BK, Bruhn T, Feineis D, Mudogo V, Brun R, Bringmann G. Cyclombandakamines A_1 and A_2, oxygen-bridged naphthylisoquinoline dimers from a Congolese *Ancistrocladus* Liana. Org Lett. 2017;19:1342–5.
112. Zhang W, Huang XJ, Zhang SY, Zhang DM, Jiang RW, Hu JY, Zhang XQ, Wang L, Ye WC. Geleganidines A–C, unusual monoterpenoid indole alkaloids from *Gelsemium elegans*. J Nat Prod. 2015;78:2036–2044.
113. Li NP, Liu M, Huang XJ, Gong XY, Zhang W, Cheng MJ, Ye WC, Wang L. Gelsecorydines A–E, five gelsedine–corynanthe-type bisindole alkaloids from the fruits of *Gelsemium elegans*. J Org Chem. 2018;83: 5707–5714.
114. Gu JH, Zhang W, Cai WY, Fu XX, Zhou HL, Li NP, Tian HY, Liu JS, Ye WC, Wang L. Gelserancines A–E, monoterpenoid indole alkaloids with unusual skeletons from *Gelsemium elegans*. Org Chem Front. 2021;8;1918–1925.
115. Jin PF. Zhan GQ, Zheng GJ, Liu JJ, Peng X, Huang L, Gao B, Yuan XH, Yao GM. Gelstriamine A, a triamino monoterpene indole alkaloid with a caged 6/5/7/6/6/5 scaffold and analgesic alkaloids from *Gelsemium elegans* stems. J Nat Prod. 2021;84:1326–1334.
116. Fu Y, Zhang Y, He H, Hou L, Di Y, Li S, Luo X, Hao X. Strynuxlines A and B, alkaloids with an unprecedented carbon skeleton from *Strychnos nux-Vomica*. J Nat Prod. 2012;75:1987–90.
117. Liu JF, Jiang ZY, Wang RR, Zheng YT, Chen JJ, Zhang XM, Ma YB. Isatisine A, a novel alkaloid with an unprecedented skeleton from leaves of *Isatis indigotica*. Org Lett. 2007;9:4127–9.
118. Chen M, Lin S, Li L, Zhu C, Wang X, Wang Y, Jiang B, Wang S, Li Y, Jiang J, Shi J. Enantiomers of an indole alkaloid containing unusual dihydrothiopyran and 1,2,4-thiadiazole rings from the root of *Isatis indigotica*. Org Lett. 2012;14:5668–71.
119. Zhang GJ, Li B, Cui HM, Chen L, Tian Y, Liu SJ, Li BW, Li M, Xia ZM, Chen XX, Hou Y, Dong JX. Orychophragines A-C, three biologically active alkaloids from *Orychophragmus violaceus*. Org Lett. 2018;20:656–9.
120. Sai CM, Li DH, Xue CM, Wang KB, Hu P, Pei YH, Bai J, Jing YK, Li ZL, Hua HM. Two pairs of enantiomeric alkaloid dimers from *Macleaya cordata*. Org Lett. 2015;17:4102–5.
121. Wang B, Yang ZF, Zhao YL, Liu YP, Deng J, Huang WY, Li XN, Wang XH, Luo XD. Anti-inflammatory isoquinoline with bis-*seco*-aporphine skeleton from *Dactylicapnos scandens*. Org Lett. 2018;20:1647–50.
122. Wang B, Zhao YJ, Zhao YL, Liu YP, Li XN, Zhang H, Luo XD. Exploring aporphine as anti-inflammatory and analgesic lead from *Dactylicapnos scandens*. Org Lett. 2020;22:257–60.

123. Morita H, Oshimi S, Hirasawa Y, Koyama K, Honda T, Ekasari W, Indrayanto G, Zaini NC. Cassiarins A and B, novel antiplasmodial alkaloids from *Cassia siamea*. Org Lett. 2007;9:3691–3.
124. Zhou M, Zhou K, Gao XM, Jiang ZY, Lv JJ, Liu ZH, Yang GY, Miao MM, Che CT, Hu QF. Fistulains A and B, new bischromones from the bark of *Cassia fistula*, and their activities. Org Lett. 2015;17:2638–41.
125. Mu ZQ, Gao H, Huang ZY, Feng XL, Yao XS. Puberunine and puberudine, two new C18-diterpenoid alkaloids from *Aconitum barbatum* var Puberulum. Org Lett. 2012;14:2758–61.
126. Guo Q, Xia H, Shi G, Zhang T, Shi J. Aconicarmisulfonine A, a sulfonated C_{20}-diterpenoid alkaloid from the lateral roots of *Aconitum carmichaelii*. Org Lett. 2018;20:816–9.
127. Zhan Z, Feng Z, Yang Y, Li L, Jiang J, Zhang P. Ternatusine A, a new pyrrole derivative with an epoxyoxepino ring from *Ranunculus ternatus*. Org Lett. 2013;15:1970–3.
128. Xue JJ, Jiang CY, Zou DL, Li BJ, Lu JC, Li DH, Lin B, Li ZL, Hua HM. Baicalensines A and B, two isoquinoline alkaloids from the roots of *Thalictrum baicalense*. Org Lett. 2020;22:7439–42.
129. Tan CH, Chen GF, Ma XQ, Jiang SH, Zhu DY. Huperzine R, a novel 15-carbon *Lycopodium* alkaloid from *Huperzia serrata*. J Nat Prod. 2002;65:1021–2.
130. Hirasawa Y, Mitsui C, Uchiyama N, Hakamatsuka T, Morita H. Hupercumines A and B, *Lycopodium* alkaloids from *Huperzia cunninghamioides*, inhibiting acetylcholinesterase. Org Lett. 2018;20:1384–7.
131. Meng WJ, Xiong J, Wang WX, Zhang HY, Zeng H, Hu JF. Phlefargesiine A, a $C_{16}N_2$ lycopodium alkaloid with an unprecedented [6/7/6/6]-tetracyclic skeleton from *Phlegmariurus fargesii*. Tetrahedron Lett. 2016;57:3218–21.
132. Zhang ZJ, Wang C, Wu XD, Huang Y, Zhou WX, Zhao QS. Phlegmadine A: a *Lycopodium* alkaloid with a unique cyclobutane ring from *Phlegmariurus phlegmaria*. J Org Chem. 2019;84:11301–5.
133. Wei K, Li W, Koike K, Chen Y, Nikaido T. Nigramides A−S, dimeric amide alkaloids from the roots of *Piper nigrum*. J Org Chem. 2005;70:1164–76.
134. Zhang XY, Li Y, Wang YY, Cai XH, Feng T, Luo XD. Tirucallane-type alkaloids from the bark of *Dysoxylum laxiracemosum*. J Nat Prod. 2010;73:1385–8.
135. Stærk D, Witt M, Oketch Rabah HA, Jaroszewski JW. A new class of spermidine-derived alkaloids. Org Lett. 2003;5:2793–6.
136. Wang W, Tang X, Hua F, Ling TJ, Wan XC, Bao GH. Camellimidazole A-C, three methylene-bridged dimeric imidazole alkaloids from *Keemun* Black Tea. Org Lett. 2018;20:2672–5.
137. Carroll AR, Arumugan G, Quinn RJ, Redburn J, Guymer G, Grimshaw P. Grandisine A and B, novel indolizidine alkaloids with human δ-opioid receptor binding affinity from the leaves of the Australian rainforest tree *Elaeocarpus grandis*. J Org Chem. 2005;70:1889–92.
138. Liu MT, Lin S, Wang YH, He WY, Li S, Wang SJ, Yang YC, Shi JG. Two novel glycosidic triterpene alkaloids from the stem barks of *Machilus yaoshansis*. Org Lett. 2007;9:129–32.
139. Toribio A, Bonfils A, Delannay E, Prost E, Harakat D, Henon E, Richard B, Litaudon M, Nuzillard JM, Renault JH. Novel *seco*-dibenzopyrrocoline alkaloid from *Cryptocarya oubatchensis*. Org Lett. 2006;8:3825–8.
140. Tong XG, Zhou LL, Wang YH, Xia C, Wang Y, Liang M, Hou FF, Cheng YX. Acortatarins A and B, two novel antioxidative spiroalkaloids with a naturally unusual morpholine motif from *Acorus tatarinowii*. Org Lett. 2010;12:1844–7.
141. Han JJ, Zhang JZ, Zhu RX, Li Y, Qiao YN, Gao Y, Jin XY, Chen W, Zhou JC, Lou HX. Plagiochianins A and B, two *ent*-2,3-*seco*-aromadendrane derivatives from the liverwort *Plagiochila duthiana*. Org Lett. 2018;20:6550–3.
142. Kobayashi J, Yoshinaga M, Yoshida N, Shiro M, Morita H. Cephalocyclidin A, a novel pentacyclic alkaloid from *Cephalotaxus harringtonia* var. *nana*. J Org Chem 2002;67:2283–6.
143. Wang YH, Gao S, Yang FM, Sun QY, Wang JS, Liu HY, Li CS, Di YT, Li SL, He HP, Hao XJ. Structure elucidation and biomimetic synthesis of hostasinine A, a new benzylphenethylamine alkaloid from *Hosta plantaginea*. Org Lett. 2007;9:5279–81.

144. Matsumoto T, Nakamura S, Ohta T, Fujimoto K, Yoshikawa M, Ogawa K, Matsuda H. A rare glutamine derivative from the flower buds of daylily. Org Lett. 2014;16:3076–8.
145. Wang KB, Di YT, Bao Y, Yuan CM, Chen G, Li DH, Bai J, He HP, Hao XJ, Pei YH, Jing YK, Li ZL, Hua HM. Peganumine A, a β-carboline dimer with a new octacyclic scaffold from *Peganum harmala*. Org Lett. 2014;16:4028–31.
146. Wu ZN, Chen NH, Tang Q, Chen S, Zhan ZC, Zhang YB, Wang GC, Li YL, Ye WC. β-Carboline alkaloids from the seeds of *Peganum harmala* and their anti-HSV-2 virus activities. Org Lett. 2020;22:7310–4.
147. Li SG, Wang YT, Zhang Q, Wang KB, Xue JJ, Li DH, Jing YK, Lin B, Hua HM. Pegaharmols A-B, axially chiral β-carboline-quinazoline dimers from the roots of *Peganum harmala*. Org Lett. 2020;22:7522–5.
148. Chen H, Kong JB, Zhang L, Wang HH, Cao YG, Zeng MN, Li M, Sun YJ, Du K, Xue GM, Wu Y, Zheng XK, Feng WS. Lycibarbarines A–C, three tetrahydroquinoline alkaloids possessing a spiro-heterocycle moiety from the fruits of *Lycium barbarum*. Org Lett. 2021;23:858–62.
149. Liu YP, Wang TW, Xie Z, Bian Y, Liu YY, Guan RQ, Liu ZY, Qiang L, Chen GY, Fu YH. Artapilosines A and B, unusual phenanthrene derivatives related to aporphine alkaloids from *Artabotrys pilosus*. J Nat Prod. 2021;84:3117–21.
150. Zhu L, Zhu DR, Zhou WX, Zhu JM, Gong LJ, Li Y, Kong LY, Luo JG. Fortuneicyclidins A and B, pyrrolizidine alkaloids with a 7-azatetracyclo[5.4.3.0.02,8]tridecane core, from *Cephalotaxus fortunei*. Org Lett. 2021;23:2807–10.

Chapter 12
Diverse Novel Steroids

Since 1999, only 17 new skeleton steroids, assigned to 3 families 4 genera, have been isolated and identified from plants (**1236–1252** Fig. 12.1; Table A11). Their architectures are characterized by containing the basic constructure cyclopentane polyhydrophenanthrene. Solanaceae produce a rich variety of novel steroids. This chapter will focus on the names, classes, structure characteristics, the types of compounds, and plant sources of these new skeleton sterides.

Subtrifloralactones A–J (**1236–1245**) are 10 novel C-18 norwithanolides based on a new C_{27} skeleton, they were isolated from *Deprea subtriflora* [1]. Physanolide A (**1246**) is a novel withasteroid with an unprecedented skeleton containing a seven-membered ring, it was isolated from *Physalis angulate* [2]. Physangulidines A–C (**1247–1249**) are three new antiproliferative withanolides with an unusual carbon framework. Of which, **1247** is the first withanolide having a disconnection between C-13 and C-17, which otherwise would have formed ring D of the ergostane skeleton [3]. Urceoloids A (**1250**) and B (**1251**) are two C_{19} steroids with a rearranged new carbon skeleton by featuring a very unique spiro[4.4]nona-3,6,8-triene system, they were isolated from *Urceola quintaretii* [4]. Bungsteroid A (**1252**) possesses an unreported carbon skeleton and represents the first carbon skeleton of a C_{34} steroid analogue featuring a unique 6/6/6/6/5-fused pentacyclic skeleton. It was isolated from *Zanthoxylum bungeanum* [5].

Y. Cheng and D. Qin, *Novel Plant Natural Product Skeletons*,
https://doi.org/10.1007/978-981-99-7329-3_12

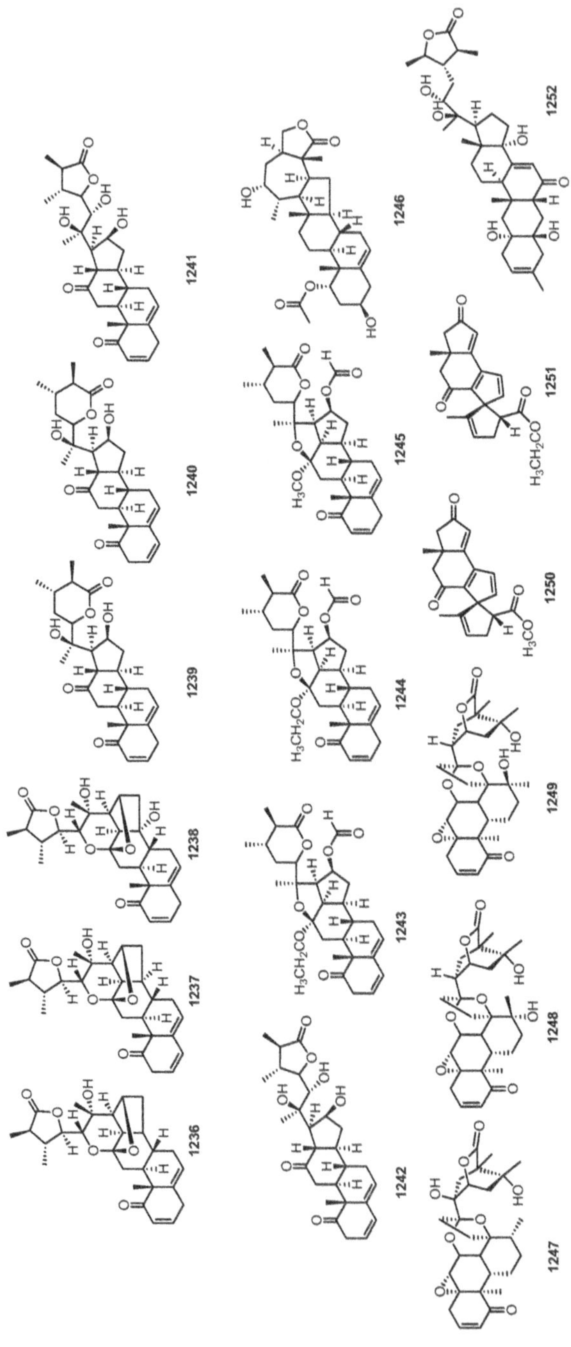

Fig. 12.1 Structures of new skeleton steroids **1236–1252**

References

1. Su BN, Park EJ, Nikolic D, Santarsiero BD, Mesecar AD, Vigo JS, Graham JG, Cabieses F, van Breemen RB, Fong HHS, Farnsworth NR, Pezzuto JM, Kinghorn AD. Activity-guided isolation of novel norwithanolides from *Deprea subtriflora* with potential cancer chemopreventive activity. J Org Chem. 2003;68:2350–61.
2. Kuo PC, Kuo TH, Damu AG, Su CR, Lee EJ, Wu TS, Shu R, Chen CM, Bastow KF, Chen TH, Lee KH. Physanolide A, a novel skeleton steroid, and other cytotoxic principles from *Physalis angulata*. Org Lett. 2006;8:2953–6.
3. Jin Z, Mashuta MS, Stolowich NJ, Vaisberg AJ, Stivers NS, Bates PJ, Lewis WH, Hammond GB. Physangulidines A, B, and C: three new antiproliferative withanolides from *Physalis angulata* L. Org Lett. 2012;14:1230–3.
4. Ren YH, Liu QF, Chen L, He SJ, Zuo JP, Fan YY, Yue JM. Urceoloids A and B, two C_{19} steroids with a rearranged spirocyclic carbon skeleton from *Urceola quintaretii*. Org Lett. 2019;21:1904–7.
5. Meng XH, Chai T, Shi YP, Yang JL. Bungsteroid A: one unusual c_{34} pentacyclic steroid analogue from *Zanthoxylum bungeanum* Maxim. J Org Chem. 2020;85:10806–12.

Chapter 13
Other Novel Skeletal Plant Natural Products

In addition to the above 11 new type skeleton compounds, there are also 113 new skeletons of other classes (1999–2021) and they contain 23 subclasses (**1253–1294** Fig. 13.1, **1295–1340** Fig. 13.2, **1341–1365** Fig. 13.3; Table A12). In view of the relatively small number of new skeleton compounds contained in each subclass, the 113 unique compounds were classified as other groups. The detailed classification of the subclasses will be described in this chapter.

13.1 Natural Pentacyclic Compound

Kingianin A (**1253**) is a new natural pentacyclic compound isolated as a racemic mixture from the barks of *Endiandra kingiana* (Lauraceae), its pentacyclic skeleton might be formed by a Diels–Alder reaction between two monomers having a bicyclo[4.2.0]octadiene backbone formed by a stereospecific electrocyclization of a linear compound of polyketide origin [1].

13.2 Thapsigargin

Transtaganolides A–D (**1254–1257**) are four novel and unusual C-19 compounds isolated from *Thapsia transtagana* (Apiaceae), this is the first time that a 7-methoxy-4,5-dihydro3*H*-oxepin-2-one ring has been found in natural products [2].

Y. Cheng and D. Qin, *Novel Plant Natural Product Skeletons*,
https://doi.org/10.1007/978-981-99-7329-3_13

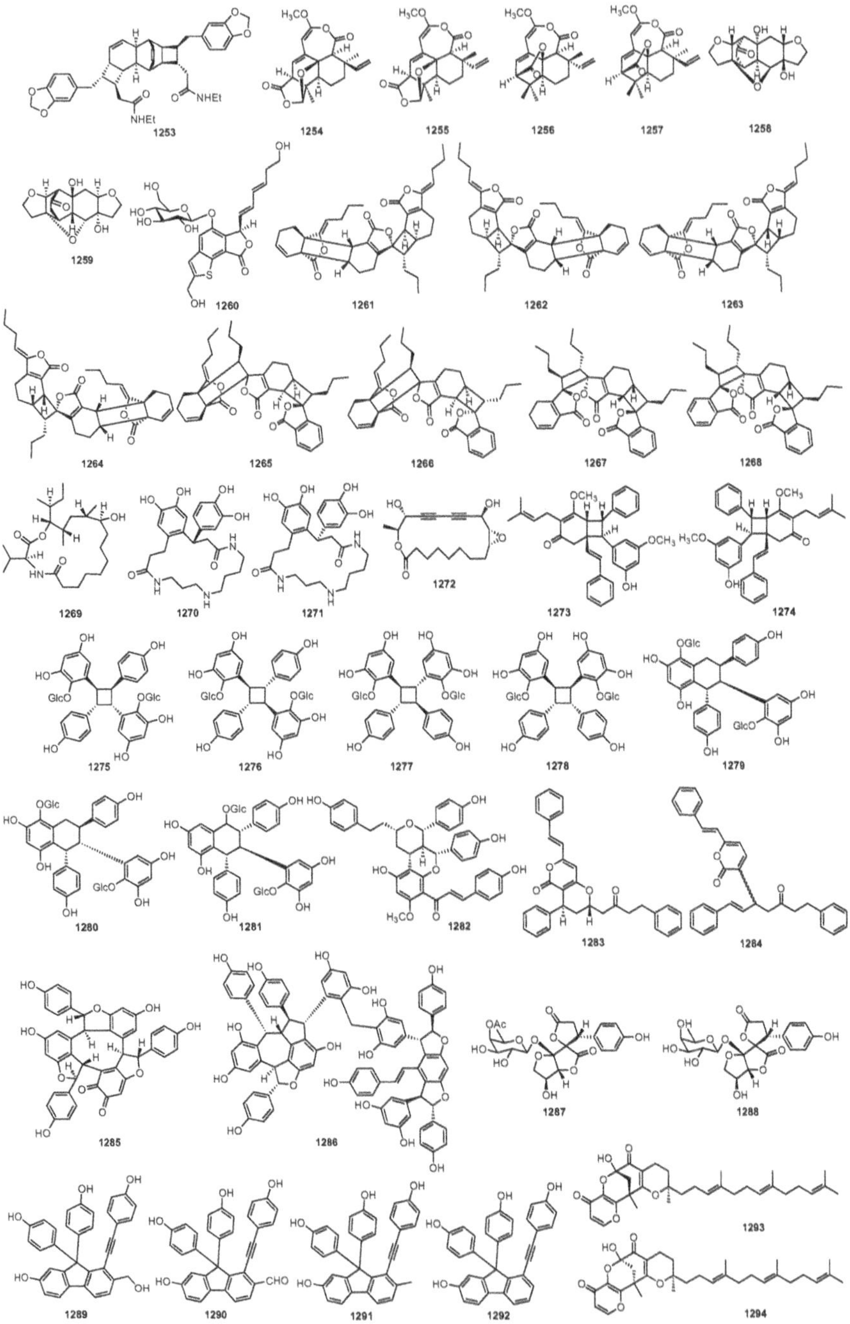

Fig. 13.1 Structures of other new skeletons **1253–1294**

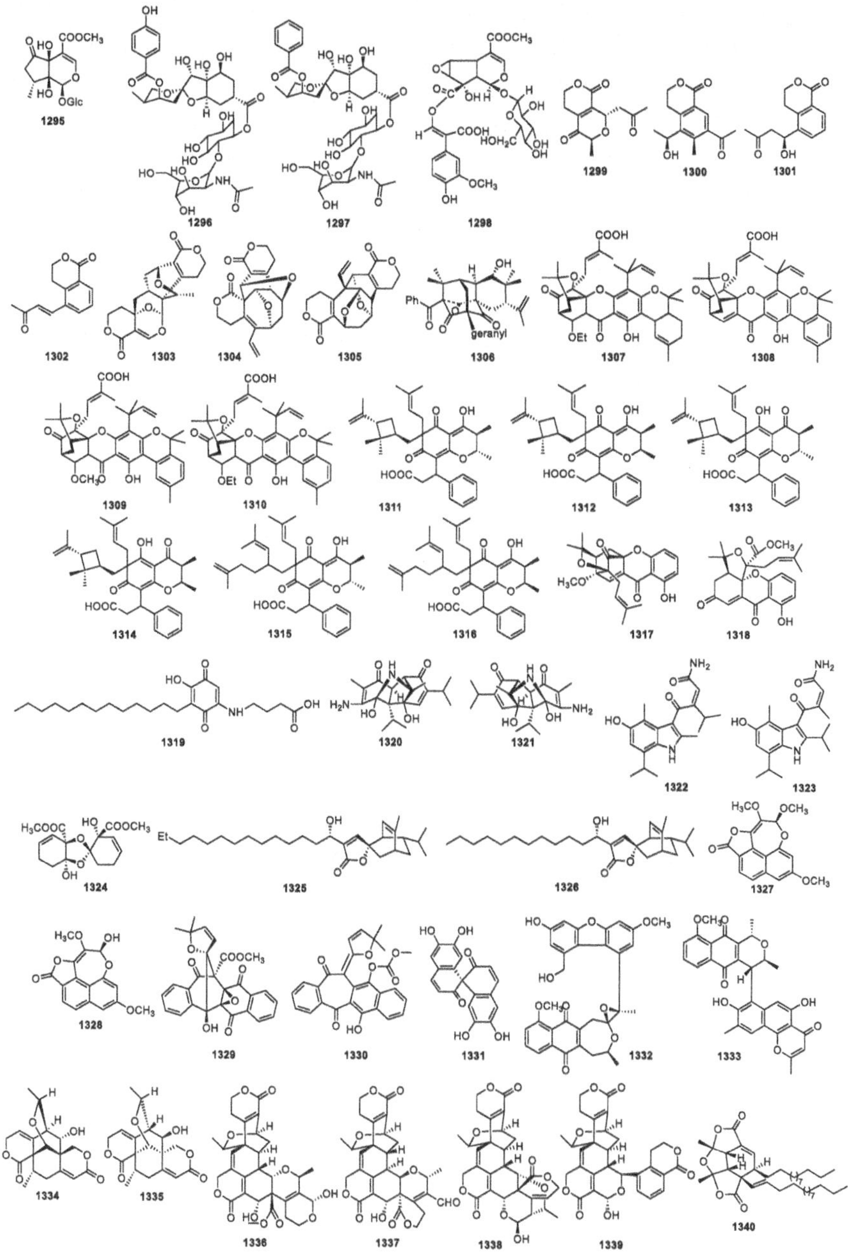

Fig. 13.2 Structures of other new skeletons **1295–1340**

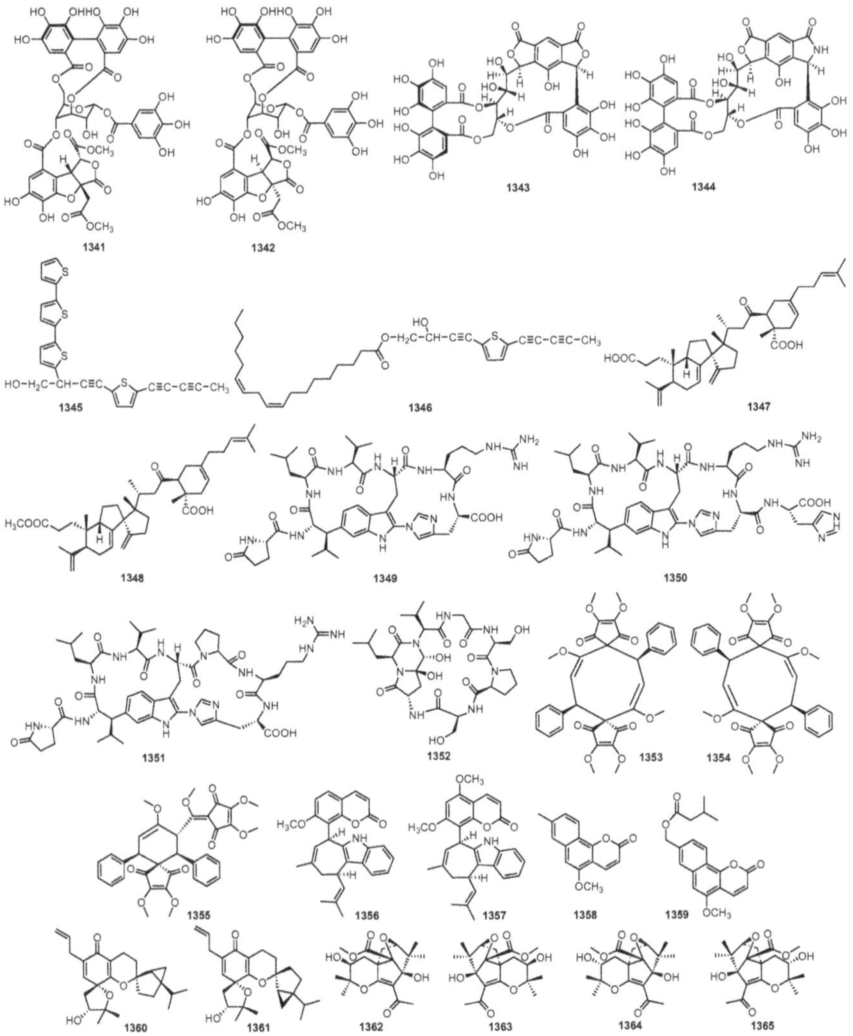

Fig. 13.3 Structures of other new skeletons **1341–1365**

13.3 Dimeric Cyclohexylethanoid Derivative

(+)-/(−)-Incarvilleatones (**1258/1259**) are an unprecedented dimeric cyclohexylethanoid analog with a racemic nature, they were isolated from the whole plant of *Incarvillea younghusbandii* (Bignoniaceae) [3].

13.4 Benzothiophene

Echinothiophene (**1260**) is a novel benzo[*b*]thiophene glycoside, it was isolated from the roots of *Echinops grijissii* (Compositae/Asteraceae) [4].

13.5 Phthalide Polymer

(−)/(+)-triligustilides A (**1261/1262**) and (−)/(+)-triligustilides B (**1263/1264**) are two pairs of enantiomeric phthalide trimers with complex polycyclic skeletons simultaneously possessing bridged, fused, and spiro ring systems. They were isolated from *Angelica sinensis* (Apiaceae) [5]. Triangeliphthalides A–D (**1265–1268**) are another four novel phthalide trimers with two new linkage styles isolated from *Angelica sinensis* [6].

13.6 Macrolactams

Dysoxylactam A (**1269**) is a 17-membered macrocyclolipopeptide comprising an unprecedented branched C19 fatty acid and *L*-valine. It was isolated from *Dysoxylum hongkongense* (Meliaceae) and its absolute configuration was determined by single-crystal X-ray diffraction [7]. (+)-(*S*)-Scocycamide and (−)-(*R*)-scocycamide (**1270** and **1271**) are a pair of new macrocyclic spermidine alkaloids. They were isolated from the roots of *Scopolia tangutica* (Solanaceae) and feature a unique 6/18 fused bicyclic framework with spermidine and catechol units [8].

13.7 Macrolide

Ivorenolide A (**1272**) is a novel 18-membered macrolide featuring conjugated acetylenic bonds and five chiral centers. It was isolated from *Khaya ivorensis* (Meliaceae) and its absolute configuration was confirmed by single-crystal X-ray diffraction [9].

13.8 Toluylene

(+)-Cajanusine (**1273**) and (−)-Cajanusine (**1274**) are a pair of new enantiomeric stilbene dimers with a unique coupling pattern isolated from the leaves of *Cajanus cajan* (Leguminosae), their absolute configurations were confirmed by single-crystal

X-ray diffraction analyses as well as CD calculations [10]. Multiflorumisides A–G (**1275–1281**) are seven new dimeric stilbene glucosides with two rare coupling patterns isolated from the roots of *Polygonum multiflorum* (Polygonaceae) [11]. Of them, **1275–1278** were presumably generated through a [2+2] cycloaddition reaction of two monomeric stilbenes to form a cyclobutane ring, which are rare in natural oligomeric stilbenes; **1279–1281** possessed identical carbon skeletons but with different configurations.

13.9 Diarylheptanoid

Calyxin I (**1282**) represents a novel carbon diarylheptanoid, which was isolated from *Alpinia blepharocalyx* (Zingiberaceae) [12]. Katsumadains A (**1283**) and B (**1284**) are also two novel diarylheptanoids representing the first example of a diarylheptanoid combined with a monocyclic α-pyrone moiety from *Alpinia katsumadai* (Zingiberaceae) [13].

13.9.1 Phenol

Hopeanolin (**1285**) is an unusual resveratral trimer with an *ortho*-quinone nucleus, it was isolated and characterized from the stem bark of *Hopea exalata* (Dipterocarpaceae) [14]. Chunganenol (**1286**) is an unusual resveratrol hexamer, it was isolated from *Vitis chunganensis* (Vitaceae) [15]. Dichotomains A (**1287**) and B (**1288**) are two new highly oxygenated phenolic derivatives that feature a spirodilactone moiety in their structures, they were isolated from the fronds of *Dicranopteris dichotoma* (Gleicheniaceae) [16]. Selaginpulvilins A–D (**1289–1292**) are new phenols with an unprecedented 9,9-diphenyl-1-(phenylethynyl)-9*H*-fluorene skeleton isolated from *Selaginella pulvinata* (Selaginellaceae) [17]. The structure of **1289** was confirmed by single-crystal X-ray diffraction. Two tocotrienol derivatives, garcipaucinones A (**1293**) and B (**1294**) were isolated from the fruit of *Garcinia paucinervis*. **1293** and **1294** are the first naturally occurring tocotrienol derivatives with a 3,10-dioxatricyclo-[7.3.1.$0^{2,7}$]tridecane skeleton incorporating an unusual γ-pyrone motif [18].

13.9.2 Iridoid

9-Hydroxy-8-epihastatoside (**1295**) is an unusual iridoid glucoside, which was isolated from *Junellia seriphioides* (Verbenaceae) [19]. Phyllanthusols A and B (**1296** and **1297**) are two norbisabolane glycosides, they were isolated from *Phyllanthus acidus* (Euphorbiaceae) [20]. Citrifolinoside (**1298**) was isolated from the leaves of *Morinda citrifolia* (Rubiaceae) [21]. There is a presence of a rearranged ferulic acid moiety in its structure. Swerilactones L–O (**1299–1302**) are four unusual secoiridoids with unprecedented C_{12} and C_{13} skeletons, they were isolated from the traditional Chinese herb *Swertia mileensis* (Gentianaceae) [22]. Sweritranslactones A–C (**1303–1305**) are three novel secoiridoid dimers, they possess a 6/6/6/6/6/6-fused hexacyclic skeleton and were obtained from swertiamarin, one of the major constituents of the genus *Swertia* (Gentianaceae) via a [4+2] cycloaddition and intramolecular nucleophilic addition under aqueous conditions [23].

13.10 Amantane

Sampsonione I (**1306**) is the first polyprenylated benzoylphloroglucinol derivative with a unique caged tetracyclo[$7.3.1.1^{3,11}.0^{3,8}$]tetradecane-2,12,14-trione skeleton. It was isolated from the aerial parts of the Chinese medicinal plant *Hypericum sampsonii* (Hypericaceae) [24].

13.11 Xanthone

Gaudichaudiic acids F–I (**1307–1310**) are a unique class of heptacyclic xanthonoids from the barks of Indonesian *Garcinia gaudichaudii* (Hypericaceae) [25]. Brasiliensophyllic acids A–F (**1311–1316**) are six novel chromanone acids, they were isolated from the bark of *Calophyllum brasiliense* (Hypericaceae) [26]. Pruniflorone T (**1317**) and pruniflorone U (**1318**) are two rare neocaged-xanthone and rearranged caged-xanthone isolated from the roots of *Cratoxylum formosum* (Hypericaceae) [27].

13.12 Quinone

1319 is an unusual 3-alkyl-1,4-benzoquinone derivative coupled through a C-N bond with γ-aminobutyric acid, it was isolated from *Embelia ribes* (Myrsinaceae) [28]. (−)-Nigegladine A, (+)-nigegladine A, nigegladine B and nigegladine C (**1320–1323**) are four thymoquinone dimers, were isolated from the seeds of *Nigella glandulifera* (Ranunculaceae). **1320** and **1321** possess a unique tricyclo[$5.4.0.1^{2,6}$]dodecane carbon skeleton, while **1322** and **1323** are two unusual diterpenoid alkaloids with indole cores [29].

13.13 Spiro Compound

Idesolide (**1324**) is a new spiro compound possessing tetrahydrobenzodioxole structure, it was isolated from the fruits of *Idesia polycarpa* (Flacourtiaceae) [30]. Yaoshanenolides A (**1325**) and B (**1326**) are two novel tricyclic spirolactones bearing long linear alkyl chains, they were formed by Diels–Alder [4+2] cycloaddition of a molecule of each butenolide with β-phellandrene and isolated from *Machilus yaoshansis* (Lauraceae) [31].

13.14 Naphthalene

Dioscorealides A (**1327**) and B (**1328**) possess naphthofuranoxepin skeleton, they were isolated from the rhizome of *Dioscorea membranacea* (Dioscoreaceae) [32]. Rubialatins A (**1329**) and B (**1330**) are two novel naphthohydroquinone dimers with unprecedented skeletons, they were isolated from the herbal plant *Rubia alata* (Rubiaceae) [33]. Spiroaxillarone A (**1331**) is the first representative of a new type of spirobisnaphthalene with a previously unknown skeletal system, it was isolated from *Cyanotis axillaris* (Commelinaceae) [34]. Eleucanainones A (**1332**) and B (**1333**) are two naphthoquinone-derived heterodimers with unprecedented carbon skeletons, they were isolated from the bulbs of *Eleutherine americana* (Iridaceae) and were determined to be the first examples of dibenzofuran- and naphthalenone-containing naphthoquinone dimers [35].

13.15 Lactone

Swerilactones A and B (**1334** and **1335**) are two novel lactones with an unprecedented 6/6/6/6/6 pentacyclic ring system, they were isolated from the traditional Chinese herb *Swertia mileensis* (Gentianaceae) [36]. Swerilactones H–K (**1336–1339**) are four novel lactones with an unprecedented C_{29} skeleton, they were isolated from *Swertia mileensis* (Gentianaceae) [37]. Paracaseolide A (**1340**) is a novel α-alkylbutenolide dimer isolated from *Sonneratia paracaseolaris* (Sonneratiaceae), it was characterized by an unusual tetraquinane oxa-cage bislactone skeleton bearing two linear alkyl chains [38].

13.16 Tannin

Punicatannins A (**1341**) and B (**1342**) are two new ellagitannins containing a rare 3-oxo-1,3,3a,8b-tetrahydrofuro[3,4-b]benzofuran moiety, they were isolated from *Punica granatum* (Punicaceae) [39]. Quercusnins A (**1343**) and B (**1344**) are two unusual ellagitannin metabolites isolated from the sapwood of *Quercus crispula* (Fagaceae) [40].

13.17 Thiophene

Echinopsacetylenes A and B (**1345** and **1346**) are two new polyacetylene thiophenes isolated from the roots of *Echinops transiliensis* (Compositae/Asteraceae) [41]. Of them, **1345** is the first natural product possessing an α-terthienyl moiety covalently linked with another thiophene moiety, while **1346** is the first natural thiophene conjugated with a fatty acid moiety.

13.18 Tetraterpene

Abibalsamins A (**1347**) and B (**1348**) are two unprecedented tetraterpenoids featuring a 3,4-*seco*-rearranged lanostane system fused with a β-myrcene lateral chain via a [4+2] Diels–Alder cycloaddition, they were isolated from the oleoresin of *Abies balsamea* (Pinaceae) [42].

13.19 Peptide

Celogentins A–C (**1349–1351**), three novel bicyclic peptides, they were isolated from the seeds of *Celosia argentea* (Amaranthaceae) [43]. Tunicyclin A (**1352**) is a novel cycloheptapeptide with a unique tricyclic ring cyclopeptide skeleton, it was isolated from *Psammosilene tunicoides* (Caryophyllaceae) [44].

13.20 Cyclopentadienedione

(+)-Linderaspirone A and (−)-linderaspirone A (**1353** and **1354**) are a pair of natural windmill-shaped enantiomers, they were isolated from the traditional Chinese medicine plant *Lindera aggregate* (Lauraceae) [45]. Bi-linderone (**1355**) was isolated as a racemate from the traditional Chinese medicinal plant *Lindera aggregata*

(Lauraceae), it has an unprecedented spirocyclopentenedione-containing carbon skeleton, no other structure with this skeleton has been reported to date [46].

13.21 Coumarin

Exotines A and B (**1356** and **1357**) are two heterodimers of isopentenyl-substituted indole and coumarin derivatives linked through a new fused heptacyclic ring system, they were isolated from the roots of *Murraya exotica* (Rutaceae) [47]. Muralatins A and B (**1358** and **1359**) are two new rare 8-methylbenzo[*h*]coumarins, they were isolated from the leaves of *Murraya alata* (Rutaceae) [48].

13.22 Isopentylphenylpropanol

Spirooliganones A (**1360**) and (**1361**) are a pair of spiro carbon epimers with a rare dioxaspiro skeleton, they were isolated from the roots of *Illicium oligandrum* (Magnoliaceae) [49]. Their absolute configurations were determined by modified Mosher's method and X-ray diffraction analysis.

13.23 Isopentenyl Acetyl Benzene

(+)- and (−)-Melicolones A (**1362** and **1363**) as well as (+)- and (−)-melicolones B (**1364** and **1365**) are a pair of rearranged prenylated acetophenone epimers with an unusual 9-oxatricyclo-[3.2.1.1^{3,8}]nonane core, they were isolated from the leaves of *Melicope ptelefolia* (Rutaceae) [50].

References

1. Leverrier A, Dau METH, Retailleau P, Awang K, Guéritte F, Litaudon M. Kingianin A: a new natural pentacyclic compound from *Endiandra kingiana*. Org Lett. 2010;12:3638–41.
2. Saouf A, Guerra FM, Rubal JJ, Moreno Dorado FJ, Akssira M, Mellouki F, López M, Pujadas AJ, Jorge ZD, Massanet GM. Transtaganolides A–D: novel metabolites from *Thapsia transtagana*. Org Lett. 2005;7:881–4.
3. Gao YP, Shen YH, Zhang SD, Tian JM, Zeng HW, Ye J, Li HL, Shan L, Zhang WD. Incarvilleatone, a new cyclohexylethanoid dimer from *Incarvillea younghusbandii* and its inhibition against nitric oxide (NO) release. Org Lett. 2012;14:1954–7.
4. Koike K, Jia Z, Nikaido T, Liu Y, Zhao Y, Guo D. Echinothiophene, a novel benzothiophene glycoside from the roots of *Echinops grijissii*. Org Lett. 1999;1:197–8.

5. Zou J, Chen GD, Zhao H, Huang Y, Luo X, Xu W, He RR, Hu D, Yao XS, Gao H. Triligustilides A and B: two pairs of phthalide trimers from *Angelica sinensis* with a complex polycyclic skeleton and their activities. Org Lett. 2018;20:884–7.
6. Zou J, Chen GD, Zhao H, Wang XX, Zhang ZJ, Qu YB, He RR, So KF, Yao XS, Gao H. Triangeliphthalides A-D: bioactive phthalide trimers with new skeletons from *Angelica sinensis* and their production mechanism. Chem Commun. 2019;55:6221–4.
7. Liu CP, Xie CY, Zhao JX, Ji KL, Lei XX, Sun H, Lou LG, Yue JM. Dysoxylactam A: a macrocyclolipopeptide reverses p-glycoprotein-mediated multidrug resistance in cancer cells. J Am Chem Soc. 2019;141:6812–6.
8. Wang JX, Zhao YP, Du NN, Han Y, Li H, Wang R, Xu Y, Liu YF, Liang XM. Scocycamides, a pair of macrocyclic dicaffeoylspermidines with butyrylcholinesterase inhibition and antioxidation activity from the roots of *Scopolia tangutica*. Org Lett. 2020;22:8240–4.
9. Zhang B, Wang Y, Yang SP, Zhou Y, Wu WB, Tang W, Zuo JP, LiY, Yue JM. Ivorenolide A, an unprecedented immunosuppressive macrolide from *Khaya ivorensis*: structural elucidation and bioinspired total synthesis. J Am Chem Soc. 2012;134:20605–8.
10. Li XL, Zhao BX, Huang XJ, Zhang DM, Jiang RW, Li YJ, Jian YQ, Wang Y, Li YL, Ye WC. (+)- and (−)-Cajanusine, a pair of new enantiomeric stilbene dimers with a new skeleton from the leaves of *Cajanus cajan*. Org Lett. 2014;16:224–7.
11. Li SG, Huang XJ, Li MM, Liu Q, Liu H, Wang Y, Ye WC. Multiflorumisides A-G, dimeric stilbene glucosides with rare coupling patterns from the roots of *Polygonum multiflorum*. J Nat Prod. 2018;81:254–63.
12. Gewali MB, Tezuka Y, Banskota AH, Ali MS, Saiki I, Dong H, Kadota S. Epicalyxin F and calyxin I: two novel antiproliferative diarylheptanoids from the seeds of *Alpinia blepharocalyx*. Org Lett. 1999;1:1733–6.
13. Yang Y, Kinoshita K, Koyama K, Takahashi K, Tai T, Nunoura Y, Watanabe K. Two novel anti-emetic principles of *Alpinia katsumadai*. J Nat Prod. 1999;62:1672–4.
14. Ge HM, Huang B, Tan SH, Shi DH, Song YC, Tan RX. Bioactive oligostilbenoids from the stem bark of *Hopea exalata*. J Nat Prod. 2006;69:1800–2.
15. He S, Jiang L, Wu B, Li C, Pan Y. Chunganenol: an unusual antioxidative resveratrol hexamer from *Vitis chunganensis*. J Org Chem. 2009;74:7966–9.
16. Li XL, Cheng X, Yang LM, Wang RR, Zheng YT, Xiao WL, Zhao Y, Xu G, Lu Y, Chang Y, Zheng QT, Zhao QS, Sun HD. Dichotomains A and B: two new highly oxygenated phenolic derivatives from *Dicranopteris dichotoma*. Org Lett. 2006;8:1937–40.
17. Liu X, Luo HB, Huang YY, Bao JM, Tang GH, Chen YY, Wang J, Yin S. Selaginpulvilins A-D, new phosphodiesterase-4 inhibitors with an unprecedented skeleton from *Selaginella pulvinata*. Org Lett. 2014;16:282–5.
18. Tan X, Han XY, Teng HD, Li QQ, Chen Yu, Lei XX, Yang GD. Structural elucidation of garcipaucinones A and B from *Garcinia paucinervis* using quantum chemical calculations. J Nat Prod 2021;84:972−8.
19. Franzyk H, Jensen SR, Olsen CE, Quiroga JM. A 9-hydroxyiridoid isolated from *Junellia seriphioides* (Verbenaceae). Org Lett. 2000;2:699–700.
20. Vongvanich N, Kittakoop P, Kramyu J, Tanticharoen M, Thebtaranonth Y. Phyllanthusols A and B, cytotoxic norbisabolane glycosides from *Phyllanthus acidus* skeels. J Org Chem. 2000;65:5420–3.
21. Sang S, He K, Liu G, Zhu N, Cheng X, Wang M, Zheng Q, Dong Z, Ghai G, Rosen RT, Ho CT. A new unusual iridoid with inhibition of activator protein-1 (ap-1) from the leaves of *Morinda citrifolia* L. Org Lett. 2001;3:1307–9.
22. Geng CA, Zhang XM, Ma YB, Luo J, Chen JJ. Swerilactones L-O, secoiridoids with C_{12} and C_{13} skeletons from *Swertia mileensis*. J Nat Prod. 2011;74:1822–5.
23. Liu L, Xu GL, Ma XX, Khan A, Tan WH, Yang ZY, Zhou ZH. Sweritranslactones A-C: unusual skeleton secoiridoid dimers via [4+2] cycloaddition from swertiamarin. J Org Chem. 2017;82:13263–7.
24. Hu LH, Sim KY. Cytotoxic polyprenylated benzoylphloroglucinol derivatives with an unusual adamantyl skeleton from *Hypericum sampsonii* (Guttiferae). Org Lett. 1999;1:879–82.

25. Xu YJ, Yip SC, Kosela S, Fitri E, Hana M, Goh SH, Sim KY. Novel cytotoxic, polyprenylated heptacyclic xanthonoids from Indonesian *Garcinia gaudichaudii* (Guttiferae). Org Lett. 2000;2:3945–8.
26. Cottiglia F, Dhanapal B, Sticher O, Heilmann J. New chromanone acids with antibacterial activity from *Calophyllum brasiliense*. J Nat Prod. 2004;67:537–41.
27. Boonnak N, Chantrapromma S, Fun HK, Yuenyongsawad S, Patrick BO, Maneerat W, Williams DE, Andersen RJ. Three types of cytotoxic natural caged-scaffolds: pure enantiomers or partial racemates. J Nat Prod. 2014;77:1562–71.
28. Lin P, Li S, Wang S, Yang Y, Shi J. A nitrogen-containing 3-alkyl-1,4-benzoquinone and a gomphilactone derivative from *Embelia ribes*. J Nat Prod. 2006;69:1629–32.
29. Tian J, Han C, Guo WH, Yin Y, Wang XB, Sun HB, Yao HQ, Yang Y, Wang C, Liu C, Yang MH, Kong LY. Nigegladines A-C, three thymoquinone dimers from *Nigella glandulifera*. Org Lett. 2017;19:6348–51.
30. Kim SH, Sung SH, Choi SY, Chung YK, Kim J, Kim YC. Idesolide: a new spiro compound from *Idesia polycarpa*. Org Lett. 2005;7:3275–7.
31. Liu M, Lin S, Gan M, Chen M, Li L, Wang S, Zi J, Fan X, Liu Y, Si Y, Yang Y, Chen X, Shi J. Yaoshanenolides A and B: new spirolactones from the bark of *Machilus yaoshansis*. Org Lett. 2012;14:1004–7.
32. Itharat A, Plubrukarn A, Kongsaeree P, Bui T, Keawpradub N, Houghton PJ. Dioscorealides and dioscoreanone, novel cytotoxic naphthofuranoxepins, and 1,4-phenanthraquinone from *Dioscorea membranacea* Pierre. Org Lett. 2003;5:2879–82.
33. Zhao SM, Wang Z, Zeng GZ, Song WW, Chen XQ, Li XN, Tan NH. New cytotoxic naphthohydroquinone dimers from *Rubia alata*. Org Lett. 2014;16:5576–9.
34. Wisetsai A, Lekphrom R, Boonmak J, Youngme S, Schevenels FT. Spiroaxillarone A, a symmetric spirobisnaphthalene with an original skeleton from *Cyanotis axillaris*. Org Lett. 2019;21:8344–8.
35. Chen D, Sun Z, Liu Y, Li Z, Liang H, Chen L, Xu X, Yang J, Ma G, Huo X. Eleucanainones A and B: two dimeric structures from the bulbs of *Eleutherine americana* with anti-MRSA activity. Org Lett. 2020;22:3449–53.
36. Geng CA, Jiang ZY, Ma YB, Luo J, Zhang XM, Wang HL, Shen Y, Zuo AX, Zhou J, Chen JJ. Swerilactones A and B, anti-HBV new lactones from a traditional Chinese herb: *Swertia mileensis* as a treatment for viral hepatitis. Org Lett. 2009;11:4120–3.
37. Geng CA, Wang LJ, Zhang XM, Ma YB, Huang XY, Luo J, Guo RH, Zhou J, Shen Y, Zuo AX, Jiang ZY, Chen JJ. Anti-hepatitis B virus active lactones from the traditional Chinese herb: *Swertia mileensis*. Chem Eur J. 2011;17:3893–903.
38. Chen XL, Liu HL, Li J, Xin GR, Guo YW. Paracaseolide A, first α-alkylbutenolide dimer with an unusual tetraquinane oxa-cage bislactone skeleton from Chinese mangrove *Sonneratia paracaseolaris*. Org Lett. 2011;13:5032–5.
39. Yuan T, Ding Y, Wan C, Li L, Xu J, Liu K, Slitt A, Ferreira D, Khan IA, Seeram NP. Antidiabetic ellagitannins from pomegranate flowers: inhibition of α-glucosidase and lipogenic gene expression. Org Lett. 2012;14:5358–61.
40. Omar M, Matsuo Y, Maeda H, Saito Y, Tanaka T. New metabolites of *C*-glycosidic ellagitannin from Japanese Oak Sapwood. Org Lett. 2014;16:1378–81.
41. Nakano H, Cantrell CL, Mamonov LK, Osbrink WLA, Ross SA. Echinopsacetylenes A and B, new thiophenes from *Echinops transiliensis*. Org Lett. 2011;13:6228–31.
42. Lavoie S, Legault J, Gauthier C, Mshvildadze V, Mercier S, Pichette A. Abibalsamins A and B, two new tetraterpenoids from *Abies balsamea* oleoresin. Org Lett. 2012;14:1504–7.
43. Kobayashi J, Suzuki H, Shimbo K, Takeya K, Morita H. Celogentins A−C, new antimitotic bicyclic peptides from the seeds of *Celosia argentea*. J Org Chem. 2001;66:6626–33.
44. Tian JM, Shen YH, Yang XW, Liang S, Tang J, Shan L, Zhang WD. Tunicyclin A, the first plant tricyclic ring cycloheptapeptide from *Psammosilene tunicoides*. Org Lett. 2009;11:1131–3.
45. Wang F, Gao Y, Zhang L, Bai B, Hu YN, Dong ZJ, Zhai QW, Zhu HJ, Liu JK. A pair of windmill-shaped enantiomers from *Lindera aggregata* with activity toward improvement of insulin sensitivity. Org Lett. 2010;12:3196–9.

46. Wang F, Gao Y, Zhang L, Liu JK. Bi-linderone, a highly modified methyl-linderone dimer from *Lindera aggregata* with activity toward improvement of insulin sensitivity in vitro. Org Lett. 2010;12:2354–7.
47. Liu BY, Zhang C, Zeng KW, Li J, Guo XY, Zhao MB, Tu PF, Jiang Y. Exotines A and B, two heterodimers of isopentenyl-substituted indole and coumarin derivatives from *Murraya exotica*. Org Lett. 2015;17:4380–3.
48. Lv HN, Wang S, Zeng KW, Li J, Guo XY, Ferreira D, Zjawiony JK, Tu PF, Jiang Y. Anti-inflammatory coumarin and benzocoumarin derivatives from *Murraya alata*. J Nat Prod. 2015;78:279–85.
49. Ma SG, Gao RM, Li YH, Jiang JD, Gong NB, Li L, Lü Y, Tang WZ, Liu YB, Qu J, Lü HN, Li Y, Yu SS. Antiviral spirooliganones A and B with unprecedented skeletons from the roots of *Illicium oligandrum*. Org Lett. 2013;15:4450–3.
50. Xu JF, Zhao HJ, Wang XB, Li ZR, Luo J, Yang MH, Yang L, Yu WY, Yao HQ, Luo JG, Kong LY. (±)-Melicolones A and B, rearranged prenylated acetophenone stereoisomers with an unusual 9-oxatricyclo[3.2.1.1^{3,8}]nonane core from the leaves of *Melicope ptelefolia*. Org Lett 2015;17:146–9.

Chapter 14
Biological Activities

New skeleton plant natural products are found in many plants belonging to various families. These compounds have been observed to possess various biological activities including cytotoxicity, anti-inflammatory, antiviral, immunomodulatory, antioxidant, neuroprotective, antibacterial, and anti-malarial activities as well as other diverse activities. In this Review, we focus mainly on the mentioned above eight activities. As for the other types of biological activities, please see the Tables.

14.1 Cytotoxic Activity

Ligulolide A (**1**) showed strong cytotoxicity (IC_{50} = 19.43 μM against HL-60 and IC_{50} > 100 μM against HO-8910) in vitro [1]. (+)-Artaboterpenoid B (**15**) exhibited cytotoxic effects against HCT-116, HepG2, A2780, NCI-H1650, and BGC-823 cell lines with IC_{50} values of 1.38 – 8.19 μM [2]. It is interesting that chlotrichene B (**54**) showed synergetic cytotoxicity with DOX in U2 OS cells (CI: 0.94 ± 0.03) [3]. Japonicone A (**55**) showed the most potent cytotoxicities against four tumor cell lines, A549, LOVO, CEM and MDA-MB-435with IC_{50} values of 1.620, 0.256, 0.001, and 0.198 μg/mL, respectively [4]. Chlorahupetone G (**64**) exhibited the most potent cytotoxicity against A549 cells with an IC_{50} value of 0.43 ± 0.12 μM, even about 4 times than paclitaxel (IC_{50} = 1.62 ± 0.13 μM). Moreover, **64** showed significant efficacy in inducing cell cycle arrest at the G0/G1 phase and apoptosis in A549 cells [5]. Dicarabrones A and B (**79** and **80**) showed moderate effects in HL-60 cells with IC_{50} values of 9.1 μM and 8.2 μM, respectively [6]. It is true that human leukemia (CCRF-CEM) cells are sensitive to carpedilactones A – D (**81** – **84**) with respective IC_{50} values of 0.14 μM, 0.32 μM, 0.35 μM, and 0.16 μM [7]. It was found that ainsliatrimers A and B (**95** and **96**) showed potent cytotoxicites against LOVO and CEM cell lines [8]. Ainsliatriolide A (**97**) displayed potent cytotoxicity with an averaged IC_{50} value of 1.17 μM against four human cancer cells [9].

Y. Cheng and D. Qin, *Novel Plant Natural Product Skeletons*,
https://doi.org/10.1007/978-981-99-7329-3_14

Ainsliatetramers A and B (**99** and **100**) exhibited potent cytotoxicity against human cancer cell lines with IC_{50} values ranging from 2 to 15 μM [10]. Cytotoxicity results showed that ainsfragolide (**101**) against five cancer cell lines with IC_{50} values in the range of 0.4 – 8.3 μM [11]. Involucratustones A and B (**120** and **121**) showed potent growth inhibitory effects (IC_{50} = 6.78–10.27 μm) against U2 OS (human osteosarcoma cell line) and SMMC-7721 (human hepatic liver carcinoma cell line) [12]. Euphorikanin A (**143**) exhibited moderate cytotoxic activities with IC50 values of 28.85 and 20.89 μM, respectively [13]. Euphomilone A (**161**) was found to inhibit nuclear factor kappa B ligand (RANKL)-induced osteoclast formation with an IC_{50} value of 12.6 μM [14]. Crokonoid A (**165**) exhibited significant cytotoxicity against HL-60 and A-549 cell lines with IC_{50} values of 1.24 μM and 1.92 μM, respectively [15]. Jatrofoliane A (**182**) showed significant multidrug resistance (MDR) reversal activity to cancer cells HepG2/ADR and HCT-15/5-FU at 10 μM [16]. The IC_{50} of fischdiabietane A (**185**) against T47D cells was about sixfold higher than that of cisplatin (the positive control). Furthermore, **185** induced apoptosis in T47D cells through the activation of caspase-3 and the degradation of poly(ADP-ribose) polymerase [17]. Excisanin H (**186**) is cytotoxic toward P388 murine leukemia cells having an IC_{50} value of 0.96 μg/m, a remarkable in vitro effect [18]. HeLa cells is much more sensitive to maoecrystal V (**187**) whose IC_{50} value is 0.02 μg/mL [19]. Cytotoxic properties of maoecrystal Z (**188**) against human cancer cells such as K562 leukemia (IC_{50} = 2.90 μg/mL), MCF7 breast (IC_{50} = 1.63 μg/mL), and A2780 ovarian (IC_{50} = 1.45 μg/mL) were evaluated [20]. Their potency was found to be comparable with that of camptothecin and paclitaxel, positive controls in the study. Bisrubescensin A (**189**) is also toxic toward several cancer cells with IC_{50} values in the range of 0.54 – 1.85 μM [21]. Hispidanin B (**203**) was found to inhibit cancer cell lines SGC7901, SMMC7721, and K562 with respective IC_{50} values of 10.7 μM, 9.8 μM, and 13.7 μM [22]. Salvileucalin B (**217**) exerted cytotoxic properties against A549 and HT-29 cells with IC_{50} values of 5.23 μg/mL and 1.88 μg/mL, respectively [23]. Using cisplatin as a positive control, spirodesertol A (**222**) was found to be more potent against A-549, SMMC-7721, and MCF-7 cancer cell lines [24]. In comparison with colon cancer (HCT-8), stomach cancer (BGC-823), lung adenocarcinoma (A549), and human ovarian cancer (A2780) cells, human hepatoma carcinoma cell line (Bel-7402) is more sensitive to secorhodomollolide B (**227**) with an IC_{50} value of 0.97 μM, indicating its selectivity toward different cancer cells [25]. HepG2 and XWLC-05 cell lines were used to evaluate the cytotoxic activities of hedychin B (**258**), exhibiting IC_{50} values of 8.0 μM and 19.7 μM, respectively [26]. Cunlanceloic acids B and C (**286** and **287**), showed remarkable cytotoxic activities against A-549 and SMMC-7721 cell lines with IC_{50} values of 5.90–9.21 μM [27]. Atisane 2 (**300**) showed slight cytotoxic activity against mouse leukemia cells (P388) with an IC_{50} value of 16 μg/mL [28]. Pallambins A (**304**) and B (**305**) were found to enable to reverse the adriamycin-induced resistance of K562/A02 cells at 10 μM, with reversal fold values of 4.3 and 1.9, respectively [29]. Cytotoxic assay revealed that tagalide A (**325**) could inhibit triple-negative breast cancer cell lines: MD-MBA-453 (IC_{50} = 1.73 μM) and MD-MBA-231 (IC_{50} = 8.12 μM) [30]. Schinalactone A (**359**) was found to show cytotoxicity against PANC-1 cell line with an

IC_{50} value of 5.9 μM [31]. Henrischinins A (**360**) and B (**361**) showed weak cytotoxicity against HL-60 cell lineswith IC_{50} values of 16.5 and 10.5 μM,respectively [32]. Schiglautone A (**368**) was tested for its inhibitory effects against HeLa, Hep G2, and SGC-7901 cell lines and showed weak cytotoxicities of 25%, 23%, and 13% respectively, with a concentration of 100 μg/mL [33]. Kadlongilactones A (**392**) and B (**393**) exerted significant inhibitory effects against human cancer cells (K562) with IC_{50} values of 1.40 μg/mL and 1.71 μg/mL, respectively [34]. Euphorol J (**426**) and euphorstranol A (**427**) exhibited significant cytotoxicities against MDA-MB-468 cells, with IC_{50} values being 3.9 and 2.9 μM, respectively [35]. Although it was reported that pseudolaridimer A (**436**) showed strong cytotoxicity against HCT116, ZR-75-30, and HL-60 human tumor cell lines, with respective IC_{50} values of 9.62, 7.84, 8.29 μg/mL, pseudolaridimer B (**437**) exhibited potent inhibition against HL-60 cell line with an IC_{50} value of 7.50 μg/mL, their biological activities are actually moderate despite their action mechanism might be new [36]. Cimicifugadine (**440**) only exhibited moderate cytotoxicity against several human cancer cell lines at GI50 values of 12.5 μM (MCF-7, breast cancer), 9.0 μM (NCI-H460, non-small cell lung cancer), and 5.0 (K-562, leukemia) [37]. Ilelic acids A (**460**) and B (**461**) showed weak inhibition on MCF-7 cell growth with IC_{50} values of 29.51 μM and 38.49 μM, respectively [38]. Machilusides A (**465**) and B (**466**) showed nonselective cytotoxic activities against ovary (A2780), colon (HCT-8), hepatoma (Bel-7402), stomach (BGC-823), and lung (A549) cell lines with IC_{50} values of 0.40–6.52 μM [39]. Turrapubesins A (**492**) showed weak activity ($IC_{50} = 12.14$ μM) against the P-388 cell line [40]. Aphanamolide A (**499**) showed cytotoxic activity against A-549 (IC_{50}: 88.1 μM) and HL-60 (IC_{50}:191.0 μM) tumor cell lines [41]. Xylogranatins A and B (**508** and **509**) are also cytotoxic in A-549 cell line with IC_{50} values of 15.7 μM and 11.3 μM, respectively [42]. Phyllanthoid A (**541**) displayed moderate cytotoxicity against breast cancer cells (MCF-7) with an IC_{50} value of 15.52 μM [43]. The IC_{50} values of perforanoid A (**547**) against HEL, K562, and CB3 cancer cells are 6.17 μM, 4.24 μM, and 3.91 μm, respectively [44]. Hyperbeanone A (**560**) exhibited specific cytotoxicities against HL-60 (human acute myeloid leukemia) and SU-DHL-4 (diffuse large B-cell lymphoma) cell lines with IC_{50} values of 7.07 and 12.49 μM, respectively. **560** also induced G1-phase cell cycle arrest and apoptosis in HL-60 and SU-DHL-4 cell lines [45]. Garciyunnanimines A–C (**570–572**) displayed cytotoxicity against three human cancer cell lines (A549, HepG2, and RPMI-8226) [46]. Hyperberins A (**573**) and B (**574**) only exhibited moderate cytotoxicity in HCT 116 cells with IC_{50} values of 14.21 μM and 16.20 μM, respectively [47]. Hyphenrones D (**578**) also exhibited moderate cytotoxic activities against five human cancer cell lines (HL-60, A-549, SMMC-7721, MCF-7, and SW480) whose IC_{50} values are 4.7–25.5 μM [48]. Although hyperhexanone A (**581**) was evaluated for its cytotoxicity, it only exhibited moderate activities against five human cancer cell lines (HL-60, SMMC-7721, A-549, MCF-7, and SW480) with IC_{50} values ranging from 11.91 μM to 16.55 μM [49]. The cytotoxicity of norascyronones A and B (**582** and **583**) was evaluated using SK-BR-3 cell line, showing IC_{50} values of 4.3 μM and 7.8 μM, respectively [50]. Hypersubone B (**601**) exhibited significant cytotoxicities against four human cancer lines in vitro (IC_{50} values: 0.07 – 7.52 μM) [51]. Dioxasampsone

B (**603**) was reported to show mild RXRα transcriptional-inhibitory activities in a dose dependent manner (5–20 μM) [52]. Hyperuralone A (**628**) was found to be cytotoxic against five human cancer cell lines in vitro (IC_{50} values: 4.6–14.4 μM) [53]. (+)-Garcimulin A, (−)-garcimulin A and garcimulin B (**630–632**) exhibited cytotoxic activities against five human cancer cell lines in vitro (IC_{50} values: 3.42–13.23 μM) [54]. Guadial C (**640**) only showed weak cytotoxic effect toward HepG2/ADM cells with an IC_{50} value of 36.70 ± 2.37 μm [55]. **656** could inhibit HepG2 and MDA-MB-231 cells with IC_{50} values of 4.39 μM and 19.92 μM, respectively, while **657** showed negligible inhibitory activity toward these two cell lines with IC_{50} values of 40.70 μM and 40.00 μM, respectively [56]. Hyperjapone D (**666**) showed activity against AGS with IC_{50} at 12.3 μM, in addition, it was also found to inhibit Hsp90 with an IC_{50} value of 21.3 μM [57]. Frutescones A and D (**670** and **671**) exhibited moderate cytotoxic activities against Caco-2 with IC_{50} values of 8.08 and 10.20 μM, respectively [58]. Littordials B, C and E (**676**, **677** and **679**) showed cytotoxic activities on two cancer cell lines (MDA-MB-231 and B16) with IC_{50} values from 6.6 to ± 1.5 to 9.2 ± 1.7 μM [59]. Psiguadials C and D (**683** and **684**) exhibited potent inhibitory effects on the growth of HepG2 cells with IC_{50} values of 104.5 ± 13.71 nM and 128.3 ± 18.2 nM, respectively, the cytotoxicity of **683** and **684** in HepG2/ADM, with IC_{50} values of 21.06 ± 1.25 μM and 23.65 ± 1.71 μM, differed significantly from that in HepG2 cells [60]. The in vitro cytotoxic activities of the eucalyptals A-C (**705–707**) were evaluated against HL-60 (human leukemia) and A-549 (human lung adenocarcinoma) tumor cell lines, **705–717** showed selective activity against HL-60 with IC_{50} values of 1.7, 6.8, and 17 μM, respectively [61]. Eucalrobusone C (**709**) showed potent cytotoxicity against HepG2, MCF-7, and U2OS cell lines, with IC_{50} values of 7.40, 8.99, and 8.50 μM, respectively [62]. Terpecurcumin Q (**777**) is toxic in human breast cancer cells (MCF-7) having IC_{50} of 3.9 μM [63]. Fischernolide D (**806**) is not only cytotoxic but can induce the apoptosis of MCF-7 cells and Bel-7402 cells via caspase activation [64]. Cryptotrione (**814**) was reported to be cytotoxic with an IC_{50} value of 6.44 μM [65]. Caesalpinnone A (**824**) is relatively a potent cytotoxic compound toward HL-60, SMMC-7721, A-549, MCF7, and SW-480 human cancer cell lines with IC_{50} values in the range of 0.54–0.87 μM [66]. (±)-Sativamides A (**894**) and B (**895**) were reported to reduced the endoplasmic reticulum (ER) stress induced cytotoxicity [67]. While (+)-pinnatifidaone C (**901**) is toxic against Hep3B cells [68]. Since the IC_{50} values of flueggine A (**907**) in MCF-7, MDA-MB-231, and MCF-7/ADR cell lines are bigger than 60 μM, their cytotoxicities are really negligible, although plant natural products except few examples such as taxol usually exert weak to moderate cytotoxicity, flueggine B (**908**) exhibited a significant inhibitory activity on the growth of MCF-7 and MDA-MB-231 cells with IC_{50} values of 135 nM and 147 nM, respectively [69]. Fluevirosines B and C (**916** and **917**) were found to inhibit the splicing of XBP1 mRNA [70]. Serratezomines A and B (**937** and **938**) exhibited cytotoxicity against murine lymphoma (L1210) cells (IC_{50}: 9.7 and 7.2 μg/mL, respectively) and human epidermoid carcinoma KB cells (IC_{50}: > 10 and 5.1 μg/mL, respectively) [71]. The cytotoxic activity of lyconadin A (**940**) is also strong against murine lymphoma L1210 cells with an IC_{50} value of 0.46 μg/mL [72]. Sieboldine A (**942**) was reported to exhibit a potent inhibitory activity against acetylcholinesterase and modest cytotoxicity [73]. Himeradine A (**943**) only exhibited moderate cytotoxicity against murine lymphoma L1210 cells (IC_{50}: 10 μg/mL) [74]. Nankakurine A (**944**) is toxic against human epidermoid carcinoma KB cells (IC_{50}: 3.1 μg/mL) [75]. While bistabercarpamine A (**968**), ervadivamine A (**970**), alasmontamine A (**972**), and tabercorymines A (**973**) and B (**974**) only exert weak to moderate cytotoxicities in the tested human cancer cells [76–79]. Bipleiophylline (**975**) was found to show appreciable in vitro cytotoxicity toward drug-sensitive and vincristine resistant (VJ300) human KB as well as Jurkat cells with respective IC_{50} values of 3.2 μg/mL, 2.0 μg/mL, and 3.7 μg/mL [80]. Cytotoxic assay found that melotenine A (**996**) are toxic against five human cancer cell lines including SK-BR-3breast, SMMC-7721 hepatocellular carcinoma, HL-60 myeloid leukemia, PANC-1 pancreatic cancer, and A-549 lung cancer [81]. Mekongenines A (**1008**) and mekongenine B (**1009**) showed moderate cell growth inhibitory activity against five human cancer cell lines (HL60, SMMC7721, A549, MCF7, and SW480) with IC_{50} values in the range of 7.15–17.05 [82]. In contrast, leucophyllidine (**1017**) showed pronounced in vitro cytotoxicity toward drug sensitive as well as vincristine-resistant (VJ300) human KB cells with respective IC_{50} values of 2.95 μg/mL and 2.92 μg/mL (5.16 μM and 5.10 μM) [83]. Murine lymphoma L1210 cells are sensitive to daphnezomine B (**1041**) with an in vitro IC_{50} value of 0.46 μg/mL, while human epidermoid carcinoma KB cells are not sensitive to daphnezomine B (IC_{50}: 8.5 μg/mL) [84]. Daphnezomines F and G (**1042** and **1043**) were found to exhibit cytotoxicity against murine lymphoma L1210 cells (IC_{50}: 8.4 μg/mL and 5.3 μg/mL, respectively) [85]. However, daphnicyclidins A–H (**1044–1051**) exhibited strong cytotoxicity against murine lymphoma L1210 cells (IC_{50}: 0.8 μg/mL, 0.1 μg/mL, 3.0 μg/mL, 1.7 μg/mL, 0.4 μg/mL, 4.3 μg/mL, 4.2 μg/mL, and 0.5 μg/mL, respectively) and moderate activities toward human epidermoid carcinoma KB cells (IC_{50}: 6.0 μg/mL, 2.6 μg/mL, 7.2 μg/mL, 4.6 μg/mL, 5.2 μg/mL, 7.6 μg/mL, > 10 μg/mL, and 0.9 μg/mL, respectively). These results indicated that murine lymphoma L1210 cells are more sensitive than human epidermoid carcinoma KB cells [86]. In addition to daphnicyclidins A–H, daphnicyclidins J (**1052**) and K (**1053**) are also toxic in murine lymphoma L1210 cells (IC_{50}: 1.9 μg/mL and 4.7 μg/mL, respectively) and human epidermoid carcinoma KB cells (IC_{50}: 2.5 μg/mL and 6.5 μg/mL, respectively) [87]. It was reported that daphcalycine (**1054**) displayed moderate cytotoxicity against human nasopharynx carcinoma KB cells with IC_{50} values of 13 μg/mL [88]. Daphniglaucins A (**1057**) and B (**1058**) were reported to exhibit cytotoxicity against murine lymphoma L1210 cells (IC_{50}: 2.7 μg/mL and 3.9 μg/mL, respectively) and human epidermoid carcinoma KB cells (IC_{50}: 2.0 μg/mL and 10.0 μg/mL, respectively) [89]. Besides, calyciphyllines A (**1059**) and B (**1060**) were also reported to inhibit murine lymphoma L1210 cells (IC_{50}: 2.1 μg/mL and 4.2 μg/mL, respectively) [90]. Calyciphylline G (**1072**) is toxic against murine leukemia L1210 cells with an IC_{50} value of 9 μg/mL [91]. Geleganidine B (**1131**) was found to have weak activity against MCF-7 cells with an IC_{50} value of 38.41 μM, PC-12 cells were used to evaluate the potential of geleganidine C (**1132**) affording an IC_{50} value of 16.10 μM [92]. The cytotoxicities of strynuxline A (**1144**) in five human cancer cell lines (HL60, SMMC7721, A549, MCF7, and SW480) are moderate (corresponding

IC_{50} values of 7.15 μM, 9.40 μM, 17.05 μM, 12.18 μM, and 13.33 μM), The IC_{50} values for strynuxline B (**1153**) in HL60, SMMC7721, A549, MCF7, and SW480 cell lines are in the range of 11.23–18.98 μM [93]. (−)-Macleayin A (**1153**) is toxic against HL-60 cell line with an IC_{50} value of 3.51 μM [94]. Baicalensine A (**1167**) was evaluated for antiproliferative activities against Caco-2 and HL-60 cancer cells and which exerted moderate activities with IC_{50} values of 9.24 ± 0.57 and 10.15 ±0.79 μM, respectively (5-FU was used as the positive control with IC_{50} values of 20.24 ± 0.86 and 3.15 ± 0.39 μM, respectively [95]. Peganumine A (**1220**) only showed moderate cytotoxic activity against MCF-7, PC-3, and HepG2 cells, but exerted selective effects on HL-60 cells with an IC_{50} value of 5.8 μM [96]. Although pegaharmol A (**1227**) was reported to be active against HL60 and A549 cells with IC_{50} values of 39.02 μM and 55.69 μM, respectively., whereas their activities are really negligible [97]. (+)-Cajanusine (**1273**) and (+)-Cajanusine (**1274**) was reported to inhibit HepG2 and HepG2/ADM cells with an IC_{50} value of 16.23 μM and 20.45 μM, respectively [98]. Phyllanthusols A and B (**1296** and **1297**) were found to exert cytotoxicity against BC (EC_{50} at 4.2 μg/mL and 4.0 μg/mL for **1296** and **1297**, respectively) and KB (EC_{50} at 14.6 μg/mL and 8.9 μg/mL for **1296** and **1297**, respectively) cell lines [99]. Sampsonione I (**1306**) has been tested for their cytotoxicity in P388 cell line, where sampsonione I was found to be active with an ED_{50} value of 6.9 μg/mL [100]. Gaudichaudiic acids F–I (**1307–1310**) showed cytotoxicity in P388 cell line with IC_{50} values of 4.6 μg/m, 3.4 μg/m, 2.0 μg/m, and 1.7 μg/mL, respectively, **1308–1310** were also cytotoxic to P388/DOX cell line with IC_{50} values of 3.4 μg/m, 2.2 μg/m, and 1.4 μg/mL, respectively, finally, **1308–1310** respectively exhibited IC_{50} values of 3.8 μg/m, 3.0 μg/m, and 3.0 μg/mL in Messa cell line [101]. Yaoshanenolides A and B (**1325** and **1326**) showed nonselective cytotoxicity to A549 cell lines with IC_{50} values of 5.1–6.6 μM [102]. Although dioscorealide A (**1327**) showed slight activity only against MCF-7, whereas dioscorealide B (**1328**) exhibited the best potency, especially against MCF-7 and COR-L23 [103]. (+)-Rubialatin A (**1329**) was found to be cytotoxic and could inhibit NF-kappa B pathway, while rubialatin B (**1330**) is not only cytotoxic but exerted a synergistic effect with TNF-α on NF-kappa B activation [104]. Abibalsamin A (**1347**) was reported to selectively inhibit the growth of A549 cells (IC_{50}: 22 ± 4 μM), while abibalsamin B (**1348**) only showed moderate cytotoxicity against A549 and DLD-1 cancer cell lines with IC_{50} values of 8.5 μM and 15 μM, respectively [105].

14.2 Anti-inflammatory Activity

(±)-eugenunilones B, D and F (**125**, **31** and **33**) (10 μM) reduced the neutrophil number in inflammatory sites in zebrafish acute inflammatory models which were induced by $CuSO_4$ or tail fin injury [106]. (±)-pinnatanoid A (**35**) showed anti-inflammatory activity against NO production in RAW264.7 macrophage cells, with an IC_{50} value of 79.45 μM [107]. Daphnenoids B and C (**39** and **40**) exhibited potential inhibitory activities on the production of NO against LPS-induced BV2 microglial cells [108]. Tricarabrols A and B (**85** and **86**) both have the ability of reducing NO production in LPS-stimulated RAW264.7 macrophages with IC_{50} values of 2.90 μM and 4.52 μM, respectively, comparable to that of positive control (indomethacin) [109]. In contrast, artesin A (**88**) only exhibited a moderate inhibition on NO production in LPS-induced BV-2 microglial cells with an IC_{50} value of 38.78 μM [110]. Of note, macrocephadiolide A (**102**) showed a potent inhibitory effect on NO production with an IC_{50} value of 0.99 μM, whereas, the potency of macrocephadiolide B (**103**) is relatively weaker (IC_{50}: 6.13 μM) in LPS-stimulated RAW264.7 macrophages [111]. Inulajaponicolide A (**106**) possessed significant inhibitory potency against NO production in LPS-induced RAW264.7 cells with IC_{50} value of 1.9 μM [112]. Nardochinoid C (**119**) was found to could dose-dependently inhibit LPS-induced iNOS and COX-2 expression and increase HO-1 expression at 10 μM [113]. Involucratustone C (**122**) has an IC_{50} value of 7.32 μM in LPS-activated RAW 264.7 macrophages using NO production as an indicator [12]. It was reported that euphopias A–C (**133–135**) could significantly inhibit NLRP3 inflammasome activation and block NLRP3 inflammasome-induced pyroptosis, additionally, a mechanistic study revealed that euphopia B (**134**) could ameliorate mitochondria damage, thereby interrupting NLRP3 inflammasome activation [114]. Euphopia D (**136**) could restrain the maturation of caspase-1 and regulate the secretion of cytokine interleukin-1β (IL-1β), thereby blocking GSDMD-mediated cell pyroptosis [115]. Officinalin A (**212**) was reported to exhibit potent nitric oxide (NO) inhibitory activity with an IC_{50} value of 2.02 μM [116]. Kravanhin B (**249**) was found to have inhibitory activity on NO production in lipopolysaccharide (LPS)-induced RAW264.7 macrophages with an IC_{50} value of 36.2 μM [117]. Amomaxin B (**256**) showed NO inhibition in LPS-activated RAW264.7 cells with an IC_{50} value of 31.33 μM, comparable to that of *N*-monomethyl-*L*-arginine (positive control) at 40.45 μM [118]. Aphanamenes A (**277**) and B (**278**) also showed significant inhibition on NO generation in LPS-activated RAW264.7 cells with IC_{50} values of 9.72 μM and 7.98 μM [119]. Populusene A (**302**) and populusin A (**303**) were found to significantly inhibit the production of TNF-α and IL-6 and the expression of iNOS, COX-2, and p-NF-κ B at 125 nM in RAW264.7 cells [120]. Hapmnioide A (**311**) was found to greatly suppress the expression of IL-6 in LPS-induced cells, suggesting a potential anti-inflammatory activity [121]. 5–50 μM pseudolarenone (**438**) showed a moderate inhibition against NO release without cytotoxicity [122]. Longipetalol A (**487**) exhibited inhibitory effects on nitric oxide production in lipopolysaccharide-induced RAW264.7 macrophages (IC_{50} 11.4 ± 0.6 μM) [123]. A report showed that walrobsin A (**489**) exhibited inhibition on the expression of iNOS and IL-1β. NO production [124]. RAW264.7 cell line was used to evaluate anti-inflammatory property of aphananoid A (**491**), showing inhibition rate of 46.80 ± 1.93% at 50 μM [125]. NO production of thaixylomolin B (**514**) in LPS and IFN-γ-induced RAW264.7 murine macrophages was conducted, giving an IC_{50} value of 84.3 μM for thaixylomolin B [126]. Elodeoidins E and F (**588** and **589**) was also found to have significant anti-inflammatory activity (IC_{50}: 6.06–10.46 μM) [127]. Soniiglucinols A and B (**592** and **593**) are the first reported PPAPs with cyclooxygenase-2 (COX-2) inhibitory activity [128]. Longisglucinol A (**625**) showed remarkable anti-inflammatory activity by inducing macrophage M2 polarization through the suppression of NF-kappa B [129]. In zebrafish acute inflammatory models induced by CuSO4, tail injury, or LPS, (−)-leptosperol A (**691**), (+)-leptosperol A (**692**) and leptosperol B (**708**) (at 2.5 μM) were found to significantly reduced the neutrophil number in inflammatory sites [130]. (±)-Fissisternoids A (**760**) and B (**761**), exhibit anti-inflammatory activity (IC_{50}: 13.2 μM, **760**; 9.8 μM, **761**) via suppression of inflammatory cytokines IL-1β, IL-6, and iNOS [131]. Anti-inflammatory assay revealed that arteannoides B and C (**779** and **780**) could inhibit NO production in LPS-induced RAW 264.7 mouse macrophages with IC_{50} values of 4.5 μM and 2.9 μM, respectively [132]. Sarglaperoxide A (**793**) was found to display 53.6% inhibitory effects on NO production in LPS-induced RAW264.7 cells at 25 μM [133]. Nicotabin A (**795**) was also active on NO production in LPS-activated RAW264.7 macrophages with an IC_{50} of 22.1 μM [134]. Perovsfolin B (**799**) demonstrated an inhibitory effect on IL-1β expression from LPS-stimulated microglial cells with an EC_{50} value of 38.4 μM [135]. (±)-Cajanusflavanol A and B (**825–828**) displayed favorable inhibitory effects on NO production with IC_{50} values of 13.62 μM and 17.52 μM, respectively [136]. Saffloquinoside A (**831**) was reported to have anti-inflammatory activity, [137] and carthorquinosides A (**835**) and B (**836**) was found to exhibit anti-inflammatory activities in LPS-stimulated HUVEC cells by regulating IL-1, IL-6, IL-10, and IFN-γ mRNA expression at concentrations as low as 4 μM [138]. Likewise, using LPS-induced RAW264.7 macrophages, abiesanol A (**858**) was found to possess potent effects on NO production with the inhibition rate of 43.0% at 50 μg/mL [139]. It has been described that dragonins A and B (**862** and **863**) could exhibit inhibition of fMLP/CB-induced superoxide anion and elastase [140]. Likely, using LPS-induced RAW264.7 mouse macrophages, (±)-subaveniumins A (**887**) and (±)-subaveniumins B (**888**) only exhibited moderate inhibition against NO production [141]. Piperhancin A (**904**) exhibited significant inhibitory effects on the iNOS and IL-6 level at 10 μM [142]. Although lycojaponicumins A–C (**947–949**) and lycojaponicumin D (**950**) were reported to inhibit LPS-induced pro-inflammatory factors in BV2 macrophages, their potency are almost negligible in when compared with the positive drug [143, 144]. Alsmaphorazine A (**976**) was found to dose-dependently inhibit NO production in LPS-stimulated J774.1 cells [145]. Rauvomine B (**1007**) and parvifloranine A (**1121**) exhibited inhibition on NO generation in RAW264.7 macrophages with IC_{50} values of 39.6 μM and 23.4 μM, respectively [146, 147]. Gelsecorydines A and C–E (**1133**, **1135–1137**) exhibited were also found to inhibit LPS-induced NO production with IC_{50} values of 14.7 μM, 16.2 μM, 13.7 μM, and 4.2 μM, respectively, while the IC_{50} value of the positive control indomethacin in this study is 21.0 μM [148]. Gelserancines B–D (**1139–1141**) (at 50 μM) significantly reduced the neutrophil number in inflammatory sites in zebrafish acute inflammatory models which were induced by tail fin injury or $CuSO_4$ [149]. Anti-inflammatory assay found that dactyllactone A (**1156**) could inhibit IL-1β and PGE2 production in a dose-dependent manner, and its inhibitory effect (2.4–12 μM) was even superior to positive control (dexamethasone at 20 μM) [150]. Dactylicapnosine A (**1157**) was reported to have anti-inflammatory effect by inhibition of the expression of TNF-α, IL-1β, and PGE2 [151]. (−)-Triligustilide A and (+)-triligustilide B (**1261** and **1262**) exhibited different levels of anti-inflammatory effects against the production of the proinflammatory cytokines (TNF-α, IL-6) and showed inhibition in a dose-dependent manner [152]. Triangeliphthalides C and D (**1267** and **1268**) exhibited the anti-inflammatory effects against the production of the pro-inflammatory cytokines (IL-6) [153]. At 20 μM, multiflorumisides A–D (**1275–1278**) could inhibit NO production in lipopolysaccharide-stimulated RAW264.7 cells with inhibition rates of 22.4%, 16.7%, 22.0%, and 16.8%, respectively, indicating their anti-inflammatory potency is not strong [154]. Anti-inflammatory activity of idesolide (**1324**) and exotine A (**1356**) were also found using LPS-induced NO production in BV2 microglial cells, with the latter having an IC_{50} value of 9.2 μM [155, 156]. Finally, NO production observation in lipopolysaccharide-induced RAW 264.7 macrophages revealed the IC_{50} values were 12.4 μM and 9.1 μM for muralatins A and B (**1358** and **1359**), respectively [157].

14.3 Antiviral Activity

Illihenin A (**37**) showed potent antiviral activity against coxsackievirus B3 with an IC_{50} value of 2.87 μM [158]. Antiviral assay showed that secoheliosphane B (**131**) showed modest activity against HSV-1 with IC_{50} value of 6.41 μM [159]. Euphorbactin (**151**) showed activity against HIV-1 replication with an IC_{50} value of 28.6 μM with selective index (SI: CC_{50}/IC_{50}) 86.1 [160]. Trigochinin C (**174**) was found to inhibit MET tyrosine kinase activity (IC_{50}: 1.95 μM) [161]. Przewalskin B (**206**) only exhibited modest anti-HIV-1 activity with $EC_{50} = 30$ μg/mL [162]. Lancifodilactone F (**355**) also showed anti-HIV activity with $EC_{50} = 20.69$ μg/mL and a selectivity index > 6.62 [163]. Lancifodilactone G (**356**) exerted negligible anti-HIV activity with $EC_{50} = 95.47$ μg/mL and a small SI [164]. Rubriflordilactone B (**358**) was reported to exhibit an EC_{50} value of 9.75 μg/mL (SI = 12.39) against HIV-1 replication with low cytotoxicity [165]. It was reported that sphenadilactone A (**363**) showed anti-HIV-1 activity with EC_{50} of 137.0 μg/mL and exerted minimal cytotoxicity against C8166 cells ($CC_{50} > 200$ μg/mL) [166]. Pre-schisanartanin (**365**) demonstrated anti-HIV-1 activity with an EC_{50} value of 13.81 μg/mL (AZT: $EC_{50} = 2.26$ μg/mL) [167]. While schinarisanlactone A (**367**) showed inhibition (11.8% survival rate) against HIV virus at 10 μM [168]. Schilancitrilactone C (**372**) was found to inhibit HIV-1 activity with an EC_{50} value of 27.54 μg/mL [169]. Kadcoccitone A (**458**) could also inhibit HIV-1 (EC_{50}: 47.91 μg/mL) [170]. Whereas, hyperisampsins A and D (**596** and **599**) showed strong anti-HIV activities with EC_{50} of 2.97 μM and 0.97 μM (SI: 4.80 for hyperisampsin A and hyperisampsin D for 7.70), respectively [171]. (±)-Cleistocaltone A (**650/651**) and cleistocaltone B (**652**) were evaluated for their in vitro antiviral activities against RSV using a cytopathic effect (CPE) reduction assay, it was found that they are more potent than that of ribavirin, as the test compounds showed IC_{50} values of 6.75 μM and 2.81 μM, respectively [172]. It is interesting that (±)-spirotriscoumarin A (**715/716**) and (±)-spirotriscoumarin B (**717/718**) exhibit 3-to-sixfold stronger antiviral activity against influenza virus A (H3N2) (IC_{50}: 3.13 and 2.87 μM, respectively) than their corresponding optically pure enantiomers [173]. Rhodomentosones A and B (**731** and **732**) exhibited good anti-RSV activities with IC_{50} values of 12.50 ± 1.42 and 15.00 ± 1.36 μM, respectively [174]. (±)-Cleistoperlone A (**733/734**) exhibited obvious anti-HSV-1 activity dose-dependently and with an IC_{50} value of 7.50 μM [175]. (+)-Rhodonoid C (**752**) showed weak inhibitory activity against herpes simplex virus type 1 (HSV-1) in vitro with an IC_{50} value of 80.6 μM (SI = 2.7) [176]. Biyouyanagin A (**766**) was also found to inhibit HIV replication in H9 lymphocytes with an EC_{50} value of 0.798 μg/mL and has no influence on H9 cell growth with IC_{50} values of > 25 μg/mL, giving a calculated therapeutic index (TI) value of > 31.3.[177] Przewalskin A (**796**) only showed modest anti-HIV-1 activity with EC_{50} = 41 μg/mL [178]. Houttuynoids A–E (**844–848**) were reported to exhibit inhibitory activities against HSV with respective IC_{50} values of 23.50 μM, 57.71 μM, 50.75 μM, 59.89 μM, and 42.03 μM, the SI of anti-HSV activity of **844–858** was 7.08, 3.15, 10.47, 3.02, and 3.21 respectively [179]. Although trigonoliimine A (**921**) was reported to show modest anti-HIV-1 activity, its EC_{50} value is much small (0.95 μg/mL) [180]. Sophalines B–D (**1019–1021**) were found to significantly inhibit HbsAg secretion by more than 50.0% at non-cytotoxic concentrations of 0.2 or 0.4 mM, suggesting that these natural products are more potent than the positive control (lamivudine) (31.5% at 1.0 mM) [181]. A study showed that flavesines A–F (**1027–1032**) exhibited inhibitory effects on the expression of HBV DNA in HepG2.2.15 cells with IC_{50} values of 44.85 μM, 86.60 μM, 32.11 μM, 74.28 μM, 70.62 μM, and 17.16 μM, respectively, better than that of positive control PFA (foscarnet) with an IC_{50} value of 105.53 μM [182]. Logeracemin A (**1086**) was found to possess significant anti-HIV activity with an EC_{50} of 4.5 μM (SI: 6.2) [183]. Antiviral assay revealed that myriberine A (**1089**) showed inhibitory effects on the hepatitis C virus (HCV) life cycle with a good therapeutic index (CC_{50}/EC_{50}: 12.0) [184]. Clausanisumine (**1123**) exhibited remarkable anti-HIV-1 reverse transcriptase effects showing an EC_{50} value of 18.58 μM [185]. Isatisine A (**1146**) only showed moderate anti-HIV-1 activity with EC_{50} = 37.8 μM and SI = 7.98 [186]. In contrast, pegaharine D (**1224**) exhibited strong antiviral activity against HSV-2 with an IC_{50} value of 2.12 μM [187]. Artapilosines A (**1232**) and B (**1233**) showed notable anti-HIV reverse transcriptase affects, with EC_{50} values of 20.93 and 125.29 μM, respectively [188]. Swerilactones L and M (**1299** and **1300**) were reported to have moderate inhibitory activities against the secretion of hepatitis B virus surface antigen (IC_{50} = 1.47 mM and 1.20 mM, respectively) and hepatitis B virus e antigen (IC_{50} = 0.88 mM and > 2.69 mM, respectively) using Hep G 2.2.15 cell line as an in vitro model [189]. Anti-HBV assay revealed that swerilactone A (**1334**) could inhibit HBsAg and HBeAg secretion with IC_{50} values of 3.66 and 3.58 mM, respectively [190]. While, swerilactones H–K (**1336–1339**) were found to inhibit HBV DNA replication with IC_{50} values ranging from 1.53 μM to 5.34 μM, equivalent to that of lamivudine [191].

14.4 Immunomodulatory Activity

A study showed that pepluacetal (**152**) and pepluanol A (**153**) could inhibit Kv1.3 by 24.9% and 46.0%, respectively, at a concentration of 30 μM, pepluanol B (**153**) shows strong effect against Kv1.3 with an IC_{50} value of 9.50 μM [192]. Scospirosin B (**221**) exhibited selective immunosuppressive activity against the proliferation of T lymphocytes (IC_{50} = 1.42 μM) [193]. Cinnamomols A (**291**) and B (**292**) exhibited significant in vitro immunostimulative activities compared to those of Tα1 and Tp5 at concentrations over 0.3906 μM [194]. Cassiabudanols A (**293**) and B (**294**) were reported to significantly promote the proliferation of ConA-induced murine T cells with enhancement rates up to 39.99% at a concentration of 0.0015 μM and enhance the proliferation of LPS-induced murine B cells with enhancement rates up to 92.36% [195]. Cordifolide A (**297**) was revealed to inhibit LPS-induced upregulation of costimulatory molecules with the suppressing effect on CD40 being more significant than that for CD80 and CD86 [196]. Colquhounoid D (**343**) and 14-epi-colquhounoid D (**344**) showed significant immunosuppressive activity on the cytokine IFN-γ secretion of mouse splenocytes induced by anti-CD3/CD4 monoclonal antibodies, with IC_{50} of 8.38 and 5.79 μM, respectively [197]. Gentianelloids A (**350**) and B (**351**) exhibited a remarkable suppression effect on IFN-γ production stimulated with anti-CD3/CD28 for 48 h (IC_{50} = 5.64 μM and 3.93 μM for **350** and **351**, respectively) [198]. Eurysoloids A (**352**) and B (**353**) could inhibit the IFN-γ production, with IC_{50} = 17.40 μM and 15.94 μM, respectively [199]. In a ConA-induced T lymphocyte proliferation assay, spiroschincarins A and B (**382** and **383**) were disclosed to possess weak inhibition with the inhibitory rate in proliferation of 56.4% and 43.5% at 50 μg/mL, respectively [200]. It is particularly interesting that phainanoid F (**421**) could inhibit T cell proliferation with an IC_{50} value of 2.04 nM (positive control CsA = 14.21 nM) and B cells with IC_{50} value of < 1.60 nM (CsA = 352.87 nM), which is much stronger than CsA [201]. Phainanolide A (**422**) also exhibited pronounced immunosuppressive activities at nM levels. Natural products with potency like phainanoids A and F are really scarce [202]. Colqueleganoid A (**479**) significantly enhanced the production of TNF-α and IL-6 in a dose-dependent manner, and the effect could be observed at a concentration as low as 5 μM and reached 5.6-fold (TNF-α) and 1.6-fold (IL-6), respectively, at 40 μM. Colqueleganoid B (**480**) was found to be only active for the TNF-α production and less potent than **479**, with an observed effect at 10 μM and 1.8-fold increase at 40 μM [203]. Hypaluton A (**558**), showed potential inhibitory activity against lipopolysaccharide (LPS)-stimulated B lymphocyte proliferation, with an IC_{50} value of (6.86 ± 0.72) μM [204]. Norwilsonnol A (**559**) significantly suppressed the proliferation of murine splenocytes, with an IC_{50} value of 1.86 ± 0.16 μM [205]. Wilsonglucinols A and C (**611** and **613**) exhibited moderate immunosuppressive activities against the proliferation of murine splenocytes stimulated by anti-CD3/anti-CD28 monoclonal antibodies with IC_{50} values of 9.7 and 9.3 μM, respectively [206]. It was reported that scopariusic acid (**797**) exhibited moderate immunosuppressive activity against mouse T cell proliferation in vitro with IC_{50} value of 2.6 μM [207]. Cascarinoids B and C (**935** and **936**) only exhibited moderate activities against ConA-induced proliferation of T lymphocyte cells and/or LPS-induced proliferation of B lymphocyte cells with IC_{50} values ranging from 6.14 μM to 16.27 μM [208]. Ophiorrhine A (**1105**) was found to inhibit LPS-induced proliferation of B lymphocyte cells with an IC_{50} value 18.6 μM, whereas ophiorrhine B (**1106**) was revealed to suppress concanavalin A (Con A) induced T-cell proliferation and LPS induced B lymphocyte cell proliferation with IC_{50} values of 13.3 μM and 7.5 μM, respectively [209]. A study also revealed that ivorenolide A (**1272**) could remarkably inhibit both ConA-induced T-cell proliferation and LPS-induced B-cell proliferation [210].

14.5 Antioxidant Activity

A study revealed that acortatarins A (**1213**) and B (**1214**) exerted antioxidant potency, in particular, **1213** could inhibit high-glucose-induced ROS generation in mesangial cells in a time-dependent fashion, indicating its potential in kidney disease [211]. Chunganenol (**1286**) is a potent singlet oxygen quencher but not an effective scavenger of hydroxyl radical and superoxide anion. At 1.42 μM, **1286** could inhibit 50% of singlet oxygen generation, which is much stronger than that of epigallocatechin gallate (EGCG, $IC_{50} = 14.5$ μM). Therefore, **1286** could be considered a selective singlet oxygen quencher, applicable to those singlet oxygen-mediated diseases such as erythropoietic protoporphyria, pellagra, and cataractogenesis [212].

14.6 Neuroprotective Activity

Illisimonin A (**9**) was disclosed to have neuroprotective effects against oxygen–glucose deprivation (OGD)-induced cell injury in SH-SY5Y cells with an EC_{50} value of 27.72 μM [213]. Jiadifenolide (**10**) and jiadifenoxolane A (**11**) was found to strongly promote neurite outgrowth in primary cultured rat cortical neurons at concentrations ranging from 0.01 to 10 μM [214]. Taxodikaloids A (**289**) and B (**290**) were evaluated for their neuroprotective activity against $A\beta 25-35$-induced damage in SH-SY5Y cells and found that taxodikaloids A and B (at 10 μM) could improve the cell viability to 78.15% and 78.88%, respectively [215]. (*m*)-Bicelaphanol A (**321**) was found to exhibit a significant in vitro neuroprotective effect against a hydrogen peroxide-induced cell viability decrease in PC12 cells at 1 μM, while (*p*)-bicelaphanol A (**322**) only showed such effects at 10 μM [216]. Incarnolide A (**390**) exerted negligible neuroprotective effect improving cell viabilities by 12.6% at the concentration of 50 μM [217]. Spirioiridotectals A, B and F (**395**, **396**, and **400**) were reported to exhibit moderate neuroprotective activities against serum-deprivation-induced PC12 cell damage at 10 μM [218]. Norfriedelanes A and B (**443** and **444**) could inhibit acetylcholinesterase with IC_{50} values of 10.3 μM and 28.7 μM, respectively [219]. At 1.0 μM, nototroneside B (**455**) was found to increase the cell viability in the free serum-treated group with cell viability of 79.33% with the negative control group [220]. Hyphenrones A, C, and D (**575**, **577**, and **578**) exhibited significant negative inhibitory activities (approximately–100% for each) at 50 μM [48]. Eucalyptusdimer A (**644**) was found to exhibit AChE inhibitory activities with IC_{50} value of 17.71 μM, which make it an excellent scaffold for further chemical synthesis and pharmacological investigations [62]. At 10 μM, both saffloflavonesides A (**833**) and B (**834**) showed strong inhibitory activity against PC12 cell damage induced by rotenone [221]. Forsythoneosides B and D (**852** and **854**) could not only inhibit PC12 cell damage induced by rotenone but increase cell viability from 53.9% to 70.1% and 67.9% at lower concentration of 0.1 μM, respectively [222]. **878** and **879** exhibited statistically significant neuroprotective effects against $A\beta_{25-35}$-induced SH-SY5Y cell death compared with the $A\beta_{25-35}$-treated group [223]. Flueggeacosine B (**919**) was found to not only induce Neuro-2a differentiation, but also promote the extension of neurites [224]. Lycoperine A (**945**) and lycojapodine A (**946**) only exerted negligible inhibition on acetylcholinesterase with an IC_{50} value of 60.9 μM and 90.3 μM, respectively [225, 226]. Lycoplanine A (**955**) has proven to be a potent Cav3.1 T-type calcium channel (TTCC) inhibitor with an IC_{50} value of 6.06 μM [227]. Studies showed that alstolarine B (**990**), hupercumines A (**1170**) and B (**1171**) showed weak to moderate acetylcholinesterase inhibitory activity with IC_{50} values of 19.3 μM, 41.9 μM, and 92.3 μM, respectively [228, 229]. Camellimidazoles A and B (**1205** and **1206**) exhibited remarkable protection against H_2O_2-induced neuronal damage at the concentration of 1.0 μM [230]. Lycibarbarines A and C (**1229** and **1231**) exhibited neuroprotective activity when evaluated for corticosterone-induced injury by reducing the apoptosis of PC12 cells through the inhibition of caspase-3 and caspase-9 [231]. (+)-(*S*)-scocycamide (**1270**) and (−)-(*R*)-scocycamide (**1271**) inhibited butyrylcholinesterase capacity, suggesting beneficial constituents against Alzheimer's disease [232].

14.7 Antibacterial Activity

Teotihuacanin (**219**) showed potent modulatory activity of multidrug resistance in vinblastine-resistant MCF-7 cancer cell line (reversal fold, $RF_{MCF\text{-}7/Vin+}$ >10,703) at 25 μg/mL [233]. The tautomer of pallamolides D and E (**309** and **310**) inhibited the hyphal morphogenesis, adhesion, and biofilm formation of DSY654 [234]. Dysoxyhainanin A (**428**) showed significant activities against four Gram-positive bacteria, *Staphylococcus aureus* ATCC 25,923 (MIC 12.5 μg/mL), *Staphylococcus epidermidis* ATCC 12,228 (MIC 6.25 μg/mL), *Micrococcus luteus* ATCC 9341 (MIC 12.5 μg/mL), and *Bacillus subtilis* CMCC 63,501 (MIC 6.25 μg/mL) [235]. (±)-Xanthchrysone B (**739/740**) showed moderate antibacterial activity against Gram-positive bacteria with the minimal inhibition concentration (MIC) value of 16 μg/mL [236]. Sarglaperoxide A (**793**) exhibited 64.5% inhibitory effect against *Staphylococcus* aureus at 25 μg/mL [133]. Erchinines A (**966**) and B (**967**) exhibited a potent antibacterial activity against *Bacillus subtilis*, which was roughly comparable to the first line antibiotic cefotaxime and much better than the plant-derived antibacterial drugs, berberine and fibraurtine, additionally, **967** showed equal bioactivity to antifungal drug griseofulvin against T. rubrum. It is noteworthy that the infection induced by the most spread dermatophyte *Trichophyton rubrum* was difficult to cure, which indicates that commonly generating reinfection by other microbials might become life threatening [237]. Antibacterial testing showed that Alstoscholarisine K (**984**) is a significant compound with an MIC value of 18.75 μg/mL against *Escherichia coli*, which is better than that of berberine (37.50 μg/mL), a well-known botanical antibacterial drug [238]. Hopeanolin (**1285**) demonstrated antifungal activity in the MIC value range 0.1–22.5 μg/mL [239]. Brasiliensophyllic acid A and B (**1311** and **1312**) showed strong antibacterial activity against the Gram-positive bacteria *B. cereus* and *S. epidermidis* [240].

14.8 Anti-malarial Activity

Anthecularin (**4**) was reported to exert antitrypanosomal (IC_{50} = 10.1 μg/mL) and antiplasmodial activity (IC_{50} = 23.3 μg/mL) and inhibit two key enzymes of the plasmodial type II fatty acid biosynthesis pathway, *Pf*FabI and *Pf*FabG (IC_{50} values = 14 μg/mL and 28.3 μg/mL, respectively) [241]. The potency for fortunoids A and B (**47** and **48**) in antimalarial activities is quite different with IC_{50} values of 10.2 μM for the former and 0.5 μM for the latter [242]. Trichloranoid A (**67**) exhibited a moderate inhibitory growth effect with IC_{50} ranges of 2.50–5.00 μM [243]. At the concentration of 40 μM, 40% inhibition of malaria parasite *P. falciparum* could be reached by walsuronoid A (**488**) [244]. It is also interesting that tomentosone A (**725**) could inhibit the growth of chloroquine-resistant and -sensitive strains of the malaria parasite *P. falciparum*, with IC_{50} values of 1.49 μM and 1.0 μM, respectively [245]. Of note, hydrangenone (**813**) was found to show in vitro antiplasmodial activity with an IC_{50} value of 1.4 μM against *P. falciparum* [246]. It was found that myristicyclin A (**849**) exhibited IC_{50} values of 35 μM, 43 μM, and 54 μM, respectively, against ring, trophozoite, and schizont stages of *Plasmodium falciparum* [247]. Flinderoles A–C (**1109–1111**) were found to have selective antimalarial activities with IC_{50} values between 0.15–1.42 μM [248]. Mbandakamine A (**1124**) exhibited significant in vitro antiplasmodial activities against the chloroquine-sensitive *Plasmodium falciparum* strain NF54. Moreover, its diacetate salt (IC_{50} = 0.043 μM) is even more effective than the free base (IC_{50} = 0.13 μM) [249]. It is rather exciting that spiromban-dakamines A1 (**1126**) and A2 (**1127**) were found to possess very good activities against both the chloroquineresistant *P. falciparum* K1 strain (IC_{50} = 7 nM and 94 nM, respectively) and the nonresistant NF54 strain (IC_{50} = 40 nM and 226 nM, respectively) [250]. Another exciting thing is that cyclombandakamines A1 (**1128**) and A2 (**1129**) were also disclosed to be strong against *P. falciparum* NF54 with IC_{50} values of 0.043 and 0.055 μM, respectively. Besides, **1129** was found to be very active against *Trypanosoma brucei rhodesiense* with an IC_{50} of 0.01 μM (SI > 1000), making it a promising candidate for further investigations [251]. Cassiarin A

(**1159**) was also revealed to show promising in vitro antiplasmodial activity against *P. falciparum* ($IC_{50} = 0.005$ μg/mL) [252]. Finally, a study revealed that spiroaxillarone A (**1331**) is active against resistant *P. falciparum* with an IC_{50} value of 2.32 μM [253].

References

1. Wu QX, Shi YP, Yang L. Unusual sesquiterpene lactones from *Ligularia virgaurea* spp. *oligocephala*. Org Lett. 2004;6:2313–6.
2. Xi FM, Ma SG, Liu YB, Li L, Yu SS. Artaboterpenoids A and B, bisabolene-derived sesquiterpenoids from *Artabotrys hexapetalus*. Org Lett. 2016;18:3374–7.
3. Chi J, Xu W, Wei S, Wang X, Li J, Gao H, Kong L, Luo J. Chlotrichenes A and B, two lindenane sesquiterpene dimers with highly fused carbon skeletons from *Chloranthus holostegius*. Org Lett. 2019;21:789–92.
4. Yan H, Qin XJ, Li X H, Yu Q, Ni W, He L, Liu HY. Japonicones A-C: three lindenane sesquiterpenoid dimers with a 12-membered ring core from Chloranthus japonicus. Tetrahedron Lett 2019;60:713–717
5. Zhang DY, Wang XX, Wang YN, Wang M, Zhuang PY, Jin Y, Liu H. Nine sesquiterpenoid dimers with four unprecedented types of carbon skeleton from *Chloranthus henryi* var. *hupehensis*. Org Chem Front. 2021;8:4374–86.
6. Wu JW, Tang CP, Chen L, Qiao Y, Geng MY, Ye Y. Dicarabrones A and B, a pair of new epimers dimerized from sesquiterpene lactones via a [3 + 2] cycloaddition from *Carpesium abrotanoides*. Org Lett. 2015;17:1656–9.
7. Yang YX, Shan L, Liu QX, Shen YH, Zhang JP, Ye J, Xu XK, Li HL, Zhang WD. Carpedilactones A–D, four new isomeric sesquiterpene lactone dimers with potent cytotoxicity from *Carpesium faberi*. Org Lett. 2014;16:4216–9.
8. Wang Y, Shen YH, Jin HZ, Fu JJ, Hu XJ, Qin JJ, Liu JH, Chen M, Yan SK, Zhang WD. Ainsliatrimers A and B, the first two guaianolide trimers from *Ainsliaea fulvioides*. Org Lett. 2008;10:5517–20.
9. Zhang R, Tang C, Liu HC, Ren Y, Xu C, Ke CQ, Yao S, Huang X, Ye Y. Ainsliatriolides A and B, two guaianolide trimers from *Ainsliaea fragrans* and their cytotoxic activities. J Org Chem. 2018;83:14175–80.
10. Zhang R, Tang C, Liu HC, Ren Y, Ke CQ, Yao S, Cai Y, Zhang N, Ye Y. Tetramerized sesquiterpenoid ainsliatetramers A and B from *Ainsliaea fragrans* and their cytotoxic activities. Org Lett. 2019;21:8211–4.
11. Ding N, Wang JY, Liu J, Zhu YJ, Hou SR, Zhao HM, Yang YS, Chen XB, Hu LH, Wang XC. Cytotoxic guaianolide sesquiterpenoids from *Ainsliaea fragrans*. J Nat Prod. 2021;84:2568–74.
12. Li QM, Luo JG, Zhang YM, Li ZR, Wang XB, Yang MH, Luo J, Sun HB, Chen YJ, Kong LY. Involucratustones A–C: unprecedented sesquiterpene dimers containing multiple contiguous quaternary carbons from *Stahlianthus involucratus*. Chem Eur J. 2015;21:13206–9.
13. Fei DQ, Dong LL, Qi FM, Fan GX, Li HH, Li ZY, Zhang ZX. Euphorikanin A, a diterpenoid lactone with a fused 5/6/7/3 ring system from Euphorbia kansui. Org Lett 2016;18:2844–2847.
14. Liu SN, Huang D, Morris Natschke SL, Ma H, Liu Z, Seeram NP, Xu J, Lee KH, Gu Q. Euphomilones A and B, ent-rosane diterpenoids with 7/5/6 and 5/7/6 skeletons from *Euphorbia milii*. Org Lett. 2016;18:6132–6135.
15. Fan YY, Shi SQ, Deng GZ, Liu HC, Xu CH, Ding J, Wang GW, Yue JM. Crokonoids A–C, a highly rearranged and dual-bridged spiro diterpenoid and two other diterpenoids from *Croton kongensis*. Org Lett. 2020;22:929–33.
16. Li W, Tang YQ, Sang J, Fan RZ, Tang GH, Yin, S. Jatrofolianes A and B: two highly modified lathyrane diterpenoids from Jatropha gossypiifolia. Org Lett 2020;22:106–109.

17. He Jun, Xu JK, Guo LB, Xia CY, Lian WW, Tian HY, Zhang J, Shi YX, Zhang WK. Fischdiabietane A, an antitumoral diterpenoid dimer featuring an unprecedented carbon skeleton from *Euphorbia fischeriana. J Org Chem.* 2021;**86**:5894−900.
18. Gui MY, Aoyagi Y, Jin YR, Li XW, Hasuda T, Takeya K. Excisanin H, a novel cytotoxic 14,20-epoxy-*ent*-kaurene diterpenoid, and three new *ent*-kaurene diterpenoids from *Rabdosia excisa*. J Nat Prod. 2004;67:373–6.
19. Li SH, Wang J, Niu XM, Shen YH, Zhang HJ, Sun HD, Li ML, Tian QE, Lu Y, Cao P, Zheng QT. Maoecrystal V, cytotoxic diterpenoidwith a novel C_{19} skeleton from *Isodon eriocalyx* (Dunn.) Hara. Org Lett. 2004;6:4327–4330.
20. Han QB, Cheung S, Tai J, Qiao CF, Song JZ, Tso TF, Sun HD, Xu HX. Maoecrystal Z, a cytotoxic diterpene from *Isodon eriocalyx* with a unique skeleton. Org Lett. 2006;8:4727–30.
21. Huang SX, Xiao WL, Li LM, Li SH, Zhou Y, Ding LS, Lou LG, Sun HD. Bisrubescensins A–C: three new dimeric *e nt*-kauranoids isolated from *Isodon rubescens*. Org Lett. 2006;8:1157–60.
22. Huang B, Xiao CJ, Huang ZY, Tian XY, Cheng X, Dong X, Jiang B. Hispidanins A-D: four new asymmetric dimeric diterpenoids from the rhizomes of *Isodon hispida*. Org Lett. 2014;16:3552–5.
23. Aoyagi Y, Yamazaki A, Nakatsugawa C, Fukaya H, Takeya K, Kawauchi S, Izumi H. Salvileucalin B, a novel diterpenoid with an unprecedented rearranged neoclerodane skeleton from *Salvia Leucantha* Cav. Org Lett. 2008;10:4429–32.
24. Zheng GJ, Kadir A, Zheng XF, Jin PF, Liu JJ. Maiwulanjiang M, Yao G, Aisa HA. Spirodesertols A and B, two highly modified spirocyclic diterpenoids with an unprecedented 6-isopropyl-3*H*-spiro[benzofuran-2,1′-cyclohexane] motif from *Salvia deserta*. Org Chem Front. 2020;7:3137–45.
25. Wang SJ, Lin S, Zhu CG, Yang YC, Li S, Zhang JJ, Chen XG, Shi JG. Highly acylated diterpenoids with a new 3,4-secograyanane skeleton from the flower buds of *Rhododendron* molle. Org Lett. 2010;12:1560–3.
26. Zhao Q, Gao JJ, Qin XJ, Hao XJ, He HP, Liu HY. Hedychins A and B, 6,7-dinorlabdane diterpenoids with a peroxide bridge from *Hedychium Forrestii*. Org Lett. 2018;20:704–7.
27. Wu XD, Ding LF, Chen B, Li XN, Peng LY, Zhao QS. Cunlanceloic acids A-D: unprecedented labdane diterpenoid dimers with AChE inhibitory and cytotoxic activities from Cunninghamia lanceolata. Org Chem Front. 2021;8:5777–84.
28. Perry NB, Burgess EJ, Baek SH, Weavers RT. The first atisane diterpenoids from a liverwort: polyols from *Lepidolaena clavigera*. Org Lett. 2001;3:4243–5.
29. Wang LN, Zhang JZ, Li X, Wang XN, Xie CF, Zhou JC, Lou HX. Pallambins A and B, unprecedented hexacyclic 19-nor-secolabdane diterpenoids from the chinese liverwort *Pallavicinia ambigua*. Org Lett 2012;14:1102–1105.
30. Zhang XH, Yang Y, Liu JJ, Shen L, Shi Z, Wu J. Tagalide A and tagalol A, naturally occurring 5/6/6/6- and 5/6/6-fused cyclic dolabrane-type diterpenes: a new insight into the anti-breast cancer activity of the dolabrane scaffold. Org Chem Front. 2018;5:1176–83.
31. He F, Pu JX, Huang SX, Wang YY, Xiao WL, Li LM, Liu JP, Zhang HB, Li Y, Sun HD. Schinalactone A, a new cytotoxic triterpenoid from *Schisandra sphenanthera*. Org Lett. 2010;12:1208–11.
32. Xue YB, Yang JH, Li XN, Du X, Pu JX, Xiao WL, Su J, Zhao W, Li Y, Sun HD. Henrischinins A−C: three new triterpenoids from *Schisandra henryi*. Org Lett 2011;13:1564–1567.
33. Meng, FY, Sun JX, Li X, Yu HY, Li SM, Ruan HL. Schiglautone A, a new tricyclic triterpenoid with a unique 6/7/9-fused skeleton from the stems of *Schisandra glaucescens*. Org Lett 2011;13:1502–1505.
34. Pu JX, Xiao WL, Lu Y, Li RT, Li HM, Zhang L, Huang SX, Li X, Zhao QS, Zheng QT, Sun HD. Kadlongilactones A and B, two novel triterpene dilactones from *Kadsura longipedunculata* possessing a unique skeleton. Org Lett. 2005;7:5079–82.
35. Yuan FY, Xu F, Fan RZ, Li W, Huang D, Tang GH, Yuan T, Gan LS, Yin S. Structural elucidation of three 9,11-seco tetracyclic triterpenoids enables the structural revision of euphorol J. J Org Chem 2021;86:7588−7593.

36. Li B, Kong DY, Shen YH, Yuan H, Yue RC, He YR, Lu L, Shan L, Li HL, Ye J, Yang XW, Su J, Liu RH, Zhang WD. Pseudolaridimers A and B, hetero-cycloartane–labdane diels-alder adducts from the cone of *Pseudolarix amabilis*. Org Lett. 2012;14:5432–5.
37. Dan C, Zhou Y, Ye D, Peng S, Ding L, Gross ML, Qiu SX. Cimicifugadine from *Cimicifuga foetida*, a new class of triterpene alklaoids with novel reactivity. Org Lett. 2007;9:1813–6.
38. Wang CQ, Wang L, Fan CL, Zhang DM, Huang XJ, Jiang RW, Bai LL, Shi JM, Wang Y, Ye WC. Ilelic acids a and b, two unusual triterpenes with a seven-membered ring from *Ilex latifolia*. Org Lett. 2012;14:4102–5.
39. Liu MT, Gan ML, Lin S, Zhang YL, Zi JC, Song WX, Fan XN, Liu Y, Yang YC, Shi JG. Machilusides A and B: structurally unprecedented homocucurbitane glycosides from the stem bark of *Machilus Yaoshansis*. Org Lett. 2011;13:2856–9.
40. Wang XN, Yin S, Fan CQ, Wang FD, Lin LP, Ding J, Yue JM. Turrapubesins A and B, first examples of halogenated and maleimide-bearing limonoids in nature from *Turraea pubescens*. Org Lett 2006;8:3845–3848.
41. Yang SP, Chen HD, Liao SG, Xie BJ, Miao ZH, Yue JM. Aphanamolide A, a new limonoid from *Aphanamixis polystachya*. Org Lett 2011;13:150–153.
42. Yin S, Fan CQ, Wang XN, Lin LP, Ding J, Yue JM. Xylogranatins A–D: novel tetranortriterpenoids with an unusual 9,10-*seco* scaffold from marine mangrove *Xylocarpus granatum*. Org Lett. 2006;8:4935–8.
43. Zhao JQ, Wang YM, He HP, Li SH, Li XN, Yang CR, Wang D, Zhu HT, Xu M, Zhang YJ. Two new highly oxygenated and rearranged limonoids from *Phyllanthus cochinchinensis*. Org Lett. 2013;15:2414–7.
44. Lv C, Yan XH, Tu Q, Di YT, Yuan CM, Fang X, Ben-David Y, Xia L, Gong JX, Shen YM, Yang Z, Hao XJ. Isolation and asymmetric total synthesis of perforanoid A. Angew Chem Int Ed 2016;55:7539–7543.
45. Yang BY, Huang JZ, Lin S, Tong QY, Yao ZY, Li FL, Ye Y, Hu ZX, Zhang YH. Hyperbeanone A, a 5,6-seco-spirocyclic polycyclic polyprenylated acylphloroglucinol derivative with an unprecedented skeleton from *Hypericum beanii*. Org Chem Front. 2021;8:6411–8.
46. Zheng D, Zhang H, Zheng CW, Lao YZ, Xu DQ, Xiao LB, Xu HX. Garciyunnanimines A–C, novel cytotoxic polycyclic polyprenylated acylphloroglucinol imines from *Garcinia yunnanensis*. Org Chem Front. 2017;4:2102–8.
47. Xu WJ, Tang PF, Lu WJ, Zhang YQ, Wang XB, Zhang H, Luo J, Kong LY. Hyperberins A and B, type B polycyclic polyprenylated acylphloroglucinols with bicyclo[5.3.1]hendecane core from *Hypericum beanii*. Org Lett. 2019;21:8558–62.
48. Yang XW, Ding Y, Zhang JJ, Liu X, Yang LX, Li XN, Ferreira D, Walker LA, Xu G. New acylphloroglucinol derivatives with diverse architectures from *Hypericum henryi*. Org Lett. 2014;16:2434–7.
49. Zhu HC, Chen CM, Yang J, Li DY, Zhang JW, Guo Y, Wang JP, Luo ZW, Xue YB, Zhang YH. Hyperhexanone A, a crucial intermediate from bicyclo[3.3.1]—to cyclohexanone monocyclic-polycyclic polyprenylated acylphloroglucinols. Tetrahedron. 2016;72:4655–9.
50. Hu YL, Hu K, Kong LM, Xia F, Yang XW, Xu G. Norascyronones A and B, 2,3,4-*nor*-polycyclic polyprenylated acylphloroglucinols from *Hypericum ascyron*. Org Lett. 2019;21:1007–10.
51. Liao Y, Liu X, Yang J, Lao YZ, Yang XW, Li XN, Zhang JJ, Ding Z, Xu HX, Xu G. Hypersubones A and B, new polycyclic acylphloroglucinols with intriguing adamantane type cores from *Hypericum subsessile*. Org Lett. 2015;17:1172–5.
52. Tian WJ, Qiu YQ, Yao XJ, Chen HF, Dai Y, Zhang XK, Yao XS. Dioxasampsones A and B, two polycyclic polyprenylated acylphloroglucinols with unusual epoxy-ring-fused skeleton from *Hypericum sampsonii*. Org Lett. 2014;16:6346–9.
53. Zhang JJ, Yang J, Liao Y, Yang XW, Ma JZ, Xiao QL, Yang LX, Xu G. Hyperuralones A and B, new acylphloroglucinol derivatives with intricately caged cores from *Hypericum uralum*. Org Lett. 2014;16:4912–5.
54. Fan YM, Yi P, Li Y, Yan C, Huang T, Gu W, Ma Y, Huang LJ, Zhang JX, Yang CL, Li Y, Yuan CM, Hao XJ. Two unusual polycyclic polyprenylated acylphloroglucinols, including a pair of enantiomers from *Garcinia multiflora*. Org Lett. 2015;17:2066–9.

55. Jian YQ, Huang XJ, Zhang DM, Jiang RW, Chen MF, Zhao BX, Wang Y, Ye WC. Guapsidial A and guadials B and C: three new meroterpenoids with unusual skeletons from the leaves of *Psidium guajava*. Chem Eur J 2015;21:9022–9027.
56. Liu C, Ang S, Huang XJ, Tian HY, Deng YY, Zhang DM, Wang Y, Ye WC, Wang L. Meroterpenoids with new skeletons from *Myrtus communis* and structure revision of myrtucommulone K. Org Lett. 2016;18:4004–7.
57. Yang XW, Li YP, Su J, Ma WG, Xu G. Hyperjapones A–E, terpenoid polymethylated acylphloroglucinols from *Hypericum japonicum*. Org Lett 2016;18:1876–1879.
58. Hou JQ, Guo C, Zhao JJ, He QW, Zhang BB, Wang H. Frutescone A–G, tasmanone-based meroterpenoids from the aerial parts of *Baeckea frutescens*. J Org Chem. 2017;82:1448–57.
59. Xu J, Zhu HL, Zhang J, Liu WY, Luo JG, Pan K, Cao WY, Bi QR, Feng F, Qu W. Littordials A–E, novel formyl-phloroglucinol-β-caryophyllene meroterpenoids from the leaves of *Psidium littorale*. Org Chem Front. 2019;6:1667–73.
60. Shao M, Wang Y, Jian YQ, Huang XJ, Zhang DM, Tang QF, Jiang RW, Sun XG, Lv ZP, Zhang XQ, Ye WC. Guadial A and psiguadials C and D, three unusual meroterpenoids from *Psidium guajava*. Org. Lett. 2012;14:5262–5265.
61. Yin S, Xue JJ, Fan CQ, Miao ZH, Ding J, Yue JM. Eucalyptals A–C with a new skeleton isolated from *Eucalyptus globulus*. Org Lett. 2007;9:5549–52.
62. Qin XJ, Feng MY, Liu H, Ni W, Rauwolf T, Porco JA, Yan H, He L, Liu HY. Eucalyptusdimers A–C, dimeric phloroglucinol–phellandrene meroterpenoids from *Eucalyptus robusta*. Org Lett. 2018;20:5066–70.
63. Lin XH, Ji S, Qiao X, Hu HB, Chen N, Dong YH, Huang Y, Guo D, Tu PF, Ye M. Density functional theory calculations in stereochemical determination of terpecurcumins J–W, cytotoxic terpene-conjugated curcuminoids from *Curcuma longa* L. J Org Chem. 2013;78:11835–48.
64. Zhang J, He J, Cheng YC, Zhang PC, Yan Y, Zhang HJ, Zhang WK, Xu JK. Fischernolides A–D, four novel diterpene-based meroterpenoid scaffolds with antitumor activities from *Euphorbia fischeriana*. Org Chem Front. 2019;6:2312–8.
65. Chen CC, Wu JH, Yang NS, Chang JY, Kuo CC, Wang SY, Kuo YH. Cytotoxic C_{35} terpenoid cryptotrione from the bark of *Cryptomeria japonica*. Org Lett. 2010;12:2786–9.
66. Zhang LJ, Bi DW, Hu J, Mu WH, Li YP, Xia GH, Yang L, Li XN, Liang XS, Wang LQ. Four hybrid flavan–chalcones, caesalpinnone a possessing a 10,11-dioxatricyclic [5.3.3.$0^{1,6}$]tridecane-bridged system and caesalpinflavans A–C from *Caesalpinia enneaphylla*. Org Lett 2017;19:4315–4318.
67. Zhu GY, Yang J, Yao XJ, Yang X, Fu J, Liu X, Bai LP, Liu L, Jiang ZH. (±)-Sativamides A and B, two pairs of racemic nor-lignanamide enantiomers from the fruits of *Cannabis sativa*. J Org Chem. 2018;83:2376–81.
68. Guo R, Zhao P, Yu X, Yao G, Lin B, Huang X, Song S. (±)-Pinnatifidaones A–D, four pairs of highly modified neolignan enantiomers with a rare spirocyclohexenone skeleton from *Crataegus pinnatifida*. Org Chem Front. 2021;8:953–60.
69. Zhao BX, Wang Y, Zhang DM, Jiang RW, Wang GC, Shi JM, Huang XJ, Chen WM, Che CT, Ye WC. Flueggines A and B, two new dimeric indolizidine alkaloids from *Flueggea virosa*. Org Lett. 2011;13:3888–91.
70. Zhang H, Zhang CR, Zhu KK, Gao AH, Luo C, Li J, Yue JM. Fluevirosines A–C: a biogenesis inspired example in the discovery of new bioactive scaffolds from *Flueggea virosa*. Org Lett. 2013;15:120–3.
71. Morita H, Arisaka M, Yoshida N, Kobayashi J. Serratezomines A–C, new alkaloids from *Lycopodium serratum* Var. *s Erratum*. J Org Chem. 2000;65:6241–5.
72. Kobayashi J, Hirasawa Y, Yoshida N, Morita H, Lyconadin A. A novel alkaloid from *Lycopodium complanatum*. J Org Chem. 2001;66:5901–4.
73. Hirasawa Y, Morita H, Shiro M, Kobayashi J, Sieboldine A. A novel tetracyclic alkaloid from *Lycopodium sieboldii*, inhibiting acetylcholinesterase. Org Lett. 2003;5:3991–3.
74. Morita H, Hirasawa Y, Kobayashi J, Himeradine A. A novel $C_{27}N_3$-type alkaloid from *Lycopodium chinense*. J Org Chem. 2003;68:4563–6.

75. Hirasawa Y, Morita H, Kobayashi J, Nankakurine A. A novel $C_{16}N_2$-type alkaloid from *Lycopodium hamiltonii*. Org Lett. 2004;6:3389–91.
76. Ma K, Wang JS, Luo J, Yang MH, Yao HQ, Sun HB, Kong LY. Bistabercarpamines A and B, first vobasinyl-chippiine-type bisindole alkaloid from *Tabernaemontana corymbosa*. Tetrahedron Lett. 2014;55:101–4.
77. Liu ZW, Zhang J, Li ST, Liu MQ, Huang XJ, Ao YL, Fan CL, Zhang DM, Zhang QW, Ye WC, Zhang XQ. Ervadivamines A and B, two unusual trimeric monoterpenoid indole alkaloids from *Ervatamia divaricata*. J Org Chem. 2018;83:10613–8.
78. Hirasawa Y, Miyama S, Hosoya T, Koyama K, Rahman A, Kusumawati I, Zaini NC, Morita H. Alasmontamine A, a first tetrakis monoterpene indole alkaloid from *Tabernaemontana elegans*. Org Lett. 2009;11:5718–21.
79. Yuan YX, Zhang Y, Guo LL, Wang YH, Goto M, Morris Natschke SL, Lee KH, Hao XJ. Tabercorymines A and B, two vobasinyl–ibogan-type bisindole alkaloids from *Tabernaemontana corymbosa*. Org Lett. 2017;19:4964–7.
80. Kam TS, Tan SJ, Ng SW, Komiyama K. Bipleiophylline, an unprecedented cytotoxic bisindole alkaloid constituted from the bridging of two indole moieties by an aromatic spacer unit. Org Lett. 2008;10:3749–52.
81. Feng T, Li Y, Liu YP, Cai XH, Wang YY, Luo XD. Melotenine A, a cytotoxic monoterpenoid indole alkaloid from *Melodinus tenuicaudatus*. Org Lett. 2010;12:968–71.
82. Fu Y, He H, Di Y, Li S, Zhang Y, Hao X. Mekongenines A and B, two new alkaloids from *Bousigonia mekongensis*. Tetrahedron Lett. 2012;53:3642–6.
83. Gan CY, Robinson WT, Etoh T, Hayashi M, Komiyama K, Kam TS. Leucophyllidine, a cytotoxic bisindole alkaloid constituted from the union of an eburnan and a new vinylquinoline alkaloid. Org Lett. 2009;11:3962–5.
84. Morita H, Yoshida N, Kobayashi J. Daphnezomines A and B, novel alkaloids with an aza-adamantane core from *Daphniphyllum humile*. J Org Chem. 1999;64:7208–12.
85. Morita H, Yoshida N, Kobayashi J. Daphnezomines F and G: novel alkaloids with 1-azabicyclo[5.2.2]undecane moiety from Daphniphyllum humile. J Org Chem. 2000;65:3558–62.
86. Kobayashi J, Inaba Y, Shiro M, Yoshida N, Morita H. Daphnicyclidins A–H, novel hexa- or pentacyclic alkaloids from two species of *Daphniphyllum*. J Am Chem Soc. 2001;123:11402–8.
87. Morita H, Yoshida N, Kobayashi J. Daphnicyclidins J and K, unique polycyclic alkaloids from *Daphniphyllum humile*. J Org Chem. 2002;67:2278–82.
88. Jossang A, Bitar HE, Pham VC, Sévenet T. Daphcalycine, a novel heptacycle fused ring system alkaloid from *Daphniphyllum calycinum*. J Org Chem. 2003;68:300–4.
89. Kobayashi J, Takatsu H, Shen YC, Morita H. Daphniglaucins A and B, novel polycyclic quaternary alkaloids from *Daphniphyllum glaucescens*. Org Lett. 2003;5:1733–6.
90. Morita H, Kobayashi J. Calyciphyllines A and B, two novel hexacyclic alkaloids from *Daphniphyllum calycinum*. Org Lett. 2003;5:2895–8.
91. Saito S, Kubota T, Kobayashi J, Calyciphylline G. A novel alkaloid with an unprecedented fused-hexacyclic skeleton from *Daphniphyllum calycinum*. Tetrahedron Lett. 2007;48:5693–5.
92. Zhang W, Huang XJ, Zhang SY, Zhang DM, Jiang RW, Hu JY, Zhang XQ, Wang L, Ye WC. Geleganidines A–C, unusual monoterpenoid indole alkaloids from *Gelsemium elegans*. J Nat Prod 2015;78:2036–2044.
93. Fu YH, Zhang Y, He HP, Hou L, Di YT, Li SL, Luo XD, Hao XJ. Strynuxlines A and B, alkaloids with an unprecedented carbon skeleton from *Strychnos nux-Vomica*. J Nat Prod. 2012;75:1987–90.
94. Sai CM, Li DH, Xue CM, Wang KB, Hu P, Pei YH, Bai J, Jing YK, Li ZL, Hua HM. Two pairs of enantiomeric alkaloid dimers from *Macleaya cordata*. Org Lett. 2015;17:4102–5.
95. Xue JJ, Jiang CY, Zou DL, Li BJ, Lu JC, Li DH, Lin B, Li ZL, Hua HM. Baicalensines A and B, two isoquinoline alkaloids from the roots of *Thalictrum baicalense*. Org Lett 2020;22:7439–7442.

96. Wang KB, Di YT, Bao Y, Yuan CM, Chen G, Li DH, Bai J, He HP, Hao XJ, Pei YH, Jing YK, Li ZL, Hua HM. Peganumine A, a β-carboline dimer with a new octacyclic scaffold from *Peganum harmala*. Org Lett. 2014;16:4028–31.
97. Li SG, Wang YT, Zhang Q, Wang KB, Xue JJ, Li DH, Jing YK, Lin B, Hua HM. Pegaharmols A–B, axially chiral β-carboline-quinazoline dimers from the roots of *Peganum harmala*. Org Lett. 2020;22:7522–5.
98. Li XL, Zhao BX, Huang XJ, Zhang DM, Jiang RW, Li YJ, Jian YQ, Wang Y, Li YL, Ye WC. (+)- and (−)-Cajanusine, a pair of new enantiomeric stilbene dimers with a new skeleton from the leaves of *Cajanus cajan*. Org Lett. 2014;16:224–7.
99. Vongvanich N, Kittakoop P, Kramyu J, Tanticharoen M, Thebtaranonth Y. Phyllanthusols A and B, cytotoxic norbisabolane glycosides from *Phyllanthus acidus* skeels. J Org Chem. 2000;65:5420–3.
100. Hu LH, Sim KY. Cytotoxic polyprenylated benzoylphloroglucinol derivatives with an unusual adamantyl skeleton from *Hypericum sampsonii* (Guttiferae). Org Lett. 1999;1:879–82.
101. Xu YJ, Yip SC, Kosela S, Fitri E, Hana M, Goh SH, Sim KY. Novel cytotoxic, polyprenylated heptacyclic xanthonoids from indonesian *Garcinia gaudichaudii* (Guttiferae). Org Lett. 2000;2:3945–8.
102. Liu MT, Lin S, Gan ML, Chen M, Li L, Wang SJ, Zi JC, Fan XN, Liu Y, Si YK, Yang YC, Chen XG, Shi JG. Yaoshanenolides A and B: new spirolactones from the bark of *Machilus yaoshansis*. Org Lett. 2012;14:1004–7.
103. Itharat A, Plubrukarn A, Kongsaeree P, Bui T, Keawpradub N, Houghton PJ. Dioscorealides and dioscoreanone, novel cytotoxic naphthofuranoxepins, and 1,4-phenanthraquinone from *Dioscorea membranacea* Pierre. Org Lett. 2003;5:2879–82.
104. Zhao SM, Wang Z, Zeng GZ, Song WW, Chen XQ, Li XN, Tan NH. New cytotoxic naphthohydroquinone dimers from *Rubia alata*. Org Lett. 2014;16:5576–9.
105. Lavoie S, Legault J, Gauthier C, Mshvildadze V, Mercier S, Pichette A. Abibalsamins A and B, two new tetraterpenoids from *Abies balsamea* Oleoresin. Org Lett. 2012;14:1504–7.
106. Chen M, Cao JQ, Ang S, Zeng TN, Li NP, Yang TJ, Liu JS, Wu Y, Ye WC, Wang L. Eugenunilones A-H: rearranged sesquiterpenoids from *Eugenia uniflora*. Org Chem Front. 2022;9:667–75.
107. Jiao SG, Su GZ, Zhou XC, Ge FX, Liu CX, Zhang RF, Peng B, Chen S, Huang LQ, Tu PF, Chai XY. Three pairs of enantiomeric sesquiterpenoids from *Syringa pinnatifolia*. J Org Chem 2021;86:7263−7270.
108. Zhao P, Li ZY, Qin SY, Xin BS, Liu YY, Lin B, Yao GD, Huang XX, Song SJ. Three unusual sesquiterpenes with distinctive ring skeletons from *Daphne penicillata* uncovered by molecular networking strategies. J Org Chem. 2021;86:15298–306.
109. Yuan J, Wen X, Ke CQ, Zhang T, Lin L, Yao S, Goodpaster JD, Tang C, Ye Y. Tricarabrols A–C, three anti-inflammatory sesquiterpene lactone trimers featuring a methylene-tethered linkage from *Carpesium faberi*. Org Chem Front. 2020;7:1374–82.
110. Zhou XD, Chai XY, Zeng KW, Zhao MB, Jiang Y, Tu PF. Artesin A, a new cage-shaped dimeric guaianolide from *Artemisia sieversiana*. Tetrahedron Lett. 2015;56:1141–3.
111. Ren YM, Zhou SZ, Zhang T, Qian M, Zhang R, Yao S, Zhu H, Tang C, Lin L, Ye Y. Targeted isolation of two disesquiterpenoid macrocephadiolides A and B from *Ainsliaea macrocephala* using a molecular networking-based dereplication strategy. Org Chem Front. 2020;7:1481–9.
112. Jin Q, Lee JW, Jang H, Lee HL, Kim JG, Wu W, Lee D, Kim EH, Kim Y, Hong JT, Lee MK, Hwang BY. Dimeric- and trimeric sesquiterpenes from the flower of *Inula japonica*. Phytochemistry. 2018;155:107–13.
113. Shen XY, Qin DP, Zhou H, Luo JF, Yao YD, Lio CK, Li HB, Dai Y, Yu Y, Yao XS. Nardochinoids A–C, three dimeric sesquiterpenoids with specific fused-ring skeletons from *Nardostachys chinensis*. Org Lett. 2018;20:5813–6.
114. Shi QQ, Zhang XJ, Wang TT, Wang Q, Sun TT, Amin M, Zhang RH, Li XL, Xiao WL. Euphopias A–C: three rearranged jatrophane diterpenoids with tricyclo[8.3.0.$0^{2,7}$]tridecane and tetracyclo[11.3.0.$0^{2,10}$.0 ^{3,7}]hexadecane cores from *Euphorbia Helioscopia*. *Org Lett* 2020;22:7820–24.

115. Shi QQ, Zhang Y, Wang TT, Xiong F, Zhang RH, Li XL, Ji X, Zhang XJ, Wang WG, Xiao WL. Euphopias D–F from *Euphorbia L.*: quantum chemical calculation-based structure elucidation and their bioactivity of inhibiting NLRP3 inflammasome activation. *Org Chem Front* 2021;8:3041–46.
116. Li LN, Liu XQ, Zhu DR, Chen C, Lin YL, Wang WL, Zhu L, Luo JG, Kong LY. Officinalins A and B, a pair of C_{23} terpenoid epimers with a tetracyclic 6/7/5/5 system from *Salvia officinalis*. Org Chem Front. 2019;6:3369–73.
117. Yin H, Luo JG, Kong LY. Tetracyclic diterpenoids with isomerized isospongian skeleton and labdane diterpenoids from the fruits of *Amomum kravanh*. J Nat Prod. 2013;76:237–42.
118. Yin H, Luo JG, Shan SM, Wang XB, Luo J, Yang MH, Kong LY. Amomaxins A and B, two unprecedented rearranged labdane norditerpenoids with a nine-membered ring from *Amomum maximum*. Org Lett. 2013;15:1572–5.
119. Zhang HJ, Luo J, Shan SM, Wang XB, Luo JG, Yang MH, Kong LY. Aphanamenes A and B, two new acyclic diterpene [4 + 2]-cycloaddition adducts from *Aphanamixis grandifolia*. Org Lett. 2013;15:5512–5.
120. Liu YY, Yan YM, Wang DW, Cheng YX. Populusene A, an anti-inflammatory diterpenoid with a bicyclo[8,4,1]pentadecane scaffold from *Populus euphratica* Resins. Org Lett. 2021;23:8657–61.
121. Zhou JC, Zhang JZ, Li RJ, Liu J, Fan PH, Li Y, Ji M, Dong YW, Yuan HQ, Lou HX. Hapmnioides A–C, rearranged labdane-type diterpenoids from the chinese liverwort *Haplomitrium mnioides*. Org Lett. 2016;18:4274–6.
122. Li B, Kong DY, Shen YH, Fu KL, Yue RC, Han ZZ, Yuan H, Liu QX, Shan L, Li HL, Yang XW, Zhang WD. Pseudolarenone, an unusual nortriterpenoid lactone with a fused 5/11/5/6/5 ring system featuring an unprecedented bicyclo[8.2.1]tridecane core from Pseudolarix amabilis. Chem Commun. 2013;49:1187.
123. Zhang DL, Li M, Han GF, Li SY, Jin DJ, Tang SA. Longipetalol A: a highly modified triterpenoid from *Dichapetalum longipetalum*. J Nat Prod. 2021;84:1556–62.
124. An FL, Sun DM, Li RJ, Zhou MM, Yang MH, Yin Y, Kong LY, Luo J. Walrobsins A and B, two anti-inflammatory limonoids from root barks of *Walsura robusta*. Org Lett. 2017;19:4568–71.
125. Yang BJ, Fan SR, Cai JY, Wang YT, Jing C, Guo JJ, Chen DZ, Hao XJ. Aphananoid A is an anti-inflammatory limonoid with a new 5/6/5 fused ring featuring a C_{24} carbon skeleton from *Aphanamixis polystachya*. J Org Chem. 2020;85:8597–602.
126. Li J, Li MY, Bruhn T, Katele FZ, Xiao Q, Pedpradab P, Wu J, Bringmann G. Thaixylomolins A–C: limonoids featuring two new motifs from the thai *Xylocarpus moluccensis*. Org Lett. 2013;15:3682–5.
127. Li QJ, Tang PF, Zhou X, Lu WJ, Xu WJ, Luo J, Kong LY. Elodeoidins A–H, acylphloroglucinol meroterpenoids possessing diverse rearranged skeletons from *Hypericum elodeoides*. Org Chem Front. 2021;8:1409–14.
128. Xie S, Qi C, Duan Y, Xu Q, Liu Y, Huang Y, Yin X, Sun W, Zhou Y, Zhang Y. Discovery of new polycyclic polyprenylated acylphloroglucinols with diverse architectures as potent cyclooxygenase-2 inhibitors. Org Chem Front. 2020;7:1349–57.
129. Zhang N, Shi Z, Xu Q, Sun W, Gu L, Xie S, Guo Y, Duan Y, Zhang K, Qi C, Zhang Y. Longisglucinols A–C, structurally intriguing polycyclic polyprenylated acylphloroglucinols with anti-inflammatory activity from *Hypericum longistylum*. Org Lett. 2020;22:7926–9.
130. Gu JH, Wang WJ, Chen JZ, Liu JS, Li NP, Cheng MJ, Hu LJ, Li CC, Ye WC, Wang L. Leptosperols A and B, two cinnamoylphloroglucinol–sesquiterpenoid hybrids from *Leptospermum scoparium*: structural elucidation and biomimetic synthesis. Org Lett. 2020;22:1796–800.
131. Xue GM, Zhao CG, Xue JF, Chen H, Zhao ZZ, Si YY, Du K, Zhi YL, Feng WS. Fissisternoids A and B, two 2′,5′-quinodihydrochalcone-based meroterpenoid enantiomers with unusual carbon skeletons from *Fissistigma bracteolatum*. Org Chem Front. 2022;9:190–6.
132. Qin DP, Pan DB, Xiao W, Li HB, Yang B, Yao XJ, Dai Y, Yu Y, Yao XS. Dimeric cadinane sesquiterpenoid derivatives from *Artemisia annua*. Org Lett. 2018;20:453–6.

133. Wang P, Li RJ, Liu RH, Jian KL, Yang MH, Yang L, Kong LY, Luo J. Sarglaperoxides A and B, sesquiterpene–normonoterpene conjugates with a peroxide bridge from the seeds of *Sarcandra glabra*. Org Lett. 2016;18:832–5.
134. Feng T, Li XM, He J, Ai HL, Chen HP, Li XN, Li ZH, Liu JK. Nicotabin A, a sesquiterpenoid derivative from *Nicotiana tabacum*. Org Lett. 2017;19:5201–3.
135. Tanaka N, Niwa K, Kajihara S, Tsuji D, Itoh K, Mamadalieva NZ, Kashiwada Y. C_{28} terpenoids from lamiaceous plant *Perovskia scrophulariifolia*: their structures and anti-neuroinflammatory activity. Org Lett. 2020;22:7667–70.
136. He QF, Wu ZL, Huang XJ, Zhong YL, Li MM, Jiang RW, Li YL, Ye WC, Wang Y. Cajanusflavanols A–C, three pairs of flavonostilbene enantiomers from *Cajanus cajan*. Org Lett. 2018;20:876–9.
137. Jiang JS, He J, Feng ZM, Zhang PC. Two new quinochalcones from the florets of *Carthamus tinctorius*. Org Lett. 2010;12:1196–9.
138. Yue SJ, Qu C, Zhang PX, Tang YP, Jin Y, Jiang JS, Yang YN, Zhang PC, Duan JA. Carthorquinosides A and B, quinochalcone *C*-glycosides with diverse dimeric skeletons from *Carthamus tinctorius*. J Nat Prod. 2016;79:2644–51.
139. Yang XW, Li SM, Feng L, Shen YH, Tian JM, Zeng HW, Liu XH, Shan L, Su J, Zhang C, Zhang WD. Abiesanol A, a novel biflavanol with unique six connective hexacyclic rings isolated from *Abies georgei*. Tetrahedron Lett. 2008;49:3042–4.
140. Kuo PC, Hung HY, Hwang TL, Du WK, Ku HC, Lee EJ, Tai SH, Chen FA, Wu TS. Anti-inflammatory flavan-3-ol-dihydroretrochalcones from *Daemonorops draco*. J Nat Prod. 2017;80:783–9.
141. Lai Y, Liu T, Sa R, Wei X, Xue Y, Wu Z, Luo Z, Xiang M, Zhang Y, Yao G. Neolignans with a rare 2-oxaspiro[4.5]deca-6,9-dien-8-one motif from the stem bark of *Cinnamomum subavenium*. J Nat Prod. 2015;78:1740–4.
142. Yang F, Su BJ, Hu YJ, Liu JL, Li H, Wang YQ, Liao HB, Liang D. Piperhancins A and B, two pairs of antineuroinflammatory cycloneolignane enantiomers from *Piper hancei*. J Org Chem. 2021;86:5284–91.
143. Wang XJ, Zhang GJ, Zhuang PY, Zhang Y, Yu SS, Bao XQ, Zhang D, Yuan YH, Chen NH, Ma S, Qu J, Li Y. Lycojaponicumins A–C, three alkaloids with an unprecedented skeleton from *Lycopodium japonicum*. Org Lett. 2012;14:2614–7.
144. Wang XJ, Liu YB, Li L, Yu SS, Lv HN, Ma SG, Bao XQ, Zhang D, Qu J, Li Y. Lycojaponicumins D and E: two new alkaloids from *Lycopodium japonicum*. Org Lett. 2012;14:5688–91.
145. Koyama K, Hirasawa Y, Nugroho AE, Hosoya T, Hoe TC, Chan KL, Morita H. Alsmaphorazines A and B, novel indole alkaloids from *Alstonia pneumatophora*. Org Lett. 2010;12:4188–91.
146. Zeng J, Zhang DB, Zhou PP, Zhang QL, Zhao L, Chen JJ, Gao K. Rauvomines A and B, two monoterpenoid indole alkaloids from *Rauvolfia vomitoria*. Org Lett. 2017;19:3998–4001.
147. Shou Q, Banbury LK, Renshaw DE, Smith JE, He X, Dowell A, Griesser HJ, Heinrich M, Wohlmuth H. Parvifloranines A and B, two 11-carbon alkaloids from *Geijera parviflora*. J Nat Prod. 2013;76:1384–7.
148. Li NP, Liu M, Huang XJ, Gong XY, Zhang W, Cheng MJ, Ye WC, Wang L. Gelsecorydines A–E, five gelsedine–corynanthe-type bisindole alkaloids from the fruits of *Gelsemium elegans*. J Org Chem. 2018;83:5707–14.
149. Gu JH, Zhang W, Cai WY, Fu XX, Zhou HL, Li NP, Tian HY, Liu JS, Ye WC, Wang L. Gelserancines A–E, monoterpenoid indole alkaloids with unusual skeletons from *Gelsemium elegans*. Org Chem Front 2021;8;1918–1925.
150. Wang B, Yang ZF, Zhao YL, Liu YP, Deng J, Huang WY, Li XN, Wang XH, Luo XD. Anti-inflammatory isoquinoline with bis-*seco*-aporphine skeleton from *Dactylicapnos scandens*. Org Lett. 2018;20:1647–50.
151. Wang B, Zhao YJ, Zhao YL, Liu YP, Li XN, Zhang H, Luo XD. Exploring aporphine as anti-inflammatory and analgesic lead from *Dactylicapnos scandens*. Org Lett. 2020;22:257–60.

152. Zou J, Chen GD, Zhao H, Huang Y, Luo X, Xu W, He RR, Hu D, Yao XS, Gao H. Triligustilides A and B: two pairs of phthalide trimers from *Angelica sinensis* with a complex polycyclic skeleton and their activities. Org Lett. 2018;20:884–7.
153. Zou J, Chen GD, Zhao H, Wang XX, Zhang ZJ, Qu YB, He RR, So KF, Yao XS, Gao H. Triangeliphthalides A–D: bioactive phthalide trimers with new skeletons from *Angelica sinensis* and their production mechanism. Chem Commun. 2019;55:6221–4.
154. Li SG, Huang XJ, Li MM, Liu Q, Liu H, Wang Y, Ye WC. Multiflorumisides A-G, dimeric stilbene glucosides with rare coupling patterns from the roots of *Polygonum multiflorum*. J Nat Prod. 2018;81:254–63.
155. Kim SH, Sung SH, Choi SY, Chung YK, Kim J, Kim YC. Idesolide: a new spiro compound from *Idesia polycarpa*. Org Lett. 2005;7:3275–7.
156. Liu BY, Zhang C, Zeng KW, Li J, Guo XY, Zhao MB, Tu PF, Jiang Y. Exotines A and B, two heterodimers of isopentenyl-substituted indole and coumarin derivatives from *Murraya exotica*. Org Lett. 2015;17:4380–3.
157. Lv HN, Wang S, Zeng KW, Li J, Guo XY, Ferreira D, Zjawiony JK, Tu PF, Jiang Y. Anti-inflammatory coumarin and benzocoumarin derivatives from *Murraya alata*. J Nat Prod. 2015;78:279–85.
158. Yong JY, Li WR, Wang XJ, Su GZ, Li M, Zhang JP, Jia HL, Li YH, Wang RB, Gan ML, Ma SG. Illihenin A: an antiviral sesquiterpenoid with a cage-like tricyclo[6.2.2.$0^{1,5}$]dodecane skeleton from Illicium henryi. J Org Chem. 2021;86:2017–22.
159. Mai ZP, Ni G, Liu YF, Li YH, Li L, Li JY, Yu DQ. Secoheliosphanes A and B and secoheliospholane A, three diterpenoids with unusual *seco* -jatrophane and *seco*-jatropholane skeletons from *Euphorbia helioscopia*. J Org Chem. 2018;83:167–73.
160. Tian Y, Guo QL, Xu WD, Zhu CG, Yang YC, Shi JG. A minor diterpenoid with a new 6/5/7/3 fused-ring skeleton from *Euphorbia micractina*. Org Lett. 2014;16:3950–3.
161. Chen HD, Yang SP, He XF, Ai J, Liu ZK, Liu HB, Geng MY, Yue JM. Trigochinins A–C: three new daphnane-type diterpenes from *Trigonostemon chinensis*. Org Lett. 2010;12:1168–71.
162. Xu G, Hou AJ, Zheng YT, Zhao Y, Li XL, Peng LY, Zhao QS. Przewalskin B, a novel diterpenoid with an unprecedented skeleton from *Salvia przewalskii* Maxim. Org Lett. 2006;9:291–3.
163. Xiao WL, Li RT, Li SH, Li XL, Sun HD, Zheng YT, Wang RR, Lu Y, Wang C, Zheng QT. Lancifodilactone F: a novel nortriterpenoid possessing a unique skeleton from *Schisandra lancifolia* and its anti-hiv activity. Org Lett. 2005;7:1263–6.
164. Xiao WL, Zhu HJ, Shen YH, Li RT, Li SH, Sun HD, Zheng YT, Wang RR, Lu Y, Wang C, Zheng QT. Lancifodilactone G: a unique nortriterpenoid isolated from *Schisandra lancifolia* and its anti-hiv activity. Org Lett. 2005;7:2145–8.
165. Xiao WL, Yang LM, Gong NB, Wu L, Wang RR, Pu JX, Li XL, Huang SX, Zheng YT, Li RT, Lu Y, Zheng QT, Sun HD. Rubriflordilactones A and B, two novel bisnortriterpenoids from *Schisandra rubriflora* and their biological activities. Org Lett. 2006;8:991–4.
166. Xiao WL, Pu JX, Chang Y, Li XL, Huang SX, Yang LM, Li LM, Lu Y, Zheng YT, Li RT, Zheng QT, Sun HD. Sphenadilactones A and B, two novel nortriterpenoids from *Schisandra sphenanthera*. Org Lett. 2006;8:1475–8.
167. Huang SX, Li RT, Liu JP, Lu Y, Chang Y, Lei C, Xiao WL, Yang LB, Zheng QT, Sun HD. Isolation and characterization of biogenetically related highly oxygenated nortriterpenoids from *Schisandra chinensis*. Org Lett. 2007;9:2079–82.
168. Lin YC, Lo IW, Chen SY, Lin PH, Chien CT, Chang S, Shen YC. Schinarisanlactone A, a new bisnortriterpenoid from *Schisandra arisanensis*. Org Lett. 2011;13:446–9.
169. Xiao WL, Huang SX, Zhang L, Tian RR, Wu L, Li L, Pu JX, Zheng YT, Lu Y, Li RT, Zheng QT, Sun HD. Nortriterpenoids from *Schisandra lancifolia*. J Nat Prod. 2006;69:650–3.
170. Liang CQ, Shi YM, Luo RH, Li XY, Gao ZH, Li XN, Yang LM, Shang SZ, Li Y, Zheng YT, Zhang HB, Xiao WL, Sun HD. Kadcoccitones A and B, two new 6/6/5/5-fused tetracyclic triterpenoids from *Kadsura coccinea*. Org Lett. 2012;14:6362–5.
171. Zhu HC, Chen CM, Yang J, Li XN, Liu JJ, Sun B, Huang SX, Li DY, Yao GM, Luo ZW, Li Y, Zhang JW, Xue YB, Zhang YH. Bioactive acylphloroglucinols with adamantyl skeleton from *Hypericum Sampsonii*. Org Lett. 2014;16:6322–5.

172. Song JG, Su JC, Song QY, Huang RL, Tang W, Hu LJ, Huang XJ, Jiang RW, Li YL, Ye WC, Wang Y. Cleistocaltones A and B, antiviral phloroglucinol–terpenoid adducts from *Cleistocalyx operculatus*. Org Lett. 2019;21:9579–83.
173. Tang ZH, Liu YB, Ma SG, Li L, Li Y, Jiang JD, Qu J, Yu SS. Antiviral spirotriscoumarins A and B: two pairs of oligomeric coumarin enantiomers with a spirodienone–sesquiterpene skeleton from *Toddalia asiatica*. Org Lett. 2016;18:5146–9.
174. Deng LM, Hu LJ, Bai Yang TZ, Wang J, Qin GQ, Song QY, Su JC, Huang XJ, Jiang RW, Tang W, Li YL, Li CC, Ye WC, Wang Y. Rhodomentosones A and B: two pairs of enantiomeric phloroglucinol trimers from rhodomyrtus tomentosa and their asymmetric biomimetic synthesis. Org Lett. 2021;23:4499–504.
175. Su JC, Wang S, Cheng W, Huang XJ, Li MM, Jiang RW, Li YL, Wang L, Ye WC, Wang Y. Phloroglucinol derivatives with unusual skeletons from *Cleistocalyx operculatus* and their in vitro antiviral activity. J Org Chem. 2018;83:8522–32.
176. Liao HB, Huang GH, Yu MH, Lei C, Hou AJ. Five pairs of meroterpenoid enantiomers from *Rhododendron capitatum*. J Org Chem. 2017;82:1632–7.
177. Tanaka N, Okasaka M, Ishimaru Y, Takaishi Y, Sato M, Okamoto M, Oshikawa T, Ahmed SU, Consentino LM, Lee KH. Biyouyanagin A, an anti-HIV agent from *Hypericum chinense* L. var. *salicifolium*. Org Lett. 2005;7:2997–9.
178. Xu G, Hou AJ, Wang RR, Liang GY, Zheng YT, Liu ZY, Li XL, Zhao Y, Huang SX, Peng LY, Zhao QS. Przewalskin A: a new c_{23} terpenoid with a 6/6/7 carbon ring skeleton from *Salvia przewalskii* Maxim. Org Lett. 2006;8:4453–6.
179. Chen SD, Gao H, Zhu QC, Wang YQ, Li T, Mu ZQ, Wu HL, Peng T, Yao XS. Houttuynoids A–E, anti-herpes simplex virus active flavonoids with novel skeletons from *Houttuynia cordata*. Org Lett. 2012;14:1772–5.
180. Tan CJ, Di YT, Wang YH, Zhang Y, Si YK, Zhang Q, Gao S, Hu XJ, Fang X, Li SF, Hao XJ. Three new indole alkaloids from *Trigonostemon lii*. Org Lett. 2010;12:2370–3.
181. Zhang YB, Zhang XL, Chen NH, Wu ZN, Ye WC, Li YL, Wang GC. Four matrine-based alkaloids with antiviral activities against HBV from the seeds of *Sophora alopecuroides*. Org Lett. 2017;19:424–7.
182. Zhang YB, Zhan LQ, Li GQ, Wang F, Wang Y, Li YL, Ye WC, Wang GC. Dimeric matrine-type alkaloids from the roots of *Sophora flavescens* and their anti-hepatitis B virus activities. J Org Chem. 2016;81:6273–80.
183. Xu JB, Zhang H, Gan LS, Han YS, Wainberg MA, Yue JM. Logeracemin A, an anti-HIV *Daphniphyllum* alkaloid dimer with a new carbon skeleton from *Daphniphyllum longeracemosum*. J Am Chem Soc. 2014;136:7631–3.
184. Huang SD, Zhang Y, Cao MM, Di YT, Tang GH, Peng ZG, Jiang JD, He HP, Hao XJ. Myriberine A, a new alkaloid with an unprecedented heteropentacyclic skeleton from *Myrioneuron faberi*. Org Lett. 2013;15:590–3.
185. Liu YP, Guo JM, Xie Z, Suo XY, Liu ZY, Qiao ZH, Guan RQ, Bian Y, Qiang L, Fu YH. Clausanisumine, a prenylated bicarbazole alkaloid from the fruits of *Clausena anisumolens* and its potential anti-HIV activity. J Org Chem. 2021;86:17722–6.
186. Liu JF, Jiang ZY, Wang RR, Zheng YT, Chen JJ, Zhang XM, Ma YB. Isatisine A, a novel alkaloid with an unprecedented skeleton from leaves of *Isatis indigotica*. Org Lett. 2007;9:4127–9.
187. Wu ZN, Chen NH, Tang Q, Chen S, Zhan ZC, Zhang YB, Wang GC, Li YL, Ye WC. β-Carboline alkaloids from the seeds of *Peganum harmala* and their anti-HSV-2 virus activities. Org Lett. 2020;22:7310–4.
188. Liu YP, Wang TW, Xie Z, Bian Y, Liu YY, Guan RQ, Liu ZY, Qiang L, Chen GY, Fu YH. Artapilosines A and B, unusual phenanthrene derivatives related to aporphine alkaloids from *Artabotrys pilosus*. J Nat Prod. 2021;84:3117–21.
189. Geng CA, Zhang XM, Ma YB, Luo J, Chen JJ. Swerilactones L–O, secoiridoids with C_{12} and C_{13} skeletons from *Swertia mileensis*. J Nat Prod. 2011;74:1822–5.
190. Geng CA, Jiang ZY, Ma YB, Luo J, Zhang XM, Wang HL, Shen Y, Zuo AX, Zhou J, Chen JJ. Swerilactones A and B, anti-HBV new lactones from a tradtional chinese herb: *Swertia mileensis* as a treatment for viral hepatitis. Org Lett. 2009;11:4120–3.

191. Geng CA, Wang LJ, Zhang XM, Ma YB, Huang XY, Luo J, Guo RH, Zhou J, Shen Y, Zuo AX, Jiang ZY, Chen JJ. Anti-hepatitis B virus active lactones from the traditional chinese herb: *Swertia mileensis*. Chem Eur J. 2011;17:3893–903.
192. Wan LS, Nian Y, Ye CJ, Shao LD, Peng XR, Geng CA, Zuo ZL, Li XN, Yang J, Zhou M, Qiu MH. Three minor diterpenoids with three carbon skeletons from *Euphorbia peplus*. Org Lett. 2016;18:2166–9.
193. Li XR, Yan BC, Hu K, He SJ, Sun HD, Zuo JP, Puno PT. Spiro *ent*-clerodane dimers: discovery and green approaches for a scalable biomimetic synthesis. Org Lett. 2021;23:5647–51.
194. Zhou L, Tuo YL, Hao Y, Guo XM, Tang W, Xue YB, Zeng JF, Zhou Y, Xiang M, Zuo JP, Yao GM, Zhang YH. Cinnamomols A and B, immunostimulative diterpenoids with a new carbon skeleton from the leaves of *Cinnamomum cassia*. Org Lett. 2017;19:3029–32.
195. Zhou HF, Guoruoluo YDZ, Tuo Y, Zhou J, Zhang H, Wang W, Xiang M, Aisa HA, Yao G. Cassiabudanols A and B, immunostimulative diterpenoids with a cassiabudane carbon skeleton featuring a 3-oxatetracyclo[6.6.1.$0^{2,6}$.$0^{10,14}$]pentadecane scaffold from cassia buds. *Org Lett*. 2019;21:549–53.
196. Pan L, Terrazas C, Lezama-Davila CM, Rege N, Gallucci JC, Satoskar AR, Kinghorn AD. Cordifolide A, a sulfur-containing clerodane diterpene glycoside from *Tinospora cordifolia*. Org Lett. 2012;14:2118–21.
197. Jing SX, Fu R, Li CH, Zhou TT, Liu YC, Liu Y, Luo SH, Li XN, Zeng F, Li S. Immunosuppresive sesterterpenoids and norsesterterpenoids from *Colquhounia coccinea* var. *mollis*. J Org Chem 2021;86:11169−11176.
198. Guo K, Liu X, Zhou TT, Liu YC, Liu Y, Shi QM, Li XN, Li SH. Gentianelloids A and B: immunosuppressive 10,11-seco-gentianellane sesterterpenoids from the traditional uighur medicine *Gentianella turkestanorum*. J Org Chem 2020;85;5511–5515.
199. Teng LL, Mu RF, Liu YC, Xiao CJ, Li DS, Gao JX, Guo K, Li XN, Liu Y, Zeng F, Li SH. Immunosuppressive and adipogenesis inhibitory sesterterpenoids with a macrocyclic ether system from *Eurysolen gracilis*. Org Lett. 2021;23:2232–7.
200. Song J, Liu Y, Zhou M, Cao H, Peng XG, Liang JJ, Zhao XY, Xiang M, Ruan HL. Spiroschincarins A-E: five spirocyclic nortriterpenoids from the fruit of *Schisandra incarnata*. Org Lett. 2017;19:1196–9.
201. Fan YY, Zhang H, Zhou Y, Liu HB, Tang W, Zhou B, Zuo JP, Yue JM. Phainanoids A–F, a new class of potent immunosuppressive triterpenoids with an unprecedented carbon skeleton from *Phyllanthus hainanensis*. J Am Chem Soc. 2015;137:138–41.
202. Fan YY, Gan LS, Liu HC, Li H, Xu CH, Zuo JP, Ding J, Yue JM. Phainanolide A, highly modified and oxygenated triterpenoid from *Phyllanthus hainanensis*. Org Lett. 2017;19:4580–3.
203. Hua J, Liu YC, Luo SH, Liu Y, Xiao CJ, Li XN, Li SH. Immunostimulatory 6/6/6/6 tetracyclic triterpenoid saponins with the methyl-30 incorporated cyclization from the root of *Colquhounia elegans*. Org Lett. 2021;23:7462–6.
204. Duan YL, Xie SS, Bu PF, Guo Y, Shi ZY, Guo Y, Cao YF, Sun WG, Qi CX, Zhang YH. Hypaluton A, an immunosuppressive 3,4-nor-polycyclic polyprenylated acylphloroglucinol from *Hypericum patulum*. J Org Chem. 2021;86:6478–85.
205. Xie SS, Zhou Y, Tan XS, Sun WG, Duan YL, Feng H, Sun LJ, Guo Y, Shi Z, Hao XC, Chen G, Qi CX, Zhang YH. Norwilsonnol A, an immunosuppressive polycyclic polyprenylated acylphloroglucinol with a spiro[5-oxatricyclo[6.4.0.03,7]dodecane-6′,1–1′,2′-dioxane] system from *Hypericum wilsonii*. Org Chem Front. 2021;8:2280–6.
206. Xie SS, Qi CX, Duan YL, Hao XC, Guo Y, Deng MY, Qiao YB, Shi ZY, Tao L, Cao YF, Gu LH, Zhou Y, Zhang YH. Wilsonglucinols A–C, homoadamantane-type polycyclic polyprenylated acylphloroglucinols with unusual fused epoxy ring skeletons from *Hypericum wilsonii*. Org Chem Front. 2020;7:464–71.
207. Zhou M, Zhang HB, Wang WG, Gong NB, Zhan R, Li XN, Du X, Li LM, Li Y, Lu Y, Pu JX, Sun HD. Scopariusic acid, a new meroditerpenoid with a unique cyclobutane ring isolated from *Isodon scoparius*. Org Lett. 2013;15:4446–9.

208. Gao XH, Xu YS, Fan YY, Gan LS, Zuo JP, Yue JM. Cascarinoids A–C, a class of diterpenoid alkaloids with unpredicted conformations from *Croton cascarilloides*. Org Lett. 2018;20:228–31.
209. Feng T, Duan KT, He SJ, Wu B, Zheng YS, Ai HL, Li ZH, He J, Zuo JP, Liu JK. Ophiorrhines A and B, two immunosuppressive monoterpenoid indole alkaloids from *Ophiorrhiza japonica*. Org Lett. 2018;20:7926–8.
210. Zhang B, Wang Y, Yang SP, Zhou Y, Wu WB, Tang W, Zuo JP, Li Y, Yue JM. Ivorenolide A, an unprecedented immunosuppressive macrolide from *Khaya ivorensis*: structural elucidation and bioinspired total synthesis. J Am Chem Soc. 2012;134:20605–8.
211. Tong XG, Zhou LL, Wang YH, Xia C, Wang Y, Liang M, Hou FF, Cheng YX. Acortatarins A and B, two novel antioxidative spiroalkaloids with a naturally unusual morpholine motif from *Acorus tatarinowii*. Org Lett. 2010;12:1844–7.
212. He S, Jiang L, Wu B, Li C, Pan Y. Chunganenol: an unusual antioxidative resveratrol hexamer from *Vitis chunganensis*. J Org Chem. 2009;74:7966–9.
213. Ma SG, Li M, Lin MB, Li L, Liu YB, Qu J, Li Y, Wang XJ, Wang RB, Xu S, Hou Q, Yu SS. Illisimonin A, a caged sesquiterpenoid with a tricyclo[5.2.1.0^{1,6}]decane skeleton from the fruits of *Illicium simonsii*. Org Lett. 2017;19:6160–3.
214. Kubo M, Okada C, Huang JM, Harada K, Hioki H, Fukuyama Y. Novel pentacyclic *seco*-prezizaane-type sesquiterpenoids with neurotrophic properties from *Illicium jiadifengpi*. Org Lett. 2009;11:5190–3.
215. Huang XH, Tao LX, Ke CQ, Tang C, Zhang HY, Ye Y, Lin LG, Yao S. Taxodikaloids A and B, two dimeric abietane-type diterpenoids from *Taxodium ascendens* possessing an oxazoline ring linkage. Org Lett. 2017;19:556–9.
216. Wang LY, Wu J, Yang Z, Wang XJ, Fu Y, Liu SZ, Wang HM, Zhu WL, Zhang HY, Zhao WM. (*m*)- and (*p*)-Bicelaphanol A, dimeric trinorditerpenes with promising neuroprotective activity from *Celastrus orbiculatus*. J Nat Prod. 2013;76:745–9.
217. Chen J, Peng XG, Liu J, Gao Y, Zhou M, Zhou J, Song J, Ruan HL. Incarnolides A and B, two schinortriterpenoids with a tricyclo[9.2.1.0^{2,8}]tetradecane-bridged system from *Schisandra incarnata*. Org Lett. 2020;22:1071–5.
218. Zhang CL, Wang Y, Liu YF, Ni G, Liang D, Luo H, Song XY, Zhang WQ, Chen RY, Chen NH, Yu DQ. Iridal-type triterpenoids with neuroprotective activities from *Iris tectorum*. J Nat Prod. 2014;77:411–5.
219. Liu JQ, Peng XR, Li XY, Li TZ, Zhang WM, Shi L, Han J, Qiu MH. Norfriedelins A–C with acetylcholinesterase inhibitory activity from acerola tree (*Malpighia emarginata*). Org Lett. 2013;15:1580–3.
220. Liu XY, Li CJ, Chen FY, Ma J, Wang S, Yuan YH, Li L, Zhang DM. Nototronesides A–C, three triterpene saponins with a 6/6/9 fused tricyclic tetranordammarane carbon skeleton from the leaves of *Panax notoginseng*. Org Lett. 2018;20:4549–53.
221. He J, Yang YN, Jiang J, Feng ZM, Zhang PC. Saffloflavonesides A and B, two rearranged derivatives of flavonoid *C*-glycosides with a furan–tetrahydrofuran ring from *Carthamus tnctorius*. Org Lett. 2014;16:5714–7.
222. Zhang F, Yang YN, Song XY, Shao SY, Feng ZM, Jiang JS, Li L, Chen NH, Zhang PC. Forsythoneosides A–D, neuroprotective phenethanoid and flavone glycoside heterodimers from the fruits of *Forsythia suspensa*. J Nat Prod. 2015;78:2390–7.
223. Yu HY, Chen ZY, Sun B, Liu J, Meng FY, Liu Y, Tian T, Jin A, Ruan HL. Lignans from the fruit of *Schisandra glaucescens* with antioxidant and neuroprotective properties. J Nat Prod. 2014;77:1311–20.
224. Wu ZL, Huang XJ, Xu MT, Ma X, Li L, Shi L, Wang WJ, Jiang RW, Ye WC, Wang Y. Flueggeacosines A–C, dimeric securinine-type alkaloid analogues with neuronal differentiation activity from *Flueggea suffruticosa*. Org Lett. 2018;20:7703–7.
225. Hirasawa Y, Kobayashi J, Morita H, Lycoperine A. A novel $C_{27}N_3$-type pentacyclic alkaloid from *Lycopodium hamiltonii*, inhibiting acetylcholinesterase. Org Lett. 2006;8:123–6.
226. He J, Chen XQ, Li MM, Zhao Y, Xu G, Cheng X, Peng LY, Xie MJ, Zheng YT, Wang YP, Zhao QS. Lycojapodine A, a novel alkaloid from *Lycopodium japonicum*. Org Lett. 2009;11:1397–400.

227. Zhang ZJ, Nian Y, Zhu QF, Li XN, Su J, Wu XD, Yang J, Zhao QS. Lycoplanine A, a $C_{16}N$ *Lycopodium* alkaloid with a 6/9/5 tricyclic skeleton from *Lycopodium complanatum*. Org Lett. 2017;19:4668–71.
228. Zhang J, Song M, Ao YL, Li Y, Zou XY, Xu J, Wang Y, Zhang DM, Zhang XQ, Ye WC. Alstolarines A and B, two unusual monoterpenoid indole alkaloids with an acetal moiety from *Alstonia scholaris*. Org Chem Front. 2020;7:3468–73.
229. Hirasawa Y, Mitsui C, Uchiyama N, Hakamatsuka T, Morita H. Hupercumines A and B, *Lycopodium* alkaloids from *Huperzia cunninghamioides*, inhibiting acetylcholinesterase. Org Lett. 2018;20:1384–7.
230. Wang W, Tang X, Hua F, Ling TJ, Wan XC, Bao GH. Camellimidazole A–C, three methylene-bridged dimeric imidazole alkaloids from *Keemun Black* Tea. Org Lett. 2018;20:2672–5.
231. Chen H, Kong JB, Zhang L, Wang HH, Cao YG, Zeng MN, Li M, Sun YJ, Du K, Xue GM, Wu Y, Zheng XK, Feng WS. Lycibarbarines A–C, three tetrahydroquinoline alkaloids possessing a spiro-heterocycle moiety from the fruits of *Lycium barbarum*. Org Lett. 2021;23:858–62.
232. Wang JX, Zhao YP, Du NN, Han Y, Li H, Wang R, Xu Y, Liu YF, Liang XM. Scocycamides, a pair of macrocyclic dicaffeoylspermidines with butyrylcholinesterase inhibition and antioxidation activity from the roots of *Scopolia tangutica*. Org Lett. 2020;22:8240–4.
233. Bautista E, Fragoso-Serrano M, Toscano RA, García-Peña MdelR, Ortega A. Teotihuacanin, a diterpene with an unusual spiro-10/6 system from *Salvia amarissima* with potent modulatory activity of multidrug resistance in cancer cells. *Org Lett.* 2015;17:3280–82.
234. Li Y, Xu Z, Zhu R, Zhou J, Zong Y, Zhang J, Zhu M, Jin X, Qiao Y, Zheng H, Lou H. Probing the interconversion of labdane lactones from the Chinese liverwort *Pallavicinia ambigua*. Org Lett. 2020;22:510–4.
235. He XF, Wang XN, Gan LS, Yin S, Dong L, Yue JM. Two novel triterpenoids from *Dysoxylum hainanense*. Org Lett. 2008;10:4327–30.
236. Liu F, Tian HY, Huang XL, Wang WJ, Li NP, He J, Ye WC, Wang L. Xanthchrysones A–C: rearranged phenylpropanoyl–phloroglucinol dimers with unusual skeletons from *Xanthostemon chrysanthus*. J Org Chem. 2019;84:15355–61.
237. Yu HF, Qin XJ, Ding CF, Wei X, Yang J, Luo JR, Liu L, Khan A, Zhang LC, Xia CF, Luo XD. Nepenthe-like indole alkaloids with antimicrobial activity from *Ervatamia chinensis*. Org Lett. 2018;20:4116–20.
238. Yu HF, Ding CF, Zhang LC, Wei X, Cheng GG, Liu YP, Zhang RP, Luo XD. Alstoscholarisine K, an antimicrobial indole from gall-induced leaves of *Alstonia scholaris*. Org Lett. 2021;23:5782–6.
239. Ge HM, Huang B, Tan SH, Shi DH, Song YC, Tan RX. Bioactive oligostilbenoids from the stem bark of *Hopea exalata*. J Nat Prod. 2006;69:1800–2.
240. Cottiglia F, Dhanapal B, Sticher O, Heilmann J. New chromanone acids with antibacterial activity from *Calophyllum brasiliense*. J Nat Prod. 2004;67:537–41.
241. Karioti A, Skaltsa H, Linden A, Perozzo R, Brun R, Tasdemir D. Anthecularin: a novel sesquiterpene lactone from *anthemis auriculata* with antiprotozoal activity. J Org Chem. 2007;72:8103–6.
242. Zhou B, Liu QF, Dalal S, Cassera MB, Yue JM. Fortunoids A-C, three sesquiterpenoid dimers with different carbon skeletons from *Chloranthus fortunei*. Org Lett. 2017;19:734–7.
243. Zhou JS, Liu QF, Zimbres Flavia M, Butler Joshua H, Cassera Maria B, Zhou B, Yue JM. Trichloranoids A–D, antimalarial sesquiterpenoid trimers from *Chloranthus spicatus*. Org Chem Front. 2021;8:1795–801.
244. Yin S, Wang XN, Fan CQ, Liao SG, Yue JM. The first limonoid peroxide in the meliaceae family: walsuronoid A from *Walsura robusta*. Org Lett. 2007;9:2353–6.
245. Hiranrat A, Mahabusarakam W, Carroll AR, Duffy S, Avery VM. Tomentosones A and B, hexacyclic phloroglucinol derivatives from the Thai shrub *Rhodomyrtus tomentosa*. J Org Chem. 2012;77:680–3.
246. Farimani MM, Taheri S, Ebrahimi SN, Bahadori MB, Khavasi HR, Zimmermann S, Brun R, Hamburger M. Hydrangenone, a new isoprenoid with an unprecedented skeleton from *Salvia hydrangea*. Org Lett. 2012;14:166–9.

247. Lu ZY, Van Wagoner RM, Pond CD, Pole AR, Jensen JB, Blankenship D, Grimberg BT, Kiapranis R, Matainaho TK, Barrows LR, Ireland CM. Myristicyclins A and B: antimalarial procyanidins from *Horsfieldia spicata* from Papua New Guinea. Org Lett. 2014;16:346–9.
248. Fernandez LS, Buchanan MS, Carroll AR, Feng YJ, Quinn RJ, Avery VM. Flinderoles A–C: antimalarial bis-indole alkaloids from *Flindersia* species. Org Lett. 2009;11:329–32.
249. Bringmann G, Lombe BK, Steinert C, Ioset KN, Brun R, Turini F, Heubl G, Mudogo V. Mbandakamines A and B, unsymmetrically coupled dimeric naphthylisoquinoline alkaloids, from a congolese *Ancistrocladus* species. Org Lett. 2013;15:2590–3.
250. Lombe BK, Bruhn T, Feineis D, Mudogo V, Brun R, Bringmann G. Antiprotozoal spirombandakamines A_1 and A_2, fused naphthylisoquinoline dimers from a congolese *Ancistrocladus* plant. Org Lett. 2017;19:6740–3.
251. Lombe BK, Bruhn T, Feineis D, Mudogo V, Brun R, Bringmann G. Cyclombandakamines A_1 and A_2, oxygen-bridged naphthylisoquinoline dimers from a congolese *Ancistrocladus* Liana. Org Lett. 2017;19:1342–5.
252. Morita H, Oshimi S, Hirasawa Y, Koyama K, Honda T, Ekasari W, Indrayanto G, Zaini NC. Cassiarins A and B, novel antiplasmodial alkaloids from *Cassia siamea*. Org Lett. 2007;9:3691–3.
253. Wisetsai A, Lekphrom R, Boonmak J, Youngme S, Schevenels FT. Spiroaxillarone A, a symmetric spirobisnaphthalene with an original skeleton from *Cyanotis axillaris*. Org Lett. 2019;21:8344–8.

Chapter 15
Conclusions and Outlook

Based on the information included in this Review, it is easy to conclude that the discovery of NSNPs that possess unique skeletons in the past 23 years has mainly uncovered members derived from the plant families Hypericaceae, Meliaceae, Euphorbiaceae, Myrtaceae, Asteraceae/Compositae and Apocynaceae (Fig. 15.1). Although the Burseraceae family contains *Canarium*, *Garuga* and *Protium* genera, the NSNPs arise mainly from the genus *Canarium*. As for the structure types of the NSNPs isolated during 1999–2021, it is clear that alkaloids particularly are dominant, and diterpenoids, triterpenoids and phloroglucinols are the second most abundant. In contrast, the numbers of structural types lignins and sterides are small (Fig. 15.2). In the context of publications per year, the numbers of articles describing NSNPs in the 1999–2011 period are small, whereas the numbers of publications about structurally novel NPs in the other years are near constant averaging 68–137 per year (Fig. 15.3). A compilation of journals shows that most publications describing NPs with new skeletons appeared in Organic Letters, followed by Journal of Organic Chemistry, and then Journal of Natural Products and Organic Chemistry Frontiers (Fig. 15.4). It is interesting that the most contributions in this area are from scientists in China, whose studies uncovered 75.3% of NPs with new skeletons from plants from 1999 to 2021 (Fig. 15.5). This large percentage might reflect the fact that many Chinese researchers are involved in investigations of herbal medicines.

Although NSNP structures included in this Review are complex and intriguing, their biological activities for the most part are not generally high even though a wide variety of biological assays are available. It is obvious that in most publications bioactivity appears to be embellish. Among all the biological evaluations, the most frequently used has been an anti-inflammation assessment determined by measuring nitric oxide (NO) release. However, the potency of most NSNPs is weak or not that strong. Screening of HIV inhibitors has also attracted considerable attention during 1999–2021. Thus, it is surprising that not many results of this assay of the NSNPs have been reported. Theoretically, plant derived NPs with complex new skeletons are created in Nature at a high energy cost and, as a result, they need to play essential roles

Y. Cheng and D. Qin, *Novel Plant Natural Product Skeletons*,
https://doi.org/10.1007/978-981-99-7329-3_15

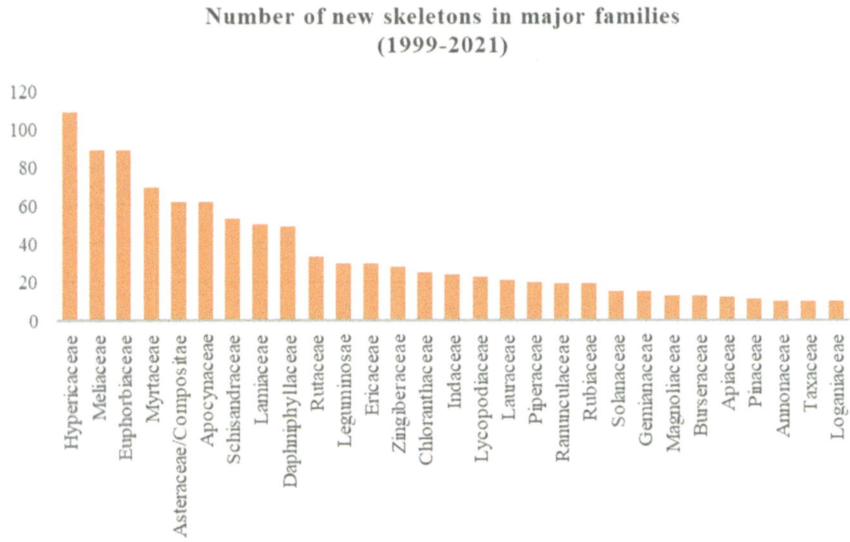

Fig. 15.1 Number of plant NPs with new skeletons in major families (1999–2021)

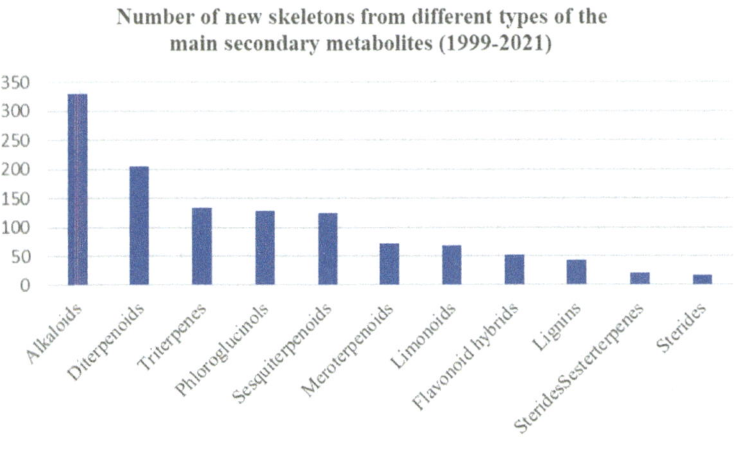

Fig. 15.2 Number of plant NPs with new skeletons in different types of the main secondary metabolites (1999–2021)

in the evolution of plants and, perhaps these properties translate into to biological activities in Human pathogens. Azadirachtin, an excellent pesticide, and paclitaxel, an antineoplastic agent acting via stabilization of tubulin polymerization, are textbook examples of this phenomenon. Some striking biological activities are also revealed in this Review. For example, anti-malarial activities of spirombandakamines A1 and A2, cyclombandakamines A1 and A2, and immunosuppressive activities of phainanoid F and phainanolide A occur at nanomolar levels. Possible reasons exist for why

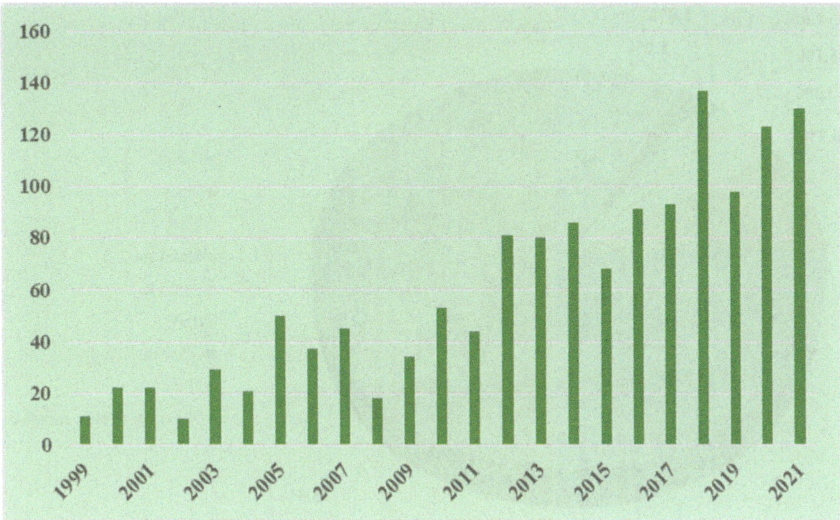

Fig. 15.3 Number of plant NPs with new skeletons from 1999 to 2021

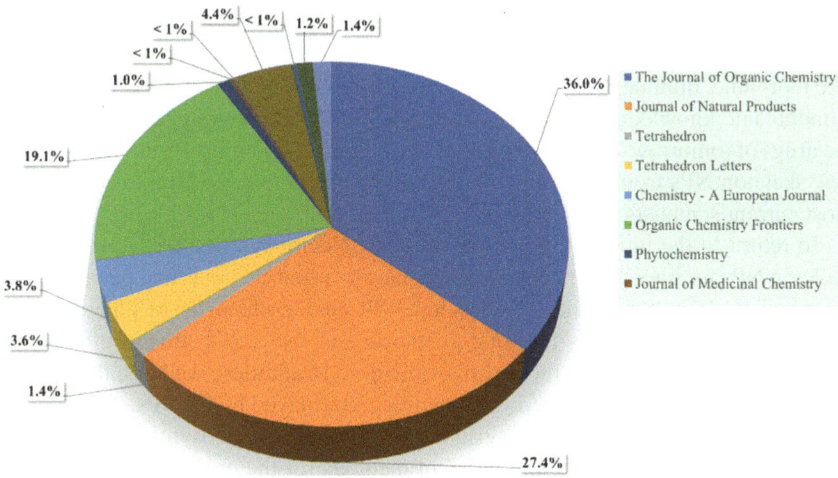

Fig. 15.4 The proportion of the number of plant NPs with new skeletons published in different journals apart from organic letters (1999–2021)

most of the NPs with new skeletons do not have interesting or strong biological activities. The most important factor could be that the limited quantities isolated from plant sources hamper biological evaluation especially in the cases of NSNPs that are isolated as minor components. Although numerous biological assays and high throughput screening methods are available for use on small sample sizes, they are not well shared. Furthermore, in retrospect, nearly all the classical drugs

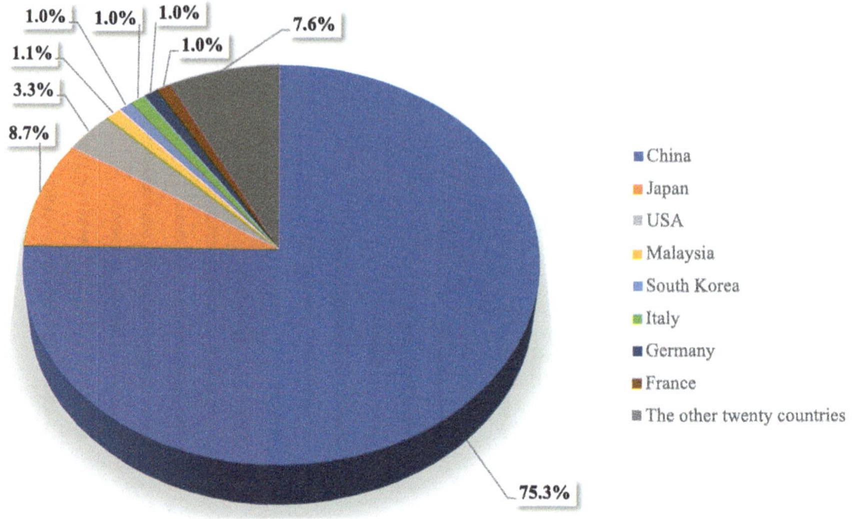

Fig. 15.5 The proportion of plant NPs with new skeletons contributed by scientists in different countries

like morphine, quinine and aspirin were discovered through inspiration provided by ethnobotanic knowledge or cultural practices. This "tradition of yesterday could be the drugs of tomorrow" philosophy should be helpful for guiding evaluation of minor new skeleton NPs from medicinal plants but it appears to have been overlooked by most current scientists.

To return to the introduction of this Review, Nature synthesizes numerous new skeleton NPs for important purposes which have amazed chemists and puzzled pharmacologists. It is hoped that this Review will enable chemists to identify more complex NSNPs as targets for total synthesis and that the synthetic efforts will provide larger quantities for biological evaluation. In addition, we hope the Review aids pharmacologists in uncovering medicinal roles played by NSNPs. This Review might also benefit synthetic biologists in identifying new biosynthetic pathways, which might in turn lead to more efficient methods for the preparation of complex NSNPs or NP-like compounds.

Acknowledgements This work was supported by grants from the National Natural Science Foundation of China (81903508, 81525026, U1702287).

Appendix

See Tables A1, A2, A3, A4, A5, A6, A7, A8, A9, A10, A11 and A12.

Y. Cheng and D. Qin, *Novel Plant Natural Product Skeletons*,
https://doi.org/10.1007/978-981-99-7329-3

Table A1 Sesquiterpenoids from plants

Compound	Trivial name	Source	Family	Structure characteristics	Activities if reported	References
1	Ligulolide A	*Ligularia virgaurea*	Asteraceae/ compositae	Represent a novel sesquiterpene carbon framework	Cytotoxic activity	[30]
2	Heliespirone B	*Helianthus annuus*	Asteraceae/ compositae	Contain six-membered spiroheterocyclic skeleton	Insecticidal action	[31]
3	Heliespirone C	*Helianthus annuus*	Asteraceae/ compositae	Contain five-membered spiroheterocyclic skeleton	Insecticidal action	[31]
4	Anthecularin	*Anthemis auriculata*	Asteraceae/ compositae	A minor sesquiterpene lactone with a novel ring system	Antiprotozoal activity	[32]
5	Artarborol	*Artemisia arborescens*	Asteraceae/ compositae	Fused cyclobutane and cyclononane rings	–	[33]
6	Wedelolide A	*Wedelia trilobata*	Asteraceae/ compositae	The first example of the novel sesquiterpene framework (9*R*)-eudesman-9,12-olide	–	[34]
7	Wedelolide B	*Wedelia trilobata*	Asteraceae/ compositae	The first example of the novel sesquiterpene framework (9*R*)-eudesman-9,12-olide	–	[34]
8	Cycloparvifloralone	American *Illicium*	Magnoliaceae	A hitherto unknown ring system with a cagelike acetal/hemiketal structure	–	[35]
9	Illisimonin A	*Illicium simonsii*	Magnoliaceae	A caged 2-oxatricyclo [3,3,0,$1^{4,7}$] nonane ring system fused to a five-membered carbocyclic ring and a five-membered lactone ring	Neuroprotective effect	[36]
10	Jiadifenolide	*Illicium jiadifengpi*	Magnoliaceae	The first example of a *seco*-prezizaane-type sesquiterpenoid with a γ-lactone formed between the C-11 carbonyl and the C-4 hydroxy group	Neurite outgrowth-promoting activity	[37]
11	Jiadifenoxolane A	*Illicium jiadifengpi*	Magnoliaceae	Novel seco-prezizaane-type sesquiterpenoid	Neurite outgrowth-promoting activity	[37]
12	Jiadifenoxolane B	*Illicium jiadifengpi*	Magnoliaceae	Novel seco-prezizaane-type sesquiterpenoid	–	[37]
13	Artaboterpenoid A	*Artabotrys hexapetalus*	Annonaceae	A novel carbon skeleton with a new C-2–C-10 linkage	–	[38]
14	(−)-artaboterpenoid B	*Artabotrys hexapetalus*	Annonaceae	The first example of 1,2-*seco*-bisabolene-type sesquiterpene lactone	–	[38]

(continued)

Table A1 (continued)

Compound	Trivial name	Source	Family	Structure characteristics	Activities if reported	References
15	(+)-artaboterpenoid B	*Artabotrys hexapetalus*	Annonaceae	The first example of 1,2-*seco*-bisabolene-type sesquiterpene lactone	Cytotoxicity	[38]
16	Volvalerelactone A	*Valeriana officinalis*	Valerianaceae	An unprecedented 3/7/6 tricyclic ring system	–	[39]
17	Narjatamolide	*Narjatamolide jatamansi*	Valerianaceae	Possessing an additional acetate unit spiro-fused with C-4 and C-15 to form a cyclopropane ring	Antiproliferative effect	[40]
18	Urechitol A	*Pentalinon andrieuxii*	Apocynaceae	The first report of the novel "campechane" trinorsesquiterpenoid skeleton	–	[41]
19	Urechitol B	*Pentalinon andrieuxii*	Apocynaceae	The first report of the novel "campechane" trinorsesquiterpenoid skeleton	–	[41]
20	Toxicodenane A	*Toxicodendron vernicifluum*	Anacardiaceae	A 6/7 ring system, a tetrahedron furan ring moiety in the molecule which made it cage like structure	–	[42]
21	Toxicodenane B	*Toxicodendron vernicifluum*	Anacardiaceae	A 6/7 ring system	Extracellular matrix inhibitor	[42]
22	Toxicodenane C	*Toxicodendron vernicifluum*	Anacardiaceae	A 5/8ringsystem, a tetrahedron furan ring moiety in the molecule which made it cage like structure	Extracellular matrix inhibitor	[42]
23	Coriatone	*Coriaria nepalensis*	Coriariaceae	Novel highly oxygenated picrotoxane-type sesquiterpene	–	[43]
24	Coriane	*Coriaria nepalensis*	Coriariaceae	The first example with the coriane-type sesquiterpene skeleton	–	[43]
25	Plagiochianin A	*Plagiochila duthiana*	Plagiochilaceae	Unprecedented 2,3:6,7-di-seco-6,8-cycloaromadendrane carbon scaffold conjugated with three cyclic acetals	Anti-AChE activity	[44]
26	Plagiochianin B	*Plagiochila duthiana*	Plagiochilaceae	An exceptional pyridine type aromadendrane alkaloid	–	[44]
27	Hodgsonox	*Lepidolaena hodgsoniae*	Lepidolaenaceae	A cyclopenta[5,1-c]pyran ring system fused to an oxirane ring	Insecticidal activity	[45]
28	Curcumane A	*Curcuma longa*	Zingiberaceae	A dicyclo[3.2.1]octane moiety	Vasorelaxant activity	[46]
29	Curcumane B	*Curcuma longa*	Zingiberaceae	A dicyclo[3.3.1]nonane moiety	Vasorelaxant activity	[46]

(continued)

Table A1 (continued)

Compound	Trivial name	Source	Family	Structure characteristics	Activities if reported	References
30	(±)-eugenunilone C	*Eugenia uniflora*	Myrtaceae	The first example of sesquiterpenoid possessing a caged tricyclo[4.3.1.03,7]decane core	–	[47]
31	(±)-eugenunilone D	*Eugenia uniflora*	Myrtaceae	The first example of sesquiterpenoid possessing an isopentyl substituted bicyclo[3.2.1]octane backbone	Anti-inflammatory effect	[47]
32	(±)-eugenunilone E	*Eugenia uniflora*	Myrtaceae	A tricyclo[4.4.0.0^{2,10}]decane scaffold which is found in nature for the first time	–	[47]
33	(±)-eugenunilone F	*Eugenia uniflora*	Myrtaceae	A tricyclo[4.4.0.0^{2,10}]decane scaffold which is found in nature for the first time	Anti-inflammatory effect	[47]
34	(±)-syringanoid A	*Syringa pinnatifolia*	Oleaceae	Represent an unprecedented 5/4/6 tricyclic backbone	Protective effects Against hypoxia-induced injury to H9c2 cells	[48]
35	(±)-pinnatanoid A	*Syringa pinnatifolia*	Oleaceae	Represent a rare 6/7 bicyclic backbone	Protective effects Against hypoxia-induced injury to H9c2 cells; anti-inflammatory activity	[48]
36	(±)-pinnatanoid B	*Syringa pinnatifolia*	Oleaceae	Represent a rare 6/7 bicyclic backbone	Protective effects Against hypoxia-induced injury to H9c2 cells	[48]
37	Illihenin A	*Illicium henryi*	Illiciaceae	Represent a class of novel 5/7/6 tricyclic sesquiterpenoids featuring a rare cage-like tricyclo[6.2.2.0^{1,5}]dodecane core	Antiviral activity	[49]
38	Daphnenoid A	*Daphne penicillata*	Thymelaeaceae	Possesses a unique caged tetracyclo [5.3.2.0^{1,6}.0^{4,11}] dodecane scaffold by unexpected cyclizations of C-1/C-11 and C-2/C-14	–	[50]
39	Daphnenoid B	*Daphne penicillata*	Thymelaeaceae	The first discovered natural sesquiterpenes with unique 5/5 spirocyclic systems in nature	Anti-inflammatory effect	[50]
40	Daphnenoid C	*Daphne penicillata*	Thymelaeaceae	The first discovered natural sesquiterpenes with unique 5/5 spirocyclic systems in nature	Anti-inflammatory effect	[50]
41	Chloraholide A	*Chloranthus Holostegius*	Chloranthaceae	A highly complex sesquiterpenoid dimer	Blocker of the potassium channel	[51]

(continued)

Table A1 (continued)

Compound	Trivial name	Source	Family	Structure characteristics	Activities if reported	References
42	Chloraholide B	*Chloranthus Holostegius*	Chloranthaceae	A highly complex, decacyclic lindenane-type sesquiterpenoid dimer with a unique 18-membered macrocyclic trilactone ring	Blocker of the potassium channel	[51]
43	Sarcanolide A	*Sarcandra hainanensis*	Chloranthaceae	An unprecedented carbon framework via the formation of C-11–C-7' bond	–	[52]
44	Sarcanolide B	*Sarcandra hainanensis*	Chloranthaceae	An unprecedented carbon framework via the formation of C-11–C-7' bond	–	[52]
45	Serratustone A	*Chloranthus serratus*	Chloranthaceae	A novel dimerization pattern of two different types of sesquiterpenoids of an elemane and an eudesmane	–	[53]
46	Serratustone B	*Chloranthus serratus*	Chloranthaceae	A novel dimerization pattern of two different types of sesquiterpenoids of an elemane and an eudesmane	–	[53]
47	Fortunoid A	*Chloranthus fortunei*	Chloranthaceae	A new carbon skeleton of rearranged lindenane dimer	Antimalarial activity	[54]
48	Fortunoid B	*Chloranthus fortunei*	Chloranthaceae	First example of heterodimeric frameworks of a lindenane and an eudesmane sesquiterpenoid	Antimalarial activity	[54]
49	Fortunoid C	*Chloranthus fortunei*	Chloranthaceae	First example of heterodimeric frameworks of a lindenane and an eudesmane sesquiterpenoid	–	[54]
50	Hedyorienoid A	*Hedyosmum orientale*	Chloranthaceae	An unprecedented heterodimeric framework of a lindenane and an aromadendrane sesquiterpenoid furnished by forming an unusual 1,3dioxolane ring	NF-κB inhibitory activity	[55]
51	Hedyorienoid B	*Hedyosmum orientale*	Chloranthaceae	A new dimerization pattern of two guaiane-type sesquiterpenoids tied together by forming a new carbon–carbon bond	NF-κB inhibitory activity	[55]
52	Chloraserrtone A	*Chloranthus serratus*	Chloranthaceae	A sesquiterpenoid dimer with two lindenane-type sesquiterpenoid monomers bridged by two six-membered rings	–	[56]
53	Chlotrichene A	*Chloranthus holostegius*	Chloranthaceae	A unique 3/5/6/6/6/6/5/3-fused octacyclic skeleton	–	[57]
54	Chlotrichene B	*Chloranthus holostegius*	Chloranthaceae	A highly fused-ring skeleton	Cytotoxicity	[57]

(continued)

Table A1 (continued)

Compound	Trivial name	Source	Family	Structure characteristics	Activities if reported	References
55	Japonicone A	*Chloranthus japonicus*	Chloranthaceae	A rare 12-membered ring framework	Cytotoxicity	[58]
56	Japonicone B	*Chloranthus japonicus*	Chloranthaceae	A rare 12-membered ring framework	–	[58]
57	Japonicone C	*Chloranthus japonicus*	Chloranthaceae	A rare 12-membered ring framework	–	[58]
58	Chlorahupetone A	*Chloranthus henryi*	Chloranthaceae	A unique 3/5/6/5/4/7/7/5-fused octacyclic carbon skeleton	–	[59]
59	Chlorahupetone B	*Chloranthus henryi*	Chloranthaceae	An unusual 3/5/6/5/4/7/6/5-fused carbon skeleton	–	[59]
60	Chlorahupetone C	*Chloranthus henryi*	Chloranthaceae	An unusual 3/5/6/5/4/7/6/5-fused carbon skeleton	–	[59]
61	Chlorahupetone D	*Chloranthus henryi*	Chloranthaceae	Possess an unprecedented 3/5/6/6/4/6/6/3/5-fused carbon framework	–	[59]
62	Chlorahupetone E	*Chloranthus henryi*	Chloranthaceae	Possess an unprecedented 3/5/6/6/4/6/6/3/5-fused carbon framework	–	[59]
63	Chlorahupetone F	*Chloranthus henryi*	Chloranthaceae	A unique C_2-symmetric cyclic rearranged lindenane-type sesquiterpenoid dimer	–	[59]
64	Chlorahupetone G	*Chloranthus henryi*	Chloranthaceae	A carbon framework via the formation of a C-11–C-7' bond with an aromatized ring D and a five-membered lactone ring fused at C-11 and C-7'	Cytotoxicity	[59]
65	Chlorahupetone H	*Chloranthus henryi*	Chloranthaceae	A carbon framework via the formation of a C-11–C-7' bond with an aromatized ring D and a five-membered lactone ring fused at C-11 and C-7'	–	[59]
66	Chlorahupetone I	*Chloranthus henryi*	Chloranthaceae	A carbon framework via the formation of a C-11–C-7' bond with an aromatized ring D and a five-membered lactone ring fused at C-11 and C-7'	–	[59]
67	Trichloranoid A	*Chloranthus spicatus*	Chloranthaceae	A sesquiterpenoid trimer incorporating a new skeleton of rearranged lindenane sesquiterpenoid	Antimalarial activity	[60]

(continued)

Table A1 (continued)

Compound	Trivial name	Source	Family	Structure characteristics	Activities if reported	References
68	Trichloranoid B	*Chloranthus spicatus*	Chloranthaceae	Contains the first lindenane sesquiterpenoid with a stereo reverse-fused cyclopropane ring	–	[60]
69	Linderalide A	*Lindera aggregata*	Lauraceae	First example of disesquiterpenoid—geranylbenzofuranone hybrids directly linked by two C–C bonds	–	[61]
70	Linderalide B	*Lindera aggregata*	Lauraceae	First example of disesquiterpenoid—geranylbenzofuranone hybrids directly linked by two C–C bonds	–	[61]
71	Linderalide C	*Lindera aggregata*	Lauraceae	The first example of disesquiterpenoid—geranylbenzofuranone hybrids directly linked by two C–C bonds	–	[61]
72	Linderalide D	*Lindera aggregata*	Lauraceae	An unprecedented carbon skeleton featuring an unusual linearly 6/ 6/5/6/6 pentacyclic ring system fused by a sesquiterpenoid unit and a geranylbenzofuranone moiety	NF-κB inhibitory activity	[61]
73	Aggreganoid A	*Lindera aggregata*	Lauraceae	Unprecedented methin-bridged sesquiterpenoid trimer possessing a unique C_{46} skeleton	Transforming growth factor (TGF)-β inhibitory activity	[62]
74	Aggreganoid B	*Lindera aggregata*	Lauraceae	Unprecedented methine- and methylene-bridged sesquiterpenoid trimer possessing a unique C_{46} skeleton	–	[62]
75	Aggreganoid C	*Lindera aggregata*	Lauraceae	The first example of carbon-bridged disesquiterpenoids with a C_{33} skeleton in the plant kingdom	–	[62]
76	Aggreganoid D	*Lindera aggregata*	Lauraceae	The first example of carbon-bridged disesquiterpenoids with a C_{31} skeleton in the plant kingdom	–	[62]
77	Aggreganoid E	*Lindera aggregata*	Lauraceae	The first example of carbon-bridged disesquiterpenoids with a C_{31} skeleton in the plant kingdom	–	[62]

(continued)

Table A1 (continued)

Compound	Trivial name	Source	Family	Structure characteristics	Activities if reported	References
78	Aggreganoid F	*Lindera aggregata*	Lauraceae	The first example of carbon-bridged disesquiterpenoids with a C_{31} skeleton in the plant kingdom	–	[62]
79	Dicarabrone A	*Carpesium abrotanoides*	Asteraceae/ compositae	A new skeleton with a cyclopentane ring connecting two monomeric units	Cytotoxicity	[63]
80	Dicarabrone B	*Carpesium abrotanoides*	Asteraceae/ compositae	A new skeleton with a cyclopentane ring connecting two monomeric units	Cytotoxicity	[63]
81	Carpedilactones A	*Carpesium faberi*	Asteraceae/ compositae	The first 2,4-linked exo-Diels–Alder adducts between a eudesmanolide dienophile and a guaianolide diene	Cytotoxicity	[64]
82	Carpedilactones B	*Carpesium faberi*	Asteraceae/ compositae	The first 2,4-linked exo-Diels–Alder adducts between a eudesmanolide dienophile and a guaianolide diene	Cytotoxicity	[64]
83	Carpedilactones C	*Carpesium faberi*	Asteraceae/ compositae	The first 2,4-linked exo-Diels–Alder adducts between a eudesmanolide dienophile and a guaianolide diene	Cytotoxicity	[64]
84	Carpedilactones D	*Carpesium faberi*	Asteraceae/ compositae	The first 1,3-linked exo-Diels–Alder adducts between a eudesmanolide dienophile and a guaianolide diene	Cytotoxicity	[64]
85	Tricarabrol A	*Carpesium faberi*	Asteraceae/ compositae	Possessing a novel C_{44} skeleton featuring a methylene-tethered linkage among the sesquiterpene scaffolds	Anti-inflammatory activity	[65]
86	Tricarabrol B	*Carpesium faberi*	Asteraceae/ compositae	Possessing a novel C_{44} skeleton featuring a methylene-tethered linkage among the sesquiterpene scaffolds	Anti-inflammatory activity	[65]
87	Tricarabrol C	*Carpesium faberi*	Asteraceae/ compositae	Possessing a novel C44 skeleton featuring a methylene-tethered linkage among the sesquiterpene scaffolds, along with a methylene bridge	–	[65]
88	Artesin A	*Artemisia sieversiana*	Asteraceae/ compositae	A cage-shaped dimeric guaianolide with a highly symmetrical skeleton	Anti-neuroinflammatory activity	[66]
89	Artemisian A	*Artemisia argyi*	Asteraceae/ compositae	The first example of [4+2] Diels–Alder type adducts presumably biosynthesized from a rare 1, 10–4, 5-diseco-guaianolide and a guaianolide diene	–	[67]

(continued)

Table A1 (continued)

Compound	Trivial name	Source	Family	Structure characteristics	Activities if reported	References
90	Artemisian B	*Artemisia argyi*	Asteraceae/ compositae	The first example of [4+2] Diels–Alder type adducts presumably biosynthesized from a rare 1, 10–4, 5-diseco-guaianolide and a guaianolide diene	Neuroprotective activity	[67]
91	Artemisian C	*Artemisia argyi*	Asteraceae/ compositae	The first example of [4+2] Diels–Alder type adducts presumably biosynthesized from a rare 1, 10–4, 5-diseco-guaianolide and a guaianolide diene	–	[67]
92	Artemisian D	*Artemisia argyi*	Asteraceae/ compositae	The first example of [4+2] Diels–Alder type adducts presumably biosynthesized from a rare 1, 10–4, 5-diseco-guaianolide and a guaianolide diene	–	[67]
93	Arteannoide A	*Artemisia annua*	Asteraceae/ compositae	A rare fused 6,8-dioxabicyclo[3.2.1]octan-7-one ring system	–	[68]
94	Ainsliadimer A	*Ainsliaea macrocephala*	Asteraceae/ compositae	A unique cyclopentane system connecting the two sesquiterpene lactone units	Inhibitory effect against the production of NO	[69]
95	Ainsliatrimer A	*Ainsliaea fulvioides*	Asteraceae/ compositae	The first two guaianolide-type sesquiterpene lactone trimer	Cytotoxicity	[70]
96	Ainsliatrimer B	*Ainsliaea fulvioides*	Asteraceae/ compositae	The first two guaianolide-type sesquiterpene lactone trimer	Cytotoxicity	[70]
97	Ainsliatriolide A	*Ainsliaea fragrans*	Asteraceae/ compositae	The first example of compound trimerized from guaianolide sesquiterpenoids through two different C–C linkages	Cytotoxicity	[71]
98	Ainsliatriolide B	*Ainsliaea fragrans*	Asteraceae/ compositae	The first example of compound trimerized from guaianolide sesquiterpenoids through two different C–C linkages	–	[71]
99	Ainsliatetramer A	*Ainsliaea fragrans*	Asteraceae/ compositae	Complex skeleton constructed from four sesquiterpene units via three different linkages	Cytotoxicity	[72]
100	Ainsliatetramer B	*Ainsliaea fragrans*	Asteraceae/ compositae	Complex skeleton constructed from four sesquiterpene units via three different linkages	Cytotoxicity	[72]
101	Ainsfragolide	*Ainsliaea fragrans*	Asteraceae/ compositae	An unusual guaianolide sesquiterpene trimer generated with a novel C–C linkage at $C_{2'}$–$C_{15''}$	Cytotoxicity	[73]

(continued)

Table A1 (continued)

Compound	Trivial name	Source	Family	Structure characteristics	Activities if reported	References
102	Macrocephadiolide A	*Ainsliaea macrocephala*	Asteraceae/ compositae	Featuring a rare 5,6-spirocyclic ketal lactone core and a C-15/C-15′ linkage between a guaianolide and a 3,4-*seco*-guaianolide monomer	Anti-inflammatory activity	[74]
103	Macrocephadiolide B	*Ainsliaea macrocephala*	Asteraceae/ compositae	Featuring a rare 5,6-spirocyclic ketal lactone core and a C-15/C-15′ linkage between a guaianolide and a 3,4-seco-guaianolide monomer	Anti-inflammatory activity	[74]
104	Macrocephatriolide A	*Ainsliaea macrocephala*	Asteraceae/ compositae	A trimeric architecture features a cyclohexene linkage and a methylene bridge	–	[75]
105	Macrocephatriolide B	*Ainsliaea macrocephala*	Asteraceae/ compositae	A new scaffold of insulin sensitizers	PTP1B inhibitory and insulin sensitizing effects	[75]
106	Inulajaponicolide A	*Inula japonica*	Asteraceae/ compositae	An undescribed carbon skeleton comprising of one xanthanolide and two guaianolide units with the linkage mode of C-11/C-3′ and C-11′/C-1′′ via a Diels–Alder cycloaddition reaction	Anti-inflammatory activity	[76]
107	Vlasoulamine A	*Vladimiria souliei*	Asteraceae/ compositae	A fully hydrogenated pyrrolo[2,1,5-cd]indolizine core	Neuroprotective activity	[77]
108	Commiphoroid A	*Resina Commiphora*	Burseraceae	Sesquiterpene dimer formed via [4+2]-cycloaddition reaction	Against adiposederived stem cell proliferation and differentiation	[78]
109	Commiphoroid B	*Resina Commiphora*	Burseraceae	Sesquiterpene dimer formed via [4+2]-cycloaddition reaction	–	[78]
110	Commiphoroid C	*Resina Commiphora*	Burseraceae	A 6/6/5/6/6/6 hexacyclic framework	–	[78]
111	Commiphoroid D	*Resina Commiphora*	Burseraceae	A 8-oxabicyclo[3.2.1]oct-6-ene skeletal core	–	[78]
112	Commiphoratone A	*Resina Commiphora*	Burseraceae	A unique 6/6/5/5/6(5)/6 heptacyclic architecture, a unique saddle-like architecture	Block lipid metabolism	[79]
113	Commiphoratone B	*Resina Commiphora*	Burseraceae	A 5(5)/7/5/6/6 ring system	Block lipid metabolism	[79]

(continued)

Table A1 (continued)

Compound	Trivial name	Source	Family	Structure characteristics	Activities if reported	References
114	Commiphorine A	*Resina Commiphora*	Burseraceae	An unprecedented sesquiterpenoid heterodimer featuring a 5/7(5)/4/10/5 ring system	Lipogenesis inhibition	[80]
115	Commiphorine B	*Resina Commiphora*	Burseraceae	An unusual 1,4-dioxin motif and a 6/6/5/6/6/6 polycyclic system	Lipogenesis inhibition, anti-inflammatory activity	[80]
116	Volvalerelactone B	*Valeriana officinalis*	Valerianaceae	A unique five-member ring system connecting the two sesquiterpenoid units	–	[39]
117	Nardochinoid A	*Nardostachys chinensis*	Valerianaceae	A rare fused 3,8-dioxatricyclo[7.2.1.$0^{1,6}$]dodecane-11-one ring system	–	[81]
118	Nardochinoid B	*Nardostachys chinensis*	Valerianaceae	The first nitrogen-containing nornardosinane–aristolane sesquiterpene conjugate	–	[81]
119	Nardochinoid C	*Nardostachys chinensis*	Valerianaceae	The first example of the dimer of a nornardosinane and a nardosinane sesquiterpene	Anti-inflammatory activity	[81]
120	Involucratustone A	*Stahlianthus involucratus*	Zingiberaceae	A rearranged homodimer of cadinane sesquiterpene fused with aunique fully substituted 1-oxaspiro[4.4]nonane core observed for the first time in natural products	Cytotoxicity	[82]
121	Involucratustone B	*Stahlianthus involucratus*	Zingiberaceae	A rearranged homodimer of cadinane sesquiterpene fused with aunique fully substituted 1-oxaspiro[4.4]nonane core observed for the first time in natural products	Cytotoxicity	[82]
122	Involucratustone C	*Stahlianthus involucratus*	Zingiberaceae	Anovel 3′,4′-seco-cadinane-dimer	Anti-inflammatory activity	[82]
123	Xylopiana A	*Xylopia vielana*	Annonaceae	An unprecedented pentacyclo[5.2.1.$0^{1,2}$.$0^{4,5'}$.$0^{5,4'}$]decane-3,2′-dione core	–	[83]
124	Eugenunilone A	*Eugenia uniflora*	Myrtaceae	Unprecedented dimer with caged tricyclo[4.4.0.$0^{2,8}$] decane	–	[47]
125	(±)-eugenunilone B	*Eugenia uniflora*	Myrtaceae	Unprecedented dimer with caged tricyclo[4.4.0.$0^{2,8}$]decane	Anti-inflammatory effect	[47]

Table A2 Diterpenoids from plants

Compound	Trivial name	Source	Family	Types	Structure characteristics	Activities if reported	References
126	Lagaspholone A	*Euphorbia lagascae*	Euphorbiaceae	Jatropholane	5/6/7/3 fused ring	–	[84]
127	Lagaspholone B	*Euphorbia lagascae*	Euphorbiaceae	Jatropholane	5/6/7/3 fused ring	–	[84]
128	Heliojatrone A	*Euphorbia helioscopia*	Euphorbiaceae	Jatrophane	*Trans*-bicyclo[8.3.0]tridecane core	–	[85]
129	Heliojatrone B	*Euphorbia helioscopia*	Euphorbiaceae	Jatrophane	*Trans*-bicyclo[8.3.0]tridecane core	Pglycoprotein inhibitory activity	[85]
130	Secoheliosphane A	*Euphorbia helioscopia*	Euphorbiaceae	Jatrophane	7,8-*seco*-jatrophane	–	[86]
131	Secoheliosphane B	*Euphorbia helioscopia*	Euphorbiaceae	Jatrophane	7,8-*seco*-jatrophane	Against HSV-1	[86]
132	Secoheliospho-lane A	*Euphorbia helioscopia*	Euphorbiaceae	Jatrophane	9,10-*seco*-7,10-epoxyjatropholane, 5/6/7/7-fused tetracyclic ring	–	[86]
133	Euphopia A	*Euphorbia helioscopia*	Euphorbiaceae	Jatrophane	Tricyclo[8.3.0.0^{2,7}]tridecane	Anti-inflammatory activity	[87]
134	Euphopia B	*Euphorbia helioscopia*	Euphorbiaceae	Jatrophane	Tricyclo[8.3.0.0^{2,7}]tridecane	Anti-inflammatory activity	[87]
135	Euphopia C	*Euphorbia helioscopia*	Euphorbiaceae	Jatrophane	Tetracyclo[11.3.0.0^{2,10}.0^{3,7}]hexadecane	Anti-inflammatory activity	[87]
136	Euphopia D	*Euphorbia helioscopia*	Euphorbiaceae	Jatrophane	A unique caged tetracyclic[10.2.1.0^{2,10}.0^{5,9}]pentadecane core skeleton	Anti-inflammatory activity	[88]
137	Euphopia E	*Euphorbia helioscopia*	Euphorbiaceae	Jatrophane	A tricyclic [8.3.0.0^{2,8}]tridecane core skeleton	–	[88]
138	Euphopia F	*Euphorbia helioscopia*	Euphorbiaceae	Jatrophane	A tricyclic[10.3.0.0^{2,8}] pentadecane core skeleton	–	[88]
139	Excolide A	*Excoecaria agallocha*	Euphorbiaceae	Labdane	Tetracyclic 2,3-secolabdanoid	–	[89]
140	11-*epi*-excolide A	*Excoecaria agallocha*	Euphorbiaceae	Labdane	Tetracyclic 2,3-secolabdanoid	–	[89]

(continued)

Table A2 (continued)

Compound	Trivial name	Source	Family	Types	Structure characteristics	Activities if reported	References
141	11,13-di-*epi*-excolide A	*Excoecaria agallocha*	Euphorbiaceae	Labdane	Tetracyclic 2,3-secolabdanoid	–	[89]
142	Excolide B	*Excoecaria agallocha*	Euphorbiaceae	Labdane	Tricyclic 2,3-secolabdanoid	–	[89]
143	Euphorikanin A	*Euphorbia kansui*	Euphorbiaceae	Ingenane	5/6/7/3-fused tetracyclic ring	Cytotoxicity	[90]
144	Pepluanol C	*Euphorbia peplus*	Euphorbiaceae	Ingenane	5/5/10 with *out*,*out*-[7.2.1]bicylcododecane core	Kv1.3 inhibitory effects	[91]
145	Pepluanol D		Euphorbiaceae	ingenane	6/6/7/3 fused-ring	Kv1.3 inhibitory effects	[91]
146	Euphorkanlide A	*Euphorbia kansuensis*	Euphorbiaceae	Ingenane	A highly modified ingenane diterpenoid with a C24 appendage forming an additional hexahydroisobenzofuran-fused 19-membered macrocyclic bis-lactone ring system	Cytotoxicity	[92]
147	Euphorstranoid A	*Euphorbia stracheyi*	Euphorbiaceae	Ingenane	Highly rearranged ingenane diterpenoids with an unusual 5/6/7/3 carbon ring system	TG-lowering Activity	[93]
148	Euphorstranoid B	*Euphorbia stracheyi*	Euphorbiaceae	Ingenane	Highly rearranged ingenane diterpenoids with an unusual 5/6/7/3 carbon ring system	TG-lowering Activity	[93]
149	Lathyranoic acid A	*Euphorbia lathyris*	Euphorbiaceae	Lathyrane	Secolathyrane	–	[94]
150	Lathyranone A	*Euphorbia lathyris*	Euphorbiaceae	Lathyrane	Cyclohexanone moiety	–	[95]
151	Euphorbactin	*Euphorbia micractina*	Euphorbiaceae	Lathyrane	6/5/7/3 fused-ring	Against HIV-1	[96]
152	Pepluacetal	*Euphorbia peplus*	Euphorbiaceae	Lathyrane	5/4/7/3 fused-ring	Immunosuppressive activity	[97]
153	Pepluanol A	*Euphorbia peplus*	Euphorbiaceae	Lathyrane	5/6/7/3 fused-ring	Immunosuppressive activity	[97]
154	Pepluanol B	*Euphorbia peplus*	Euphorbiaceae	Lathyrane	5/5/8/3 fused-ring	Immunosuppressive activity	[97]
155	Wallichanol A	*Euphorbia wallichii*	Euphorbiaceae	Wallichane	Cyclobutane ring	Inhibit osteoclastogenesis	[98]

(continued)

Table A2 (continued)

Compound	Trivial name	Source	Family	Types	Structure characteristics	Activities if reported	References
156	Wallichanol B	*Euphorbia wallichii*	Euphorbiaceae	Wallichane	Cyclobutane ring	Inhibit osteoclastogenesis	[98]
157	Quorumolide A	*Euphorbia antiquorum*	Euphorbiaceae	Cembrane	Cembranoid embedding an α,β-unsaturated-γ-lactone and a tetrahydro-2*H*-pyran motif within the 14-membered ring	–	[99]
158	Euphordraculoate A	*Euphorbia dracunculoides*	Euphorbiaceae	Tigliane	6/6/3-fused ring system with a 2-methyl-2-cyclopentenone moiety	–	[100]
159	Euphordraculoate B	*Euphorbia dracunculoides*	Euphorbiaceae	Tigliane	5/5/6/3-fused tetracyclic ring core	Inhibit Wnt pathway	[100]
160	Pedrolide	*Euphorbia pedroi*	Euphorbiaceae	Tigliane	Containing a bicycle[2.2.1]heptane system	Reversed multidrug resistance	[101]
161	Euphomilone A	*Euphorbia milii*	Euphorbiaceae	Rosane	7/5/6 tricyclic system	Inhibition of osteoclastogenesis	[102]
162	Euphomilone B	*Euphorbia milii*	Euphorbiaceae	Rosane	5/7/6 fused ring system	–	[102]
163	Pre-segetanin	Sea Spurge *Euphorbia paralias*	Euphorbiaceae	Segetane	Provides a first insight into the biosynthesis of diterpenoids with a segetane skeleton	–	[103]
164	Gaditanone	*Euphorbia gaditana*	Euphorbiaceae	Gaditanane	5/6/4/6 fused-ring skeleton	–	[104]
165	Crokonoid A	*Croton kongensis*	Euphorbiaceae	Kaurene	A dual-bridged tricyclo[4.4.1.1^{1,4}]dodecane-2,11-dione ring	Cytotoxicity	[105]
166	Laevinoid A	*Croton laevigatus*	Euphorbiaceae	Clerodane	Rearranged *ent*-clerodane scaffold incorporating an unusual 3/5 bicyclic motif	–	[106]
167	Laevinoid B	*Croton laevigatus*	Euphorbiaceae	Clerodane	Rearranged ent-clerodane scaffold incorporating an unusual 3/5 bicyclic motif, first chlorinated member of the clerodane	–	[106]
168	Crotonpenoid A	*Croton yanhuii*	Euphorbiaceae	Clerodane	A new 10-(butan-2-yl)-1,6,12-trimethyltricyclo[7.2.1.0^{2,7}]dodecane skeleton	Agonistic effect on pregnane X receptor	[107]

(continued)

Table A2 (continued)

Compound	Trivial name	Source	Family	Types	Structure characteristics	Activities if reported	References
169	Crotonpenoid B	*Croton yanhuii*	Euphorbiaceae	Clerodane	A new 10-(butan-2-yl)-1,6,12-trimethyltricyclo[7.2.1.$0^{2,7}$]dodecane skeleton	Agonistic effect on pregnane X receptor	[107]
170	Mangelonoid A	*Croton mangelong*	Euphorbiaceae	Cembrane	Bicyclo[9.3.1]pentadecane core and a rare bridgehead double bond	NF-κB inhibition	[108]
171	Mangelonoid B	*Croton mangelong*	Euphorbiaceae	Cembrane	Bicyclo[9.3.1]pentadecane core and a rare bridgehead double bond	NF-κB inhibition	[108]
172	Trigochinin A	*Trigonostemon chinensis*	Euphorbiaceae	Daphnane	Oxetane ring that formed between two oxygenated quaternary carbons	Against MET tyrosine kinase activity	[109]
173	Trigochinin B	*Trigonostemon chinensis*	Euphorbiaceae	Daphnane	Oxetane ring that formed between two oxygenated quaternary carbons	Against MET tyrosine kinase activity	[109]
174	Trigochinin C	*Trigonostemon chinensis*	Euphorbiaceae	Daphnane	Oxetane ring that formed between two oxygenated quaternary carbons	Against MET tyrosine kinase activity	[109]
175	Trigonothyrin A	*Trigonostemon thyrsoideum*	Euphorbiaceae	Daphnane	Oxygen-bridged four-membered-ring system, linkage mode of 12,13,14-orthoester	–	[110]
176	Trigonothyrin B	*Trigonostemon thyrsoideum*	Euphorbiaceae	Daphnane	Oxygen-bridged four-membered-ring system, linkage mode of 12,13,14-orthoester	–	[110]
177	Trigonothyrin C	*Trigonostemon thyrsoideum*	Euphorbiaceae	Daphnane	Oxygen-bridged four-membered-ring system, linkage mode of 12,13,14-orthoester	Inhibit HIV-1	[110]
178	Trigocherrin A	*Trigonostemon cherrieri*	Euphorbiaceae	Daphnane	α,β-unsaturated dichlorovinyl moiety	Against Chikungunya virus	[111]
179	Jatrophalactam	*Jatropha curcas*	Euphorbiaceae	Casbane	5/13/3 tricyclic skeleton	–	[112]
180	Spirocurcasone	*Jatropha curcas*	Euphorbiaceae	Rhamnofolane	"spirorhamnofolane" skeleton	–	[113]
181	Jatrofoliane A	*Jatropha gossypiifolia*	Euphorbiaceae	Lathyrane	An unusual transannular 1,3-dioxolane moiety, a unique 5/6/5/8/3 ring system	–	[114]
182	Jatrofoliane B	*Jatropha gossypiifolia*	Euphorbiaceae	Lathyrane	A new 10,11:13,14-diseco-lathyrane skeleton with a 12-membered macrocyclic lactone ring	Antitumor activity	[114]

(continued)

Table A2 (continued)

Compound	Trivial name	Source	Family	Types	Structure characteristics	Activities if reported	References
183	Euphohyrisnoid A	*Euphorbia lathyris*	Euphorbiaceae	Lathyrane	Featuring a unique tetracyclo-[10.2.2.$0^{1,10}$.$0^{3,7}$]cetane skeleton	Lipid-lowering activity	[115]
184	Euphohyrisnoid B	*Euphorbia lathyris*	Euphorbiaceae	Lathyrane	Featuring a unique tricyclo[8.4.1.$0^{3,7}$]pentadecane skeleton	–	[115]
185	Fischdiabietane A	*Euphorbia fischeriana*	Euphorbiaceae	Abietane	A novel asymmetric diterpenoid dimer with a unique nonacyclic 6/6/6/5/7/6/6/6/6 ring system	Cytotoxicity	[116]
186	Excisanin H	*Rabdosia excisa*	Lamiaceae	Kaurane	14,20-epoxy-*ent*-kaurene	Cytotoxicity	[117]
187	Maoecrystal V	*Isodon eriocalyx*	Lamiaceae	Kaurane	6,7-*seco*-6-*nor*-15(8 → 9)-abeo-5,8-epoxy-*ent*-kaurane	Cytotoxicity	[118]
188	Maoecrystal Z	*Isodon eriocalyx*	Lamiaceae	Kaurane	Tetracyclic 6,7:8,15-di-seco-7,20-olide-6,8-cycloent-kaurane	Anti-tumor activity	[119]
189	Bisrubescensin A	*Isodon rubescens*	Lamiaceae	Kaurane	C_{23} ent-kaurane dimer	Cytotoxicity	[120]
190	Bisrubescensin B	*Isodon rubescens*	Lamiaceae	Kaurane	Dimer linked through single carbon–carbon bond	–	[120]
191	Bisrubescensin C	*Isodon rubescens*	Lamiaceae	Kaurane	Dimer linked through single carbon–carbon bond	–	[120]
192	Neolaxiflorin A	*Isodon eriocalyx* var. *laxiflora*	Lamiaceae	Kaurane	Bicyclo[3.1.0]hexane unit	–	[121]
193	Neolaxiflorin B	*Isodon eriocalyx* var. *laxiflora*	Lamiaceae	Kaurane	An α,β-unsaturated ketone in its five-membered ring A	–	[121]
194	Ternifolide A	*Isodon ternifolius*	Lamiaceae	Kaurane	10-memberedlactone ring	–	[122]
195	Laxiflorolide A	*Isodon eriocalyx*	Lamiaceae	Kaurane	*Ent*-kauranoids bearing a unique C_{22} carbon framework	–	[123]
196	Laxiflorolide B	*Isodon eriocalyx*	Lamiaceae	Kaurane	*Ent*-kauranoids bearing a unique C_{22} carbon framework	–	[123]
197	Pharicusin A	*Isodon pharicus*	Lamiaceae	Kaurane	Composed of a benzoyl group coupled to a 7α,20-epoxy-*ent*-kauranoid, represents an unprecedented C_{27} meroditerpenoid	–	[124]

(continued)

Table A2 (continued)

Compound	Trivial name	Source	Family	Types	Structure characteristics	Activities if reported	References
198	Maoeriocalysin A	*Isodon eriocalyx*	Lamiaceae	Kaurane	A novel rearranged *ent*-kaurane diterpenoid with an unprecedented 4,5-*seco*-3,5-cyclo-7,20-epoxy-*ent*-kaurane scaffold	–	[125]
199	Maoeriocalysin B	*Isodon eriocalyx*	Lamiaceae	Kaurane	A rare 9,10-*seco*-7,20-epoxy-*ent*-kaurane diterpenoid	Anti-AChE activity, anticoagulant activity	[125]
200	Maoeriocalysin C	*Isodon eriocalyx*	Lamiaceae	Kaurane	A rare 9,10-*seco*-7,20-epoxy-*ent*-kaurane diterpenoid	–	[125]
201	Maoeriocalysin D	*Isodon eriocalyx*	Lamiaceae	Kaurane	A rare 9,10-*seco*-7,20-epoxy-*ent*-kaurane diterpenoid	–	[125]
202	Hispidanin A	*Isodon hispida*	Lamiaceae	Totarane and labdane	Asymmetric dimer	–	[126]
203	Hispidanin B	*Isodon hispida*	Lamiaceae	Totarane and labdane	Asymmetric dimer	Cytotoxicity	[126]
204	Hispidanin C	*Isodon hispida*	Lamiaceae	Totarane and labdane	Asymmetric dimer	–	[126]
205	Hispidanin D	*Isodon hispida*	Lamiaceae	Totarane and labdane	Asymmetric dimer	–	[126]
206	Przewalskin B	*Salvia przewAlskii* Maxim	Lamiaceae	Abietane	Five-membered spiro ring and α-hydroxy-β-ketone lactone moiety	Anti-HIV-1_{IIIB} activity	[127]
207	Rubesanolide A	*Isodon rubescens*	Lamiaceae	Abietane	Three six-member rings form chair, boat, and twisted-chair conformations, β-lactone group formed between C-9 and C-20	–	[128]
208	Rubesanolide B	*Isodon rubescens*	Lamiaceae	Abietane	Three six-member rings form chair, boat, and twisted-chair conformations, β-lactone group formed between C-9 and C-20	Against aromatase	[128]
209	Salviyunnanone A	*Salvia yunnanensis*	Lamiaceae	Abietane	An unprecedented 7/5/6/3 ring system	–	[129]

(continued)

Table A2 (continued)

Compound	Trivial name	Source	Family	Types	Structure characteristics	Activities if reported	References
210	Salpratlactone A	*Salvia prattii*	Lamiaceae	Abietane	Unique 6/5 carbocyclic rings bearing a γ-lactone ring through an exocyclic double bond	$Ca_V3.1$ TTCC agonist	[130]
211	Salpratlactone B	*Salvia prattii*	Lamiaceae	Abietane	Unique 6/5 carbocyclic rings bearing a γ-lactone ring through an exocyclic double bond	–	[130]
212	Officinalin A	*Salvia officinalis*	Lamiaceae	Abietane	An unprecedented carbon skeleton with a tetracycline-[9.6.0.0^{3,8}.0^{12,16}]-heptadecane core and a peroxide bridge	Anti-inflammatory activity	[131]
213	Officinalin B	*Salvia officinalis*	Lamiaceae	Abietane	Tetracycline-[9.6.0.0^{3,8}.0^{12,16}]-heptadecane core and a peroxide bridge	–	[131]
214	Salvinorin C	*Salvia divinorum*	Lamiaceae	Clerodane	*Trans*-neoclerodane	–	[132]
215	Salvinicin A	*Salvia divinorum*	Lamiaceae	Clerodane	3,4-dihydroxy-2,5-dimethoxytetrahydrofuran ring	Antagonist activity	[133]
216	Salvinicin B	*Salvia divinorum*	Lamiaceae	Clerodane	3,4-dihydroxy-2,5-dimethoxytetrahydrofuran ring	Antagonist activity	[133]
217	Salvileucalin B	*Salvia leucantha* Cav	Lamiaceae	Clerodane	Tricyclo[3.2.1.02,7]octane substructure	Cytotoxicity	[134]
218	Microphyllandi-olide	*Salvia microphylla*	Lamiaceae	Clerodane	9/3 bicyclic ring	–	[135]
219	Teotihuacanin	*Salvia amarissima*	Lamiaceae	Clerodane	Spiro-10/6 bicyclic system	Modulatory activity of multidrug resistance	[136]
220	Scospirosin A	*Isodon scoparius*	Lamiaceae	Clerodane	Unprecedented spiro ent-clerodane dimers with 6/6/10/6 ring system	–	[137]
221	Scospirosin B	*Isodon scoparius*	Lamiaceae	Clerodane	Unprecedented spiro ent-clerodane dimers with 6/6/6/6/6 ring system	Immunosuppressive activity	[137]
222	Spirodesertol A	*Salvia deserta*	Lamiaceae	Icetexane	An unprecedented 6-isopropyl-3*H*-spiro[benzofuran2,1'-cyclohexane] motif	Cytotoxicity	[138]

(continued)

Table A2 (continued)

Compound	Trivial name	Source	Family	Types	Structure characteristics	Activities if reported	References
223	Spirodesertol B	*Salvia deserta*	Lamiaceae	Icetexane	An unprecedented 6-isopropyl-3*H*-spiro[benzofuran2,1'-cyclohexane] motif	–	[138]
224	Leonuketal	*Leonurus japonicus*	Lamiaceae	Labdane	Bridged spiroketal moiety fused with a ketal-γ-lactone unit	Vasorelaxant activity	[139]
225	Hyptisolide A	*Hyptis crenata* Pohl ex Benth	Lamiaceae	Abietane	7,8;11,12-bis-*seco*-abietane	Neuroprotective effects	[140]
226	Secorhodomol-lolide A	*Rhododendron molle*	Ericaceae	Grayanane	Highly acylated diterpenoids with a new 3,4-secograyanane skeleton	–	[141]
227	Secorhodomol-lolide B	*Rhododendron molle*	Ericaceae	Grayanane	Highly acylated diterpenoids with a new 3,4-secograyanane skeleton	Cytotoxicity	[141]
228	Secorhodomol-lolide C	*Rhododendron molle*	Ericaceae	Grayanane	Highly acylated diterpenoids with a new 3,4-secograyanane skeleton	–	[141]
229	Secorhodomol-lolide D	*Rhododendron molle*	Ericaceae	Grayanane	Highly acylated diterpenoids with a new 3,4-secograyanane skeleton	Analgesic and sedative effects	[141]
230	Mollolide A	*Rhododendron molle*	Ericaceae	Grayanane	1,10:2,3-disecograyanane	Analgesic effect	[142]
231	Micranthanone A	*Rhododendron micranthum*	Ericaceae	Grayanane	New tetracyclic diterpene carbon skeleton	–	[143]
232	Mollanol A	Rhododendron molle	Ericaceae	Grayanane	C-nor-D-homograyanane	Transcriptional activation effect	[144]
233	Rhodomollein XXV	Rhododendron molle	Ericaceae	Grayanane	Caged oxa-tricyclo[3.3.1.0^{3,7}]nonane ring system	–	[144]
234	Rhodomollacetal A	*Rhododendron molle*	Ericaceae	Grayanane	*cis/cis/cis/cis*-fused 6/6/6/6/5 pentacyclic ring system, 11,13,18-trioxa-pentacyclo [8.7.1.1^{5,8}.0^{2,8}.0^{12,17}]nonadecane scaffold	PTP1B inhibitory activities	[145]

(continued)

Table A2 (continued)

Compound	Trivial name	Source	Family	Types	Structure characteristics	Activities if reported	References
235	Rhodomollacetal B	*Rhododendron molle*	Ericaceae	Grayanane	4-oxatricyclo[7.2.1.$0^{1,6}$]dodecane moiety and a 2,3dihydro-4*H*-pyran-4-one unit	PTP1B inhibitory activities	[145]
236	Rhodomollacetal C	*Rhododendron molle*	Ericaceae	Grayanane	4-oxatricyclo[7.2.1.$0^{1,6}$]dodecane moiety and a 2,3dihydro-4*H*-pyran-4-one unit	PTP1B inhibitory activities	[145]
237	Rhodomollanol A	*Rhododendron molle*	Ericaceae	Grayanane	*cis*/*trans*/*trans*/*cis*/*cis*-fused 3/5/7/5/5/5 hexacyclic ring system, 7-oxabicyclo[4.2.1]nonane core decorated with three cyclopentane units	PTP1B inhibitory activity	[146]
238	Mollebenzylanol A	*Rhododendron molle*	Ericaceae	Grayanane	9-benzyl-8,10-dioxatricyclo[5.2.1.$0^{1,5}$]decanecore	PTP1B inhibitory activity	[147]
239	Mollebenzylanol B	*Rhododendron molle*	Ericaceae	Grayanane	9-benzyl-8,10-dioxatricyclo[5.2.1.$0^{1,5}$]decanecore	PTP1B inhibitory activity	[147]
240	Mollactone A	*Rhododendron molle*	Ericaceae	Grayanane	A unique 3-oxa-tricyclo[4,3,2,$0^{2,6}$]undecane motif	PTP1B inhibitory activity	[148]
241	Mollactone B	*Rhododendron molle*	Ericaceae	Grayanane	A unique 3-oxa-tricyclo[4,3,2,$0^{2,6}$]undecane motif	PTP1B inhibitory activity	[148]
242	Mollactone C	*Rhododendron molle*	Ericaceae	Grayanane	A unique 3-oxa-tricyclo[4,3,2,$0^{2,6}$]undecane motif	PTP1B inhibitory activity	[148]
243	Pierisoid A	*Pieris formosa*	Ericaceae	Grayanane	3,4-seco-grayanane	Antifeedant effect	[149]
244	Pierisoid B	*Pieris formosa*	Ericaceae	Grayanane	3,4-seco-grayanane	Antifeedant effect	[149]
245	Pierisketolide A	*Pieris formosa*	Ericaceae	Kaurene	A-homo-B-nor-*ent*-kaurane carbon skeleton with a pentacyclic 7/ 5/5/6/5 ring system	Analgesic activity	[150]
246	Pierisketone B	*Pieris formosa*	Ericaceae	Kaurene	A-homo-B-nor-*ent*-kaurane carbon skeleton with a tetracyclic 7/5/ 6/5 ring system	–	[150]
247	Pierisketone C	*Pieris formosa*	Ericaceae	Kaurene	A-homo-B-nor-*ent*-kaurane carbon skeleton with a tetracyclic 7/5/ 6/5 ring system	–	[150]
248	Kravanhin A	*Amomum kravanh*	Zingiberaceae	Spongian	*Trans-anti-cis* fused tricyclic ring system	–	[151]
249	Kravanhin B	*Amomum kravanh*	Zingiberaceae	Spongian	*Trans-anti-cis* fused tricyclic ring system	Anti-inflammatory	[151]
250	Kravanhin C	*Amomum kravanh*	Zingiberaceae	Spongian	*Trans-anti-cis* fused tricyclic ring system	–	[151]

(continued)

Table A2 (continued)

Compound	Trivial name	Source	Family	Types	Structure characteristics	Activities if reported	References
251	Maximumin A	*Amomum maximum*	Zingiberaceae	Abdane	A unique 6/6/6/6-fused ring system	Against nuclear factor kappa B (NF-κB)	[152]
252	Maximumin B	*Amomum maximum*	Zingiberaceae	Abdane	A 16(13 $\rightarrow$ 12)-*abeo*-γ-pyrone motif	Against nuclear factor kappa B (NF-κB)	[152]
253	Maximumin C	*Amomum maximum*	Zingiberaceae	Abdane	A 16(13 $\rightarrow$ 12)-*abeo*-γ-pyrone motif	Against nuclear factor kappa B (NF-κB)	[152]
254	Maximumin D	*Amomum maximum*	Zingiberaceae	Abdane	The first example of 13(12 $\rightarrow$ 17)-*abeo*-12,17-cyclolabdane diterpenoid	Against nuclear factor kappa B (NF-κB)	[152]
255	Amomaxin A	*Amomum maximum*	Zingiberaceae	Labdane	Nine-membered ring	–	[153]
256	Amomaxin B	*Amomum maximum*	Zingiberaceae	Labdane	Nine-membered ring	Anti-inflammatory	[153]
257	Hedychin A	*Hedychium forrestii*	Zingiberaceae	Labdane	6,7-dinorlabdane ditepenoids with a peroxide bridge	Cytotoxicity	[154]
258	Hedychin B	*Hedychium forrestii*	Zingiberaceae	Labdane	6,7-dinorlabdane ditepenoids with a peroxide bridge	Cytotoxicity	[154]
259	Mannolide A	*Cephalotaxus Troponoids*	Taxaceae	Cephalotane	Intact Diterpenoid Skeleton	–	[155]
260	Mannolide B	*Cephalotaxus Troponoids*	Taxaceae	Cephalotane	Intact Diterpenoid Skeleton	–	[155]
261	Mannolide C	*Cephalotaxus Troponoids*	Taxaceae	Cephalotane	Intact Diterpenoid Skeleton	–	[155]
262	Amentoditaxone	*Amentotaxus Formosana*	Taxaceae		5/5/9 fused ring	–	[156]
263	Grandione	*Torreya grandis*	Taxaceae	Abietane	A heptacyclic rearranged abietane-type dimer	–	[157]
264	Canataxpropel-lane	*Taxus canadensis*	Taxaceae	Taxane	An unprecedented 5/5/4/6/6/6-membered hexacyclic skeleton containing [3.3.2]propellane	–	[158]
265	Neovibsanin F	*Viburnum suspensum*	Caprifoliaceae	Vibsane	Bicyclo[3.3.1]nonane ring	–	[159]

(continued)

Table A2 (continued)

Compound	Trivial name	Source	Family	Types	Structure characteristics	Activities if reported	References
266	14-*epi*-neovibsanin F	*Viburnum suspensum*	Caprifoliaceae	Vibsane	Bicyclo[3.3.1]nonane ring	–	[159]
267	14-*epi*-18-oxoneovibsanin F	*Viburnum suspensum*	Caprifoliaceae	Vibsane	Bicyclo[3.3.1]nonane ring	–	[159]
268	Vibsatin A	*Viburnum tinus* cv. variegatus	Caprifoliaceae	Vibsane	Bicyclo[4.2.1]nonane moiety	Enhanced the neurite outgrowth	[160]
269	Vibsatin B	*Viburnum tinus* cv. variegatus	Caprifoliaceae	Vibsane	Bicyclo[4.2.1]nonane moiety	Enhanced the neurite outgrowth	[160]
270	(+)-paeoveitol	*Paeonia veitchii*	Ranunculaceae	Menthane	6/5/6/6fused tetracyclic ring	–	[161]
271	(−)-paeoveitol	*Paeonia veitchii*	Ranunculaceae	Menthane	6/5/6/6fused tetracyclic ring	–	[161]
272	Vulgarisin A	*Prunella vulgaris*	Ranunculaceae	Vulgarisane	5/6/4/5 fused tetracyclic ring	Cytotoxicity	[162]
273	Nigellamine A_1	*Nigella sativa*	Ranunculaceae	Dolabellane	Diterpene alkaloid	Lipid metabolism promoting activity	[163]
274	Nigellamine B_1	*Nigella sativa*	Ranunculaceae	Dolabellane	Diterpene alkaloid	–	[163]
275	Nigellamine C_1	*Nigella sativa*	Ranunculaceae	Dolabellane	Diterpene alkaloid	Lipid metabolism promoting activity	[163]
276	Nigellamine D_1	*Nigella sativa*	Ranunculaceae	Dolabellane	Diterpene alkaloid	Lipid metabolism promoting activity	[163]
277	Aphanamene A	*Aphanamixis grandifolia*	Meliaceae	Acyclic diterpene	[4+2]-cycloaddition dimer	Anti-inflammatory	[164]
278	Aphanamene B	*Aphanamixis grandifolia*	Meliaceae	Acyclic diterpene	[4+2]-cycloaddition dimer	Anti-inflammatory	[164]
279	Hongkonoid A	*Dysoxylum hongkongense*	Meliaceae	Phytane	A unique 5,5,5-fused tricyclic spiroketal butyrolactone moiety and diterpenoid-derived long chain	–	[165]

(continued)

Table A2 (continued)

Compound	Trivial name	Source	Family	Types	Structure characteristics	Activities if reported	References
280	Hongkonoid B	*Dysoxylum hongkongense*	Meliaceae	Phytane	A unique 5,5,5-fused tricyclic spiroketal butyrolactone moiety and diterpenoid-derived long chain	–	[165]
281	Hongkonoid C	*Dysoxylum hongkongense*	Meliaceae	Phytane	A unique 5,5,5-fused tricyclic spiroketal butyrolactone moiety and diterpenoid-derived long chain	Selective inhibition against mouse 11β-HSD1, against the sterol synthesis in HepG2 cells	[165]
282	Hongkonoid D	*Dysoxylum hongkongense*	Meliaceae	Phytane	A unique 5,5,5-fused tricyclic spiroketal butyrolactone moiety and diterpenoid-derived long chain	Selective inhibition against mouse 11β-HSD1	[165]
283	Bicunningine A	*Cunninghamia lanceolata*	Taxodiaceae	Abietane	A 2,3-dihydrofuran ring fusing an abietane and a 4,5-seco-abietane diterpene	–	[166]
284	Bicunningine B	*Cunninghamia lanceolata*	Taxodiaceae	Abietane	A 2,3-dihydrofuran ring fusing an abietane and a 4,5-seco-abietane diterpene	–	[166]
285	Cunlanceloic acid A	*Cunninghamia lanceolata*	Taxodiaceae	Labdane	An unprecedented dimeric labdane backbone linked through C-16/C-16′ bonds	AChE inhibitory activity	[167]
286	Cunlanceloic acid B	*Cunninghamia lanceolata*	Taxodiaceae	Labdane	Asymmetric diterpenoid dimer with a new carbon skeleton linked through C12/C-13′ and C-15/C-16′ bonds between two labdane units	Cytotoxicity	[167]
287	Cunlanceloic acid C	*Cunninghamia lanceolata*	Taxodiaceae	Labdane	Asymmetric diterpenoid dimer with a new carbon skeleton linked through C12/C-13′ and C-15/C-16′ bonds between two labdane units	Cytotoxicity	[167]
288	Cunlanceloic acid D	*Cunninghamia lanceolata*	Taxodiaceae	Labdane	Asymmetric diterpenoid dimer with a new carbon skeleton linked through C12/C-13′ and C-15/C-16′ bonds between two labdane units	–	[167]
289	Taxodikaloid A	*Taxodium ascendens*	Taxodiaceae	Abietane	Oxazoline ring linkage between two monomers	Neuroprotective activity	[168]

(continued)

Table A2 (continued)

Compound	Trivial name	Source	Family	Types	Structure characteristics	Activities if reported	References
290	Taxodikaloid B	*Taxodium ascendens*	Taxodiaceae	Abietane	Oxazoline ring linkage between two monomers	Neuroprotective activity	[168]
291	Cinnamomol A	*Cinnamomum cassia*	Lauraceae	Isoryanodane	Cage-like, rigid, 5/5/5/5/5/6-fused hexacyclic ring system	Immunomodulatoryactivity	[169]
292	Cinnamomol B	*Cinnamomum cassia*	Lauraceae	Isoryanodane	Cage-like, rigid, 5/5/5/5/5/6-fused hexacyclic ring system	Immunomodulatoryactivity	[169]
293	Cassiabudanol A	*Cinnamomum cassia*	Lauraceae	Isoryanodane (cassiabudane)	An unprecedented 11,14-*cyclo*-8,14:12,13-di-*seco*-isoryanodane (cassiabudane) carbon skeleton featuring a unique 3-oxatetracyclo[6.6.1.$0^{2,6}$.$0^{10,14}$]pentadecane bridged system	Immunostimulative activity	[170]
294	Cassiabudanol B	*Cinnamomum cassia*	Lauraceae	Isoryanodane (cassiabudane)	An unprecedented 11,14-*cyclo*-8,14:12,13-di-*seco*-isoryanodane (cassiabudane) carbon skeleton featuring a unique 3-oxatetracyclo[6.6.1.$0^{2,6}$.$0^{10,14}$]pentadecane bridged system	Immunostimulative activity	[170]
295	Mitrephorone A	*Mitrephora glabra* Scheff	Annonaceae	Trachylobane	Hexacyclic ring system with adjacent ketone moieties and oxetane ring	Cytotoxicity	[171]
296	Annoglabayin	*Annona glabra*	Annonaceae	Kaurane	Carbon bridge between two nor-*ent*-kaurane monomeric units	–	[172]
297	Cordifolide A	*Tinospora cordifolia*	Menispermaceae	Clerodane	Sulfur-containing clerodane	Immunomodulatoryactivity	[173]
298	Scaparvin A	*Scapania parva*	Scapaniaceae	Clerodane	C-6/C-11 bond and a ketal ring	–	[174]
299	Atisane 1	*Lepidolaena clavigera*	Lepidolaenaceae	Cembrane	Poly-oxygenated	–	[175]
300	Atisane 2	*Lepidolaena clavigera*	Lepidolaenaceae	Cembrane	Poly-oxygenated	Cytotoxic and insecticidal activity	[175]
301	Populusone	*Populus euphratica*	Salicaceae	Cembrane	An unprecedented, snail-shaped trinorditerpenoid skeleton	–	[176]

(continued)

Table A2 (continued)

Compound	Trivial name	Source	Family	Types	Structure characteristics	Activities if reported	References
302	Populusene A	*Populus euphratica*	Salicaceae	Cembrane	Featuring a bicyclo[8.4.1]-pentadecane nucleus and a bridgehead double bond (anti-Bredt system)	Anti-inflammatory	[177]
303	Populusin A	*Populus euphratica*	Salicaceae	Cembrane	Possessing an uncommon dioxatricyclo[6.6.1.$1^{2,5}$]hexadecane scaffold	Anti-inflammatory	[177]
304	Pallambin A	*Pallavicinia ambigua*	Pallaviciniaceae	Labdane	Tetracyclo[4.4.$0^{3,5}$.$0^{2,8}$]decane skeleton	Cytotoxicity	[178]
305	Pallambin B	*Pallavicinia ambigua*	Pallaviciniaceae	Labdane	Tetracyclo[4.4.$0^{3,5}$.$0^{2,8}$]decane skeleton	Cytotoxicity	[178]
306	Pallamolide A	*Pallavicinia ambigua*	Pallaviciniaceae	Labdane	A bicyclo[2.2.2]octane moiety	–	[179]
307	Pallamolide B	*Pallavicinia ambigua*	Pallaviciniaceae	Labdane	A bicyclo[2.2.2]octane moiety	–	[179]
308	Pallamolide C	*Pallavicinia ambigua*	Pallaviciniaceae	Labdane	A bicyclo[2.2.2]octane moiety	–	[179]
309	Pallamolide D	*Pallavicinia ambigua*	Pallaviciniaceae	Labdane	A bicyclo[2.2.2]octane moiety	Bacteriostatic action	[179]
310	Pallamolide E	*Pallavicinia ambigua*	Pallaviciniaceae	Labdane	A bicyclo[2.2.2]octane moiety	Bacteriostatic action	[179]
311	Hapmnioide A	*Haplomitrium mnioides*	Haplomitriaceae	Labdane	1,10:5,6-di-*seco*-12,20olide-1,5:6,10-dicyclolabdane	Anti-inflammatory activity	[180]
312	Hapmnioide B	*Haplomitrium mnioides*	Haplomitriaceae	Labdane	1,10:5,6-di-*seco*-12,20:6,18diolide-1,5:18,8-dicyclolabdane	–	[180]
313	Hapmnioide C	*Haplomitrium mnioides*	Haplomitriaceae	Labdane	1,10:5,6-di-*seco*-12,20:6,18diolide-1,5:18,8-dicyclolabdane	–	[180]
314	Hypophyllin A	*Hypoestes phyllostachya*	Acanthaceae	Labdane	7,8-seco-6,11-cyclolabdane	–	[181]
315	Hypophyllin B	*Hypoestes phyllostachya*	Acanthaceae	Labdane	6(7 → 8)-abeolabdane	–	[181]
316	Hypophyllin C	*Hypoestes phyllostachya*	Acanthaceae	Labdane	6(7 → 8)-abeolabdane	–	[181]
317	Hypophyllin D	*Hypoestes phyllostachya*	Acanthaceae	Labdane	8,9-dioxatricyclic[4.2.1.$1^{3,7}$]decane moiety	Vasorelaxant activity	[181]
318	Schaffnerine	*Acacia schaffneri*	Mimosaceae	Cassane	12-membered heterocyclic ring with a C_2 symmetry axis	–	[182]

(continued)

Table A2 (continued)

Compound	Trivial name	Source	Family	Types	Structure characteristics	Activities if reported	References
319	Caesalminaxin A1	*Caesalpinia minax*	Leguminosae	Cassane	Cleavage of the C-13–C-14 bond	–	[183]
320	Caesalminaxin A2	*Caesalpinia minax*	Leguminosae	Cassane	Cleavage of the C-13–C-14 bond	–	[183]
321	(*M*)-bicelaphanol A	*Celastrus orbiculatus*	Celastraceae	Podocarpane	Dimeric podocarpane-type trinorditerpene	Neuroprotective effect	[184]
322	(*P*)-bicelaphanol A	*Celastrus orbiculatus*	Celastraceae	Podocarpane	Dimeric podocarpane-type trinorditerpene	Neuroprotective effect	[184]
323	Tagalsin I	*Ceriops tagal*	Rhizophoraceae	Bisdolabrane	Bisdolabrane backbone tetraditerpenoid	–	[185]
324	Tagalsin J	*Ceriops tagal*	Rhizophoraceae	Bisdolabrane	Bisdolabrane backbone tetraditerpenoid	–	[185]
325	Tagalide A	*Ceriops tagal*	Rhizophoraceae	Dolabrane	The first C22 skeletal dolabrane with an unprecedented 5/6/6/6-fused tetracyclic core	Anti-breast cancer activity	[186]
326	Tagalol	*Ceriops tagal*	Rhizophoraceae	Dolabrane	The first 3-nordolabrane with a novel 5/6/6-fused tricarbocyclic scaffold	–	[186]
327	Premnafulvol A	*Premna fulva*	Verbenaceae	Isopimarane	6/5/7/3-fused tetracyclic core	–	[187]
328	Fraxinuacidoside	*Fraxinus sieboldiana*	Oleaceae		Norditerpene glucopyranoside	–	[188]
329	Oliviformislac-tone A	*Icacina oliviformis*	Icacinaceae	Pimarane	An unprecedented 4,12-dioxatetracyclo[8.6.0.0^{2,7}.0^{10,14}]hexadecane core	PTP1B inhibitory activity	[189]
330	Oliviformislac-tone B	*Icacina oliviformis*	Icacinaceae	Pimarane	An unprecedented 4,12-dioxatetracyclo[8.6.0.0^{2,7}.0^{10,14}]hexadecane core	–	[189]
331	Trichanthol A	*Icacina trichantha*	Icacinaceae	Pimarane	The first example of a pimarane-derived diterpenoid dimer furnished by forming an undescribed C-16–C-7′ linkage	–	[190]
332	Commiphorane A	*Resina Commiphora*	Burseraceae	–	A 6/6/6/6 ring system	–	[191]
333	Commiphorane B	*Resina Commiphora*	Burseraceae	–	A 6/6/6/6 ring system	–	[191]

Table A3 Sesterterpenoids from plants

Compound	Trivial name	Source	Family	Structure characteristics	Activities if reported	References
334	Bolivianine	*Hedyosmum Angustifolium*	Chloranthaceae	Polycyclic structure	–	[192]
335	Genepolide	*Artemisia umbelliformis*	Compositae/asteraceae	γ-lactone with a novel carbon skeleton	–	[193]
336	Leucosceptrine	*Leucosceptrum canum*	Lamiaceae	Novel skeleton	Prolylendopeptidase inhibitory activity	[194]
337	Norleucosceptroid A	*Leucosceptrum canum*	Lamiaceae	Highly oxygenated pentanor-sesterterpenoids	Antifeedant activity	[195]
338	Norleucosceptroid B	*Leucosceptrum canum*	Lamiaceae	Highly oxygenated pentanor-sesterterpenoids	Antifeedant activity	[195]
339	Norleucosceptroid C	*Leucosceptrum canum*	Lamiaceae	Highly oxygenated pentanor-sesterterpenoids	Antifeedant activity	[195]
340	Colquhounoid A	*Colquhounia coccinea*	Lamiaceae	Highly oxygenated cyclopropyl containing sesterterpenoid	–	[196]
341	Colquhounoid B	*Colquhounia coccinea*	Lamiaceae	3,21-*seco* derivative of colquhounoid A	Antifeedant activity	[196]
342	Colquhounoid C	*Colquhounia coccinea*	Lamiaceae	1,6-dioxaspiro[4.4] substructure	Antifeedant activity	[196]
343	Colquhounoid D	*Colquhounia coccinea*	Lamiaceae	New C-14 epimeric sesterterpenoid	Immunosuppressive activity	[197]
344	14-*epi*-colquhounoid D	*Colquhounia coccinea*	Lamiaceae	New C-14 epimeric sesterterpenoid	Immunosuppressive activity	[197]
345	Norcolquhounoid A	*Colquhounia coccinea*	Lamiaceae	New C_{21} framework	–	[197]
346	Norcolquhounoid B	*Colquhounia coccinea*	Lamiaceae	New C_{21} framework	–	[197]
347	Norcolquhounoid C	*Colquhounia coccinea*	Lamiaceae	New C_{20} framework	–	[197]
348	Norcolquhounoid D	*Colquhounia coccinea*	Lamiaceae	New C_{20} framework	–	[197]
349	Norcolquhounoid E	*Colquhounia coccinea*	Lamiaceae	New C_{20} framework	–	[197]
350	Gentianelloid A	*Gentianella turkestanorum*	Gemianaceae	An unusual 10,11-*seco*-gentianellane skeleton	Immunosuppressive activity	[198]
351	Gentianelloid B	*Gentianella turkestanorum*	Gemianaceae	An unusual 10,11-*seco*-gentianellane skeleton	Immunosuppressive activity	[198]

(continued)

Table A3 (continued)

Compound	Trivial name	Source	Family	Structure characteristics	Activities if reported	References
352	Eurysoloid A	*Eurysolen gracilis*	Papilionaceae	Possessing a pentacyclic 5/6/5/10/5 framework with an unusual macrocyclic ether system	Immunosuppressive activity	[199]
353	Eurysoloid B	*Eurysolen gracilis*	Papilionaceae	Possessing a pentacyclic 5/6/5/10/5 framework with an unusual macrocyclic ether system	Immunosuppressive activity, adipogenesis inhibitory activity	[199]

Table A4 Triterpenoids from plants

Compound	Trivial name	Source	Family	Structure characteristics	Activities if reported	References
354	Micrandilactone A	*Schisandra micrantha*	Schisandraceae	Highly oxidized, rearranged cycloartane skeleton	–	[200]
355	Lancifodilactone F	*Schisandra lancifolia*	Schisandraceae	Rearranged pentanortriterpenoid backbone	Cytotoxicity, anti-HIV activity	[201]
356	Lancifodilactone G	*Schisandra lancifolia*	Schisandraceae	Partial enol structure and a spirocyclic moiety	Cytotoxicity, anti-HIV activity	[202]
357	Rubriflordilactone A	*Schisandra Rubriflora*	Schisandraceae	Highly unsaturated rearranged bisnortriterpenoids possessing a biosynthetically modified aromatic D-ring	Anti-HIV-1 activity	[203]
358	Rubriflordilactone B	*Schisandra Rubriflora*	Schisandraceae	Highly unsaturated rearranged bisnortriterpenoids possessing a biosynthetically modified aromatic D-ring	Anti-HIV-1 activity	[203]
359	Schinalactone A	*Schisandra sphenanthera*	Schisandraceae	Five membered carbon ring featuring C-30 connected to C-1	Cytotoxicity	[204]
360	Henrischinin A	*Schisandra henryi*	Schisandraceae	3-one-2-oxabicyclo[3.2.1]-octane	Cytotoxicity	[205]
361	Henrischinin B	*Schisandra henryi*	Schisandraceae	3-one-2-oxabicyclo[3.2.1]-octane	Cytotoxicity	[205]
362	Henrischinin C	*Schisandra henryi*	Schisandraceae	*3-one-2-oxabicyclo[3.2.1]-octane*	–	[205]
363	Sphenadilactone A	*Schisandra Sphenanthera*	Schisandraceae	Nortriterpenoid with a diversity of highly oxygenated structure	Anti-HIV-1 activity	[206]
364	Sphenadilactone B	*Schisandra sphenanthera*	Schisandraceae	Nortriterpenoid with a diversity of highly oxygenated structure	–	[206]
365	Pre-schisanartanin	*Schisandra chinensis*	Schisandraceae	Highly oxygenated nortriterpene skeleton including a unique 7/8/3 consecutive carbocycle	Anti-HIV-1 activity	[207]
366	Arisandilactone A	*Schisandra arisanensis*	Schisandraceae	5/5/7/5/8/5-fused hexacyclic ring system	Anti-HSV-1 virus activity	[208]

(continued)

Table A4 (continued)

Compound	Trivial name	Source	Family	Structure characteristics	Activities if reported	References
367	Schinarisanlactone A	*Schisandra arisanensis*	Schisandraceae	5/7/7/5/7/5/6/5-fused octacyclic ring system	Anti-HIV activity	[209]
368	Schiglautone A	*Schisandra glaucescens*	Schisandraceae	6/7/9-fused tricyclic carbon backbone	Cytotoxicity	[210]
369	Schicagenin A	*Schisandra chinensis*	Schisandraceae	Tetracyclic oxa-cage motif and C_9 side chain	–	[211]
370	Schicagenin B	*Schisandra chinensis*	Schisandraceae	Tetracyclic oxa-cage motif and C_9 side chain	–	[211]
371	Schicagenin C	*Schisandra chinensis*	Schisandraceae	Tetracyclic oxa-cage motif and C_9 side chain	–	[211]
372	Schilancitrilactone A	*Schisandra lancifolia*	Schisandraceae	5/5/7/5/5/5-fused hexacyclic ring system with a C_{29} backbone	–	[212]
373	Schilancitrilactone B	*Schisandra lancifolia*	Schisandraceae	C_{27} skeleton with a 5/7/5/5/5-fused pentacyclic ring system	–	[212]
374	Schilancitrilactone C	*Schisandra lancifolia*	Schisandraceae	C_{27} skeleton with a 5/7/5/5/5-fused pentacyclic ring system	Anti-HIV-1 activity	[212]
375	Lancolide A	*Schisandra lancifolia*	Schisandraceae	Tricyclo[6.3.0.0^{2,11}]undecane-bridged system	Antiplatelet aggregation	[213]
376	Lancolide B	*Schisandra lancifolia*	Schisandraceae	Tricyclo[6.3.0.0^{2,11}]undecane-bridged system	–	[213]
377	Lancolide C	*Schisandra lancifolia*	Schisandraceae	Tricyclo[6.3.0.0^{2,11}]undecane-bridged system	–	[213]
378	Lancolide D	*Schisandra lancifolia*	Schisandraceae	Tricyclo[6.3.0.0^{2,11}]undecane-bridged system	Antiplatelet aggregation	[213]
379	Lancifonin E	*Schisandra lancifolia*	Schisandraceae	7/7 fused carbocyclic core with an internal ester bridge between C-9 and C-14	Antioxidative activity	[214]
380	Lancifonin F	*Schisandra lancifolia*	Schisandraceae	7/7 fused carbocyclic core with an internal ester bridge between C-9 and C-14	–	[214]

(continued)

Table A4 (continued)

Compound	Trivial name	Source	Family	Structure characteristics	Activities if reported	References
381	Schincalide A	*Schisandra incarnate*	Schisandraceae	Tricyclo[5.2.1.$0^{1,6}$]decane-bridged system	Immunosuppressive activity	[215]
382	Spiroschincarin A	*Schisandra incarnata*	Schisandraceae	1-oxaspiro[6.6]tridecane motif	Immunosuppressive activity	[216]
383	Spiroschincarin B	*Schisandra incarnata*	Schisandraceae	1-oxaspiro[6.6]tridecane motif	Immunosuppressive activity	[216]
384	Spiroschincarin C	*Schisandra incarnata*	Schisandraceae	1-oxaspiro[6.6]tridecane motif	–	[216]
385	Spiroschincarin D	*Schisandra incarnata*	Schisandraceae	1-oxaspiro[6.6]tridecane motif	–	[216]
386	Spiroschincarin E	*Schisandra incarnata*	Schisandraceae	1-oxaspiro[6.6]tridecane motif	–	[216]
387	Schilancidilactone C	*Schisandra lancifolia*	Schisandraceae	The first 3-norlancischiartane with unusual configurational inversions occurring at C-1 and C-10	–	[217]
388	Schincalactone A	*Schisandra incarnata*	Schisandraceae	5/5/6/11/3 ring system	–	[218]
389	Schincalactone B	*Schisandra incarnata*	Schisandraceae	5/5/6/11/3 ring system	–	[218]
390	Incarnolide A	*Schisandra incarnata*	Schisandraceae	A tricyclo[9.2.1.$0^{2,8}$]tetradecane-bridged system	Antiviral and Neuroprotective activities	[219]
391	Incarnolide B	*Schisandra incarnata*	Schisandraceae	A tricyclo[9.2.1.02,8]tetradecane-bridged system	–	[219]
392	Kadlongilactone A	*Kadsura longipeduncu-lata*	Schisandraceae	Rearranged hexacyclic backbone	Cytotoxicity	[220]
393	Kadlongilactone B	*Kadsura Longipeduncu-lata*	Schisandraceae	Rearranged hexacyclic backbone	Cytotoxicity	[220]

(continued)

Table A4 (continued)

Compound	Trivial name	Source	Family	Structure characteristics	Activities if reported	References
394	Kadsuphilactone A	*Kadsura Philippinensis*	Schisandraceae	Eleven-membered system	–	[221]
395	Spirioiridotectal A	*Iris tectorum*	Indaceae	3,6-dihydro-2*H*-pyran moiety	Neuroprotective activity	[222]
396	Spirioiridotectal B	*Iris tectorum*	Indaceae	3,6-dihydro-2*H*-pyran moiety	Neuroprotective activity	[222]
397	Spirioiridotectal C	*Iris tectorum*	Indaceae	3,6-dihydro-2*H*-pyran moiety	–	[222]
398	Spirioiridotectal D	*Iris tectorum*	Indaceae	3,6-dihydro-2*H*-pyran moiety	–	[222]
399	Spirioiridotectal E	*Iris tectorum*	Indaceae	3,6-dihydro-2*H*-pyran moiety	–	[222]
400	Spirioiridotectal F	*Iris tectorum*	Indaceae	3,6-dihydro-2*H*-pyran moiety	Neuroprotective activity	[222]
401	Polycycloiridal A	*Iris tectorum*	Indaceae	α terpineol moiety resulting from cyclization of the homofarnesylside chain	Hepatoprotective activity	[223]
402	Polycycloiridal B	*Iris tectorum*	Indaceae	α terpineol moiety resulting from cyclization of the homofarnesylside chain	Hepatoprotective activity	[223]
403	Polycycloiridal C	*Iris tectorum*	Indaceae	α terpineol moiety resulting from cyclization of the homofarnesylside chain	Hepatoprotective activity	[223]
404	Polycycloiridal D	*Iris tectorum*	Indaceae	α terpineol moiety resulting from cyclization of the homofarnesylside chain	Hepatoprotective activity	[223]
405	Polycycloiridal E	*Iris tectorum*	Indaceae	Cyclopentane ring	–	[224]
406	Polycycloiridal F	*Iris tectorum*	Indaceae	Cyclopentane ring	–	[224]
407	Polycycloiridal G	*Iris tectorum*	Indaceae	Cyclopentane ring	–	[224]
408	Polycycloiridal H	*Iris tectorum*	Indaceae	Cyclopentane ring	–	[224]
409	Polycycloiridal I	*Iris tectorum*	Indaceae	Cyclopentane ring	–	[224]
410	Polycycloiridal J	*Iris tectorum*	Indaceae	Cyclopentane ring	–	[224]
411	Dibelamcandal A	*Belamcanda chinensis*	Indaceae	Six-membered ring linking two iridal type triterpenoid nuclei	Molluscicide activity	[225]

(continued)

Table A4 (continued)

Compound	Trivial name	Source	Family	Structure characteristics	Activities if reported	References
412	Belamchinane A	*Belamcanda chinensis*	Indaceae	4/6/6/6/5 polycyclic system	Against age-related renal fibrosis	[226]
413	Belamchinane B	*Belamcanda chinensis*	Indaceae	4/6/6/6/5 polycyclic system	Against age-related renal fibrosis	[226]
414	Belamchinane C	*Belamcanda chinensis*	Indaceae	4/6/6/6/5 polycyclic system	Against age-related renal fibrosis	[226]
415	Belamchinane D	*Belamcanda chinensis*	Indaceae	4/6/6/6/5 polycyclic system	Against age-related renal fibrosis	[226]
416	Phainanoid A	*Phyllanthus hainanensis*	Euphorbiaceae	Incorporating two unique motifs of a 4,5-and a 5,5-spirocyclic systems	–	[227]
417	Phainanoid B	*Phyllanthus hainanensis*	Euphorbiaceae	Incorporating two unique motifs of a 4,5-and a 5,5-spirocyclic systems	–	[227]
418	Phainanoid C	*Phyllanthus hainanensis*	Euphorbiaceae	Incorporating two unique motifs of a 4,5-and a 5,5-spirocyclic systems	–	[227]
419	Phainanoid D	*Phyllanthus hainanensis*	Euphorbiaceae	Incorporating two unique motifs of a 4,5-and a 5,5-spirocyclic systems	–	[227]
420	Phainanoid E	*Phyllanthus hainanensis*	Euphorbiaceae	Incorporating two unique motifs of a 4,5-and a 5,5-spirocyclic systems	–	[227]
421	Phainanoid F	*Phyllanthus hainanensis*	Euphorbiaceae	Incorporating two unique motifs of a 4,5- and a 5,5-spirocyclic systems	Immunosuppressive activity	[227]
422	Phainanolide A	*Phyllanthus hainanensis*	Euphorbiaceae	6/9/6 heterotricyclic system in the down-left and a highly oxygenated 5,5-spirocyclic ketal lactone motif in the up-right	Cytotoxic and immunosuppressive activities	[228]
423	Spiropedroxodiol	*Euphorbia pedroi*	Euphorbiaceae	Spiro scaffold	Multidrug-resistance reversers	[229]
424	Ebracpene A	*Euphorbia ebracteolata*	Euphorbiaceae	Unusual ring C-*seco* and ring D-aromatic nor-triterpenoids	–	[230]
425	Ebracpene B	*Euphorbia ebracteolata*	Euphorbiaceae	Unusual ring C-*seco* and ring D-aromatic nor-triterpenoids	Lipase Inhibitory	[230]

(continued)

Table A4 (continued)

Compound	Trivial name	Source	Family	Structure characteristics	Activities if reported	References
426	Euphorol J	*Euphorbia stracheyi*	Euphorbiaceae	The first instances of 9,11-*seco*-euphane	Cytotoxicity	[231]
427	Euphorstranol A	*Euphorbia stracheyi*	Euphorbiaceae	The first instances of 9,11-*seco*-euphane	Cytotoxicity	[231]
428	Dysoxyhainanin A	*Dysoxylum hainanense*	Meliaceae	1,3-cyclo-2,3-seco A ring with a formamido containing appendage	Antibacterial activity	[232]
429	Dysoxyhainanin B	*Dysoxylum hainanense*	Meliaceae	1,2-dinor-3,10:9,10-*bisseco* skeleton	–	[232]
430	Cyclodammarane	*Aglaia odorata*	Meliaceae	A new type of natural five-membered-ring triterpenoid	–	[233]
431	Walsucochin A	*Walsura cochinchinensis*	Meliaceae	Phenylacetylene moiety fused into a contracted five-membered C-ring	Cell protecting activity	[234]
432	Walsucochin B	*Walsura cochinchinensis*	Meliaceae	Phenylacetylene moiety fused into a contracted five-membered C-ring	Cell protecting activity	[234]
433	Turraenine	*Turraea sp.*	Meliaceae	Nitrogen-containing dimeric nor-multiflorane triterpene	–	[235]
434	Spirochensilide A	*Abies chensiensis*	Pinaceae	8,10-cyclo-9,10-seco and methyl-rearranged carbon skeleton	Anti-inflammatory activity	[236]
435	Spirochensilide B	*Abies chensiensis*	Pinaceae	8,10-cyclo-9,10-seco and methyl-rearranged carbon skeleton	–	[236]
436	Pseudolaridimer A	*Pseudolarix amabilis*	Pinaceae	Heterodimer formed via a [4+2] Diels–Alder cycloaddition	Cytotoxicity	[237]
437	Pseudolaridimer B	*Pseudolarix amabilis*	Pinaceae	Heterodimer formed via a [4+2] Diels–Alder cycloaddition	Cytotoxicity	[237]
438	Pseudolarenone	*Pseudolarix amabilis*	Pinaceae	Featuring a unique bicyclo[8.2.1]tridecane core	Anti-inflammatory activity	[238]
439	Podocarpaside	*Actaea podocarpa*	Ranunculacea	A new class of triterpenoids	Anticomplement activity	[239]

(continued)

Table A4 (continued)

Compound	Trivial name	Source	Family	Structure characteristics	Activities if reported	References
440	Cimicifugadine	*Cimicifuga foetida*	Ranunculacea	Pyridine ring incorporated to a cycloartane triterpenoid nucleus	Cytotoxicity	[240]
441	Lygodipenoid A	*Lygodium Japonicum*	Lygodiaceae	Novel C33 tetracyclic triterpenoids with a new 9,19:24,32-dicyclopropane skeleton	Agonist activity	[241]
442	Lygodipenoid B	*Lygodium Japonicum*	Lygodiaceae	Novel C_{33} tetracyclic triterpenoids with a new 9,19:24,32-dicyclopropane skeleton	–	[241]
443	Norfriedelane A	*Malpighia emarginata*	Malpighiaceae	3-norfriedelane, possessing the α-oxo-β-lactone group	Anti-AchE activity	[242]
444	Norfriedelane B	*Malpighia emarginata*	Malpighiaceae	1,2-dinorfriedelane with a keto-lactone group	Anti-AchE activity	[242]
445	Norfriedelane C	*Malpighia emarginata*	Malpighiaceae	1,2,3-trinorfriedelane	–	[242]
446	(+)-(3*S*,7*S*,8*S*,13*R*,14*S*,17*R*,18*R*,19*R*,21*S*)-25-norfern-5(10),9(11)diene-3,7,19,28-tetraol	*Sinocalamus affinis*	Poaceae	25-norfern carbon skeleton	Anti-PTP1B activity	[243]
447	(+)-(3*S*,7*S*,8*S*,13*R*,14*S*,17*R*,18*R*,19*S*,20*S*,21*S*)-25-norfern-5(10),9(11)-diene-3,7,19,20,28-pentaol	*Sinocalamus affinis*	Poaceae	25-norfern carbon skeleton	Anti-PTP1B activity	[243]
448	(+)-(3*S*,7*S*,8*S*,13*R*,14*S*,17*R*,18*R*,21*S*)-25-Norfern-5(10),9(11)-diene19-oxo-3,7,28-triol	*Sinocalamus affinis*	Poaceae	25-norfern carbon skeleton	Anti-PTP1B activity	[243]
449	(+)-(3*S*,4*R*,7*S*,8*S*,13*R*,14*S*,17*R*,18*R*,19*R*,21*S*)-25-norfern-5(10),9(11)-diene-3,7,19,23,28-pentaol	*Sinocalamus affinis*	Poaceae	25-norfern carbon skeleton	Anti-PTP1B activity	[243]
450	(+)-(3*S*,5*S*,7*S*,8*S*,13*R*,14*S*,17*R*,18*R*,19*R*,21*S*)-25-norfern-1(10),9(11)-diene-3,7,19,28-tetraol	*Sinocalamus affinis*	Poaceae	25-norfern carbon skeleton	–	[243]
451	Alstonic acid B	*Alstonia scholaris*	Apocynaceae	2,3-secofernanes with the cyclization between C-3 and C-9	–	[244]
452	Alstoscholarinoid A	*Alstonia scholaris*	Apocynaceae	New subtypes of pentacyclic triterpenoids, with unique 6/6/6/7/5 ring system	Antihyperuricemic bioactivity	[245]

(continued)

Table A4 (continued)

Compound	Trivial name	Source	Family	Structure characteristics	Activities if reported	References
453	Alstoscholarinoid B	*Alstonia scholaris*	Apocynaceae	New subtypes of pentacyclic triterpenoids, with unique 6/6/5/6/6/6 ring system	Antihyperuricemic bioactivity	[245]
454	Nototroneside A	*Panax notoginseng*	Araliaceae	6/6/9 fused tricyclic tetranordammarane carbon skeleton	–	[246]
455	Nototroneside B	*Panax notoginseng*	Araliaceae	6/6/9 fused tricyclic tetranordammarane carbon skeleton	Neuroprotective effect	[246]
456	Nototroneside C	*Panax notoginseng*	Araliaceae	6/6/9 fused tricyclic tetranordammarane carbon skeleton	–	[246]
457	Alismanin A	*Alisma orientale*	Alismaceae	C_{34} skeleton with four six-membered and one fivemembered rings	Agonistic effects on PXR	[247]
458	Kadcoccitone A	*Kadsura coccinea*	Magnoliaceae	6/6/5/5-fused tetracyclic ring system unit and a C_9 side chain	Anti-HIV-1 activity	[248]
459	Kadcoccitone B	*Kadsura coccinea*	Magnoliaceae	6/6/5/5-fused tetracyclic ring system unit and a C_9 side chain	–	[248]
460	Ilelic acid A	*Ilex latifolia*	Aquifoliaceae	Seven membered ring	Cytotoxicity	[249]
461	Ilelic acid B	*Ilex latifolia*	Aquifoliaceae	Seven membered ring	Cytotoxicity	[249]
462	Phyteumoside A	*Phyteuma orbiculare*	Campanu-laceae	Two additional tetrahydropyran rings	–	[250]
463	Phyteumoside B	*Phyteuma orbiculare*	Campanu-laceae	17-polypodene aglycon	–	[250]
464	Canarene	*Canarium schweinfurthii*	Burseraceae	Novel skeleton	α-glucosidase inhibitory activity	[251]
465	Machiluside A	*Machilus yaoshansis*	Lauraceae	Cage-like tricyclic ring moiety	Cytotoxicity	[252]
466	Machiluside B	*Machilus yaoshansis*	Lauraceae	Cage-like tricyclic ring moiety	Cytotoxicity	[252]
467	Cucurbalsaminone A	*Momordica balsamina*	Cucurbitaceae	Featuring a unique 5/6/3/6/5-fused pentacyclic carbon skeleton	MDR-reversing activity	[253]

(continued)

Table A4 (continued)

Compound	Trivial name	Source	Family	Structure characteristics	Activities if reported	References
468	Cucurbalsaminone B	*Momordica balsamina*	Cucurbitaceae	Featuring a unique 5/6/3/6/5-fused pentacyclic carbon skeleton	MDR-reversing activity	[253]
469	Cucurbalsaminone C	*Momordica balsamina*	Cucurbitaceae	Featuring a unique 5/6/3/6/5-fused pentacyclic carbon skeleton	MDR-reversing activity	[253]
470	Duboscic acid	*Duboscia macrocarpa*	Tiliaceae	First member of a new class of triterpenoids	α-glucosidase inhibition	[254]
471	8,14-*seco*-oleana-8(26),13-dien-3β-ol	*Stevia viscida*	Compositae/ Asteraceae	*Seco*-C oleanane	–	[255]
472	8,14-*seco*-oleana-8(26),13-dien-3β-ol acetate	*Stevia eupatoria*	Compositae/ Asteraceae	*Seco*-C oleanane	–	[255]
473	3β-acetoxy-11α-hydroxy-11(12 $\rightarrow$ 3)abeooleanan-12-al	*Ficus microcarpa*	Moraceae	Five-membered C ring	–	[256]
474	3β-hydroxy-20-oxo-29(20 $\rightarrow$ 9)abeolupane	*Ficus microcarpa*	Moraceae	Unique skeleton	–	[256]
475	29,30-dinor-3β-acetoxy-18,19-dioxo-18,19-secolupane	*Ficus microcarpa*	Moraceae	Unique skeleton	–	[256]
476	Salvadione-A	*Salvia bucharica*	Labiatae	Novel carbon skeleta	–	[257]
477	Salvadione-B	*Salvia bucharica*	Labiatae	Novel carbon skeleta	–	[257]
478	Teuviscin A	*Teucrium viscidum*	Labiatae	A rare 7(8 $\rightarrow$ 9)abeo-9*R*-D:C-friedo-B′:A’-*neo*-gammacerane skeleton,	–	[258]
479	Colqueleganoid A	*Colquhounia elegans*	Labiatae	The first methyl-30 incorporated 6/6/6/ 6-cyclized carbon skeleton	Immunosuppressive action	[259]
480	Colqueleganoid B	*Colquhounia elegans*	Labiatae	The first methyl-30 incorporated 6/6/6/ 6-cyclized carbon skeleton	Immunosuppressive action	[259]
481	*Ent*-epicatechinoceanothic acid A	*Zizyphus jujuba*	Rhamnaceae	C–C bond linkage with catechin moiety	Antiproliferative activity	[260]
482	*Ent*-epicatechinoceanothic acid B	*Zizyphus jujuba*	Rhamnaceae	C–C bond linkage with catechin moiety	–	[260]
483	Epicatechino-3-deoxyceanothetric acid A	*Zizyphus jujuba*	Rhamnaceae	C–C bond linkage with catechin moiety	–	[260]

(continued)

Table A4 (continued)

Compound	Trivial name	Source	Family	Structure characteristics	Activities if reported	References
484	Picraquassin A	*Picrasma quassioides*	Simarubaceae	21,24-cycloapotirucallane skeleton	–	[261]
485	Mispyric acid	*Mischocarpus pyriformis*	Sapindaceae	Monocyclic Triterpenoid with a novel skeleton	Inhibiting DNA polymerase β activity	[262]
486	Volvalerenol A	*Valeriana hardwickii*	Valerianaceae	7/12/7 tricyclic ring system	–	[263]
487	Longipetalol A	*Dichapetalum longipetalum*	Dichapeta-laceae	An unprecedented highly modified triterpenoid with a unique 1,2-*seco*-3-(2-oxo-phenylethyl)-17α-13,30-cyclodammarane skeleton	Anti-inflammatory activity	[264]

Table A5 Limonoids from plants

Compound	Trivial name	Source	Structure characteristics	Activities if reported	References
488	Walsuronoid A	*Walsura robusta*	3,4-peroxide-bridged A-seco skeleton	Antimalarial activity	[265]
489	Walrobsin A	*Walsura robusta*	5-oxatricyclo[5.4.1^{1,4}]hendecane ring system	Anti-inflammatory activity	[266]
490	Walrobsin B	*Walsura robusta*	5-oxatricyclo[5.4.1^{1,4}]hendecane ring system	–	[266]
491	Aphananoid A	*Aphanamixis polystachya*	A rare C_{24} appendage and new 5/6/5 fused-ring framework	Anti-inflammatory activity	[267]
492	Turrapubesin A	*Turraea pubescens*	The first examples of halogenated limonoids	Cytotoxicity	[268]
493	Turrapubesin B	*Turraea pubescens*	The first examples of maleimide-bearing limonoids	–	[268]
494	Ciliatonoid A	*Toona ciliata*	*cis*-fused central motif of methylhexahydro-3*H*,6*H*-furo[3,4-*c*]oxepin-6-one	–	[269]
495	Ciliatonoid B	*Toona ciliata*	*cis*-fused central motif of methylhexahydro-3*H*,6*H*-furo[3,4-*c*]oxepin-6-one	–	[269]
496	Tooniliatone A	*Toona ciliata*	6/5/6/5 tetracarbocyclic skeleton	Modulation of adriamycin susceptibility	[270]
497	Perforalactone A	*Harrisonia perforata*	Aunique cagelike 2,4-dioxaadamantane ring system and amigrated side chain	Insecticidal effect	[271]
498	Musidunin	*Croton jatrophoides*	Acetal annulation of A, A′, and B′ rings	Antifeedant activity	[272]
499	Aphanamolide A	*Aphanamixis polystachya*	Unprecedented carbon skeleton formed by the key linkage between C-3 and C-6	Cytotoxity	[273]
500	Aphanamixoid A	*Aphanamixis polystachya*	Ring A, B-seco limonoid with a unique C-2–C-30 bond	Antifeedent activity	[274]
501	Aphanaonoid A	*Aphanamixis polystachya*	An unprecedented 7/6/5 tricyclic skeleton	–	[275]
502	Aphanaonoid B	*Aphanamixis polystachya*	2,6-dioxabicyclo[3.2.2]nonan-3-one caged ring A system	–	[275]
503	Cipadesin A	*Cipadessa cinerascens*	Rings A and C were joined via C-10 and C-11	–	[276]
504	Cipadesin B	*Cipadessa cinerascens*	Rings A and C were joined via C-10 and C-11	–	[276]
505	Cipadesin C	*Cipadessa cinerascens*	Rings A and C were joined via C-10 and C-11	–	[276]
506	Cipacinoid A	*Cipadessa cinerascens*	Rearranged tetrahydropyranyl ring B incorporating usually exocyclic C-30	–	[277]
507	Trichiconin A	*Trichilia connaroides*	Rearranged A, B-ring system	–	[278]

(continued)

Table A5 (continued)

Compound	Trivial name	Source	Structure characteristics	Activities if reported	References
508	Xylogranatin A	*Xylocarpus granatum*	9,10-*seco* skeleton, 1,9-oxygen bridge	Cytotoxicity	[279]
509	Xylogranatin B	*Xylocarpus granatum*	9,10-*seco* skeleton	Cytotoxicity	[279]
510	Xylogranatin C	*Xylocarpus granatum*	9,10-*seco* skeleton	Cytotoxicity	[279]
511	Xylogranatin D	*Xylocarpus granatum*	C-30–C-9 linkage	Cytotoxicity	[279]
512	Khayalenoid A	*Khaya senegalensis*	8-oxa-tricyclo[4.3.2.$0^{2,7}$]undecane motif	–	[280]
513	Khayalenoid B	*Khaya senegalensis*	8-oxa-tricyclo[4.3.2.$0^{2,7}$]undecane motif	–	[280]
514	Thaixylomolin B	*Xylocarpus moluccensis*	Pentasubstituted pyridine scaffold	Antiinflammatory activity	[281]
515	Thaixylomolin C	*Xylocarpus moluccensis*	Pentasubstituted pyridine scaffold	–	[281]
516	Xylomexicanin I	*Xylocarpus granatum*	A bridged skeleton between the B- and C-rings	–	[282]
517	Triconoid A	*Trichilia connaroides*	Rearranged mexicanolide skeleton	–	[283]
518	Triconoid B	*Trichilia connaroides*	Rearranged mexicanolide skeleton	–	[283]
519	Triconoid C	*Trichilia connaroides*	Rearranged mexicanolide skeleton	–	[283]
520	Xyloccensin O	*Xylocarpus granatum*	8,9,30-phragmalin ortho ester	Antifeedant activity	[284]
521	Xyloccensin P	*Xylocarpus granatum*	8,9,30-phragmalin ortho ester	Antifeedant activity	[284]
522	Chuktabularin A	*Chukrasia tabularis*	2,7-dioxabicyclo[2.2.1]heptane system	–	[285]
523	Chuktabularin B	*Chukrasia tabularis*	2,7-dioxabicyclo[2.2.1]heptane system	–	[285]
524	Chuktabularin C	*Chukrasia tabularis*	2,7-dioxabicyclo[2.2.1]heptane system	–	[285]
525	Chuktabularin D	*Chukrasia tabularis*	2,7-dioxabicyclo[2.2.1]heptane system	–	[285]
526	Chuktabrin A	*Chukrasia tabularis*	1,3-dioxolan-2-one and a 3,4-dihydro-2*H*-pyran	–	[286]
527	Chuktabrin B	*Chukrasia tabularis*	Polycyclic skeleton with a biosynthetically extended C2 unit (acetyl) at C-15	–	[286]
528	Chukvelutin A	*Chukrasia tabularis*	2,7-dioxabicyclo[2.2.1]heptane moiety	–	[287]
529	Chukvelutin B	*Chukrasia tabularis*	2,7-dioxabicyclo[2.2.1]heptane moiety	–	[287]
530	Chukvelutin C	*Chukrasia tabularis*	2,7-dioxabicyclo[2.2.1]heptane moiety	–	[287]
531	Chukrasone B	*Chukrasia tabularis*	16,19-dinor limonoid backbone with an extended C3 unit at C-15	Inhibition of the delayed rectifier (IK)Kþ current	[288]

(continued)

Table A5 (continued)

Compound	Trivial name	Source	Structure characteristics	Activities if reported	References
532	Chukfuransin A	*Chukrasia tabularis*	C-15/C-20 linkage, 2-oxaspiro[4.4]non-3-ene fragment	Cytotoxity	[289]
533	Chukfuransin B	*Chukrasia tabularis*	C-15/C-20 linkage, 2-oxaspiro[4.4]non-3-ene fragment	–	[289]
534	Chukfuransin C	*Chukrasia tabularis*	C-15/C-21 bonding	–	[289]
535	Chukfuransin D	*Chukrasia tabularis*	C-15/C-21 bonding	–	[289]
536	Guianolide A	*Carapa guianensis*	C-11–C-21 bond	Cytotoxity	[290]
537	Guianolide B	*Carapa guianensis*	C-11–C-21 bond	–	[290]
538	Thaixylomolin A	*Xylocarpus moluccensis*	6-oxabicyclo[3.2.1]octan-3-one motif	–	[281]
539	Walsucochinoid A	*Walsura cochinchinensis*	A five-membered C ring fused with a six-membered aromatic D ring	–	[291]
540	Walsucochinoid B	*Walsura cochinchinensis*	A five-membered C ring fused with a six-membered aromatic D ring	–	[291]
541	Phyllanthoid A	*Phyllanthus cochinchinensis*	6/6/5/6-fused ring system, both 19/30 and 19/29 oxygen bridges	Antifeedant activity, cytotoxity	[292]
542	Phyllanthoid B	*Phyllanthus cochinchinensis*	6/6/5/6-fused ring system	–	[292]
543	Grandifotane A	*Khaya grandifoliola*	Complex hexacyclic carbon skeleton	–	[293]
544	Chukrasone A	*Chukrasia tabularis*	Highly rearranged A/B ring system	Inhibition of the delayed rectifier (IK)Kþ current	[294]
545	Trichiconin B	*Trichilia connaroides*	A, B, D-*seco* skeleton	Anti-HIV activity	[278]
546	Trichiconin C	*Trichilia connaroides*	A, B, D-*seco* skeleton	Anti-HIV activity	[278]
547	Perforanoid A	*Harrisonia perforata*	An all-carbon quaternary stereocenter at C13 and anovel BCD tricyclic ring system	Cytotoxic activity	[295]
548	Triconoid D	*Trichilia connaroides*	Rearranged 1,2-*seco*-phragmalin skeleton	–	[283]
549	Harpertrioate A	*Harrisonia perforata*	A rearranged ring B incorporating exocyclic C-30	Against Alzheimer's disease	[296]
550	Cipadonoid A	*Cipadessa cinerasecns*	17*S*-configuration, spirocyclic skeleton	Anti-PTP1B activity	[297]
551	Cipacinoid B	*Cipadessa cinerascens*	17*S*-configuration, spirocyclic skeleton	–	[297]
552	Cipacinoid C	*Cipadessa cinerascens*	17*S*-configuration, spirocyclic skeleton	–	[297]
553	Cipacinoid D	*Cipadessa cinerascens*	spirocyclic skeleton	–	[297]

Table A6 Phloroglucinols from plants

Compound	Trivial name	Source	Family	Structure characteristics	Activities if reported	References
554	Garsubelone A	*Garcinia subelliptica*	Hypericaceae	The first dimeric polycyclic polyprenylated acylphloroglucinols type metabolite featuring a complicated 6/6/6/6/6/6/6 heptacyclic architecture containing 10 stereogenic centers	–	[298]
555	Hyperforone A	*Hypericum forrestii*	Hypericaceae	A unique C-1 H-substituted bicyclo[5.3.1]hendecane scaffold	–	[299]
556	Hyperforone B	*Hypericum forrestii*	Hypericaceae	A unique C-1 H-substituted bicyclo[5.3.1]hendecane scaffold	–	[299]
557	Hyperforone C	*Hypericum forrestii*	Hypericaceae	A unique C-1 H-substituted bicyclo[5.3.1]hendecane scaffold	–	[299]
558	Hypaluton A	*Hypericum patulum*	Hypericaceae	The first PPAP possessing an unparalleled 3,4-nor-bicyclic polyprenylated acylphloroglucinol (BPAP) scaffold	Immunosuppressive activity	[300]
559	Norwilsonnol A	*Hypericum wilsonii*	Hypericaceae	A new 6/6/5/5 tetracyclic system based on a spiro[5-oxatricyclo[6.4.0.0^{3,7}]dodecane-6′,1–1′,2’-dioxane] core	Immunosuppressive activity	[301]
560	Hyperbeanone A	*Hypericum beanii*	Hypericaceae	An undescribed benz[*f*]indene-1,9(4*H*)-dione ring system fused to a tricyclic γ-lactone unit via a ketone carbonyl	Cytotoxicity	[302]
561	Hymoin A	*Hypericum monogynum*	Hypericaceae	The first report on PPAPs with rare caged pentacyclic 5/6/6/5/5 architecture	–	[303]
562	Hymoin B	*Hypericum monogynum*	Hypericaceae	The first report on PPAPs with rare caged pentacyclic 5/6/6/5/5 architecture	–	[303]
563	Hymoin C	*Hypericum monogynum*	Hypericaceae	The first report on PPAPs with rare caged pentacyclic 5/6/6/6/5 ring system	Anti-coagulant activity	[303]
564	Hymoin D	*Hypericum monogynum*	Hypericaceae	The first report on PPAPs with rare caged pentacyclic 5/6/6/6/5 ring system	–	[303]
565	Hypermonin A	*Hypericum monogynum*	Hypericaceae	The first examples of highly modified norPPAPs characterized by a rare 7/6/6/5-tetracyclic system	–	[304]
566	Hypermonin B	*Hypericum monogynum*	Hypericaceae	The first examples of highly modified norPPAPs characterized by a rare 7/6/6/5-tetracyclic system	–	[304]
567	Hypermonin C	*Hypericum monogynum*	Hypericaceae	The first examples of highly modified norPPAPs characterized by a rare 7/6/6/5-tetracyclic system	–	[304]

(continued)

Table A6 (continued)

Compound	Trivial name	Source	Family	Structure characteristics	Activities if reported	References
568	Hypermonin D	*Hypericum monogynum*	Hypericaceae	The first examples of highly modified norPPAPs characterized by a rare 7/6/6/5-tetracyclic system	–	[304]
569	Garcicowin A	*Garcinia cowa*	Hypericaceae	Unusual polyprenylated acylphloroglucinol derivative unsubstituted at C-2 and C-6	–	[305]
570	Garciyunnanimine A	*Garcinia yunnanensis*	Hypericaceae	Contain unprecedented imine functionality at C-10	Cytotoxicity	[306]
571	Garciyunnanimine B	*Garcinia yunnanensis*	Hypericaceae	Contain unprecedented imine functionality at C-10	Cytotoxicity	[306]
572	Garciyunnanimine C	*Garcinia yunnanensis*	Hypericaceae	Contain unprecedented imine functionality at C-10	Cytotoxicity	[306]
573	Hyperberin A	*Hypericum beanii*	Hypericaceae	Bicyclo[5.3.1]hendecane core	Cytotoxic and anti-inflammatory activities	[307]
574	Hyperberin B	*Hypericum beanii*	Hypericaceae	Bicyclo[5.3.1]hendecane core	Cytotoxic and anti-inflammatory activities	[307]
575	Hyphenrone A	*Hypericum henryi*	Hypericaceae	Cleavage of the C-1/C-9 bond	AChE agonist	[308]
576	Hyphenrone B	*Hypericum henryi*	Hypericaceae	Cleavage of the C-1/C-9 bond	–	[308]
577	Hyphenrone C	*Hypericum henryi*	Hypericaceae	5/8/5 and 6/6/5/8/5 fused ring systems	AChE agonist	[308]
578	Hyphenrone D	*Hypericum henryi*	Hypericaceae	5/8/5 and 6/6/5/8/5 fused ring systems	AChE agonist, cytotoxicity	[308]
579	Ascyronone A	*Hypericum ascyron*	Hypericaceae	Unusual seven-membered carbon core fused with a c-lactone ring	–	[309]
580	Ascyronone B	*Hypericum ascyron*	Hypericaceae	Unusual seven-membered carbon core fused with a c-lactone ring	–	[309]
581	Hyperhexanone A	*Hypericum sampsonii*	Hypericaceae	Represent the first PPAP with a 1,2-seco-bicyclo[3.3.1]-PPAP core that resulting from the cleavage of C-1/C-2 bond via a retro-Claisen condensation reaction	Cytotoxicity	[310]

(continued)

Table A6 (continued)

Compound	Trivial name	Source	Family	Structure characteristics	Activities if reported	References
582	Norascyronone A	*Hypericum ascyron*	Hypericaceae	An intriguing 6/6/5/6 ring system via a [4+2] intramolecular radical cyclization	Cytotoxicity	[311]
583	Norascyronone B	*Hypericum ascyron*	Hypericaceae	An intriguing 6/6/5/6 ring system via a [4+2] intramolecular radical cyclization	Cytotoxicity	[311]
584	Elodeoidin A	*Hypericum elodeoides*	Hypericaceae	Key cyclizations with C-1 or C-3 established a 2-cyclopentyltetrahydrofuran moiet	–	[312]
585	Elodeoidin B	*Hypericum elodeoides*	Hypericaceae	Key cyclizations with C-1 or C-3 established a 2-cyclopentyltetrahydrofuran moiet	–	[312]
586	Elodeoidin C	*Hypericum elodeoides*	Hypericaceae	A [5,5]-spiroketal-fused 5–5–5–5 skeleton	–	[312]
587	Elodeoidin D	*Hypericum elodeoides*	Hypericaceae	A [5,5]-spiroketal-fused 5–5–5–5 skeleton	–	[312]
588	Elodeoidin E	*Hypericum elodeoides*	Hypericaceae	A 1,2-dioxonane-bridged 5–9–5 framework	Anti-inflammatory activity	[312]
589	Elodeoidin F	*Hypericum elodeoides*	Hypericaceae	A 1,2-dioxonane-bridged 5–9–5 framework	Anti-inflammatory activity	[312]
590	Elodeoidin G	*Hypericum elodeoides*	Hypericaceae	A 2,5-dioxabicyclo[2.2.2]octane caged structure	–	[312]
591	Elodeoidin H	*Hypericum elodeoides*	Hypericaceae	A 2,5-dioxabicyclo[2.2.2]octane caged structure	–	[312]
592	Soniiglucinol A	*Hypericum wilsonii*	Hypericaceae	A fascinating tricyclo-[7.3.1.0^{3,7}]tridecane core bearing an unusual 5/7/6 carbon skeleton	Anti-inflammatory activity	[313]
593	Soniiglucinol B	*Hypericum wilsonii*	Hypericaceae	A fascinating tricyclo-[7.3.1.0^{3,7}]tridecane core bearing an unusual 5/7/6 carbon skeleton	Anti-inflammatory activity	[313]
594	Soniiglucinol C	*Hypericum wilsonii*	Hypericaceae	Tricyclo-[7.3.1.0^{2,7}]tridecane and tricyclo-[6.3.1.0^{2,6}]dodecane moieties	–	[313]
595	Soniiglucinol D	*Hypericum wilsonii*	Hypericaceae	Tricyclo-[7.3.1.0^{2,7}]tridecane and tricyclo-[6.3.1.0^{2,6}]dodecane moieties	–	[313]

(continued)

Table A6 (continued)

Compound	Trivial name	Source	Family	Structure characteristics	Activities if reported	References
596	Hyperisampsin A	*Hypericum sampsonii*	Hypericaceae	Tetracyclo[6.3.1.1^{3,10}.0^{3,7}]tridecane skeleton	Anti-HIV activity	[314]
597	Hyperisampsin B	*Hypericum sampsonii*	Hypericaceae	Tetracyclo[6.3.1.1^{3,10}.0^{3,7}]tridecane skeleton	–	[314]
598	Hyperisampsin C	*Hypericum sampsonii*	Hypericaceae	Tetracyclo[6.3.1.1^{3,10}.0^{3,7}]tridecane skeleton	–	[314]
599	Hyperisampsin D	*Hypericum sampsonii*	Hypericaceae	Tetracyclo[6.3.1.1^{3,10}.0^{3,7}]tridecane skeleton	Anti-HIV activity	[314]
600	Hypersubone A	*Hypericum subsessile*	Hypericaceae	*Seco*-adamantane architecture and a tetracyclo-[6.3.1.1^{3,10}.0^{4,8}]-tridecane core combined with a peroxide ring	–	[315]
601	Hypersubone B	*Hypericum subsessile*	Hypericaceae	*Seco*-adamantane architecture and a tetracyclo-[6.3.1.1^{3,10}.0^{4,8}]-tridecane core combined with a peroxide ring	Cytotoxicity	[315]
602	Dioxasampsone A	*Hypericum sampsonii*	H ypericaceae	Unexpected hexacyclic skeleton with a rare 2,7-dioxabicyclo[2.2.1]heptane moiety	Cytotoxicity	[316]
603	Dioxasampsone B	*Hypericum sampsonii*	Hypericaceae	Tetrahydrofuro[3,4-*b*]furan-fused tricyclo[4.3.1.1^{5,7}]undecane skeleton	RXRα transcriptional–inhibitory activity	[316]
604	Hypatulone A	*Hypericum patulum*	Hypericaceae	A tricyclo-[4.3.1.1^{3,8}]-undecane core and a unique 5/5/7/6/ 6 pentacyclic ring system	Nitric oxide (NO) inhibitory activity	[317]
605	Norsampsone A	*Hypericum sampsonii*	Hypericaceae	Loss of C-2 carbonyl in the phloroglucinol ring	–	[318]
606	Norsampsone B	*Hypericum sampsonii*	Hypericaceae	Loss of C-2 carbonyl in the phloroglucinol ring	–	[318]
607	Norsampsone C	*Hypericum sampsonii*	Hypericaceae	Loss of C-2 carbonyl in the phloroglucinol ring	RXRα transcriptional–inhibitory activity	[318]
608	Norsampsone D	*Hypericum sampsonii*	Hypericaceae	Loss of C-2 carbonyl in the phloroglucinol ring	–	[318]
609	Hypatulin A	*Hypericum patulum*	Hypericaceae	Densely substituted tricyclic octahydro-1,5-methanopentalene core	Antimicrobacterial activity	[319]

(continued)

Table A6 (continued)

Compound	Trivial name	Source	Family	Structure characteristics	Activities if reported	References
610	Hypatulin B	*Hypericum patulum*	Hypericaceae	Bicyclo[3.2.1]octane moiety	–	[319]
611	Wilsonglucinol A	*Hypericum wilsonii*	Hypericaceae	A new hexahydrofuro[2,3-c][1,2]dioxine motif	Immunosuppressive activity	[320]
612	Wilsonglucinol B	*Hypericum wilsonii*	Hypericaceae	Decahydro-1,8:6a,10-dimethanocycloocta[1,2-*b*:1,8-*b*']difuran fragment	–	[320]
613	Wilsonglucinol C	*Hypericum wilsonii*	Hypericaceae	Decahydro-1,8:6a,10-dimethanocycloocta[1,2-*b*:1,8-*b*']difuran fragment	Immunosuppressive activity	[320]
614	Garcibractinone A	*Garcinia bracteata*	Hypericaceae	An unprecedented caged tricyclo-[4.4.1.1^{1,4}] dodecane skeleton	Anti-inflammatory activity	[321]
615	Garcibractinone B	*Garcinia bracteata*	Hypericaceae	An unprecedented caged tricyclo-[4.4.1.1^{1,4}] dodecane skeleton	Anti-inflammatory activity	[321]
616	Biyouyanagiol	*Hypericum chinense*	Hypericaceae	Cyclopenta-1,3-dione moiety	–	[322]
617	Biyoulactone A	*Hypericum chinense*	Hypericaceae	Dilactone structure containing C–C bonded bi- and tricyclic γ-lactone moiety	–	[323]
618	Biyoulactone B	*Hypericum chinense*	Hypericaceae	Dilactone structure containing C–C bonded bi- and tricyclic γ-lactone moiety	–	[323]
619	Biyoulactone C	*Hypericum chinense*	Hypericaceae	Dilactone structure containing C–C bonded bi- and tricyclic γ-lactone moiety	–	[323]
620	Furanmonogone A	*Hypericum monogynum*	Hypericaceae	Unprecedented 4,5-seco-3(2*H*)-furanone skeleton	Anti-inflammatory activity	[324]
621	Furanmonogone B	*Hypericum monogynum*	Hypericaceae	Unprecedented 4,5-seco-3(2*H*)-furanone skeleton	–	[324]
622	Hyperilongenol A	*Hypericum longistylum*	Hypericaceae	The first examples of naturally occurring 12,13-*seco*-spirocyclic PPAPs with an enolizable β,β'-tricarbonyl system	–	[325]
623	Hyperilongenol B	*Hypericum longistylum*	Hypericaceae	First examples of naturally occurring 12,13-*seco*-spirocyclic PPAPs with an enolizable β,β'-tricarbonyl system	Antibacterial activity	[325]

(continued)

Table A6 (continued)

Compound	Trivial name	Source	Family	Structure characteristics	Activities if reported	References
624	Hyperilongenol C	*Hypericum longistylum*	Hypericaceae	First examples of naturally occurring 12,13-*seco*-spirocyclic PPAPs with an enolizable β,β'-tricarbonyl system	–	[325]
625	Longisglucinol A	*Hypericum longistylum*	Hypericaceae	Featuring an unparalleled 6/6/6/5 Fused ring skeleton based on a unique 8-oxa-tetracyclo-[8.3.3.$0^{1,9}$.$0^{3,7}$]-cetane core	Anti-inflammatory Activity	[326]
626	Longisglucinol B	*Hypericum longistylum*	Hypericaceae	Featuring a 6/ 5/5 spirocyclic skeleton system	–	[326]
627	Longisglucinol C	*Hypericum longistylum*	Hypericaceae	Featuring a 6/ 5/5 spirocyclic skeleton system	–	[326]
628	Hyperuralone A	*Hypericum uralum*	Hypericaceae	Tetracyclo[5.3.1.$1^{4,9}$.$0^{4,11}$]-dodecane core	Cytotoxicity	[327]
629	Hyperuralone B	*Hypericum uralum*	Hypericaceae	Complex caged skeleton	–	[327]
630	(+)-garcimulin A	*Garcinia multiflora*	Hypericaceae	Caged tetracyclo[5.4.1.$1^{1,5}$.$0^{9,13}$]tridecane skeleton	Cytotoxicity	[328]
631	(−)-garcimulin A	*Garcinia multiflora*	Hypericaceae	Caged tetracyclo[5.4.1.$1^{1,5}$.$0^{9,13}$]tridecane skeleton	Cytotoxicity	[328]
632	Garcimulin B	*Garcinia multiflora*	Hypericaceae	Caged tetracyclo[5.4.1.$1^{1,5}$.$0^{9,13}$]tridecane skeleton	Cytotoxicity	[328]
633	(+)-garmultin A	*Garcinia multiflora*	Hypericaceae	2,11-dioxatricyclo[4.4.1.$0^{3,9}$]undecane and tricyclo[4.3.1.$0^{3,7}$]decane	–	[329]
634	(−)-garmultin A	*Garcinia multiflora*	Hypericaceae	2,11-dioxatricyclo[4.4.1.$0^{3,9}$]undecane and tricyclo[4.3.1.$0^{3,7}$]decane	–	[329]
635	(−)-garmultin B	*Garcinia multiflora*	Hypericaceae	2,11-dioxatricyclo[4.4.1.$0^{3,9}$]undecane and tricyclo[4.3.1.$0^{3,7}$]decane	–	[329]
636	(+)-garmultin F	*Garcinia multiflora*	Hypericaceae	31,35-γ-lactone ring formed through oxidative cleavage of the C-35 – C-36 bond	–	[329]
637	(−)-garmultin F	*Garcinia multiflora*	Hypericaceae	31,35-γ-lactone ring formed through oxidative cleavage of the C-35 – C-36 bond	–	[329]

(continued)

Table A6 (continued)

Compound	Trivial name	Source	Family	Structure characteristics	Activities if reported	References
638	(−)-garmultin G	*Garcinia multiflora*	Hypericaceae	31,35-γ-lactone ring formed through oxidative cleavage of the C-35 − C-36 bond	–	[329]
639	Guadial B	*Psidium guajava*	Myrtaceae	Monoterpene-based meroterpenoids with unprecedented skeleton	–	[330]
640	Guadial C	*Psidium guajava*	Myrtaceae	Monoterpene-based meroterpenoids with unprecedented skeleton	Cytotoxicity	[330]
641	Operculatol A	*Cleistocalyx operculatus*	Myrtaceae	Bearing a 2,4-dimethyl-cinnamyl-phloroglucinol moiety	–	[331]
642	Operculatol B	*Cleistocalyx operculatus*	Myrtaceae	Bearing a 2,4-dimethyl-cinnamyl-phloroglucinol moiety	–	[331]
643	Eucalrobusone F	*Eucalyptus robusta*	Myrtaceae	Novel adduct formed between aformyl-derived carbonatom on the phloroglucinol ring and monoterpene	–	[332]
644	Eucalyptusdimer A	*Eucalyptus robusta*	Myrtaceae	Fused skeleton between two phellandrene and two acylphloroglucinol subunits	Anti-AChE activity	[333]
645	Eucalyptusdimer B	*Eucalyptus robusta*	Myrtaceae	Fused skeleton between two phellandrene and two acylphloroglucinol subunits	–	[333]
646	Eucalyptusdimer C	*Eucalyptus robusta*	Myrtaceae	Fused skeleton between two phellandrene and two acylphloroglucinol subunits	–	[333]
647	Hypatone A	*Hypericum patulum*	Hypericaceae	Spiro[bicyclo[3.2.1]octane-6,1′-cyclohexan]-2′,4′,6′-trione core	Opposite effects on $Ca_V3.1$ low voltage-gated Ca^{2+} channel	[334]
648	Guadial A	*Psidium guajava*	Myrtaceae	The first monoterpene-based meroterpenoid	–	[335]
649	Oliganthin M	*Garcinia oligantha*	Hypericaceae	The first novel hybrid monoterpene-tetrahydroxanthone	–	[336]
650	(+)-cleistocaltone A	*Cleistocalyx operculatus*	Myrtaceae	Tricyclo[11.3.1.$0^{3;8}$]heptadecane bridged ring system with an unusual bridgehead enol	Anti-RSV activity	[337]
651	(−)-cleistocaltone A	*Cleistocalyx operculatus*	Myrtaceae	Tricyclo[11.3.1.$0^{3;8}$]heptadecane bridged ring system with an unusual bridgehead enol	Anti-RSV activity	[337]

(continued)

Table A6 (continued)

Compound	Trivial name	Source	Family	Structure characteristics	Activities if reported	References
652	Cleistocaltone B	*Cleistocalyx operculatus*	Myrtaceae	2,2,4-trimethyl-cinnamyl-β-triketone unit	Anti-RSV activity	[337]
653	Guajadial	*Psidium guajava*	Myrtaceae	Caryophyllene-based meroterpenoid	–	[338]
654	Psiguadial B	*Psidium guajava*	Myrtaceae	Globulol-based and a caryolane-based meroterpenoid	–	[339]
655	Guapsidial A	*Psidium guajava*	Myrtaceae	Possessing an unusual coupling pattern	–	[331]
656		*Myrtus communis*	Myrtaceae	Octahydrospiro[bicyclo[7.2.0]undecane-2,2′-chromene] tetracyclic ring system	cytotoxicity	[340]
657		*Myrtus communis*	Myrtaceae	Sesquiterpene-based meroterpenoid	cytotoxicity	[340]
658		*Myrtus communis*	Myrtaceae	Sesquiterpene-based meroterpenoid	–	[340]
659	Rhodomyrtial A	*Rhodomyrtus tomentosa*	Myrtaceae	Triketone-sesquiterpene-triketone adduct	–	[341]
660	Rhodomyrtial B	*Rhodomyrtus tomentosa*	Myrtaceae	Triketone-sesquiterpene-triketone adduct	–	[341]
661	Guajavadimer A	*Psidium guajava*	Myrtaceae	Most complex dimeric sesquiterpene-based meroterpenoids	Hepatoprotective activity	[342]
662	(+)-hyperjapone A	*Hypericum japonicum*	Hypericaceae	11/6/6 fused ring system	Antitumor activity	[343]
663	(−)-hyperjapone A	*Hypericum japonicum*	Hypericaceae	11/6/6 fused ring system	Antitumor activity	[343]
664	Hyperjapone B	*Hypericum japonicum*	Hypericaceae	Containing a caryophyllane-type sesquiterpenoid moiety in the molecule	Antitumor activity	[343]
665	Hyperjapone C	*Hypericum japonicum*	Hypericaceae	Containing a caryophyllane-type sesquiterpenoid moiety in the molecule	–	[343]

(continued)

Table A6 (continued)

Compound	Trivial name	Source	Family	Structure characteristics	Activities if reported	References
666	Hyperjapone D	*Hypericum japonicum*	Hypericaceae	Containing a caryophyllane-type sesquiterpenoid moiety in the molecule	Antitumor activity	[343]
667	Hyperjapone E	*Hypericum japonicum*	Hypericaceae	Containing a caryophyllane-type sesquiterpenoid moiety in the molecule	–	[343]
668	Drychampone A	*Dryopteris championii*	Dryopteridaceae	11/6/6 ring system coupled with a pyronone moiety	–	[344]
669	Drychampone C	*Dryopteris championii*	Dryopteridaceae	11/6/6 ring system coupled with a pyronone moiety	–	[344]
670	Frutescone A	*Baeckea frutescens*	Myrtaceae	Oxa-spiro[5.8] tetradecadiene ring system	Cytotoxicity	[345]
671	Frutescone D	*Baeckea frutescens*	Myrtaceae	Oxa-spiro[5.8] tetradecadiene ring system	Cytotoxicity	[345]
672	Rhodomyrtusial A	*Rhodomyrtus tomentosa*	Myrtaceae	Featuring a unique6/5/5/9/4 fused pentacyclic ring system	–	[346]
673	Rhodomyrtusial B	*Rhodomyrtus tomentosa*	Myrtaceae	Featuring a unique6/5/5/9/4 fused pentacyclic ring system	–	[346]
674	Rhodomyrtusial C	*Rhodomyrtus tomentosa*	Myrtaceae	Featuring a unique6/5/5/9/4 fused pentacyclic ring system	–	[346]
675	Littordial A	*Psidium littorale*	Myrtaceae	Unusual acyl phloroglucinol units	–	[347]
676	Littordial B	*Psidium littorale*	Myrtaceae	Unusual acyl phloroglucinol units	Cytotoxicity	[347]
677	Littordial C	*Psidium littorale*	Myrtaceae	Unusual acyl phloroglucinol units	Cytotoxicity	[347]
678	Littordial D	*Psidium littorale*	Myrtaceae	Unusual acyl phloroglucinol units	–	[347]
679	Littordial E	*Psidium littorale*	Myrtaceae	Unusual acyl phloroglucinol units, 6/7/9/4-fused tetracyclic ring system	Cytotoxicity	[347]

(continued)

Table A6 (continued)

Compound	Trivial name	Source	Family	Structure characteristics	Activities if reported	References
680	Psidial B	*Psidium guajava*	Myrtaceae	3,5-diformylbenzyl phloroglucinol-coupled sesquiterpenoid	Anti-PTP1B activity	[348]
681	Psidial C	*Psidium guajava*	Myrtaceae	3,5-diformylbenzyl phloroglucinol-coupled sesquiterpenoid	Anti-PTP1B activity	[348]
682	Psiguadial A	*Psidium guajava*	Myrtaceae	Globulol-based and a caryolane-based meroterpenoid	–	[339]
683	Psiguadial C	*Psidium guajava*	Myrtaceae	Sesquiterpene-based meroterpenoid	Cytotoxicity	[335]
684	Psiguadial D	*Psidium guajava*	Myrtaceae	Sesquiterpene-based meroterpenoid	Cytotoxicity	[335]
685	Hyperjaponol D	*Hypericum japonicum*	Hypericaceae	6/6/10 ring system	–	[349]
686	Hyperjaponol E	*Hypericum japonicum*	Hypericaceae	6/6/10 ring system	–	[349]
687	Hyperjaponol F	*Hypericum japonicum*	Hypericaceae	6/6/10 ring system	–	[349]
688	Hyperjaponol G	*Hypericum japonicum*	Hypericaceae	6/6/10 ring system	–	[349]
689	Eucalrobusone D	*Eucalyptus robusta*	Myrtaceae	Connecting a chroman ring to a bicyclogermacrane nucleus at the C-3/C-4 position	–	[332]
690	Eucalrobusone E	*Eucalyptus robusta*	Myrtaceae	Connecting a chroman ring to a bicyclogermacrane nucleus at the C-3/C-4 position	–	[332]
691	(−)-leptosperol A	*Leptospermum scoparium*	Myrtaceae	Featuring unprecedented 1-benzyl-2-(2-phenylethyl) cyclodecane and 2-benzyl-3-phenylethyl decahydronaphthalene backbone	Anti-inflammatory activity	[350]
692	(+)-leptosperol A	*Leptospermum scoparium*	Myrtaceae	Featuring unprecedented 1-benzyl-2-(2-phenylethyl) cyclodecane and 2-benzyl-3-phenylethyl decahydronaphthalene backbone	Anti-inflammatory activity	[350]
693	(+)-hyperjaponol A	Hypericum japonicum	Hypericaceae	6/6/11 ring system	–	[349]

(continued)

Table A6 (continued)

Compound	Trivial name	Source	Family	Structure characteristics	Activities if reported	References
694	(−)-hyperjaponol A	*Hypericum japonicum*	Hypericaceae	6/6/11 ring system	–	[349]
695	(+)-hyperjaponol B	*Hypericum japonicum*	Hypericaceae	6/6/11 ring system	–	[349]
696	(−)-hyperjaponol B	*Hypericum japonicum*	Hypericaceae	6/6/11 ring system	–	[349]
697		*Myrtus communis*	Myrtaceae	Sesquiterpene-based meroterpenoid	–	[340]
698		*Myrtus communis*	Myrtaceae	Sesquiterpene-based meroterpenoid	–	[340]
699	(−)-hypulatone A	*Hypericum patulum*	Myrtaceae	An unprecedented spiro[benzofuran2,1′-cycloundecan]-4′-ene-4,6(5*H*)-dione core	inhibition of late I_{Na}	[351]
700	(+)-hypulatone A	*Hypericum patulum*	Myrtaceae	An unprecedented spiro[benzofuran2,1′-cycloundecan]-4′-ene-4,6(5*H*)-dione core	Inhibition of late I_{Na}	[351]
701	(−)-hypulatone B	*Hypericum patulum*	Myrtaceae	An unprecedented spiro[benzofuran2,1′-cycloundecan]-4′-ene-4,6(5*H*)-dione core	Inhibition of late I_{Na}	[351]
702	(+)-hypulatone B	*Hypericum patulum*	Myrtaceae	An unprecedented spiro[benzofuran2,1′-cycloundecan]-4′-ene-4,6(5*H*)-dione core	Inhibition of late I_{Na}	[351]
703	(+)-hyperjaponol C	*Hypericum japonicum*	Hypericaceae	6/6/7/5 ring system	–	[349]
704	(−)-hyperjaponol C	*Hypericum japonicum*	Hypericaceae	6/6/7/5 ring system	–	[349]
705	Eucalyptal A	*Eucalyptus globulus*	Myrtaceae	3,5-diformyl-isopentyl phloroglucinol-coupled cadinane	Cytotoxicity	[352]
706	Eucalyptal B	*Eucalyptus globulus*	Myrtaceae	3,5-diformyl-isopentyl phloroglucinol-coupled cadinane	Cytotoxicity	[352]

(continued)

Table A6 (continued)

Compound	Trivial name	Source	Family	Structure characteristics	Activities if reported	References
707	Eucalyptal C	*Eucalyptus globulus*	Myrtaceae	3,5-diformyl-isopentyl phloroglucinol-coupled cadinane	Cytotoxicity	[352]
708	Leptosperol B	*Leptospermum scoparium*	Myrtaceae	Featuring unprecedented 1-benzyl-2-(2-phenylethyl) cyclodecane and 2-benzyl-3-phenylethyl decahydronaphthalene backbone	Anti-inflammatory activity	[350]
709	Eucalrobusone C	*Eucalyptus robusta*	Myrtaceae	Formyl-phloroglucinol meroterpenoid	Cytotoxicity	[332]
710	Eucalrobusone A	*Eucalyptus robusta*	Myrtaceae	Amaaliane sesquiterpene moiety is attached at a formyl-derived positionon the phloroglucinol ring	–	[332]
711	Eucalrobusone B	*Eucalyptus robusta*	Myrtaceae	Amaaliane sesquiterpene moiety is attached at a formyl-derived positionon the phloroglucinol ring	–	[332]
712	Eucalrobusone G	*Eucalyptus robusta*	Myrtaceae	The first examples of cubebane-based FPMs connected by an unusual 1-oxaspiro[5.5]undecane core	–	[332]
713	Eucalrobusone H	*Eucalyptus robusta*	Myrtaceae	The first examples of cubebane-based FPMs connected by an unusual 1-oxaspiro[5.5]undecane core	–	[332]
714	Eucalrobusone I	*Eucalyptus robusta*	Myrtaceae	The first examples of cubebane-based FPMs connected by an unusual 1-oxaspiro[5.5]undecane core	–	[332]
715	(+)-spirotriscoumarin A	*Toddalia asiatica*	Rutaceae	Complex coumarin trimers with a spirodienone–sesquiterpene-like hybrid skeleton	Antiviral activity	[352]
716	(−)-spirotriscoumarin A	*Toddalia asiatica*	Rutaceae	Complex coumarin trimers with a spirodienone–sesquiterpene-like hybrid skeleton	Antiviral activity	[353]
717	(+)-spirotriscoumarin B	*Toddalia asiatica*	Rutaceae	Complex coumarin trimers with a spirodienone–sesquiterpene-like hybrid skeleton	Antiviral activity	[353]
718	(−)-spirotriscoumarin B	*Toddalia asiatica*	Rutaceae	Complex coumarin trimers with a spirodienone–sesquiterpene-like hybrid skeleton	Antiviral activity	[353]

(continued)

Table A6 (continued)

Compound	Trivial name	Source	Family	Structure characteristics	Activities if reported	References
719	Chlorabietol A	*Chloranthus oldhamii*	Chloranthaceae	Ent-abietane-type diterpenoid coupled with an alkenyl phloroglucinol moiety by forming an unexpected 2,3-dihydrofuran ring	Anti-PTP1B activity	[354]
720	Chlorabietol B	*Chloranthus oldhamii*	Chloranthaceae	Ent-abietane-type diterpenoid coupled with an alkenyl phloroglucinol moiety by forming an unexpected 2,3-dihydrofuran ring	Anti-PTP1B activity	[354]
721	Chlorabietol C	*Chloranthus oldhamii*	Chloranthaceae	Ent-abietane-type diterpenoid coupled with an alkenyl phloroglucinol moiety by forming an unexpected 2,3-dihydrofuran ring	Anti-PTP1B activity	[354]
722	Lysidicin A	*Lysidice rhodostegia*	Caesalpiniaceae	Spirocyclic benzodihydrofuran skeleton	–	[355]
723	Lysidicin B	*Lysidice rhodostegia*	Caesalpiniaceae	Spirocyclic benzodihydrofuran skeleton	–	[355]
724	Lysidicin C	*Lysidice rhodostegia*	Caesalpiniaceae	Unprecedented carbon skeleton	–	[355]
725	Tomentosone A	*Rhodomyrtus tomentosa*	Myrtaceae	Hexacyclic ring system	Antimalarial activity	[356]
726	Tomentosone B	*Rhodomyrtus tomentosa*	Myrtaceae	Hexacyclic ring system	–	[356]
727	Myrtucommuacetalone	*Myrtus communis*	Myrtaceae	Unprecedented carbon skeleton	Inhibitory effect against nitric oxide (NO) production, antiproliferative activity	[357]
728	Hypercohin K	*Hypericum cohaerens*	Hypericaceae	A unique spiro-fused cyclopropane ring	Increasing the activity of AChE	[358]
729	Hyperprin A	*Hypericum przewalskii*	Hypericaceae	A new 6/6/6/6/5/5 hexacyclic system with an unprecedented tetracyclo[10.3.1.$0^{3,8}$.$0^{8,12}$]hexadecane motif	–	[359]
730	H yperprin B	*Hypericum przewalskii*	Hypericaceae	A unique 6/8/6/6 tetracyclic scaffold	PTP1B inhibition activity	[359]
731	Rhodomentosone A	*Rhodomyrtus tomentosa*	Myrtaceae	Enantiomeric phloroglucinol trimer featuring a unique 6/5/5/6/5/5/6-fused ring system	Antiviral activity	[360]

(continued)

Table A6 (continued)

Compound	Trivial name	Source	Family	Structure characteristics	Activities if reported	References
732	Rhodomentosone B	*Rhodomyrtus tomentosa*	Myrtaceae	Enantiomeric phloroglucinol trimer featuring a unique 6/5/5/6/5/5/6-fused ring system	Antiviral activity	[360]
733	(+)-cleistoperlone A	*Cleistocalyx operculatus*	Myrtaceae	Phloroglucinol dimers possessing an unprecedented polycyclic skeleton with a highly functionalized dihydropyrano[3,2-*d*]xanthene tetracyclic core	Anti-HSV-1 activity	[331]
734	(−)-cleistoperlone A	*Cleistocalyx operculatus*	Myrtaceae	Phloroglucinol dimers possessing an unprecedented polycyclic skeleton with a highly functionalized dihydropyrano[3,2-*d*]xanthene tetracyclic core	Anti-HSV-1 activity	[331]
735	(+)-cleistoperlone B	*Cleistocalyx operculatus*	Myrtaceae	Phloroglucinol dimers possessing an unprecedented polycyclic skeleton with a highly functionalized dihydropyrano[3,2-*d*]xanthene tetracyclic core	–	[331]
736	(−)-cleistoperlone B	*Cleistocalyx operculatus*	Myrtaceae	Phloroglucinol dimers possessing an unprecedented polycyclic skeleton with a highly functionalized dihydropyrano[3,2-*d*]xanthene tetracyclic core	–	[331]
737	(+)-xanthchrysone A	*Xanthostemon chrysanthus*	Myrtaceae	Bis-phenylpropanoyl-benzo[*b*]cyclopent[*e*] oxepine tricyclic backbone	–	[361]
738	(−)-xanthchrysone A	*Xanthostemon chrysanthus*	Myrtaceae	Bis-phenylpropanoyl-benzo[*b*]cyclopent[*e*] oxepine tricyclic backbone	–	[361]
739	(+)-xanthchrysone B	*Xanthostemon chrysanthus*	Myrtaceae	First examples of 1-(cyclopentylmethyl)-3-(3-phenylpropanoyl)benzene scaffold	Antibacterial activity	[361]
740	(−)-xanthchrysone B	*Xanthostemon chrysanthus*	Myrtaceae	First examples of 1-(cyclopentylmethyl)-3-(3-phenylpropanoyl)benzene scaffold	Antibacterial activity	[361]
741	(+)-xanthchrysone C	*Xanthostemon chrysanthus*	Myrtaceae	First examples of 1-(cyclopentylmethyl)-3-(3-phenylpropanoyl)benzene scaffold	–	[361]
742	(−)-xanthchrysone C	*Xanthostemon chrysanthus*	Myrtaceae	First examples of 1-(cyclopentylmethyl)-3-(3-phenylpropanoyl)benzene scaffold	–	[361]
743	(+)-melipatulinone A	*Melicope patulinervia*	Rutaceae	A novel spiro[hydrobenzofuran-2,3′-furan] 5/5/6 tricyclic ring system	Pancreatic lipase inhibitory effect	[362]

(continued)

Table A6 (continued)

Compound	Trivial name	Source	Family	Structure characteristics	Activities if reported	References
744	(−)-melipatulinone A	*Melicope patulinervia*	Rutaceae	A novel spiro[hydrobenzofuran-2,3′-furan] 5/5/6 tricyclic ring system	Pancreatic lipase inhibitory effect	[362]
745	(+)-melipatulinone B	*Melicope patulinervia*	Rutaceae	A novel spiro[hydrobenzofuran-2,3′-furan] 5/5/6 tricyclic ring system	Pancreatic lipase inhibitory effect	[362]
746	(−)-melipatulinone B	*Melicope patulinervia*	Rutaceae	A novel spiro[hydrobenzofuran-2,3′-furan] 5/5/6 tricyclic ring system	Pancreatic lipase inhibitory effect	[362]
747	(−)-melipatulinone C	*Melicope patulinervia*	Rutaceae	An unprecedented spiro[cyclopenta[*b*]hydrofuran-2,3′-furan] 5/5/5 tricyclic framework	Pancreatic lipase inhibitory effect	[362]
748	(+)-melipatulinone C	*Melicope patulinervia*	Rutaceae	An unprecedented spiro[cyclopenta[*b*]hydrofuran-2,3′-furan] 5/5/5 tricyclic framework	Pancreatic lipase inhibitory effect	[362]

Table A7 Meroterpenoids from plants

Compound	Trivial name	Source	Family	Structure characteristics	Types	Activities if reported	References
749	Littoralisone	*Verbena littoralis*	Verbenaceae	Heptacyclic skeleton including four- and nine-membered ring	Monoterpene	Enhancers of the action of nerve growth factor (NGF)	[363]
750	(+)-rhodonoid A	*Rhododendron capitatum*	Ericaceae	6/6/6/4 ring system	Monoterpenoid	–	[364]
751	(−)-rhodonoid A	*Rhododendron capitatum*	Ericaceae	6/6/6/4 ring system	Monoterpenoid	–	[364]
752	(+)-rhodonoid C	*Rhododendron capitatum*	Ericaceae	6/6/6/5 ring system	Monoterpene	Anti-HIV-1 activity	[365]
753	(−)-rhodonoid C	*Rhododendron capitatum*	Ericaceae	6/6/6/5 ring system	Monoterpene	–	[365]
754	(+)-rhodonoid D	*Rhododendron capitatum*	Ericaceae	6/6/5/5 ring system	Monoterpene	–	[365]
755	(−)-rhodonoid D	*Rhododendron capitatum*	Ericaceae	6/6/5/5 ring system	Monoterpene	–	[365]
756	Magterpenoid A	*Magnolia officinalis*	Magnoliaceae	4,6,11-trioxatricyclo[5.3.1.0^{1,5}]undecane framework with an irregular monoterpenoid moiety	Monoterpenoid	Anti-PTP1B activity	[366]
757	(+)-magterpenoid B	*Magnolia officinalis*	Magnoliaceae	6/6/6/6 polycyclic skeleton	Monoterpenoid	–	[366]
758	(−)-magterpenoid B	*Magnolia officinalis*	Magnoliaceae	6/6/6/6 polycyclic skeleton	Monoterpenoid	Anti-PTP1B activity	[366]
759	Magterpenoid C	*Magnolia officinalis*	Magnoliaceae	A novel terpenoid quinone with a C6 – C3 unit	Monoterpenoid	–	[366]
760	(±)-fissisternoid A	*Fissistigma bracteolatum*	Annonaceae	An unprecedented meroterpenoid featuring a unique tricyclo [3,3,1,0$^{1',5'}$] decane central framework	Monoterpenoid	Anti-inflammatory activity	[367]
761	(±)-fissisternoid B	*Fissistigma bracteolatum*	Annonaceae	A rare 6/6/5/4 tetracyclic carbon skeleton	Monoterpenoid	Anti-inflammatory activity	[367]
762	Fissistigmatin A	*Fissistigma bracteolatum*	Annonaceae	The first example of a new group of natural products consisting of a flavonoid and a sesquiterpene moiety	Sesquiterpene	–	[368]

(continued)

Table A7 (continued)

Compound	Trivial name	Source	Family	Structure characteristics	Types	Activities if reported	References
763	Fissistigmatin B	*Fissistigma bracteolatum*	Annonaceae	The first example of a new group of natural products consisting of a flavonoid and a sesquiterpene moiety	Sesquiterpene	–	[368]
764	Fissistigmatin C	*Fissistigma bracteolatum*	Annonaceae	The first example of a new group of natural products consisting of a flavonoid and a sesquiterpene moiety	Sesquiterpene	–	[368]
765	Fissistigmatin D	*Fissistigma bracteolatum*	Annonaceae	The first example of a new group of natural products consisting of a flavonoid and a sesquiterpene moiety	Sesquiterpene	–	[368]
766	Biyouyanagin A	*Hypericum chinense*	Hypericaceae	Contain sesquiterpene, cyclobutane, and spirolactone moiety	Sesquiterpene	Anti-HIV activity	[369]
767	Hyperdioxane A	*Hypericum ascyron*	Hypericaceae	Conjugate of dibenzo-1,4-dioxane and sesquiterpene with an unprecedented heptacyclic ring system	Sesquiterpene	–	[370]
768	Nardoaristolone A	*Nardostachys chinensis Batal*	Valerianaceae	Aristolane-type sesquiterpenoid with a chalcone moiety fused by a 2,3-dihydrofuran ring	Sesquiterpenoid	Cardiomyocytes protective activity	[371]
769	Nudibaccatumone	*Piper nudibaccatum*	Piperaceae	Trimer comprising a phenylpropanoid and two sesquiterpene moieties	Sesquiterpene	–	[372]
770	Terpecurcumin J	*Curcuma longa*	Zingiberaceae	Hydrobenzannulated[6,6]-spiroketal	Sesquiterpene	–	[373]
771	Terpecurcumin K	*Curcuma longa*	Zingiberaceae	Hydrobenzannulated[6,6]-spiroketal	Sesquiterpene	–	[373]
772	Terpecurcumin L	*Curcuma longa*	Zingiberaceae	Bicyclo[2.2.2]octene (3–7)	Sesquiterpene	–	[373]
773	Terpecurcumin M	*Curcuma longa*	Zingiberaceae	Bicyclo[2.2.2]octene (3 – 7)	Sesquiterpene	–	[373]
774	Terpecurcumin N	*Curcuma longa*	Zingiberaceae	Bicyclo[2.2.2]octene (3 – 7)	Sesquiterpene	–	[373]
775	Terpecurcumin O	*Curcuma longa*	Zingiberaceae	Bicyclo[2.2.2]octene (3 – 7)	Sesquiterpene	–	[373]
776	Terpecurcumin P	*Curcuma longa*	Zingiberaceae	Bicyclo[2.2.2]octene (3 – 7)	Sesquiterpene	–	[373]
777	Terpecurcumin Q	*Curcuma longa*	Zingiberaceae	Bicyclo[3.1.3]octene	Sesquiterpene	Cytotoxicity	[373]
778	Terpecurcumin V	*Curcuma longa*	Zingiberaceae	Bicyclo[3.1.3]octene	Sesquiterpene	–	[373]
779	Arteannoide B	*Artemisia annua*	Compositae/ Asteraceae	The first examples of cadinene sesquiterpene–phenylpropanoid conjugate	Sesquiterpene	Anti-inflammatory activity	[68]

(continued)

Table A7 (continued)

Compound	Trivial name	Source	Family	Structure characteristics	Types	Activities if reported	References
780	Arteannoide C	*Artemisia annua*	Compositae/ Asteraceae	The first examples of cadinene sesquiterpene–phenylpropanoid conjugate	Sesquiterpene	Anti-inflammatory activity	[68]
781	Parasubindole A	*Parasenecio albus*	Compositae/ Asteraceae	12*H*-cyclopentane[*b*]naphthalenespiro-1,3′-indole skeleton	Sesquiterpene	Neuroprotective effect	[374]
782	Parasubindole B	*Parasenecio albus*	Compositae/ Asteraceae	12*H*-cyclopentane[*b*]naphthalenespiro-1,3′-indole skeleton	Sesquiterpene	–	[374]
783	Parasubindole C	*Parasenecio albus*	Compositae/ Asteraceae	12*H*-cyclopentane[*b*]naphthalenespiro-1,3′-indole skeleton	Sesquiterpene	–	[374]
784	Parasubindole D	*Parasenecio albus*	Compositae/ Asteraceae	12*H*-cyclopentane[*b*]naphthalenespiro-1,3′-indole skeleton	Sesquiterpene	Neuroprotective effect	[374]
785	Parasubindole E	*Parasenecio albus*	Compositae/ Asteraceae	12*H*-cyclopentane[*b*]naphthalenespiro-1,3′-indole skeleton	Sesquiterpene	–	[374]
786	Parasubindole F	*Parasenecio albus*	Compositae/ Asteraceae	12*H*-cyclopentane[*b*]naphthalenespiro-1,3′-indole skeleton	Sesquiterpene	–	[374]
787	Parasubindole G	*Parasenecio albus*	Compositae/ Asteraceae	12*H*-cyclopentane[*b*]naphthalenespiro-1,3′-indole skeleton	Sesquiterpene	–	[374]
788	Spiroalanpyrroid A	*Inula helenium*	Compositae/ Asteraceae	An unprecedented eudesmanolide–pyrrolizidine spiro[5.5] framework	Sesquiterpene	–	[375]
789	Spiroalanpyrroid B	*Inula helenium*	Compositae/ Asteraceae	An unprecedented eudesmanolide–pyrrolizidine spiro[5.5] framework	Sesquiterpene	–	[375]
790	Xanthanoltrimer A	*Xanthium italicum*	Compositae/ asteraceae	The first discovered xanthanolide sesquiterpene trimer	Sesquiterpene	–	[376]
791	Xanthanoltrimer A	*Xanthium italicum*	Compositae/ asteraceae	The first discovered xanthanolide sesquiterpene trimer	Sesquiterpene	–	[376]
792	Xanthanoltrimer A	*Xanthium italicum*	Compositae/ asteraceae	The first discovered xanthanolide sesquiterpene trimer	Sesquiterpene	–	[376]
793	Sarglaperoxide A	*Sarcandra glabra*	Chloranthaceae	A pair of unusual sesquiterpene–normonoterpene conjugates with a peroxide bridge	Sesquiterpene	Antibacterial and anti-inflammatory activities	[377]
794	Sarglaperoxide B	*Sarcandra glabra*	Chloranthaceae	A pair of unusual sesquiterpene–normonoterpene conjugates with a peroxide bridge	Sesquiterpene	–	[377]

(continued)

Table A7 (continued)

Compound	Trivial name	Source	Family	Structure characteristics	Types	Activities if reported	References
795	Nicotabin A	*Nicotiana tabacum*	Solanaceae	Possessing a fused 5/6/5/5/5 ring system	Sesquiterpene	Anti-inflammatory activity	[378]
796	Przewalskin A	*Salvia przewalskii*	Lamiaceae	6/6/7 carbon ring skeleton	Diterpenoid	Anti-HIV-1 activity	[379]
797	Scopariusic acid	*Isodon scoparius*	Lamiaceae	Cyclobutane ring and 1-octen-3-ol substituent	Diterpene	Cytotoxicity and immunosuppressive activity	[380]
798	Perovsfolin A	*Perovskia scrophulariifolia*	Lamiaceae	An unprecedented 6/8/6/6/6 pentacyclic carbon skeleton with a $C_6 - C_3$ ester moiety	Diterpene	–	[381]
799	Perovsfolin B	*Perovskia scrophulariifolia*	Lamiaceae	An unprecedented 6/8/6/6/6 pentacyclic carbon skeleton with a C6 – C3 ester moiety	Diterpene	Anti-neuroinflammatory activity	[381]
800	Pimelotide A	*Pimelea elongata*	Thymelaeaceae	Daphnane ketallactone-type diterpenoid orthoesters	Diterpene	–	[382]
801	Pimelotide B	*Pimelea elongata*	Thymelaeaceae	Daphnane ketallactone-type diterpenoid orthoesters	Diterpene	–	[382]
802	Dysohonin A	*Dysoxylum hongkongense*	Meliaceae	An unprecedented 6,15,6-fused heterotricyclic ring system	Diterpene	PTP1B inhibitory activity	[383]
803	Fischernolide A	*Euphorbia fischeriana*	Euphorbiaceae	An unprecedented 28-carbon skeleton with a novel scaffold	Diterpene	–	[384]
804	Fischernolide B	*Euphorbia fischeriana*	Euphorbiaceae	An unprecedented 28-carbon skeleton with a novel scaffold	Diterpene	–	[384]
805	Fischernolide C	*Euphorbia fischeriana*	Euphorbiaceae	An unprecedented 28-carbon skeleton with a novel scaffold	Diterpene	–	[384]
806	Fischernolide D	*Euphorbia fischeriana*	Euphorbiaceae	An unprecedented 28-carbon skeleton with a novel scaffold	Diterpene	Cytotoxicity	[384]
807	Pycnanthuquinone A	*Pycnanthus angolensis*	Myristicaceae	Terpenoid-quinone structure containing a fused 6,6,5-ring skeleton	Diterpene	Antihyperglycemic activity	[385]
808	Pycnanthuquinone B	*Pycnanthus angolensis*	Myristicaceae	Terpenoid-quinone structure containing a fused 6,6,5-ring skeleton	Diterpene	Antihyperglycemic activity	[385]
809	Celamonol A	*Celastrus monospermus*	Celastraceae	Featuring an unusual pattern of conjunction between a 24,29-dinorfriedelanetype triterpenoid and a catechin	Triterpene	Against the proliferation of B lymphocytes	[386]

(continued)

Table A7 (continued)

Compound	Trivial name	Source	Family	Structure characteristics	Types	Activities if reported	References
810	Celamonol B	*Celastrus monospermus*	Celastraceae	Featuring an unusual pattern of conjunction between a 24,29-dinorfriedelanetype triterpenoid and a catechin	Triterpene	Against the proliferation of B lymphocytes	[386]
811	Celamonol C	*Celastrus monospermus*	Celastraceae	Featuring an unusual pattern of conjunction between a 24,29-dinorfriedelanetype triterpenoid and a catechin	Triterpene	Against the proliferation of B lymphocytes	[386]
812	Celamonol D	*Celastrus monospermus*	Celastraceae	Featuring an unusual pattern of conjunction between a 24,29-dinorfriedelanetype triterpenoid and a catechin	Triterpene	Against the proliferation of B lymphocytes	[386]
813	Hydrangenone	*Salvia hydrangea*	Lamiaceae	6/7/6/5/5 membered carbon ring skeleton	Isoprenoid	Antiplasmodial activity	[387]
814	Cryptotrione	*Cryptomeria japonica*	Taxodiaceae	Abietane diterpene with a unique bicyclic sesquiterpene	Diterpene-sesquiterpene	Cytotoxicity	[388]
815	Helikaurolide A	*Helianthus annuus*	Compositae/ Asteraceae	Combining a sesquiterpene lactone and a kaurane diterpene acid	Diterpene-sesquiterpene	–	[389]
816	Helikaurolide B	*Helianthus annuus*	Compositae/ Asteraceae	Combining a sesquiterpene lactone and a kaurane diterpene acid	Diterpene-sesquiterpene	–	[389]
817	Helikaurolide C	*Helianthus annuus*	Compositae/ Asteraceae	Combining a sesquiterpene lactone and a kaurane diterpene acid	Diterpene-sesquiterpene	–	[389]
818	Helikaurolide D	*Helianthus annuus*	Compositae/ Asteraceae	Combining a sesquiterpene lactone and a kaurane diterpene acid	Diterpene-sesquiterpene	–	[389]
819	Hitorin A	*Chloranthus japonicus*	Chloranthaceae	6/5/5/5/5/3 hexacyclic skeleton including one γ-lactone ring and two tetrahydrofuran rings	Sesquiterpene-monoterpene	–	[390]
820	Hitorin B	*Chloranthus japonicus*	Chloranthaceae	6/5/5/5/5/3 hexacyclic skeleton including one γ-lactone ring and two tetrahydrofuran rings	Sesquiterpene-monoterpene	–	[390]

Table A8 Flavonoid hybrids from plants

Compound	Trivial name	Source	Family	Structure characteristics	Activities if reported	References
821	Caeruleanone A	*Millettia caerulea*	Leguminosae	The first example of a rotenoid-type compound in which the aromaticity of the D-ring is disrupted	–	[391]
822	(−)-millpuline A	*Millettia pulchra*	Leguminosae	A complicated skeleton constructed by a four-membered carbocyclic ring	Inhibitory activity on miR-144-3p expression	[392]
823	(+)-millpuline A	*Millettia pulchra*	Leguminosae	A complicated skeleton constructed by a four-membered carbocyclic ring	–	[392]
824	Caesalpinnone A	*Caesalpinia enneaphylla*	Leguminosae	10,11-dioxatricyclic [5.3.3.$0^{1,6}$]tridecane-bridged system	Cytotoxicity	[393]
825	(+)-cajanusflavanol A	*Cajanus cajan*	Leguminosae	Highly functionalized cyclopenta[1,2,3-de]isobenzopyran-1-one tricyclic core	Anti-inflammatory activity	[394]
826	(−)-cajanusflavanol A	*Cajanus cajan*	Leguminosae	Highly functionalized cyclopenta[1,2,3-de]isobenzopyran-1-one tricyclic core	Anti-inflammatory activity	[394]
827	(+)-cajanusflavanol B	*Cajanus cajan*	Leguminosae	The first example of methylene-unit-linked flavonostilbenes	Anti-inflammatory activity	[394]
828	(−)-cajanusflavanol B	*Cajanus cajan*	Leguminosae	The first example of methylene-unit-linked flavonostilbenes	Anti-inflammatory activity	[394]
829	(+)-cajanusflavanol C	*Cajanus cajan*	Leguminosae	The first example of methylene-unit-linked flavonostilbenes	–	[394]
830	(−)-cajanusflavanol C	*Cajanus cajan*	Leguminosae	The first example of methylene-unit-linked flavonostilbenes	–	[394]
831	Saffloquinoside A	*Carthamus tinctorius*	Compositae/ asteraceae	Six-five member dioxaspirocycle	Anti-inflammatory activity	[395]
832	Saffloquinoside B	*Carthamus tinctorius*	Compositae/ asteraceae	Cyclohexatrione skeleton with a benzyl group and two C-glycosyl units	–	[395]
833	Saffloflavoneside A	*Carthamus tinctorius*	Compositae/ asteraceae	Possessing a furan conjoining tetrahydrofuran ring	Neuroprotective activity	[396]
834	Saffloflavoneside B	*Carthamus tinctorius*	Compositae/ asteraceae	Possessing a furan conjoining tetrahydrofuran ring	Neuroprotective activity	[396]

(continued)

Table A8 (continued)

Compound	Trivial name	Source	Family	Structure characteristics	Activities if reported	References
835	Carthorquinoside A	*Carthamus tinctorius*	Compositae/ asteraceae	Quinochalcone–flavonol structure linked via a methylene bridge	Anti-inflammatory activitiy	[397]
836	Carthorquinoside B	*Carthamus tinctorius*	Compositae/ asteraceae	Two glucopyranosylquinochalcone moieties linked via the formyl carbon of an acyclic glucosyl unit	Anti-inflammatory activitiy	[397]
837	Morusyunnansin A	*Morus yunnanensis*	Moraceae	Two monomers were linked through the isoprenoid groups	–	[398]
838	Morusyunnansin B	*Morus yunnanensis*	Moraceae	Two monomers were linked through the isoprenoid groups	–	[398]
839	Acutifolin A	*Brosimum acutifolium*	Moraceae	Bicyclo[3.3.1]non-3-ene-2,9-dione ring	–	[399]
840	Morunigrine A	*Morus nigra*	Moraceae	A rearranged chalcone – stilbene/2-arylbenzofuran core decorated with a unique methylbiphenyl moiety	PTP1B inhibitory activity	[400]
841	Morunigrine B	*Morus nigra*	Moraceae	A rearranged chalcone–stilbene/2-arylbenzofuran core decorated with a unique methylbiphenyl moiety	PTP1B inhibitory activity	[400]
842	(+)-tephrorins A	*Tephrosia purpurea*	Papilionaceae	Tetrahydrofuran moiety	–	[401]
843	(+)-tephrorins B	*Tephrosia purpurea*	Papilionaceae	Tetrahydrofuran moiety	Cancer chemopreventive property	[401]
844	Houttuynoid A	*Houttuynia cordata*	Saururaceae	Flavonoid with houttuynin (3-oxododecanal) tethered to hyperoside	Anti-HSV activity	[402]
845	Houttuynoid B	*Houttuynia cordata*	Saururaceae	Flavonoid with houttuynin (3-oxododecanal) tethered to hyperoside	Anti-HSV activity	[402]
846	Houttuynoid C	*Houttuynia cordata*	Saururaceae	Flavonoid with houttuynin (3-oxododecanal) tethered to hyperoside	Anti-HSV activity	[402]
847	Houttuynoid D	*Houttuynia cordata*	Saururaceae	Flavonoid with houttuynin (3-oxododecanal) tethered to hyperoside	Anti-HSV activity	[402]
848	Houttuynoid E	*Houttuynia cordata*	Saururaceae	Flavonoid with houttuynin (3-oxododecanal) tethered to hyperoside	Anti-HSV activity	[402]

(continued)

Table A8 (continued)

Compound	Trivial name	Source	Family	Structure characteristics	Activities if reported	References
849	Myristicyclin A	*Horsfieldia spicata*	Myristicaceae	Procyanidin-like congeners of myristinins lacking a pendant aromatic ring	Antimalarial activity	[403]
850	Myristicyclin B	*Horsfieldia spicata*	Myristicaceae	Procyanidin-like congeners of myristinins lacking a pendant aromatic ring	–	[403]
851	Forsythoneoside A	*Forsythia suspensa*	Oleaceae	A flavonoid unit fused to a phenylethanoid glycoside through a pyran ring or carbon–carbon bond	–	[404]
852	Forsythoneoside B	*Forsythia suspensa*	Oleaceae	A flavonoid unit fused to a phenylethanoid glycoside through a pyran ring or carbon–carbon bond	Neuroprotective activity	[404]
853	Forsythoneoside C	*Forsythia suspensa*	Oleaceae	A flavonoid unit fused to a phenylethanoid glycoside through a pyran ring or carbon–carbon bond	–	[404]
854	Forsythoneoside D	*Forsythia suspensa*	Oleaceae	A flavonoid unit fused to a phenylethanoid glycoside through a pyran ring or carbon–carbon bond	Neuroprotective activity	[404]
855	Melanodiol 4″-O-protocatechuate	*Aronia melanocarpa*	Rosaceae	Fused pentacyclic core with two contiguous hemiketals	Hydroxyl radical scavenging activity, quinone reductase-inducing activity	[405]
856	Melanodiol	*Aronia melanocarpa*	Rosaceae	Fused pentacyclic core with two contiguous hemiketals	Hydroxyl radical scavenging activity, quinone reductase-inducing activity	[405]
857	Polyflavanostilbene A	*Polygonum cuspidatum*	Polygonaceae	Rearranged flavanol skeleton fused to stilbene via a hexahydrocyclopenta[*c*]furan moiety	Inhibitory activity against α-glucosidase	[406]
858	Abiesanol A	*Abies georgei*	Pinaceae	A novel biflavanol with unique six connective hexacyclic rings by cyclization	Anti-inflammatory activitiy	[407]
859	Pinutwindoublin	*Pinus massoniana*	Pinaceae	A-type-only trimer linked with ($2\alpha \rightarrow O \rightarrow 5,4\alpha \rightarrow 6$) and ($2\alpha \rightarrow O \rightarrow 7,4\alpha \rightarrow 8$) bonds	–	[408]
860	Pinuspirotetrin	*Pinus massoniana*	Pinaceae	A spirotype dimer connected with an A-type dimer	–	[408]

(continued)

Table A8 (continued)

Compound	Trivial name	Source	Family	Structure characteristics	Activities if reported	References
861	Pinumassohexin	*Pinus massoniana*	Pinaceae	A hexameric PAC with mixed A + B-type interflavanyl linkages	Increasing the modulus of elasticity of dentin	[408]
862	Dragonin A	*Daemonorops draco*	Palmaceae	A-type flavan-3-ol-dihydroretrochalcone dimer	Anti-inflammatory activitiy	[409]
863	Dragonin B	*Daemonorops draco*	Palmaceae	A-type flavan-3-ol-dihydroretrochalcone dimer	Anti-inflammatory activitiy	[409]
864	Dragonin C	*Daemonorops draco*	Palmaceae	A-type flavan-3-ol-dihydroretrochalcone dimer	–	[409]
865	Dragonin D	*Daemonorops draco*	Palmaceae	A-type flavan-3-ol-dihydroretrochalcone dimer	–	[409]
866	Rugonidine A	*Alchornea rugosa*	Euphorbiaceae	An unprecedented 1,6-dioxa-7,9-diazaspiro[4.5]dec-7-en-8-amine scaffold	Antidiabetic activity	[410]
867	Rugonidine B	*Alchornea rugosa*	Euphorbiaceae	An unprecedented 1,6-dioxa-7,9-diazaspiro[4.5]dec-7-en-8-amine scaffold	Antidiabetic activity	[410]
868	Rugonidine C	*Alchornea rugosa*	Euphorbiaceae	An unprecedented 1,6-dioxa-7,9-diazaspiro[4.5]dec-7-en-8-amine scaffold	Antidiabetic activity	[410]
869	Rugonidine D	*Alchornea rugosa*	Euphorbiaceae	An unprecedented 1,6-dioxa-7,9-diazaspiro[4.5]dec-7-en-8-amine scaffold	–	[410]
870	Rugonidine E	*Alchornea rugosa*	Euphorbiaceae	An unprecedented 1,6-dioxa-7,9-diazaspiro[4.5]dec-7-en-8-amine scaffold	–	[410]
871	Rugonidine F	*Alchornea rugosa*	Euphorbiaceae	An unprecedented 1,6-dioxa-7,9-diazaspiro[4.5]dec-7-en-8-amine scaffold	–	[410]

Table A9 Lignin from plants

Compound	Trivial name	Source	Family	Structure characteristics	Activities if reported	References
872	Taiwanschirin A	*Schizandra arisanensis*	Schisandraceae	3,4-pentano-2,3-dihydrobenzo[*b*]furan skeleton	–	[411]
873	Taiwanschirin B	*Schizandra arisanensis*	Schisandraceae	3,4-pentano-2,3-dihydrobenzo[*b*]furan skeleton	–	[411]
874	Taiwanschirin C	*Schizandra arisanensis*	Schisandraceae	3,4-pentano-2,3-dihydrobenzo[*b*]furan skeleton	–	[411]
875	Taiwankadsurin A	*Kadsura Philippinensis*	Schisandraceae	3,4-{1′-[(*Z*)-2″-methoxy-2″-oxo-ethylidene]}-pentano(2,3-dihydro-benzo[*b*]furano)-3-(2‴-methoxycarbonyl-2‴-hydroxy-2‴,3′-epoxide) skeleton	Cytotoxicity	[412]
876	Taiwankadsurin B	*Kadsura Philippinensis*	Schisandraceae	3,4-{1′-[(*Z*)-2″-methoxy-2″-oxo-ethylidene]}-pentano(2,3-dihydro-benzo[*b*]furano)-3-(2‴-methoxycarbonyl-2‴-hydroxy-2‴,3′-epoxide) skeleton	Cytotoxicity	[412]
877	Taiwankadsurin C	*Kadsura Philippinensis*	Schisandraceae	3,4-{1′-[(*Z*)-2″-methoxy-2″-oxo-ethylidene]}-pentano(2,3-dihydro-benzo[*b*]furano)-3-(2‴-methoxycarbonyl-2‴-hydroxy-2‴,3′-epoxide) skeleton	–	[412]
878	(7′*R*,8′*S*)-3,4-dimethoxy-3′,4′-methylenedioxy-7,8-*seco*-7,7′-epoxylignan-7,8-dion	*Schisandra glaucescens*	Schisandraceae	7,8-*seco*-lignan	Neuroprotective effect	[413]

(continued)

Table A9 (continued)

Compound	Trivial name	Source	Family	Structure characteristics	Activities if reported	References
879	(7′*R*,8′*S*)-3,4-methylenedioxy-3′,4′-dimethoxy-7,8-*seco*-7,7′-epoxylignan-7,8-dione	*Schisandra glaucescens*	Schisandraceae	7,8-*seco*-lignan	Neuroprotective effect	[413]
880	Kadsuraol A	*Kadsura longipedunculata*	Schisandraceae	Four-membered ring across C-1′–C-8	–	[414]
881	Kadsuraol B	*Kadsura longipedunculata*	Schisandraceae	Four-membered ring across C-1′–C-8	–	[414]
882	Kadsuraol C	*Kadsura longipedunculata*	Schisandraceae	Four-membered ring across C-1′–C-8	Hepatoprotective activity	[414]
883	(±)-torreyunlignan A	*Torreya yunnanensis*	Taxaceae	8–9′ linked neolignan enantiomers featuring a rare (*E*)-2-styryl-1,3-dioxane moiety	Phosphodiesterase-9A inhibitor	[415]
884	(±)-torreyunlignan B	*Torreya yunnanensis*	Taxaceae	8–9′ linked neolignan enantiomers featuring a rare (*E*)-2-styryl-1,3-dioxane moiety	Phosphodiesterase-9A inhibitor	[415]
885	(±)-torreyunlignan C	*Torreya yunnanensis*	Taxaceae	8–9′ linked neolignan enantiomers featuring a rare (*E*)-2-styryl-1,3-dioxane moiety	Phosphodiesterase-9A inhibitor	[415]
886	(±)-torreyunlignan D	*Torreya yunnanensis*	Taxaceae	8–9′ linked neolignan enantiomers featuring a rare (*E*)-2-styryl-1,3-dioxane moiety	Phosphodiesterase-9A inhibitor	[415]
887	(±)-subaveniumins A	*Cinnamomum subavenium*	Lauraceae	2-oxaspiro[4.5]deca-6,9-dien-8-one motif	Anti-inflammatory activity	[416]
888	(±)-subaveniumins B	*Cinnamomum subavenium*	Lauraceae	2-oxaspiro[4.5]deca-6,9-dien-8-one motif	Anti-inflammatory activity	[416]
889	Aiphanol	*Aiphanes aculeata*	Palmaceae	A stilbene moiety is linked with a phenylpropane unit through a dioxane bridge	Cyclooxygenase inhibitory	[417]

(continued)

Table A9 (continued)

Compound	Trivial name	Source	Family	Structure characteristics	Activities if reported	References
890	Chimarrhinin	*Chimarrhis turbinata*	Rubiaceae	A new C6.C3 lignan skeleton type	Antioxidant activity	[418]
891	Rufescenolide	*Cordia rufescens*	Boraginaceae	A bicyclic [2.2.2] octene skeleton	–	[419]
892	(+)-sibiricumin A	*Xanthium sibiricum*	Compositae/asteraceae	Spirodienone neolignan	Inhibitory effects on NO production	[420]
893	(−)-sibiricumin A	*Xanthium sibiricum*	Compositae/asteraceae	Spirodienone neolignan	Inhibitory effects on NO production	[420]
894	(±)-sativamide A	*Cannabis sativa*	Moraceae	Benzo-angular triquinane skeleton	Cytotoxicity	[421]
895	(±)-sativamide B	*Cannabis sativa*	Moraceae	Benzo-angular triquinane skeleton	Cytotoxicity	[421]
896	(−)-pinnatifidaone A	*Crataegus pinnatifida*	Rosaceae	An unprecedented 5/6/6 tricyclic ring system with a rare 2-oxaspiro[4.5]deca-6-en-8-one motif	–	[422]
897	(+)-pinnatifidaone A	*Crataegus pinnatifida*	Rosaceae	An unprecedented 5/6/6 tricyclic ring system with a rare 2-oxaspiro[4.5]deca-6-en-8-one motif	–	[422]
898	(−)-pinnatifidaone B	*Crataegus pinnatifida*	Rosaceae	An unprecedented 5/6/6 tricyclic ring system with a rare 2-oxaspiro[4.5]deca-6-en-8-one motif	–	[422]
899	(+)-pinnatifidaone B	*Crataegus pinnatifida*	Rosaceae	An unprecedented 5/6/6 tricyclic ring system with a rare 2-oxaspiro[4.5]deca-6-en-8-one motif	–	[422]
900	(−)-pinnatifidaone C	*Crataegus pinnatifida*	Rosaceae	A unique 2-oxaspiro[4.5]deca6-en-8-one 5/5/6/6 tetracyclic framework	–	[422]
901	(+)-pinnatifidaone C	*Crataegus pinnatifida*	Rosaceae	A unique 2-oxaspiro[4.5]deca6-en-8-one 5/5/6/6 tetracyclic framework	Cytotoxicity	[422]

(continued)

Table A9 (continued)

Compound	Trivial name	Source	Family	Structure characteristics	Activities if reported	References
902	Pibeneolignan A	*Piper betle*	Piperaceae	New naturally occurring neolignan skeleton	–	[423]
903	Pibeneolignan B	*Piper betle*	Piperaceae	New naturally occurring neolignan skeleton	–	[423]
904	(±)-piperhancin A	*Piper hancei*	Piperaceae	An unprecedented 1′,2:1,2′-dicyclo-8,3′-neolignane	Anti-inflammatory activity	[424]

Table A10 Alkaloids from plants

Compound	Trivial name	Source	Family	Structure characteristics	Activities if reported	References
905	Flueggenine A	*Flueggea virosa*	Euphorbiaceae	C-linked dimeric indolizidine alkaloid	Cytotoxicity	[425]
906	Flueggenine B	*Flueggea virosa*	Euphorbiaceae	C-linked dimeric indolizidine alkaloid	–	[425]
907	Flueggine A	*Flueggea virosa*	Euphorbiaceae	C,C-linked dimeric indolizidine alkaloid	Cytotoxic activity	[426]
908	Flueggine B	*Flueggea virosa*	Euphorbiaceae	C,C-linked dimeric indolizidine alkaloid	Cytotoxic activity	[426]
909	Suffrutine A	*Flueggea suffruticosa*	Euphorbiaceae	An unprecedented C_{20} framework constructed from an indolizidine heterocycle and a benzofuranone moiety tethered by a conjugated diene chain	Regulating the morphology of Neuro-2a	[427]
910	Suffrutine B	*Flueggea suffruticosa*	Euphorbiaceae	An unprecedented C_{20} framework constructed from an indolizidine heterocycle and a benzofuranone moiety tethered by a conjugated diene chain	–	[427]
911	Fluvirosaone A	*Flueggea virosa*	Euphorbiaceae	Pentacyclic skeleton	Lipidlowering effects	[428]
912	Fluvirosaone B	*Flueggea virosa*	Euphorbiaceae	Pentacyclic skeleton	Lipidlowering effects	[428]
913	Virosaine A	*Flueggea virosa*	Euphorbiaceae	7-oxa-1-azabicyclo[3.2.1]octane ring system	–	[429]
914	Virosaine B	*Flueggea virosa*	Euphorbiaceae	7-oxa-1-azabicyclo[3.2.1]octane ring system	–	[429]
915	Fluevirosine A	*Flueggea virosa*	Euphorbiaceae	C,C-linked trimeric Securinega alkaloid	–	[430]
916	Fluevirosine B	*Flueggea virosa*	Euphorbiaceae	C,C-linked trimeric Securinega alkaloid	Antitumor activity	[430]
917	Fluevirosine C	*Flueggea virosa*	Euphorbiaceae	C,C-linked trimeric Securinega alkaloid	Antitumor activity	[430]
918	Flueggeacosine A	*Flueggea suffruticosa*	Euphorbiaceae	Dimeric alkaloid	–	[431]
919	Flueggeacosine B	*Flueggea suffruticosa*	Euphorbiaceae	Dimeric alkaloid	Neuronal differentiation activity	[431]
920	Flueggeacosine C	*Flueggea suffruticosa*	Euphorbiaceae	Dimeric alkaloid	–	[431]
921	Trigonoliimine A	*Trigonostemon lii*	Euphorbiaceae	Polycyclic skeleton	Anti-HIV-1 activity	[432]
922	Trigonoliimine B	*Trigonostemon lii*	Euphorbiaceae	Polycyclic skeleton	–	[432]
923	Trigonoliimine C	*Trigonostemon lii*	Euphorbiaceae	Polycyclic skeleton	–	[432]

(continued)

Table A10 (continued)

Compound	Trivial name	Source	Family	Structure characteristics	Activities if reported	References
924	Trigolutesin A	*Trigonostemon lutescens*	Euphorbiaceae	Polycyclic skeleton	Anti-AChE activity	[433]
925	Trigolutesin B	*Trigonostemon lutescens*	Euphorbiaceae	Polycyclic skeleton	–	[433]
926	Trigolute A	*Trigonostemon lutescens*	Euphorbiaceae	Polycyclic skeleton	–	[433]
927	Trigolute B	*Trigonostemon lutescens*	Euphorbiaceae	Polycyclic skeleton	–	[433]
928	Trigolute C	*Trigonostemon lutescens*	Euphorbiaceae	Polycyclic skeleton	–	[433]
929	Trigolute D	*Trigonostemon lutescens*	Euphorbiaceae	Polycyclic skeleton	–	[433]
930	Voatinggine	*Tabernaemontana* species	Euphorbiaceae	Pentacyclic system	–	[434]
931	Tabertinggine	*Tabernaemontana* species	Euphorbiaceae	Pentacyclic system	Reversing multidrug-resistance in vincristine-resistant KB (VJ300) cells	[434]
932	Saludimerine A	*Croton flavens*	Euphorbiaceae	Dimericbiaryltype morphinanedienone alkaloid	–	[435]
933	Saludimerine B	*Croton flavens*	Euphorbiaceae	Dimericbiaryltype morphinanedienone alkaloid	–	[435]
934	Cascarinoid A	*Croton cascarilloides*	Euphorbiaceae	Crotofolane diterpenoid alkaloids	–	[436]
935	Cascarinoid B	*Croton cascarilloides*	Euphorbiaceae	Crotofolane diterpenoid alkaloids	Immunosuppressive activity	[436]
936	Cascarinoid C	*Croton cascarilloides*	Euphorbiaceae	Crotofolane diterpenoid alkaloids	Immunosuppressive activity	[436]
937	Serratezomine A	*Lycopodium serratum* var. *serratum*	Lycopodiaceae	A 2-oxabicyclo[3.3.1]nonan-3-one and an indolizidine ring connected through a spiro carbon	Cytotoxicity	[437]
938	Serratezomine B	*Lycopodium serratum* var. *serratum*	Lycopodiaceae	N-oxide of serratinine-type alkaloid	Cytotoxicity	[437]
939	Serratezomine C	*Lycopodium serratum* var. *serratum*	Lycopodiaceae	Lycodoline-type skeleton	–	[437]
940	Lyconadin A	*Lycopodium complanatum*	Lycopodiaceae	Three six-membered, one five-membered, and one α-pyridone ring	Cytotoxicity	[438]
941	Lycoposerramine-A	*Lycopodium serratum*	Lycopodiaceae	1,2,4-oxadiazolidin-5-one residue	–	[439]

(continued)

Table A10 (continued)

Compound	Trivial name	Source	Family	Structure characteristics	Activities if reported	References
942	Sieboldine A	*Lycopodium sieboldii*	Lycopodiaceae	Fused-tetracyclic ring system	Cytotoxic and anti-AChE activity	[440]
943	Himeradine A	*Lycopodium chinense*	Lycopodiaceae	$C_{27}N_3$-type Lycopodium alkaloid	Cytotoxicity	[441]
944	Nankakurine A	*Lycopodium hamiltonii*	Lycopodiaceae	Fused-tetracyclic skeleton	Cytotoxicity	[442]
945	Lycoperine A	*Lycopodium hamiltonii*	Lycopodiaceae	$C_{27}N_3$-type alkaloid consisting of two octahydroquinoline rings and a piperidine ring	Anti-AChE activity	[443]
946	Lycojapodine A	*Lycopodium japonicum*	Lycopodiaceae	6/6/6/7 tetracyclic ring system	Anti-AChE and anti-HIV-1 activity	[444]
947	Lycojaponicumin A	*Lycopodium japonicum*	Lycopodiaceae	5/5/5/5/6pentacyclicring system with a 1-aza-7-oxabicyclo[2.2.1]heptane moiety	Anti-neuroinflammation	[445]
948	Lycojaponicumin B	*Lycopodium japonicum*	Lycopodiaceae	5/5/5/5/6pentacyclicring system with a 1-aza-7-oxabicyclo[2.2.1]heptane moiety	Anti-neuroinflammation	[445]
949	Lycojaponicumin C	*Lycopodium japonicum*	Lycopodiaceae	6/5/5/6 tetracyclic skeketon	Anti-neuroinflammation	[445]
950	Lycojaponicumin D	*Lycopodium japonicum*	Lycopodiaceae	5/7/6/6 tetracyclic skeleton	Anti-neuroinflammation	[446]
951	Lycospidine A	*Lycopodium complanatum*	Lycopodiaceae	5/6/6/6 tetracyclicringsystem	–	[447]
952	Annotinolide A	*Lycopodium annotinum*	Lycopodiaceae	Cyclopropane ring	Anti-Alzheimer's disease	[448]
953	Annotinolide B	*Lycopodium annotinum*	Lycopodiaceae	Cyclobutane ring	Anti-Alzheimer's disease	[448]
954	Annotinolide C	*Lycopodium annotinum*	Lycopodiaceae	12-spiro-9,12-γ-lactone moiety	Anti-Alzheimer's disease	[448]
955	Lycoplanine A	*Lycopodium complanatum*	Lycopodiaceae	6/9/5 tricyclic ring skeleton	Anti-AChE activity	[449]
956	Lycoplatyrine A	*Lycopodium platyrhizoma*	Lycopodiaceae	An unusual lycodine–piperidine adduct	–	[450]
957	Palhinine A	*Palhinhaea cernua*	Lycopodiaceae	5/6/6/9 tetracyclic ring system	–	[451]
958	Lycopalhine A	*Palhinhaea cernua*	Lycopodiaceae	Possesses a fused hexacyclic (5/5/5/6/6/6) ring system comprising a 5,9-diaza-tricyclo[6.2.1.$0^{4,9}$]undecane moiety and a tricyclo[5.2.1.$0^{4,8}$]decane moiety	Weak BChE inhibitory activity	[452]
959	Isopalhinine A	*Palhinhaea cernua*	Lycopodiaceae	Pentacyclic (5/6/6/6/7)	–	[453]

(continued)

Table A10 (continued)

Compound	Trivial name	Source	Family	Structure characteristics	Activities if reported	References
960	Tabernine A	*Tabernaemontana elegans*	Apocynaceae	Unusual and intriguing ring system	Modulate multidrug resistance	[454]
961	Tabernine B	*Tabernaemontana elegans*	Apocynaceae	Unusual and intriguing ring system	–	[454]
962	Tabernine C	*Tabernaemontana elegans*	Apocynaceae	Unusual and intriguing ring system	Modulate multidrug resistance	[454]
963	Actinophyllic acid	*Alstonia actinophylla*	Apocynaceae	2,3,6,7,9,13c-hexahydro-1*H*-1,7,8-(methanetriyloxymethano)pyrrolo[1′,2′:1,2] azocino[4,3-*b*]indole-8(5*H*)-carboxylic acid skeleton	Carboxypeptidase U inhibitor	[455]
964	Criofolinine	*Tabernaemontana corymbosa*	Apocynaceae	Incorporating a pyrroloazepine motif within a pentacyclic ring system	–	[456]
965	Vernavosine	*Tabernaemontana corymbosa*	Apocynaceae	Incorporating a pyridopyrimidine moiety embedded within a pentacyclic carbon framework	–	[456]
966	Erchinine A	*Ervatamia chinensis*	Apocynaceae	Heterocyclic architecture with three unique hemiaminals	Antibacterial activity	[457]
967	Erchinine B	*Ervatamia chinensis*	Apocynaceae	Heterocyclic architecture with three unique hemiaminals	Antibacterial and antibacterial activity	[457]
968	Bistabercarpamine A	*Tabernaemontana corymbosa*	Apocynaceae	Unprecedented bis-vobasinyl-chippiine-type skeleton	Cytotoxicity	[458]
969	Bistabercarpamine B	*Tabernaemontana corymbosa*	Apocynaceae	Unprecedented bis-vobasinyl-chippiine-type skeleton	–	[458]
970	Ervadivamine A	*Ervatamia divaricata*	Apocynaceae	Vobasine-iboga-vobasine-type alkaloid with both C–C and C–N linkage patterns	Cytotoxicity	[459]
971	Ervadivamine B	*Ervatamia divaricata*	Apocynaceae	Vobasine-iboga-vobasine-type alkaloid with both C–C and C–N linkage patterns	–	[459]
972	Alasmontamine A	*Tabernaemontana elegans*	Apocynaceae	Bis-vobtusine-type skeletons	Cytotoxicity	[460]
973	Tabercorymine A	*Tabernaemontana corymbosa*	Apocynaceae	Caged heteropentacyclic ring system	Antiproliferative activity	[461]
974	Tabercorymine B	*Tabernaemontana corymbosa*	Apocynaceae	Vobasinyl–ibogan-type	Antiproliferative activity	[461]
975	Bipleiophylline	*Alstonia angustifolia*	Apocynaceae	*Two indole moieties are bridged by an aromatic spacer unit*	Cytotoxicity	[462]

(continued)

Table A10 (continued)

Compound	Trivial name	Source	Family	Structure characteristics	Activities if reported	References
976	Alsmaphorazine A	*Alstonia pneumatophora*	Apocynaceae	1,2-oxazinane and an isoxazolidine chromophore	Anti-inflammatory activity	[463]
977	Alsmaphorazine B	*Alstonia pneumatophora*	Apocynaceae	1,2-oxazinane and an isoxazolidine chromophore	–	[463]
978	Lumutinine A	*Alstonia macrophylla*	Apocynaceae	Linear, ring A/F-fused macroline–macroline-type bisindole	–	[464]
979	Lumutinine B	*Alstonia macrophylla*	Apocynaceae	Linear, ring A/F-fused macroline–macroline-type bisindole	–	[464]
980	Alistonitrine A	*Alstonia scholaris*	Apocynaceae	Caged skeleton with a unique 6/5/6/5/5/6 ring system	–	[465]
981	Alstoscholarisine H	*Alstonia scholaris*	Apocynaceae	C-3/N-1 bond	–	[466]
982	Alstoscholarisine I	*Alstonia scholaris*	Apocynaceae	C-3/N-1 bond	–	[466]
983	Alstoscholarisine J	*Alstonia scholaris*	Apocynaceae	C-3/N-1 bond	–	[466]
984	Alstoscholarisine k	*Alstonia scholaris*	Apocynaceae	A novel 6/5/6/6/6/6/6/5 octacyclic architecture	Antibacterial bioactivity	[467]
985	Alstonlarsine A	*Alstonia scholaris*	Apocynaceae	A new carbon skeleton with a cage-shaped 9-azatricyclo[4.3.1.0^{3,8}]decane motif	DRAK2 inhibitory activity	[468]
986	(19,20) *E*-alstoscholarine	*Alstonia scholaris*	Apocynaceae	Skeleton rearrangement and two additional carbons	–	[469]
987	(19,20) *Z*-alstoscholarine	*Alstonia scholaris*	Apocynaceae	Skeleton rearrangement and two additional carbons	–	[469]
988	Scholarisine A	*Alstonia scholaris*	Apocynaceae	A cage skeleton	–	[470]
989	Alstolarine A	*Alstonia scholaris*	Apocynaceae	An unprecedented 21-oxa-4-azaspiro[4,4]nonane unit	Promising vasorelaxant activity	[471]
990	Alstolarine B	*Alstonia scholaris*	Apocynaceae	A unique 3-oxa-4-azatricyclo[5.1.1.1^{3,20}]undecane ring system	Promising vasorelaxant activity, anti-AChE activity	[471]
991	Subincanadine A	*Aspidosperma subincanum*	Apocynaceae	1-azoniatricyclo[4.3.3.0^{1,5}]undecane moiety	–	[472]

(continued)

Table A10 (continued)

Compound	Trivial name	Source	Family	Structure characteristics	Activities if reported	References
992	Subincanadine B	*Aspidosperma subincanum*	Apocynaceae	1-azoniatricyclo[4.3.3.$0^{1,5}$]undecane moiety	–	[472]
993	Subincanadine C	*Aspidosperma subincanum*	Apocynaceae	1-azoniatricyclo[4.3.3.$0^{1,5}$]undecane moiety	–	[472]
994	Melohenine A	*Melodinus henryi*	Apocynaceae	Arranged compactly in eight rings	–	[473]
995	Melohenine B	*Melodinus henryi*	Apocynaceae	6/9/6/6 tetracyclic ring system	–	[473]
996	Melotenine A	*Melodinus tenuicaudatus*	Apocynaceae	6/5/5/6/7 pentacyclic rearranged ring system	Cytotoxic activity	[474]
997	Melocochine A	*Melodinus cochinchinensis*	Apocynaceae	A rare 1*H*-benzo[*b*]azepane ring system within monoterpenoid indole alkaloid categories	Enhance lysosomal biogenesis	[475]
998	Melocochine B	*Melodinus cochinchinensis*	Apocynaceae	A rare 1*H*-benzo[*b*]azepane ring system within monoterpenoid indole alkaloid	Enhance lysosomal biogenesis	[475]
999	Meloyunnanine A	*Melodinus yunnanensis*	Apocynaceae	A caged-6/6/5/6/5/5 ring system	–	[476]
1000	Meloyunnanine B	*Melodinus yunnanensis*	Apocynaceae	A caged-6/6/5/6/5/5 ring system	–	[476]
1001	Meloyunnanine C	*Melodinus yunnanensis*	Apocynaceae	A caged-6/6/5/6/5/5 ring system	–	[476]
1002	Kopsifoline A	Malayan *Kopsia* species	Apocynaceae	An unprecedented carbon skeleton, in wich C-18 is linked to C-16 instead of to C-2	–	[477]
1003	Kopsifoline B	Malayan *Kopsia* species	Apocynaceae	An unprecedented carbon skeleton, in wich C-18 is linked to C-16 instead of to C-2	–	[477]
1004	Kopsifoline C	Malayan *Kopsia* species	Apocynaceae	An unprecedented carbon skeleton, in wich C-18 is linked to C-16 instead of to C-2	–	[477]
1005	Arboflorine	*Kopsia arborea*	Apocynaceae	Pentacyclic carbon skeleton and incorporating a third nitrogen atom	–	[478]
1006	Rauvomine A	*Rauvolfia vomitoria*	Apocynaceae	C_{18} sarpagine-type normonoterpenoid indole alkaloid with a chlorine atom	–	[479]

(continued)

Table A10 (continued)

Compound	Trivial name	Source	Family	Structure characteristics	Activities if reported	References
1007	Rauvomine B	*Rauvolfia vomitoria*	Apocynaceae	6/5/6/6/3/5-fused hexcyclos sarpagine-type normonoterpenoid indole alkaloid possessing a cyclopropane ring and C_{18} skeleton	Anti-inflammatory activity	[479]
1008	Mekongenine A	*Bousigonia mekongensis*	Apocynaceae	The first member of a new structural subclass of the dimeric alkaloids with a rare 2,7-secoeburnamine half incorporating an unprecedented 6/9/6/6 tetracyclic ring system	Cytotoxicity	[480]
1009	Mekongenine B	*Bousigonia mekongensis*	Apocynaceae	Consisting of an eburnamine-aspidospermine type skeleton	Cytotoxicity	[480]
1010	Bisnicalaterine B	*Hunteria zeylanica*	Apocynaceae	Consisting of an eburnane and a corynanthe type of skeleton	Vasorelaxant activity	[481]
1011	Bisnicalaterine C	*Hunteria zeylanica*	Apocynaceae	Consisting of an eburnane and a corynanthe type of skeleton	Vasorelaxant activity	[482]
1012	Goniomedine A	*Gonioma malagasy*	Apocynaceae	Quebrachamine-pleioarpamine-type skeleton, fused via a dihydropyran unit	–	[482]
1013	Goniomedine B	*Gonioma malagasy*	Apocynaceae	Quebrachamine-pleioarpamine-type skeleton, fused via a dihydropyran unit	Anti-malarial	[482]
1014	Pleiokomenine A	*Pleiocarpa mutica*	Apocynaceae	Dimeric aspidofractinine alkaloid linked by a methylene bridge	Antiplasmodial activity	[483]
1015	Pleiokomenine B	*Pleiocarpa mutica*	Apocynaceae	Dimeric aspidofractinine alkaloid linked by a methylene bridge	Antiplasmodial activity	[483]
1016	Calophyline A	*Winchia calophylla*	Apocynaceae	7/5 ring system	–	[484]
1017	Leucophyllidine	*Leuconotis griffithii*	Apocynaceae	Tetrahydrobenzo[*b*][1,8]naphthyridine chromophore	Cytotoxicity	[485]
1018	Sophaline A	*Sophora alopecuroides*	Leguminosae	6/6/ 6/4 ring system	–	[486]
1019	Sophaline B	*Sophora alopecuroides*	Leguminosae	6/5/6/6 ring system	Antiviral activity	[486]
1020	Sophaline C	*Sophora alopecuroides*	Leguminosae	Matrine–acetophenone alkaloid	Antiviral activity	[486]

(continued)

Table A10 (continued)

Compound	Trivial name	Source	Family	Structure characteristics	Activities if reported	References
1021	Sophaline D	*Sophora alopecuroides*	Leguminosae	Matrine–acetophenone alkaloid	Antiviral activity	[486]
1022	Sophaline E	*Sophora alopecuroides*	Leguminosae	Sparteine-indolizine alkaloid	–	[487]
1023	Sophaline F	*Sophora alopecuroides*	Leguminosae	Matrine-indolizine alkaloid	Antiviral activities	[487]
1024	Sophaline G	*Sophora alopecuroides*	Leguminosae	Normatrine-julolidine alkaloid	–	[487]
1025	Sophaline H	*Sophora alopecuroides*	Leguminosae	Normatrine-julolidine alkaloid	–	[487]
1026	Sophaline I	*Sophora alopecuroides*	Leguminosae	C-14–C-10′ connection	–	[487]
1027	Flavesine A	*Sophora flavescens*	Leguminosae	Dimeric matrine-type alkaloid	Anti-HBV activity	[488]
1028	Flavesine B	*Sophora flavescens*	Leguminosae	Dimeric matrine-type alkaloid	Anti-HBV activity	[488]
1029	Flavesine C	*Sophora flavescens*	Leguminosae	Dimeric matrine-type alkaloid	Anti-HBV activity	[488]
1030	Flavesine D	*Sophora flavescens*	Leguminosae	Dimeric matrine-type alkaloid	Anti-HBV activity	[488]
1031	Flavesine E	*Sophora flavescens*	Leguminosae	Dimeric matrine-type alkaloid	Anti-HBV activity	[488]
1032	Flavesine F	*Sophora flavescens*	Leguminosae	Dimerization pattern constructed by matrine and (−)-cytisine	Anti-HBV activity	[460]
1033	Erythrivarine A	*Erythrina variegata*	Leguminosae	Spirocyclic (6/5/6/6) ring	–	[488]
1034	Erythrivarine B	*Erythrina variegata*	Leguminosae	Spiro-fused (6/5/7/6) ring	–	[489]
1035	Erythrivarine J	*Erythrina variegata*	Leguminosae	Featured a 6/6/5/6/6/5/6/6/6 ring system and super conjugated double bond system	–	[490]
1036	Erythrivarine K	*Erythrina variegata*	Leguminosae	Featured a 6/6/5/6/6/5/6/6/6 ring system and super conjugated double bond system	–	[490]
1037	Erythrivarine L	*Erythrina variegata*	Leguminosae	Featured a 6/6/5/6/6/5/6/6/6 ring system and super conjugated double bond system	–	[490]
1038	Erythrivarine M	*Erythrina variegata*	Leguminosae	Assembled by C-7/8′ connection	–	[490]
1039	Erythrivarine N	*Erythrina variegata*	Leguminosae	Assembled by C-7/8′ connection	–	[490]
1040	Daphnezomine A	*Daphniphyllum humile*	Daphniphyllaceae	Azaadamantane core containing an amino ketal bridge	–	[491]
1041	Daphnezomine B	*Daphniphyllum humile*	Daphniphyllaceae	Azaadamantane core containing an amino ketal bridge	Cytotoxicity	[491]

(continued)

Table A10 (continued)

Compound	Trivial name	Source	Family	Structure characteristics	Activities if reported	References
1042	Daphnezomine F	*Daphniphyllum humile*	Daphniphyllaceae	1-azabicyclo[5.2.2]undecane ring system	Cytotoxicity	[492]
1043	Daphnezomine G	*Daphniphyllum humile*	Daphniphyllaceae	1-azabicyclo[5.2.2]undecane ring system	Cytotoxicity	[492]
1044	Daphnicyclidin A	*Daphniphyllum teijsmanni*	Daphniphyllaceae	Novel type of natural products consisting of fused hexa ring system	Cytotoxicity	[493]
1045	Daphnicyclidin B	*Daphniphyllum teijsmanni*	Daphniphyllaceae	Novel type of natural products consisting of fused hexa ring system	Cytotoxicity	[493]
1046	Daphnicyclidin C	*Daphniphyllum teijsmanni*	Daphniphyllaceae	Novel type of natural products consisting of fused hexa ring system	Cytotoxicity	[493]
1047	Daphnicyclidin D	*Daphniphyllum teijsmanni*	Daphniphyllaceae	Novel type of natural products consisting of fused hexa ring system	Cytotoxicity	[493]
1048	Daphnicyclidin E	*Daphniphyllum teijsmanni*	Daphniphyllaceae	Novel type of natural products consisting of fused hexa ring system	Cytotoxicity	[493]
1049	Daphnicyclidin F	*Daphniphyllum teijsmanni*	Daphniphyllaceae	Novel type of natural products consisting of fused hexa ring system	Cytotoxicity	[493]
1050	Daphnicyclidin G	*Daphniphyllum teijsmanni*	Daphniphyllaceae	Novel type of natural products consisting of fused hexa ring system	Cytotoxicity	[493]
1051	Daphnicyclidin H	*Daphniphyllum teijsmanni*	Daphniphyllaceae	Novel type of natural products consisting of pentacyclic ring system	Cytotoxicity	[493]
1052	Daphnicyclidin J	*Daphniphyllum humile*	Daphniphyllaceae	Fused pentacyclic skeleton	Cytotoxicity	[494]
1053	Daphnicyclidin K	*Daphniphyllum humile*	Daphniphyllaceae	Fused hexacyclic skeleton	Cytotoxicity	[494]
1054	Daphcalycine	*Daphniphyllum calycinum*	Daphniphyllaceae	Heptacycle fused ring system	Cytotoxicity	[495]
1055	Deoxycalyciphylline B	*Daphniphyllum subverticillatum*	Daphniphyllaceae	Fused hexacyclic skeleton	–	[496]
1056	Deoxyisocalyciphylline B	*Daphniphyllum subverticillatum*	Daphniphyllaceae	Fused hexacyclic skeleton	–	[496]
1057	Daphniglaucin A	*Daphniphyllum glaucescens*	Daphniphyllaceae	1-azoniatetracyclo [5.2.2.0.^{1,6}0.^{4,9}]undecane ring system	Cytotoxicity	[497]
1058	Daphniglaucin B	*Daphniphyllum glaucescens*	Daphniphyllaceae	1-azoniatetracyclo [5.2.2.0.^{1,6}0.^{4,9}]undecane ring system	Cytotoxicity	[497]

(continued)

Table A10 (continued)

Compound	Trivial name	Source	Family	Structure characteristics	Activities if reported	References
1059	Calyciphylline A	*Daphniphyllum calycinum*	Daphniphyllaceae	Fused-hexacyclic ring system	Cytotoxicity	[498]
1060	Calyciphylline B	*Daphniphyllum calycinum*	Daphniphyllaceae	Fused-hexacyclic ring system	Cytotoxicity	[498]
1061	Daphnipaxinin	*Daphniphyllum paxianum*	Daphniphyllaceae	Heterohexacyclic skeleton	–	[499]
1062	Paxdaphnidine A	*Daphniphyllum paxianum*	Daphniphyllaceae	Pentacyclic skeleton	–	[500]
1063	Paxdaphnidine B	*Daphniphyllum paxianum*	Daphniphyllaceae	Tetracyclic skeleton	–	[500]
1064	Daphmanidin C	*Daphniphyllum teijsmanii*	Daphniphyllaceae	Fused-pentacyclic skeleton	–	[501]
1065	Daphmanidin D	*Daphniphyllum teijsmanii*	Daphniphyllaceae	Fused-pentacyclic skeleton	–	[501]
1066	Calycilactone A	*Daphniphyllum calycillum*	Daphniphyllaceae	A rearranged fused-hexacyclic ring system	–	[502]
1067	Macropodumine A	*Daphniphyllum macropodum*	Daphniphyllaceae	A fused pentacyclic system including an unusual eleven-membered macrolactone ring	–	[503]
1068	Macropodumine B	*Daphniphyllum macropodum*	Daphniphyllaceae	A rare cyclopentadienyl carbanion	–	[503]
1069	Macropodumine C	*Daphniphyllum macropodum*	Daphniphyllaceae	Novel pentacyclic alkaloid	–	[503]
1070	Calyciphylline C	*Daphniphyllum calycinum*	Daphniphyllaceae	An unprecedented fused-hexacyclic ring system	–	[504]
1071	Daphlongeramine A	*Daphniphyllum longeracemosum*	Daphniphyllaceae	A unique fused octacyclic skeleton	–	[505]
1072	Calyciphylline G	*Daphniphyllum calycinum*	Daphniphyllaceae	An unprecedented fused-hexacyclic skeleton containing a 5-azatricyclo[6.2.1.$0^{1,5}$]undecane ring	Cytotoxicity	[506]
1073	Calydaphninone	*Daphniphyllum calycillum*	Daphniphyllaceae	4-azatricyclo[5.2.2.$0^{1,4}$]undecan ring system	–	[507]
1074	Daphlongeranine A	*Daphniphyllum longeracemosum*	Daphniphyllaceae	Heptacyclic ring system	Anti-platelet aggregation	[508]
1075	Daphlongeranine B	*Daphniphyllum longeracemosum*	Daphniphyllaceae	Hexacyclic ring system	Anti-platelet aggregation	[508]
1076	Macropodumine D	*Daphniphyllum macropodum*	Daphniphyllaceae	Hexacyclic-fused ring system	–	[509]
1077	Macropodumine E	*Daphniphyllum macropodum*	Daphniphyllaceae	Pentacyclic-fused ring system	–	[509]

(continued)

Table A10 (continued)

Compound	Trivial name	Source	Family	Structure characteristics	Activities if reported	References
1078	Calyciphylline D	*Daphniphyllum calycinum*	Daphniphyllaceae	Fused-pentacyclic skeleton containing a 8-azatricyclo[4.2.1.0.^{4,8}]nonane ring system	–	[510]
1079	Calycinumine A	*Daphniphyllum calycinum*	Daphniphyllaceae	C-22-nor yuzurimine-type alkaloid	–	[511]
1080	Calycinumine B	*Daphniphyllum calycinum*	Daphniphyllaceae	Heteroatom-containing adamantane-like western hemisphere of the alkaloid	–	[511]
1081	Daphenylline	*Daphniphyllum longeracemosum*	Daphniphyllaceae	22-nor-calyciphylline skeleton	–	[512]
1082	Daphhimalenine A	*Daphniphyllum himalense*	Daphniphyllaceae	Rearrangement C-21 skeleton, containing a unique 1-azabicyclo[5.2.1]decane ring system	–	[513]
1083	Angustimine	*Daphniphyllum angustifolium*	Daphniphyllaceae	Hexacyclic fused skeleton	–	[514]
1084	Hybridaphniphylline A	*Daphniphyllum longeracemosum*	Daphniphyllaceae	*Daphniphyllum* alkaloid and iridoid hybrids with unique decacyclic fused skeletons	–	[515]
1085	Hybridaphniphylline B	*Daphniphyllum longeracemosum*	Daphniphyllaceae	*Daphniphyllum* alkaloid and iridoid hybrids with unique decacyclic fused skeletons	–	[515]
1086	Logeracemin A	*Daphniphyllum longeracemosum*	Daphniphyllaceae	The first Daphniphyllum alkaloid dimer featuring an unprecedented carbon skeleton with a unique conjugated trispiro[4,5] decane backbone	Anti-HIV activity	[516]
1087	Himalensine A	*Daphniphyllum himalense*	Daphniphyllaceae	13,14,22-trinorcalyciphylline A type backbone	–	[517]
1088	Himalensine B	*Daphniphyllum himalense*	Daphniphyllaceae	22-nor-1,13-secodaphnicyclidin framework	Marginal inhibitory activity	[517]
1089	Myriberine A	*Myrioneuron faberi*	Rubiaceae	Pentacyclic system	Anti-HCVactivity	[518]
1090	Myrifabine	*Myrioneuron faberi*	Rubiaceae	Dimer with 12 chiral centers	Antimicrobial and cytotoxic activity	[519]
1091	(±)-β-Myrifabral A	*Myrioneuron faberi*	Rubiaceae	Cyclohexane-fused octahydroquinolizine skeleton	–	[520]

(continued)

Table A10 (continued)

Compound	Trivial name	Source	Family	Structure characteristics	Activities if reported	References
1092	(±)-α-myrifabral A	*Myrioneuron faberi*	Rubiaceae	Cyclohexane-fused octahydroquinolizine skeleton	–	[520]
1093	(±)-β-myrifabral B	*Myrioneuron faberi*	Rubiaceae	Cyclohexane-fused octahydroquinolizine skeleton	–	[520]
1094	(±)-α-myrifabral B	*Myrioneuron faberi*	Rubiaceae	Cyclohexane-fused octahydroquinolizine skeleton	–	[520]
1095	Macrophyllionium	*Uncaria macrophylla*	Rubiaceae	Oxindole alkaloid inner salt	–	[521]
1096	(±)-uncarilin A	*Uncaria rhynchophylla*	Rubiaceae	Symmetric four-membered core	–	[522]
1097	(±)-uncarilin B	*Uncaria rhynchophylla*	Rubiaceae	Symmetric four-membered core	Agonistic activity	[522]
1098	Rhynchine A	*Uncaria rhynchophylla*	Rubiaceae	A unique 6/5/7/5/5 ring system	Against the $Ca_V3.1$ calcium channel	[523]
1099	Rhynchine B	*Uncaria rhynchophylla*	Rubiaceae	A unique 6/5/7/5/5 ring system	Against the $Ca_V3.1$ calcium channel	[523]
1100	Rhynchine C	*Uncaria rhynchophylla*	Rubiaceae	A unique 6/5/7/5/5 ring system	–	[523]
1101	Rhynchine D	*Uncaria rhynchophylla*	Rubiaceae	A unique 6/5/7/5/5 ring system	–	[523]
1102	Rhynchine E	*Uncaria rhynchophylla*	Rubiaceae	A unique 6/5/7/5/5 ring system	–	[523]
1103	Brachycerine	*Psychotria brachyceras*	Rubiaceae	Structure formed by the combination of a 1-epiloganin derivative and tryptamine	–	[524]
1104	Psychotripine	*Psychotria pilifera*	Rubiaceae	Hendecacyclic system bearing a hexahydro-1,3,5-triazine unit	–	[525]
1105	Ophiorrhine A	*Ophiorrhiza japonica*	Rubiaceae	Spirocyclic ring system	Immunosuppressive activity	[526]
1106	Ophiorrhine B	*Ophiorrhiza japonica*	Rubiaceae	Spirocyclic ring system	Immunosuppressive activity	[526]
1107	Robustanoid A	*Coffea canephora*	Rubiaceae	An unprecedented 1,2,3,4,8b,8c-hexahydro-2a,4a-diazapentaleno[1,6-*ab*]indene moiety	–	[527]
1108	Robustanoid B	*Coffea canephora*	Rubiaceae	An unprecedented 1,2,3,4,8b,8c-hexahydro-2a,4a-diazapentaleno[1,6-*ab*]indene moiety	α-glucosidase inhibitory activity	[527]

(continued)

Table A10 (continued)

Compound	Trivial name	Source	Family	Structure characteristics	Activities if reported	References
1109	Flinderole A	*Flindersia Species*	Rutaceae	Rearranged skeleton	Antimalarial activity	[528]
1110	Flinderole B	*Flindersia Species*	Rutaceae	Rearranged skeleton	Antimalarial activity	[528]
1111	Flinderole C	*Flindersia Species*	Rutaceae	Rearranged skeleton	Antimalarial activity	[528]
1112	Raputindole A	*Raputia simulans*	Rutaceae	Fused cyclopentyl unit	Enzyme inhibitory activity	[529]
1113	Raputindole B	*Raputia simulans*	Rutaceae	Fused cyclopentyl unit	Enzyme inhibitory activity	[529]
1114	Raputindole C	*Raputia simulans*	Rutaceae	Fused cyclopentyl unit	Enzyme inhibitory activity	[529]
1115	Raputindole D	*Raputia simulans*	Rutaceae	Fused cyclopentyl unit	Enzyme inhibitory activity	[529]
1116	Zanthomuurolanine	*Zanthoxylum nitidum*	Rutaceae	Dihydrochelerythrine and a cadinane-type sesquiterpene linked by a methylene bridge	–	[530]
1117	*Epi*-zanthomuurolanine	*Zanthoxylum nitidum*	Rutaceae	Dihydrochelerythrine and a cadinane-type sesquiterpene linked by a methylene bridge	–	[530]
1118	Zanthocadinanine A	*Zanthoxylum nitidum*	Rutaceae	Dihydrochelerythrine and a cadinane-type sesquiterpene linked by a methylene bridge	–	[530]
1119	Zanthocadinanine B	*Zanthoxylum nitidum*	Rutaceae	Dihydrochelerythrine and a cadinane-type sesquiterpene linked by a methylene bridge	–	[530]
1120	*Epi*-zanthocadinanine B	*Zanthoxylum nitidum*	Rutaceae	Dihydrochelerythrine and a cadinane-type sesquiterpene linked by a methylene bridge	–	[530]
1121	Parvifloranine A	*Geijera parviflora*	Rutaceae	11-carbon skeleton linked with amino acid	Anti-inflammatory activity	[531]
1122	Parvifloranine B	*Geijera parviflora*	Rutaceae	11-carbon skeleton linked with amino acid	–	[531]
1123	Clausanisumine	*Clausena anisum-olens*	Rutaceae	An unprecedented carbon skeleton composed of a simple carbazole alkaloid and a prenylated carbazole alkaloid	Anti-HIV-1 reverse transcriptase effect	[532]
1124	Mbandakamine A	Congolese *Ancistrocladus* Species	Ancistrocladaceae	Dimeric alkaloids with an unsymmetrically coupled central biaryl axis	Antimalarial activity	[533]
1125	Mbandakamine B	Congolese *Ancistrocladus* Species	Ancistrocladaceae	Dimeric alkaloids with an unsymmetrically coupled central biaryl axis	–	[533]
1126	Spirombandakamine A_1	Congolese *Ancistrocladus*	Ancistrocladaceae	Cage-like stereostructure	Antimalarial activity	[534]
1127	Spirombandakamine A_2	Congolese *Ancistrocladus*	Ancistrocladaceae	Cage-like stereostructure	Antimalarial activity	[534]

(continued)

Table A10 (continued)

Compound	Trivial name	Source	Family	Structure characteristics	Activities if reported	References
1128	Cyclombandakamine A_1	Congolese *Ancistrocladus*	Ancistrocladaceae	Oxygen-bridged dimeric naphthylisoquinoline alkaloid	Antimalarial activity	[535]
1129	Cyclombandakamine A_2	Congolese *Ancistrocladus*	Ancistrocladaceae	Oxygen-bridged dimeric naphthylisoquinoline alkaloid	Antimalarial activity	[535]
1130	Geleganidine A	*Gelsemium elegans*	Loganiaceae	Aromatic azo- linked dimeric monoterpenoid indole alkaloid	–	[536]
1131	Geleganidine B	*Gelsemium elegans*	Loganiaceae	Urea-linked dimeric monoterpenoid indole alkaloid	Cytotoxic activity	[536]
1132	Geleganidine C	*Gelsemium elegans*	Loganiaceae	Gelsedine–corynanthe-type bisindole alkaloid	Cytotoxic activity	[536]
1133	Gelsecorydine A	*Gelsemium elegans*	Loganiaceae	Gelsedine–corynanthe-type bisindole alkaloid with 6/5/7/6/5/6 heterohexacyclic ring system	Anti-inflammatory activity	[537]
1134	Gelsecorydine B	*Gelsemium elegans*	Loganiaceae	Gelsedine–corynanthe-type bisindole alkaloid	–	[537]
1135	Gelsecorydine C	*Gelsemium elegans*	Loganiaceae	Gelsedine–corynanthe-type bisindole alkaloid	Anti-inflammatory activity	[537]
1136	Gelsecorydine D	*Gelsemium elegans*	Loganiaceae	Gelsedine–corynanthe-type bisindole alkaloid	Anti-inflammatory activity	[537]
1137	Gelsecorydine E	*Gelsemium elegans*	Loganiaceae	Gelsedine–corynanthe-type bisindole alkaloid	Anti-inflammatory activity	[537]
1138	Gelserancine A	*Gelsemium elegans*	Loganiaceae	A new skeleton with an unusual trimethyl-dihydrofuranone unit	–	[538]
1139	Gelserancine B	*Gelsemium elegans*	Loganiaceae	Novel gelsedine-iridoid adduct constructed through unusual C_{19}–$C_{11'}$ and N_4–$C_{3'}$ linkages forming an additional pyridine ring	Anti-inflammatory effect	[538]
1140	Gelserancine C	*Gelsemium elegans*	Loganiaceae	Novel gelsedine-iridoid adduct constructed through unusual C_{19}–$C_{11'}$ and N_4–$C_{3'}$ linkages forming an additional pyridine ring	Anti-inflammatory effect	[538]

(continued)

Table A10 (continued)

Compound	Trivial name	Source	Family	Structure characteristics	Activities if reported	References
1141	Gelserancine D	*Gelsemium elegans*	Loganiaceae	Photochemical *E*/*Z* tautomeric MIA with a highly conjugated skeleton	Anti-inflammatory effect	[538]
1142	Gelserancine E	*Gelsemium elegans*	Loganiaceae	Photochemical *E*/*Z* tautomeric MIA with a highly conjugated skeleton	–	[538]
1143	Gelstriamine A	*Gelsemium elegans*	Loganiaceae	An unprecedented 6/5/7/6/6/5 heterohexacyclic scaffold bearing a unique hexahydrooxazolo[4,5-*b*]pyridin-2(3*H*)-one motif	–	[539]
1144	Strynuxline A	*Strychnos nux-vomica*	Loganiaceae	6/5/9/6/7/6 hexacyclic ring system	Cytotoxicity	[540]
1145	Strynuxline B	*Strychnos nux-vomica*	Loganiaceae	6/5/9/6/7/6 hexacyclic ring system	Cytotoxicity	[540]
1146	Isatisine A	*Isatis indigotica*	Cruciferae	An unprecedented fused-pentacyclic skeleton	Anti-HIV-1 activity	[541]
1147	–	*Isatis indigotica*	Cruciferae	Dihydrothiopyran and 1,2,4-thiadiazole rings	Antiviral activity	[542]
1148	–	*Isatis indigotica*	Cruciferae	Dihydrothiopyran and 1,2,4-thiadiazole rings	Antiviral activity	[542]
1149	Orychophragine A	*Orychophragmus violaceus*	Cruciferae	2-piperazinone-fused 2,4-dioxohexahydro-1,3,5triazine skeleton	–	[543]
1150	Orychophragine B	*Orychophragmus violaceus*	Cruciferae	2-piperazinone-fused 2,4-dioxohexahydro-1,3,5triazine skeleton	–	[543]
1151	Orychophragine C	*Orychophragmus violaceus*	Cruciferae	2-piperazinone-fused 2,4-dioxohexahydro-1,3,5triazine skeleton	^{60}Co γ radiation protection activity	[543]
1152	(+)-macleayin A	*Macleaya cordata*	Papaveraceae	6,13′-coupling pattern hints a hitherto unprecedented C–C linkage	–	[544]
1153	(−)-macleayin A	*Macleaya cordata*	Papaveraceae	6,13′-coupling pattern hints a hitherto unprecedented C–C linkage	Cytotoxic activity	[544]
1154	(+)-macleayin B	*Macleaya cordata*	Papaveraceae	6,13′-coupling pattern hints a hitherto unprecedented C–C linkage	–	[544]

(continued)

Table A10 (continued)

Compound	Trivial name	Source	Family	Structure characteristics	Activities if reported	References
1155	(−)-macleayin B	*Macleaya cordata*	Papaveraceae	6,13′-coupling pattern hints a hitherto unprecedented C–C linkage	–	[544]
1156	Dactyllactone A	*Dactylicapnos scandens*	Papaveraceae	A rearranged and reconstructed D ring	Anti-inflammatory activity	[545]
1157	Dactylicapnosine A	*Dactylicapnos scandens*	Papaveraceae	The five-membered carbon ring D	Anti-inflammatory and analgesic activity	[546]
1158	Dactylicapnosine B	*Dactylicapnos scandens*	Papaveraceae	The five-membered carbon ring D	–	[546]
1159	Cassiarin A	*Cassia siamea*	Caesalpiniaceae	Tricyclic skeleton	Antiplasmodial activity	[547]
1160	Cassiarin B	*Cassia siamea*	Caesalpiniaceae	Dimeric chromone alkaloid	–	[547]
1161	Fistulain A	*Cassia fistula*	Caesalpiniaceae	Dimeric chromone alkaloid	Cytotoxic activity	[548]
1162	Fistulain B	*Cassia fistula*	Caesalpiniaceae	Rearranged E ring	Cytotoxic activity	[548]
1163	Puberunine	*Aconitum barbatum* var. *puberulum*	Ranunculaceae	Opened A ring	–	[549]
1164	Puberudine	*Aconitum barbatum* var. *puberulum*	Ranunculaceae	Epoxyoxepino[4,5-c] pyrroleriing	–	[549]
1165	Aconicarmisulfonine A	*Aconitum carmichaelii*	Ranunculaceae	$C_{15}N$ skeleton	analgesic effect	[550]
1166	Ternatusine A	*Ranunculus ternatus*	Ranunculaceae	Sulfonated carbon skeleton	–	[551]
1167	Baicalensine A	*Thalictrum baicalense*	Ranunculaceae	A new class of alkaloid dimers containing berberine conjugated to a ring-opened isoquinoline	Cytotoxicity	[552]
1168	Baicalensine B	*Thalictrum baicalense*	Ranunculaceae	The first reported natural benzylisoquinoline bearing a formyl group at C-3	–	[552]
1169	Huperzine R	*Huperzia serrata*	Huperziaceae	$C_{38}N_4$-type alkaloid	Anti-AChE activity	[553]
1170	Hupercumine A	*Huperzia cunninghamioides*	Huperziaceae	$C_{27}N_3$-type alkaloid	Anti-AChE activity	[554]
1171	Hupercumine B	*Huperzia cunninghamioides*	Huperziaceae	An unprecedented [6/7/6/6]-tetracyclic skeleton by forming a new linkage between C-4 and C-14	Anti-AChE activity	[554]

(continued)

Table A10 (continued)

Compound	Trivial name	Source	Family	Structure characteristics	Activities if reported	References
1172	Phlefargesiine A	*Phlegmariurus fargesii*	Huperziaceae	A unique cyclobutane ring and a complex tetracyclo-[4.2.2.0^{3,8}.0^{3,10}]decane-bridged system	Weak AChE inhibitory activity	[555]
1173	Phlegmadine A	*Phlegmariurus phlegmaria*	Huperziaceae	Cyclohexene ring	–	[556]
1174	Nigramide A	*Piper nigrum*	Piperaceae	Cyclohexene ring	–	[557]
1175	Nigramide B	*Piper nigrum*	Piperaceae	Cyclohexene ring	–	[557]
1176	Nigramide C	*Piper nigrum*	Piperaceae	Cyclohexene ring	–	[557]
1177	Nigramide D	*Piper nigrum*	Piperaceae	Cyclohexene ring	–	[557]
1178	Nigramide E	*Piper nigrum*	Piperaceae	Cyclohexene ring	–	[557]
1179	Nigramide F	*Piper nigrum*	Piperaceae	Cyclohexene ring	–	[557]
1180	Nigramide G	*Piper nigrum*	Piperaceae	Cyclohexene ring	–	[557]
1181	Nigramide H	*Piper nigrum*	Piperaceae	Cyclohexene ring	–	[557]
1182	Nigramide I	*Piper nigrum*	Piperaceae	Cyclohexene ring	–	[557]
1183	Nigramide J	*Piper nigrum*	Piperaceae	Cyclohexene ring	–	[557]
1184	Nigramide K	*Piper nigrum*	Piperaceae	Cyclohexene ring	–	[557]
1185	Nigramide L	*Piper nigrum*	Piperaceae	Cyclohexene ring	–	[557]
1186	Nigramide M	*Piper nigrum*	Piperaceae	Cyclohexene ring	–	[557]
1187	Nigramide N	*Piper nigrum*	Piperaceae	Cyclohexene ring	–	[557]
1188	Nigramide O	*Piper nigrum*	Piperaceae	Cyclobutane ring	–	[557]
1189	Nigramide P	*Piper nigrum*	Piperaceae	Cyclobutane ring	–	[557]
1190	Nigramide Q	*Piper nigrum*	Piperaceae	Cyclobutane ring	–	[557]
1191	Nigramide R	*Piper nigrum*	Piperaceae	Cyclobutane ring	–	[557]
1192	Nigramide S	*Piper nigrum*	Piperaceae	Tirucallane-type alkaloid	–	[557]
1193	Laxiracemosin A	*Dysoxylum laxiracemosum*	Meliaceae	Tirucallane-type alkaloid	cytotoxicity	[558]
1194	Laxiracemosin B	*Dysoxylum laxiracemosum*	Meliaceae	Tirucallane-type alkaloid	–	[558]
1195	Laxiracemosin C	*Dysoxylum laxiracemosum*	Meliaceae	Tirucallane-type alkaloid	–	[558]

(continued)

Table A10 (continued)

Compound	Trivial name	Source	Family	Structure characteristics	Activities if reported	References
1196	Laxiracemosin D	*Dysoxylum laxiracemosum*	Meliaceae	Tirucallane-type alkaloid	–	[558]
1197	Laxiracemosin E	*Dysoxylum laxiracemosum*	Meliaceae	Tirucallane-type alkaloid	Cytotoxicity	[558]
1198	Laxiracemosin F	*Dysoxylum laxiracemosum*	Meliaceae	Tirucallane-type alkaloid	–	[558]
1199	Laxiracemosin G	*Dysoxylum laxiracemosum*	Meliaceae	Tirucallane-type alkaloid	–	[558]
1200	Laxiracemosin H	*Dysoxylum laxiracemosum*	Meliaceae	Tirucallane-type alkaloid	–	[558]
1201	Dovyalicin A	*Dovyalis macrocalyx*	Flacourtiaceae	Perhydro-1,5-diazocine or a perhydro1,4-diazepine ring system	–	[559]
1202	Dovyalicin B	*Dovyalis macrocalyx*	Flacourtiaceae	Perhydro-1,5-diazocine or a perhydro1,4-diazepine ring system	–	[559]
1203	Dovyalicin C	*Dovyalis macrocalyx*	Flacourtiaceae	Perhydro-1,5-diazocine or a perhydro1,4-diazepine ring system	–	[559]
1204	Dovyalicin D	*Dovyalis macrocalyx*	Flacourtiaceae	Perhydro-1,5-diazocine or a perhydro1,4-diazepine ring system	–	[560]
1205	Camellimidazole A	*Camellia sinensis*	Theaceae	Dimeric alkaloid	Neuroprotective effect	[560]
1206	Camellimidazole B	*Camellia sinensis*	Theaceae	Dimeric alkaloid	Neuroprotective effect	[560]
1207	Camellimidazole C	*Camellia sinensis*	Theaceae	Dimeric alkaloid	–	[560]
1208	Grandisine A	*Elaeocarpus grandis*	Elaeocarpaceae	Pyrano[2′,3′:4,5]pyrano[2,3-*g*]indolizin-4-one	Human δ-opioid receptor binding affinity	[561]
1209	Grandisine B	*Elaeocarpus grandis*	Elaeocarpaceae	Contain both an indolizidine and an isoquinuclidinone moiety	Human δ-opioid receptor binding affinity	[561]
1210	Machilaminoside A	*Machilus yaoshansis*	Lauraceae	Glycosidic triterpene alkaloid	Cytotoxicity	[562]
1211	Machilaminoside B	*Machilus yaoshansis*	Lauraceae	Glycosidic triterpene alkaloid	Cytotoxicity	[562]
1212	Oubatchensine	*Cryptocarya oubatchensis*	Lauraceae	Seco-dibenzopyrrocoline natural compound class	–	[563]
1213	Acortatarin A	*Acorus tatarinowii*	Araceae	Spiroalkaloid with a naturally unusual morpholine motif	Antioxidative effect	[564]
1214	Acortatarin B	*Acorus tatarinowii*	Araceae	Spiroalkaloid with a naturally unusual morpholine motif	Antioxidative effect	[564]

(continued)

Table A10 (continued)

Compound	Trivial name	Source	Family	Structure characteristics	Activities if reported	References
1215	Plagiochianin A	*Liverwort Plagiochila duthiana*	Plagiochilaceae	2,3:6,7-di-*seco*-6,8-cycloaromadendrane carbon scaffold conjugated with three cyclic acetals	Anti-AChE activity	[565]
1216	Plagiochianin B	Liverwort *Plagiochila duthiana*	Plagiochilaceae	A pyridine type aromadendrane alkaloid	–	[565]
1217	Cephalocyclidin A	*Cephalotaxus harringtonia* var. *nana*	Cephalotaxaceae	Fused-pentacyclic skeleton and six-consecutive asymmetric center	–	[566]
1218	Hostasinine A	*Hosta plantaginea*	Liliaceae	C-4–C-6 linkage and a nitrone moiety	Anti-AChE activity	[567]
1219	Hemerocallisamine I	*Hemerocallis fulva* var. *kwanso*	Asphodelaceae	Glutamine derivative with a pyrrole ring	–	[568]
1220	Peganumine A	*Peganum harmala*	Zygophyllaceae	Dimeric alkaloid characterized by a unique 3,9-diazatetracyclo[6.5.2.01,9.03,8]pentadec-2-one scaffold	Cytotoxic activity	[569]
1221	Pegaharine A	*Peganum harmala*	Zygophyllaceae	An azepine–indole system	–	[570]
1222	(±)-pegaharine B	*Peganum harmala*	Zygophyllaceae	A classic tricyclic β-carboline fused with an additional pyrrole ring	–	[570]
1223	(±)-pegaharine C	*Peganum harmala*	Zygophyllaceae	A rare tetracyclic β-carboline is connected to a classic tricyclic β-carboline through a C-15–C-1′ bridge	–	[570]
1224	(±)-pegaharine D	*Peganum harmala*	Zygophyllaceae	A rare tetracyclic β-carboline is connected to a classic tricyclic β-carboline through a C-15–C-1′ bridge	Antiviral activity	[570]
1225	Pegaharine E	*Peganum harmala*	Zygophyllaceae	A rare tetracyclic β-carboline is connected to a classic tricyclic β-carboline through a C-15–C-1′ bridge	–	[570]
1226	Pegaharine F	*Peganum harmala*	Zygophyllaceae	A rare tetracyclic β-carboline is connected to a classic tricyclic β-carboline through a C-15–C-1′ bridge	–	[570]
1227	Pegaharmol A	*Peganum harmala*	Zygophyllaceae	An axial chirality between an aryl unit and a nonaryl unit	Cytotoxicity	[571]

(continued)

Table A10 (continued)

Compound	Trivial name	Source	Family	Structure characteristics	Activities if reported	References
1228	Pegaharmol B	*Peganum harmala*	Zygophyllaceae	An axial chirality between an aryl unit and a nonaryl unit	–	[571]
1229	Lycibarbarine A	*Lycium barbarum*	Solanaceae	A unique tetracyclic tetrahydroquinoline–oxazine–ketohexoside fused motif	Neuroprotective activity	[572]
1230	Lycibarbarine B	*Lycium barbarum*	Solanaceae	A unique tetracyclic tetrahydroquinoline–oxazine–ketohexoside fused motif	–	[572]
1231	Lycibarbarine C	*Lycium barbarum*	Solanaceae	A unique tetracyclic tetrahydroquinoline–oxazine–ketohexoside fused motif	Neuroprotective activity	[572]
1232	Artapilosine A	*Artabotrys pilosus*	Annonaceae	The first compound representative of a new class of phenanthrene derivatives having an unprecedented carbon skeleton	Anti-HIV reverse transcriptase affect	[573]
1233	Artapilosine B	*Artabotrys pilosus*	Annonaceae	A novel phenanthrene derivative having a hydroxyethyl as a substituent on the phenanthrene ring	Anti-HIV reverse transcriptase affect	[573]
1234	Fortuneicyclidin A	*Cephalotaxus fortunei*	Taxaceae	An unprecedented 7-azatetracyclo-[5.4.3.0.0^{2,8}]tridecane core	Against α-glucosidase	[574]
1235	Fortuneicyclidin B	*Cephalotaxus fortunei*	Taxaceae	An unprecedented 7-azatetracyclo-[5.4.3.0.0^{2,8}]tridecane core	–	[574]

Table A11 Sterides from plants

Compound	Trivial name	Source	Family	Structure characteristics	Activities if reported	References
1236	Subtrifloralactone A	*Deprea subtriflora*	Solanaceae	New C_{27} skeleton	Cancer chemopreventive activity	[575]
1237	Subtrifloralactone B	*Deprea subtriflora*	Solanaceae	New C_{27} skeleton	–	[575]
1238	Subtrifloralactone C	*Deprea subtriflora*	Solanaceae	New C_{27} skeleton	–	[575]
1239	Subtrifloralactone D	*Deprea subtriflora*	Solanaceae	New C_{27} skeleton	–	[575]
1240	Subtrifloralactone E	*Deprea subtriflora*	Solanaceae	New C_{27} skeleton	–	[575]
1241	Subtrifloralactone F	*Deprea subtriflora*	Solanaceae	New C_{27} skeleton	Cancer chemopreventive activity	[575]
1242	Subtrifloralactone G	*Deprea subtriflora*	Solanaceae	New C_{27} skeleton	–	[575]
1243	Subtrifloralactone H	*Deprea subtriflora*	Solanaceae	New C_{27} skeleton	–	[575]
1244	Subtrifloralactone I	*Deprea subtriflora*	Solanaceae	New C_{27} skeleton	–	[575]
1245	Subtrifloralactone J	*Deprea subtriflora*	Solanaceae	New C_{27} skeleton	–	[575]
1246	Physanolide A	*Physalis angulata*	Solanaceae	A seven-membered ring	–	[576]
1247	Physangulidine A	*Physalis angulata*	Solanaceae	Unusual disconnection between C-13 and C-17	Antiproliferative activity	[577]
1248	Physangulidine B	*Physalis angulata*	Solanaceae	Unusual disconnection between C-13 and C-17	Antiproliferative activity	[577]
1249	Physangulidine C	*Physalis angulata*	Solanaceae	Unusual disconnection between C-13 and C-17	Antiproliferative activity	[577]
1250	Urceoloid A	*Urceola quintaretii*	Apocynaceae	A unique spiro[4.4]nona-3,6,8-triene system	–	[578]
1251	Urceoloid B	*Urceola quintaretii*	Apocynaceae	A unique spiro[4.4]nona-3,6,8-triene system	Immunosuppressive activity	[578]
1252	Bungsteroid A	*Zanthoxylum bungeanum*	Rutaceae	A unique 6/6/6/6/5-fused pentacyclic skeleton	Antiproliferative effect	[579]

Table A12 Others from plants

Compound	Trivial name	Source	Family	Structure characteristics	Activities if reported	References
Natural pentacyclic compound						
1253	Kingianin A	*Endiandra kingiana*	Lauraceae	Bicyclo[4.2.0]octadienebackbone	–	[580]
Thapsigargin						
1254	Transtaganolide A	*Thapsia transtagana*	Apiaceae	7-methoxy-4,5-dihydro-3*H*-oxepin-2-one ring	–	[581]
1255	Transtaganolide B	*Thapsia transtagana*	Apiaceae	7-methoxy-4,5-dihydro-3*H*-oxepin-2-one ring	–	[581]
1256	Transtaganolide C	*Thapsia transtagana*	Apiaceae	7-methoxy-4,5-dihydro-3*H*-oxepin-2-one ring	–	[581]
1257	Transtaganolide D	*Thapsia transtagana*	Apiaceae	7-methoxy-4,5-dihydro-3*H*-oxepin-2-one ring	–	[553]
Dimeric cyclohexylethanoid derivative						
1258	(−)-incarvilleatone	*Incarvillea younghusbandii*	Bignoniaceae	The first cyclohexylethanoid dimer connected by a six-membered ring	Inhibition against nitric oxide (NO) release	[582]
1259	(+)-incarvilleatone	*Incarvillea younghusbandii*	Bignoniaceae	The first cyclohexylethanoid dimer connected by a six-membered ring	Inhibition against nitric oxide (NO) release	[582]
Benzothiophene						
1260	Echinothiophene	*Echinops grijissii*	Compositae/asteraceae	Benzo[*b*]thiophene glycoside	–	[583]
Phthalide polymer						
1261	(−)-triligustilide A	*Angelica sinensis*	Apiaceae	Complex polycyclic skeletons simultaneously possessing bridged, fused, and spiro ring systems	Anti-inflammatory effect	[584]
1262	(+)-triligustilide A	*Angelica sinensis*	Apiaceae	Complex polycyclic skeletons simultaneously possessing bridged, fused, and spiro ring systems	–	[584]
1263	(−)-triligustilide B	*Angelica sinensis*	Apiaceae	Complex polycyclic skeletons simultaneously possessing bridged, fused, and spiro ring systems	–	[584]

(continued)

Table A12 (continued)

Compound	Trivial name	Source	Family	Structure characteristics	Activities if reported	References
1264	(+)-triligustilide B	*Angelica sinensis*	Apiaceae	Complex polycyclic skeletons simultaneously possessing bridged, fused, and spiro ring systems	Anti-inflammatory activity	[584]
1265	Triangeliphthalide A	*Angelica sinensis*	Apiaceae	Phthalide trimer with new skeleton	–	[585]
1266	Triangeliphthalide B	*Angelica sinensis*	Apiaceae	Phthalide trimer with new skeleton	–	[585]
1267	Triangeliphthalide C	*Angelica sinensis*	Apiaceae	Phthalide trimer with new skeleton	Anti-inflammatory activity	[585]
1268	Triangeliphthalide D	*Angelica sinensis*	Apiaceae	Phthalide trimer with new skeleton	Anti-inflammatory activity	[585]
Macrolactams						
1269	Dysoxylactam	*Dysoxylum hongkongense*	Meliaceae	An unprecedented branched C19 fatty acid and an L-valine	Reversing multidrug resistance	[586]
1270	(+)-(*S*)-scocycamide	*Scopolia tangutica*	Solanaceae	A unique 6/18 fused bicyclic framework with spermidine and catechol units	Antioxidant and anti-BChE activities	[587]
1271	(−)-(*R*)-scocycamide	*Scopolia tangutica*	Solanaceae	A unique 6/18 fused bicyclic framework with spermidine and catechol units	Antioxidant and anti-BChE activities	[587]
Macrolide						
1272	Ivorenolide A	*Khaya ivorensis*	Meliaceae	A novel 18-membered macrolide featuring conjugated acetylenic bonds and five chiral centers	Immunosuppressive activity	[588]
Toluylene						
1273	(+)–cajanusine	*Cajanus cajan*	Leguminosae	Bicyclo[4.2.0]oct-4-en-3-one unit	–	[589]
1274	(−)–cajanusine	*Cajanus cajan*	Leguminosae	Bicyclo[4.2.0]oct-4-en-3-one unit	Cytotoxicity	[589]
1275	Multiflorumiside A	*Polygonum multiflorum*	Polygonaceae	Two monomeric stilbenes to form a cyclobutane ring	Anti-inflammatory activity	[590]
1276	Multiflorumiside B	*Polygonum multiflorum*	Polygonaceae	Two monomeric stilbenes to form a cyclobutane ring	Anti-inflammatory activity	[590]
1277	Multiflorumiside C	*Polygonum multiflorum*	Polygonaceae	Two monomeric stilbenes to form a cyclobutane ring	Anti-inflammatory activity	[590]

(continued)

Table A12 (continued)

Compound	Trivial name	Source	Family	Structure characteristics	Activities if reported	References
1278	Multiflorumiside D	*Polygonum multiflorum*	Polygonaceae	Two monomeric stilbenes to form a cyclobutane ring	Anti-inflammatory activity	[590]
1279	Multiflorumiside E	*Polygonum multiflorum*	Polygonaceae	Two monomeric stilbene moieties coupled to form a tetralin system	–	[590]
1280	Multiflorumiside F	*Polygonum multiflorum*	Polygonaceae	Two monomeric stilbene moieties coupled to form a tetralin system	–	[590]
1281	Multiflorumiside G	*Polygonum multiflorum*	Polygonaceae	Two monomeric stilbene moieties coupled to form a tetralin system	–	[590]
Diarylheptanoid						
1282	Calyxin I	*Alpinia blepharocalyx*	Zingiberaceae	Ether linkage between C-8 and C-2′′	–	[591]
1283	Katsumadain A	*Alpinia katsumadai*	Zingiberaceae	Novel skeleton	Anti-emetic activity	[592]
1284	Katsumadain B	*Alpinia katsumadai*	Zingiberaceae	Novel skeleton	Anti-emetic activity	[592]
Phenol						
1285	Hopeanolin	*Hopea exalata*	Dipterocarpaceae	Resveratral trimer with an *ortho*-quinone nucleus	Antifungal activity	[593]
1286	Chunganenol	*Vitis chunganensis*	Vitaceae	Resveratrol hexamer, two stilbene units are connected by a methylene bridge	Antioxidant activity	[594]
1287	Dichotomain A	*Dicranopteris dichotoma*	Gleicheniaceae	Spirodilactone moiety	–	[595]
1288	Dichotomain B	*Dicranopteris dichotoma*	Gleicheniaceae	Spirodilactone moiety	Anti-HIV-1 activity	[595]
1289	Selaginpulvilin A	*Selaginella pulvinata*	Selaginellaceae	9,9-diphenyl-1-(phenylethynyl)-9*H*-fluorene skeleton	Phosphodiesterase-4 inhibitor	[596]
1290	Selaginpulvilin B	*Selaginella pulvinata*	Selaginellaceae	9,9-diphenyl-1-(phenylethynyl)-9*H*-fluorene skeleton	Phosphodiesterase-4 inhibitor	[596]
1291	Selaginpulvilin C	*Selaginella pulvinata*	Selaginellaceae	9,9-diphenyl-1-(phenylethynyl)-9*H*-fluorene skeleton	Phosphodiesterase-4 inhibitor	[596]
1292	Selaginpulvilin D	*Selaginella pulvinata*	Selaginellaceae	9,9-diphenyl-1-(phenylethynyl)-9*H*-fluorene skeleton	Phosphodiesterase-4 inhibitor	[596]

(continued)

Table A12 (continued)

Compound	Trivial name	Source	Family	Structure characteristics	Activities if reported	References
1293	Garcipaucinone A	*Garcinia paucinervis*	Hypericaceae	The first naturally occurring tocotrienol derivatives with a 3,10-dioxatricyclo-[7.3.1.$0^{2,7}$]tridecane skeleton incorporating an unusual γ-pyrone motif	–	[597]
1294	Garcipaucinone B	*Garcinia paucinervis*	Hypericaceae	The first naturally occurring tocotrienol derivatives with a 3,10-dioxatricyclo-[7.3.1.$0^{2,7}$]tridecane skeleton incorporating an unusual γ-pyrone motif	–	[597]
Iridoid						
1295	9-hydroxy-8-epihastatoside	*Junellia seriphioides*	Verbenaceae	The first ordinary iridoid glucoside with a 9-substituent	–	[598]
1296	Phyllanthusol A	*Phyllanthus acidus*	Euphorbiaceae	Norbisabolane glycoside	Cytotoxicity	[599]
1297	Phyllanthusol B	*Phyllanthus acidus*	Euphorbiaceae	Norbisabolane glycoside	Cytotoxicity	[599]
1298	Citrifolinoside	*Morinda citrifolia*	Rubiaceae	Presence of a rearranged ferulic acid moiety	Inhibition of activator protein-1 (AP-1)	[600]
1299	Swerilactone L	*Swertia mileensis*	Gentianaceae	Secoiridoids with C_{12} skeleton	Anti-HBV activity	[601]
1300	Swerilactone M	*Swertia mileensis*	Gentianaceae	Secoiridoids with C_{13} skeleton	Anti-HBV activity	[601]
1301	Swerilactone N	*Swertia mileensis*	Gentianaceae	Secoiridoids with C_{13} skeleton	–	[601]
1302	Swerilactone O	*Swertia mileensis*	Gentianaceae	Secoiridoids with C_{13} skeleton	–	[601]
1303	Sweritranslactone A	*Swertia*	Gentianaceae	6/6/6/6/6/6-fused hexacyclic skeleton	–	[602]
1304	Sweritranslactone B	*Swertia*	Gentianaceae	6/6/6/6/6/6-fused hexacyclic skeleton	–	[602]
1305	Sweritranslactone C	*Swertia*	Gentianaceae	6/6/6/6/6/6-fused hexacyclic skeleton	–	[602]
Amantane						
1306	Sampsonione I	*Hypericum sampsonii*	Hypericaceae	Caged tetracyclo[7.3.1.$1^{3,11}$.$0^{3,8}$]tetradecane-2,12,14-trione skeleton	Cytotoxicity	[603]

(continued)

Table A12 (continued)

Compound	Trivial name	Source	Family	Structure characteristics	Activities if reported	References
Xanthone						
1307	Gaudichaudiic acid F	*Garcinia gaudichaudii*	Hypericaceae	Heptacyclic xanthonoid	Cytotoxicity	[604]
1308	Gaudichaudiic acid G	*Garcinia gaudichaudii*	Hypericaceae	Heptacyclic xanthonoid contain an unusual toluene-fused dimethylpyran ring	Cytotoxicity	[604]
1309	Gaudichaudiic acid H	*Garcinia gaudichaudii*	Hypericaceae	Heptacyclic xanthonoid contain an unusual toluene-fused dimethylpyran ring	Cytotoxicity	[604]
1310	Gaudichaudiic acid I	*Garcinia gaudichaudii*	Hypericaceae	Heptacyclic xanthonoid contain an unusual toluene-fused dimethylpyran ring	Cytotoxicity	[604]
1311	Brasiliensophyllic acid A	*Calophyllum brasiliense*	Hypericaceae	Cyclobutyl ring	Antibacterial activity	[605]
1312	Brasiliensophyllic acid B	*Calophyllum brasiliense*	Hypericaceae	Cyclobutyl ring	Antibacterial activity	[605]
1313	Brasiliensophyllic acid C	*Calophyllum brasiliense*	Hypericaceae	Cyclobutyl ring	–	[605]
1314	Brasiliensophyllic acid D	*Calophyllum brasiliense*	Hypericaceae	Cyclobutyl ring	–	[605]
1315	Brasiliensophyllic acid E	*Calophyllum brasiliense*	Hypericaceae	Novel skeleton	–	[605]
1316	Brasiliensophyllic acid F	*Calophyllum brasiliense*	Hypericaceae	Novel skeleton	–	[605]
1317	Pruniflorone T	*Cratoxylum formosum*	Hypericaceae	Neocaged-xanthone	–	[606]
1318	Pruniflorone U	*Cratoxylum formosum*	Hypericaceae	Rearranged caged-xanthone	–	[606]
Quinone						
1319	3-alkyl-1,4-benzoquinone derivative, N-(3-carboxylpropyl)-5-amino-2-hydroxy-3-tridecyl1,4-benzoquinone	*Embelia ribes*	Myrsinaceae	3-alkyl-1,4-benzoquinone derivative, coupled through a C-N bond with γ-aminobutyric acid	–	[607]
1320	(−)-nigegladine A	*Nigella glandulifera*	Ranunculaceae	Tricyclo[5.4.0.$1^{2,6}$]dodecane carbon skeleton	Cardiomyocyte protection	[608]
1321	(+)-nigegladine A	*Nigella glandulifera*	Ranunculaceae	Tricyclo[5.4.0.$1^{2,6}$]dodecane carbon skeleton	Cardiomyocyte protection	[608]
1322	Nigegladine B	*Nigella glandulifera*	Ranunculaceae	Diterpenoid alkaloids with indole core	–	[608]
1323	Nigegladine C	*Nigella glandulifera*	Ranunculaceae	Diterpenoid alkaloids with indole core	–	[580]

(continued)

Table A12 (continued)

Compound	Trivial name	Source	Family	Structure characteristics	Activities if reported	References
Spirolactone						
1324	Idesolide	*Idesia polycarpa*	Flacourtiaceae	Tetrahydrobenzodioxole structure	Anti-inflammatory effect	[609]
1325	Yaoshanenolide A	*Machilus yaoshansis*	Lauraceae	Tricyclic spirolactones bearing long linear alkyl chain	Cytotoxicity	[610]
1326	Yaoshanenolide B	*Machilus yaoshansis*	Lauraceae	Tricyclic spirolactones bearing long linear alkyl chain	Cytotoxicity	[610]
Naphthalene						
1327	Dioscorealide A	*Dioscorea membranacea*	Dioscoreaceae	Naphthofuranoxepin skeleton	Cytotoxicity	[611]
1328	Dioscorealide B	*Dioscorea membranacea*	Dioscoreaceae	Naphthofuranoxepin skeleton	Cytotoxicity	[611]
1329	Rubialatin A	*Rubia alata*	Rubiaceae	6/6/ 5/6/6 carbon skeleton coupled with a spirocycloisopentene group	Cytotoxicity	[612]
1330	Rubialatin B	*Rubia alata*	Rubiaceae	6/7/6/6 tetracyclic system	Cytotoxicity	[612]
1331	Spiroaxillarone A	*Cyanotis axillaris*	Commelinaceae	A new type of spirobisnaphthalene with a previously unknown skeletal system	Antimalarial activity	[613]
1332	Eleucanainone A	*Eleutherine americana*	Iridaceae	A seven-oxygen-containing carbon ring connected to a bridged ether bond with a C-12-C-11′ linked dibenzofuran dimer	Anti-MRSA activity	[614]
1333	Eleucanainone B	*Eleutherine americana*	Iridaceae	The first example of dibenzofuran- and naphthalenone-containing naphthoquinone dimer	–	[614]
Lactone						
1334	Swerilactone A	*Swertia mileensis*	Gentianaceae	6/6/6/6/6 pentacyclic ring system	Anti-HBV activity	[615]
1335	Swerilactone B	*Swertia mileensis*	Gentianaceae	6/6/6/6/6 pentacyclic ring system	–	[615]
1336	Swerilactone H	*Swertia mileensis*	Gentianaceae	An unprecedented C29 skeleton	Anti-HBV activity	[616]
1337	Swerilactone I	*Swertia mileensis*	Gentianaceae	An unprecedented C29 skeleton	Anti-HBV activity	[616]
1338	Swerilactone J	*Swertia mileensis*	Gentianaceae	An unprecedented C29 skeleton	Anti-HBV activity	[616]
1339	Swerilactone K	*Swertia mileensis*	Gentianaceae	An unprecedented C29 skeleton	Anti-HBV activity	[616]

(continued)

Table A12 (continued)

Compound	Trivial name	Source	Family	Structure characteristics	Activities if reported	References
1340	Paracaseolide A	*Sonneratia paracaseolaris*	Sonneratiaceae	α-alkylbutenolide dimer	Inhibitory activity against dual specificity phosphatase CDC25B	[617]
Tannin						
1341	Punicatannin A	*Punica granatum*	Punicaceae	3-oxo-1,3,3a,8b-tetrahydrofuro[3,4-*b*]benzofuran moiety	Inhibition of α-glucosidase and lipogenic gene expression	[618]
1342	Punicatannin B	*Punica granatum*	Punicaceae	3-oxo-1,3,3a,8b-tetrahydrofuro[3,4-*b*]benzofuran moiety	–	[618]
1343	Quercusnin A	*Quercus crispula*	Fagaceae	Unusual ellagitannin	–	[619]
1344	Quercusnin B	*Quercus crispula*	Fagaceae	The first reported example of an ellagitannin containing a nitrogen atom in its skeleton	–	[619]
Thiophene						
1345	Echinopsacetylene A	*Echinops transiliensis*	Compositae/asteraceae	α-terthienyl moiety covalently linked with another thiophene moiety	Toxicity against the Formosoan subterranean termite	[620]
1346	Echinopsacetylene B	*Echinops transiliensis*	Compositae/asteraceae	Natural thiophene conjugated with a fatty acid moiety	–	[620]
Tetraterpene						
1347	Abibalsamin A	*Abies balsamea*	Pinaceae	3,4-*seco*-rearranged lanostane system fused with a β-myrcenelateral chain	Cytotoxicity	[621]
1348	Abibalsamin B	*Abies balsamea*	Pinaceae	3,4-*seco*-rearranged lanostane system fused with a β-myrcenelateral chain	Cytotoxicity	[621]
Peptide						
1349	Celogentin A	*Celosia argentea*	Amaranthaceae	Moroidin-type bicyclic peptide	Antimitotic activity	[622]
1350	Celogentin B	*Celosia argentea*	Amaranthaceae	Moroidin-type bicyclic peptide	Antimitotic activity	[622]
1351	Celogentin C	*Celosia argentea*	Amaranthaceae	Moroidin-type bicyclic peptide	Antimitotic activity	[622]
1352	Tunicyclin A	*Psammosilene tunicoides*	Caryophyllaceae	Tricyclic ring cyclopeptide skeleton	–	[623]

(continued)

Table A12 (continued)

Compound	Trivial name	Source	Family	Structure characteristics	Activities if reported	References
Cyclopentadienedione						
1353	(+)-linderaspirone A	*Lindera aggregata*	Lauraceae	Novel windmill-shaped molecule isolated from nature	Improvement of insulin sensitivity	[624]
1354	(+)-linderaspirone A	*Lindera aggregata*	Lauraceae	Novel windmill-shaped molecule isolated from nature	Improvement of insulin sensitivity	[624]
1355	Bi-linderone	*Lindera aggregata*	Lauraceae	Spirocyclopentenedione-containing carbon skeleton	Improvement of insulin sensitivity	[625]
Coumarin						
1356	Exotine A	*Murraya exotica*	Rutaceae	Fused heptacyclic ring system	Anti-neuroinflammatory activity	[626]
1357	Exotine B	*Murraya exotica*	Rutaceae	Fused heptacyclic ring system	–	[626]
1358	Alatin A	*Murraya alata*	Rutaceae	8-methylbenzo[*h*]coumarin	Anti-inflammatory activity	[627]
1359	Alatin B	*Murraya alata*	Rutaceae	8-methylbenzo[*h*]coumarin	Anti-inflammatory activity	[627]
Isopentylphenylpropanol						
1360	Spirooliganone A	*Illicium oligandrum*	Magnoliaceae	Dioxaspiro skeleton	Antiviral activity	[628]
1361	Spirooliganoneb	*Illicium oligandrum*	Magnoliaceae	Dioxaspiro skeleton	Antiviral activity	[628]
Isopentenyl acetyl benzene						
1362	(+)-melicolone A	*Melicope ptelefolia*	Rutaceae	9-oxatricyclo[3.2.1.1^{3,8}]nonane core	Prevent diabetic endothelial dysfunction	[629]
1363	(−)-melicolone A	*Melicope ptelefolia*	Rutaceae	9-oxatricyclo[3.2.1.1^{3,8}]nonane core	Prevent diabetic endothelial dysfunction	[629]
1364	(+)-melicolone B	*Melicope ptelefolia*	Rutaceae	9-oxatricyclo[3.2.1.1^{3,8}]nonane core	Prevent diabetic endothelial dysfunction	[629]
1365	(−)-melicolone B	*Melicope ptelefolia*	Rutaceae	9-oxatricyclo[3.2.1.1^{3,8}]nonane core	Prevent diabetic endothelial dysfunction	[629]

References

1. Newman DJ, Cragg GM, Snader KM. Natural products as sources of new drugs over the period 1981 2002. J Nat Prod 2003;66:1022–1037.
2. Newman DJ, Cragg GM. Natural products as sources of new drugs over the nearly four decades from 01/1981 to 09/2019.J Nat Prod 2020;83:770−803.
3. Ciochina R, Grossman RB. Polycyclic Polyprenylated acylphloroglucinols. Chem Rev 2006;106:3963−3986.
4. Wu YB, Ni ZY, Shi QW, Dong M, Kiyota H, Gu YC, Cong B. Constituents from Salvia species and their biological activities. Chem Rev 2012;112:5967–6026.
5. Vasas A, Hohmann J. Euphorbia diterpenes: isolation, structure, biological activity, and synthesis (2008−2012). Chem Rev 2014;114:8579−8612.
6. Wang HB, Wang XY, Liu LP, Qin GW, Kang TG. Tigliane diterpenoids from the *Euphorbiaceae* and *Thymelaeaceae* families. Chem Rev 2015;115:2975−3011.
7. Tan QG, Luo XD. Meliaceous limonoids: chemistry and biological activities. Chem Rev 2011;111:7437–7522.
8. Silva LN, Zimmer KR, Macedo AJ, Trentin DS. Plant natural products targeting bacterial virulence factors. Chem Rev 2016;116:9162−9236.
9. Yang XW, Grossman RB, Xu G. Research progress of polycyclic polyprenylated acylphloroglucinols. Chem Rev 2018;118:3508−3558.
10. Almeida A, Dong L, Appendino G, Bak S. Plant triterpenoids with bond-missing skeletons: biogenesis, distribution and bioactivity. Nat Prod Rep 2020;37:1207–1228.
11. Shen Y, Liang WJ, Shi YN, Kennelly E J, Zhao DK. Structural diversity, bioactivities, and biosynthesis of natural diterpenoid alkaloids. Nat Prod Rep 2020;37:763–796.
12. YZhang, H, Liu HB, Yue JM. Organic carbonates from natural sources. Chem Rev 2014;114:883–898.
13. Woodward RB, Edited Zagalak B, Friedrich, W. Synthetic vitamin B12. Proc. Eur. Symp. 1979;3rd:37–87.
14. Gooch JW. Encyclopedic Dictionary of polymers. Springer 2011;Haworth Projection.
15. de la Torre MC, Sierra MA. Comments on recent achievements in biomimetic organic synthesis. Angew Chem Int Ed 2004;43:160–181.
16. Hillier SG, Lathe R. Terpenes, hormones and life: isoprene rule revisited. J Endocrinol 2019;242:R9–R22.
17. Moss GP. Basic terminology of stereochemistry. Pure Appl Chem 1996;68:2193–2222
18. Corey EJ, Cheng XM. The logic of chemical synthesis. Wiley 1989.
19. Corey EJ. Retrosynthetic thinking–essentials and examples. Chem Soc Rev 1988;17:111–133.
20. Corey EJ. The logic of chemical synthesis: multistep synthesis of complex carbogenic molecules (Nobel Lecture), Angew Chem Int Ed 1991;30:455–465.

21. Kenakin T. Principles: Receptor theory in pharmacology. Trends Pharmacol Sci. 2004;25:186–192.
22. Gilani AH, Rahman AU. Trends in ethnopharmocology. J Ethnopharmacol 2005;100:43–49.
23. Ro DK, Paradise EM, Ouellet M, Fisher KJ, Newman KL, Ndungu JM, Ho KA, Eachus RA, Ham TS, Kirby J, Chang MCY, Withers ST, Shiba Y, Sarpong R, Keasling JD. Production of the antimalarial drug precursor artemisinic acid in engineered yeast. Nature 2006;440:940–943.
24. Paddon CJ, Westfall PJ, Pitera DJ, et al.. High-level semi-synthetic production of the potent antimalarial artemisinin. Nature 2013;496:528–532.
25. Leonard E, Ajikumar PK, Thayer K, Xiao WH, Mo JD, Tidor B, Stephanopoulos G, Prather KLJ. Combining metabolic and protein engineering of a terpenoid biosynthetic pathway for overproduction and selectivity control. P Natl Acad Sci USA 2010;107:13654–13659.
26. Ajikumar PK, Xiao WH, Tyo KEJ, Wang Y, Simeon F, Leonard E, Mucha O, Phon TH, Pfeifer B, Stephanopoulos G. Isoprenoid pathway optimization for taxol precursor overproduction in Escherichia Coli Science 2010;330:70–74.
27. Zhou YJ, Gao W, Rong Q, Jin G, Chu H, Liu W, Yang W, Zhu Z, Li G, Zhu G, Huang L, Zhao Z K. Modular pathway engineering of diterpenoid synthases and the mevalonic acid pathway for miltiradiene production. J Am Chem Soc 2012;134:3234–3241.
28. Galanie S, Thodey K, Trenchard IJ, Interrante MF, Smolke CD. Complete biosynthesis of opioids in yeast. Science 2015;349:1095–1100.
29. Luo X, Reiter MA, d'Espaux L, Wong J, Denby CM, Lechner A, Zhang Y, Grzybowski AT, Harth S, Lin W, Lee H, Yu C, Shin J, Deng K, Benites VT, Wang G, Baidoo EEK, Chen Y, Dev I, Petzold CJ, Keasling JD. Complete biosynthesis of cannabinoids and their unnatural analogues in Yeast. Nature 2019:567:123–126.
30. Wu QX, Shi YP, Yang L. Unusual sesquiterpene lactones from Ligularia virgaurea spp. oligocephala. Org Lett 2004;6:2313–2316.
31. Macías FA, Galindo JLG, Varela RM, Torres A, Molinillo JMG, Fronczek FR. Heliespirones B and C: Two new plant heliespiranes with a novel spiro heterocyclic sesquiterpene skeleton. Org Lett 2006;8:4513–4516.
32. Karioti A, Skaltsa H, Linden A, Perozzo R, Brun R, Tasdemir D. Anthecularin: A novel sesquiterpene lactone from anthemis auriculata with antiprotozoal activity. J Org Chem 2007;72:8103–8106.
33. Fattorusso C, Stendardo E, Appendino G, Fattorusso E, Luciano P, Romano A, Taglialatela-Scafati O. Artarborol, a nor-caryophyllane sesquiterpene alcohol from *artemisia arborescens*. stereostructure assignment through concurrence of NMR data and computational analysis. Org Lett 2007;9:2377–2380.
34. Ton That Q, Jossang J, Jossang A, Nguyen Kim P P, Jaureguiberry G. Wedelolides A and B: novel sesquiterpene δ-lactones, (9R)-eudesman-9,12-olides, from *Wedelia trilobata*. J Org Chem 2007;72:7102–7105.
35. Schmidt TJ. Novel seco-prezizaane sesquiterpenes from north American *Illicium* species. J Nat Prod 1999;62:684–687.

36. Ma SG, Li M, Lin MB, Li L, Liu YB, Qu J, Li Y, Wang XJ, Wang RB, Xu S, Hou Q, Yu SS. Illisimonin A, a caged sesquiterpenoid with a tricyclo[5.2.1.$0^{1,6}$]decane skeleton from the fruits of *Illicium simonsii*. Org Lett 2017;19:6160–6163.
37. Kubo M, Okada C, Huang JM, Harada K, Hioki H, Fukuyama Y. Novel pentacyclic seco-prezizaane-type sesquiterpenoids with neurotrophic properties from *Illicium jiadifengpi*. Org Lett 2009;11:5190–5193.
38. Xi FM, Ma SG, Liu YB, Li L, Yu SS. Artaboterpenoids A and B, bisabolene-derived sesquiterpenoids from *Artabotrys hexapetalus*. Org Lett 2016;18:3374–3377.
39. Wang PC, Ran XH, Luo HR, Hu JM, Chen R, Ma QY, Dai HF, Liu YQ, Xie MJ, Zhou J, Zhao YX. Volvalerelactones A and B, two new sesquiterpenoid lactones with an unprecedented skeleton from *Valeriana officinalis* Var. *latifolia*. Org Lett 2011;13:3036–3039.
40. Chai T, Meng XH, Wang CB, Wang K, Ma LM, Shi YP, and Yang JL. Narjatamolide, an unusual homoguaiane sesquiterpene lactone from *Nardostachys jatamansi*. J Org Chem 2021;86:11006−11010.
41. Yam-Puc A, Escalante-Erosa F, Pech-López M, Chan-Bacab MJ, Arunachalampillai A, Wendt OF, Sterner O, Peña-Rodríguez LM. Trinorsesquiterpenoids from the root extract of *Pentalinon andrieuxii*. J Nat Prod 2009;72:745–748.
42. He JB, Luo J, Zhang L, Yan YM, Cheng YX. Sesquiterpenoids with New Carbon Skeletons from the Resin of Toxicodendron vernicifluum as new types of extracellular matrix inhibitors. Org Lett 2013;15:3602–3605.
43. Shen YH, Li SH, Li RT, Han QB, Zhao QS, Liang L, Sun HD, Lu Y, Cao P, Zheng QT. Coriatone and corianlactone, two novel sesquiterpenes from Coriaria nepalensis. Org Lett 2004;6:1593–1595.
44. Han JJ, Zhang JZ, Zhu RX, Li Y, Qiao YN, Gao Y, Jin XY, Chen W, Zhou JC, Lou HX. Plagiochianins A and B, two *ent-2,3-seco*-aromadendrane derivatives from the Liverwort *Plagiochila duthiana*. Org Lett 2018;20:6550–6553.
45. Ainge GD, Gerard PJ, Hinkley SFR, Lorimer SD, Weavers RT. Hodgsonox, a new class of sesquiterpene from the Liverwort *Lepidolaena hodgsoniae*. Isolation directed by insecticidal activity. J Org Chem 2001;66:2818–2821
46. Liu Y, Liu F, Qiao MM, Guo L, Chen MH, Peng C, Xiong L. Curcumanes A and B, two bicyclic sesquiterpenoids with significant vasorelaxant activity from *Curcuma longa*. Org Lett 2019;21:1197–1201.
47. Chen M, Cao JQ, Ang S, Zeng TN, Li NP, Yang TJ, Liu JS, Wu Y, Ye WC, Wang L. Eugenunilones A–H: rearranged sesquiterpenoids from *Eugenia uniflora*. Org Chem Front 2022;9:667–675.
48. Jiao SG, Su GZ, Zhou XC, Ge FX, Liu CX, Zhang RF, Peng B, Chen S, Huang LQ, Tu PF, Chai XY. Three pairs of enantiomeric sesquiterpenoids from *Syringa pinnatifolia*. J Org Chem 2021;86:7263−7270.

49. Yong JY, Li WR, Wang XJ, Su GZ, Li M, Zhang JP, Jia HL, Li YH, Wang RB, Gan ML, Ma SG. Illihenin A: an antiviral sesquiterpenoid with a cage-like tricyclo[6.2.2.$0^{1,5}$]dodecane skeleton from *Illicium henryi*. J Org Chem 2021;86:2017−2022.
50. Zhao P, Li ZY, Qin SY, Xin BS, Liu YY, Lin B, Yao GD, Huang XX, Song SJ. Three unusual sesquiterpenes with distinctive ring skeletons from *Daphne penicillata* uncovered by molecular networking strategies. J Org Chem 2021;86:15298−15306.
51. Yang SP, Gao ZB, Wang FD, Liao SG, Chen HD, Zhang CR, Hu GY, Yue JM. Chlorahololides A and B, two potent and selective blockers of the potassium channel isolated from *Chloranthus holostegius*. Org Lett 2007;9:903–906.
52. He XF, Zhang S, Zhu RX, Yang SP, Yuan T, Yue JM. Sarcanolides A and B: two sesquiterpenoid dimers with a nonacyclic scaffold from *Sarcandra hainanensis*. Tetrahedron 2011;67:3170–3174.
53. Yuan T, Zhu RX, Yang SP, Zhang H, Zhang CR, Yue JM. Serratustones A and B representing a new dimerization pattern of two types of sesquiterpenoids from *Chloranthus serratus*. Org Lett 2012;14:3198–3201.
54. Zhou B, Liu QF, Dalal S, Cassera MB, Yue JM. Fortunoids A–C, three sesquiterpenoid dimers with different carbon skeletons from *Chloranthus fortunei*. Org Lett 2017;19:734–737.
55. Fan YY, Sun YL, Zhou B, Zhao JX, Sheng L, Li JY, Yue JM. Hedyorienoids A and B, two sesquiterpenoid dimers featuring different polycyclic skeletons from *Hedyosmum orientale*. Org Lett 2018;20:5435–5438.
56. Bai B, Ye SX, Yang DP, Zhu LP, Tang GH, Chen YY, Li GQ, Zhao ZM. Chloraserrtone A, a sesquiterpenoid dimer from *Chloranthus serratus*. J Nat Prod 2019;82:407–411.
57. Chi J, Xu WJ, Wei SS, Wang XB, Li JX, Gao HL, Kong LY, Luo J. Chlotrichenes A and B, two lindenane sesquiterpene dimers with highly fused carbon skeletons from *Chloranthus holostegius*. Org Lett 2019;21:789–792.
58. Yan H, Qin XJ, Li X H, Yu Q, Ni W, He L, Liu HY. Japonicones A-C: three lindenane sesquiterpenoid dimers with a 12-membered ring core from *Chloranthus japonicus*. Tetrahedron Lett 2019;60:713–717.
59. Zhang DY, Wang XX, Wang YN, Wang M, Zhuang PY, Jin Y, Liu H. Nine sesquiterpenoid dimers with four unprecedented types of carbon skeleton from *Chloranthus henryi* var. *hupehensis*. Org Chem Front 2021;8:4374–4386.
60. Zhou JS, Liu QF, Zimbres Flavia M, Butler Joshua H, Cassera Maria B, Zhou B, Yue JM. Trichloranoids A–D, antimalarial sesquiterpenoid trimers from *Chloranthus spicatus*. Org Chem Front 2021;8:1795–1801.
61. Liu X, Yang J, Yao XJ, Yang X, Fu J, Bai LP, Liu L, Jiang ZH, Zhu GY. Linderalides A–D, disesquiterpenoid–geranylbenzofuranone conjugates from *Lindera aggregata*. J Org Chem 2019;84:8242–8247.
62. Liu X, Yang J, Fu J, Yao XJ, Wang JR, Liu L, Jiang ZH, Zhu GY. Aggreganoids A–F, carbon-bridged sesquiterpenoid dimers and trimers from *Lindera aggregata*. Org Lett 2019;21:5753–5756.

63. Wu JW, Tang CP, Chen L, Qiao Y, Geng MY, Ye Y. Dicarabrones A and B, a pair of new epimers dimerized from sesquiterpene lactones via a [3 + 2] cycloaddition from *Carpesium abrotanoides*. Org Lett 2015;17:1656–1659.
64. Yang YX, Shan L, Liu QX, Shen YH, Zhang JP, Ye J, Xu XK, Li HL, Zhang WD. Carpedilactones A–D, four new isomeric sesquiterpene lactone dimers with potent cytotoxicity from *Carpesium faberi*. Org Lett 2014;16:4216–4219.
65. Yuan J, Wen X, Ke CQ, Zhang T, Lin L, Yao S, Goodpaster JD, Tang C, Ye Y. Tricarabrols A–C, three anti-inflammatory sesquiterpene lactone trimers featuring a methylene-tethered linkage from *Carpesium faberi*. Org Chem Front 2020;7:1374–1382.
66. Zhou XD, Chai XY, Zeng K W, Zhao MB, Jiang Y, Tu PF. Artesin A, a new cage-shaped dimeric guaianolide from *Artemisia sieversiana*. Tetrahedron Lett 2015;56:1141–1143.
67. Xue GM, Han C, Chen C, Li LN, Wang XB, Yang MH, Gu YC, Luo JG, Kong LY. Artemisians A–D, diseco-guaianolide involved heterodimeric [4 + 2] adducts from *Artemisia argyi*. Org Lett 2017;19:5410–5413.
68. Qin DP, Pan DB, Xiao W, Li HB, Yang B, Yao XJ, Dai Y, Yu Y, Yao XS. Dimeric cadinane sesquiterpenoid derivatives from *Artemisia annua*. Org Lett 2018;20:453–456.
69. Wu ZJ, Xu XK, Shen YH, Su J, Tian JM, Liang S, Li HL, Liu RH, Zhang WD. Ainsliadimer A, a new sesquiterpene lactone dimer with an unusual carbon skeleton from *Ainsliaea macrocephala*. Org Lett 2008;10:2397–2400.
70. Wang Y, Shen YH, Jin HZ, Fu JJ, Hu XJ, Qin JJ, Liu JH, Chen M, Yan SK, Zhang WD. Ainsliatrimers A and B, the first two guaianolide trimers from *Ainsliaea fulvioides*. Org Lett 2008;10:5517–5520.
71. Zhang R, Tang C, Liu HC, Ren Y, Xu C, Ke CQ, Yao S, Huang X, Ye Y. Ainsliatriolides A and B, two guaianolide trimers from *Ainsliaea fragrans* and their cytotoxic activities. J Org Chem 2018;83:14175–14180.
72. Zhang R, Tang C, Liu HC, Ren Y, Ke CQ, Yao S, Cai Y, Zhang N, Ye Y. Tetramerized sesquiterpenoid ainsliatetramers A and B from *Ainsliaea fragrans* and their cytotoxic activities. Org Lett 2019;21:8211–8214.
73. Ding N, Wang JY, Liu J, Zhu YJ, Hou SR, Zhao HM, Yang YS, Chen XB, Hu LH, Wang XC. Cytotoxic guaianolide sesquiterpenoids from *Ainsliaea fragrans*. J Nat Prod 2021;84:2568−2574.
74. Ren YM, Zhou SZ, Zhang T, Qian M, Zhang R, Yao S, Zhu H, Tang C, Lin L, Ye Y. Targeted isolation of two disesquiterpenoid macrocephadiolides A and B from *Ainsliaea macrocephala* using a molecular networking-based dereplication strategy. Org Chem Front 2020;7:1481–1489.
75. Ren YM, Zhang R, Feng ZL, Ke CQ, Yao S, Tang CP, Lin LG, and Ye Y. Macrocephatriolides A and B: two guaianolide trimers from *Ainsliaea macrocephala* as PTP1B inhibitors and insulin sensitizers. J Org Chem 2021;86:17782−17789.
76. Jin Q, Lee JW, Jang H, Lee HL, Kim JG, Wu W, Lee D, Kim EH, Kim Y, Hong JT, Lee MK, Hwang BY. Dimeric- and trimeric sesquiterpenes from the flower of *Inula japonica*. Phytochemistry 2018;155:107–113.

77. Wu ZL, Wang Q, Wang JX, Dong HY, Xu XK, Shen YH, Li HL, Zhang WD. Vlasoulamine A, a neuroprotective [3.2.2]cyclazine sesquiterpene lactone dimer from the roots of *Vladimiria souliei*. Org Lett 2018;20:7567–7570.
78. Liu JW, Zhang MY, Yan YM, Wei XY, Dong L, Zhu YX, Cheng YX. Characterization of sesquiterpene dimers from *Resina commiphora* that promote adipose-derived stem cell proliferation and differentiation. J Org Chem 2018;83:2725–2733.
79. Liu JW, Liu Y, Yan YM, Yang J, Lu XF, Cheng YX. Commiphoratones A and B, two sesquiterpene dimers from *Resina commiphora*. Org Lett 2018;20:2220–2223.
80. Dong L, Qin DP, Di QQ, Liu Y, Chen WL, Wang SM, Cheng YX. Commiphorines A and B, unprecedented sesquiterpenoid dimers from *Resina commiphora* with striking activities on anti-inflammation and lipogenesis inhibition. Org Chem Front 2019;6:3825–3833.
81. Shen XY, Qin DP, Zhou H, Luo JF, Yao YD, Lio CK, Li HB, Dai Y, Yu Y, Yao XS. Nardochinoids A–C, three dimeric sesquiterpenoids with specific fused-ring skeletons from *Nardostachys chinensis*. Org Lett 2018;20:5813–5816.
82. Li QM, Luo JG, Zhang YM, Li ZR, Wang XB, Yang MH, Luo J, Sun, H B, Chen YJ, Kong LY. Involucratustones A-C: unprecedented sesquiterpene dimers containing multiple contiguous quaternary carbons from *Stahlianthus involucratus*. Chem Eur J 2015;21:13206–13209.
83. Zhang YL, Zhou XW, Wang XB, Wu L, Yang MH, Luo J, Yin Y, Luo JG, Kong LY. Xylopiana A, a dimeric guaiane with a case-shaped core from *Xylopia vielana*: structural elucidation and biomimetic conversion. Org Lett 2017;19:3013–3016.
84. Duarte N, Ferreira MJU. Lagaspholones A and B: two new jatropholane-type diterpenes from *Euphorbia lagascae*. Org Lett 2007;9:489–492.
85. Mai ZP, Ni G, Liu YF, Li L, Li JY, Yu DQ. Heliojatrones A and B, two jatrophane-derived diterpenoids with a 5/10 fused-ring skeleton from *Euphorbia helioscopia*: structural elucidation and biomimetic conversion. Org Lett 2018;20:3124–3127
86. Mai ZP, Ni G, Liu YF, Li YH, Li L, Li JY, Yu DQ. Secoheliosphanes A and B and secoheliospholane A, three diterpenoids with unusual seco-jatrophane and seco-jatropholane skeletons from *Euphorbia helioscopia*. J Org Chem 2018;83:167–173.
87. Shi QQ, Zhang XJ, Wang TT, Wang Q, Sun TT, Amin M, Zhang RH, Li XL, Xiao WL. Euphopias A–C: three rearranged jatrophane diterpenoids with tricyclo[8.3.0.0^{2,7}]tridecane and tetracyclo[11.3.0.0^{2,10}.0^{3,7}]hexadecane cores from *Euphorbia Helioscopia*. Org Lett 2020;22:7820–7824.
88. Shi QQ, Zhang Y, Wang TT, Xiong F, Zhang RH, Li XL, Ji X, Zhang XJ, Wang WG, Xiao WL. Euphopias D–F from *Euphorbia L.*: quantum chemical calculation-based structure elucidation and their bioactivity of inhibiting NLRP3 inflammasome activation. Org Chem Front 2021;8:3041–3046.

89. Annam SCHVAR, Ankireddy M, Sura MB, Ponnapalli MG, Sarma AVS,S,J.B. Epimeric excolides from the stems of *Excoecaria agallocha* and structural revision of rhizophorin A. Org Lett 2015;17:2840–2843.
90. Fei DQ, Dong LL, Qi FM, Fan GX, Li HH, Li ZY, Zhang ZX. Euphorikanin A, a diterpenoid lactone with a fused 5/6/7/3 ring system from *Euphorbia kansui*. Org Lett 2016;18:2844–2847.
91. Wan LS, Nian Y, Peng XR, Shao LD, Li XN, Yang J, Zhou M, Qiu MH. Pepluanols C–D, two diterpenoids with two skeletons from *Euphorbia peplus*. Org Lett 2018;20:3074–3078.
92. Yan XL, Sang J, Chen SX, Li W, Tang GH, Gan LS, Yin S. Euphorkanlide A, a highly modified ingenane diterpenoid with a c24 appendage from *Euphorbia kansuensis*. Org Lett 2019;21:4128–4131.
93. Yuan FY, Pan YH, Yin AP, Li W, Huang D, Yan XL, Wu SQ, Tang GH, Pu R, Yin S. Euphorstranoids A and B, two highly rearranged ingenane diterpenoids from *Euphorbia stracheyi*: structural elucidation, chemical transformation, and lipid-lowering activity. Org Chem Front 2022;9:775–780.
94. Liao SG, Zhan ZJ, Yang SP, Yue JM. Lathyranoic Acid A: first secolathyrane diterpenoid in nature from *Euphorbia lathyris*. Org Lett 2005;7:1379–1382.
95. Gao S, Liu HY, Wang YH, He HP, Wang JS, Di YT, Li CS, Fang X, Hao XJ. Lathyranone A: a diterpenoid possessing an unprecedented skeleton from *Euphorbia lathyris*. Org Lett 2007;9:3453–3455.
96. Tian Y, Guo QL, Xu WD, Zhu CG, Yang YC, Shi JG. A minor diterpenoid with a new 6/5/7/3 fused-ring skeleton from *Euphorbia micractina*. Org Lett 2014;16:3950–3953.
97. Tian Y, Guo QL, Xu WD, Zhu CG, Yang YC, Shi JG. A minor diterpenoid with a new 6/5/7/3 fused-ring skeleton from *Euphorbia micractina*. Org Lett 2014;16:3950–3953.
98. Pan L, Zhou P, Zhang X, Peng S, Ding L, Qiu SX. Skeleton-rearranged pentacyclic diterpenoids possessing a cyclobutane ring from *Euphorbia wallichii*. Org. Lett. 2006;8:2775–2778.
99. Qi WY, Zhao JX, Wei WJ, Gao K, Yue JM. Quorumolides A–C, three cembranoids from *Euphorbia antiquorum*. J Org Chem 2018;83:1041–1045.
100. Wang L, Yang J, Kong LM, Deng J, Xiong Z, Huang J, Luo J, Yan Y, Hu Y, Li XN, Li Y, Zhao Y, Huang SX. Natural and semisynthetic tigliane diterpenoids with new carbon skeletons from *Euphorbia dracunculoides* as a wnt signaling pathway inhibitor. Org. Lett. 2017;19:3911–3914.
101. Ferreira RJ, Spengler G, Orthaber A, dos Santos DJVA, Ferreira MJU. Pedrolane, a polycyclic diterpene scaffold containing a bicyclo[2.2.1]heptane system, from *Euphorbia pedroi*. Org Lett 2021;23:274–278.
102. Liu SN, Huang D, Morris Natschke SL, Ma H, Liu Z, Seeram NP, Xu J, Lee KH, Gu Q. Euphomilones A and B, ent-rosane diterpenoids with 7/5/6 and 5/7/6 skeletons from *Euphorbia milii*. Org Lett 2016;18:6132–6135.
103. Barile E, Lanzotti V. Biogenetical related highly oxygenated macrocyclic diterpenes from sea spurge *Euphorbia paralias*. Org Lett 2007;9:3603–3606.

104. Flores Giubi ME, Durán Peña MJ, Botubol Ares JM, Escobar Montaño F, Zorrilla D, Macías Sánchez AJ, Hernández Galán R. Gaditanone, a diterpenoid based on an unprecedented carbon skeleton isolated from *Euphorbia gaditana*. J Nat Prod 2017;80:2161–2165.
105. Fan YY, Shi SQ, Deng GZ, Liu HC, Xu CH, Ding J, Wang GW, Yue JM. Crokonoids A–C, a highly rearranged and dual-bridged spiro diterpenoid and two other diterpenoids from *Croton kongensis*. Org Lett 2020;22:929–933.
106. Wang GC, Zhang H, Liu HB, Yue JM. Laevinoids A and B: two diterpenoids with an unprecedented backbone from *Croton laevigatus*. Org Lett 2013;15:4880–4883.
107. Li W, Wang RM, Pan YH, Zhao YY, Yuan FY, Huang D, Tang GH, Bi HC, Yin S. Crotonpenoids A and B, two highly modified clerodane diterpenoids with a tricyclo[7.2.1.$0^{2,7}$]dodecane core from *Croton yanhuii*: isolation, structural elucidation, and biomimetic semisynthesis. Org Lett 2020;2:4435–4439.
108. Zhang WY, Zhao JX, Sheng L, Fan YY, Li JY, Gao K, Yue JM. Mangelonoids A and B, two pairs of macrocyclic diterpenoid enantiomers from *Croton mangelong*. Org Lett 2018;20:4040–4043.
109. Chen HD, Yang SP, He XF, Ai J, Liu ZK, Liu HB, Geng MY, Yue JM. Trigochinins A−C: three new daphnane-type diterpenes from *Trigonostemon chinensis*. Org Lett 2010;12:1168–1171.
110. Zhang L, Luo RH, Wang F, Jiang MY, Dong ZJ, Yang LM, Zheng YT, Liu JK. Highly functionalized daphnane diterpenoids from *Trigonostemon thyrsoideum*. Org Lett 2010;12:152–155.
111. Allard PM, Martin MT, Tran Huu Dau ME, Leyssen P, Guéritte F, Litaudon M. Trigocherrin A, the first natural chlorinated daphnane diterpene orthoester from *Trigonostemon cherrieri*. Org Lett 2012;14:342–345.
112. Wang XC, Zheng ZP, Gan XW, Hu LH. Jatrophalactam, a novel diterpenoid lactam isolated from *Jatropha curcas*. Org Lett 2009;11:5522–5524.
113. Chianese G, Fattorusso E, Aiyelaagbe OO, Luciano P, Schröder HC, Müller WEG, Taglialatela Scafati O. Spirocurcasone, a diterpenoid with a novel carbon skeleton from *Jatropha curcas*. Org Lett 2011;13:316–319.
114. Li W, Tang YQ, Sang J, Fan RZ, Tang GH, Yin, S. Jatrofolianes A and B: two highly modified lathyrane diterpenoids from *Jatropha gossypiifolia*. Org Lett 2020;22:106–109.
115. Huang D, Pan YH, Yuan FY, Li W, Yan XL, Zou MF, Tang GH, Yin S. Euphohyrisnoids A and B, two highly rearranged lathyrane diterpenoids from *Euphorbia lathyris*. Org Lett 2021;23:9602−9605.
116. He Jun, Xu JK, Guo LB, Xia CY, Lian WW, Tian HY, Zhang J, Shi YX, and Zhang WK. Fischdiabietane A, an antitumoral diterpenoid dimer featuring an unprecedented carbon skeleton from *Euphorbia fischeriana*. J Org Chem 2021;86:5894−5900.
117. Gui MY. Aoyagi Y, Jin YR, Li XW, Hasuda T, Takeya K. Excisanin H, a novel cytotoxic 14,20-epoxy-*ent*-kaurene diterpenoid, and three new *ent*-kaurene diterpenoids from *Rabdosia excisa*. J Nat Prod 2004;67:373–376.

118. Li SH, Wang J, Niu XM, Shen YH, Zhang HJ, Sun HD, Li ML, Tian QE, Lu Y, Cao P, Zheng QT. Maoecrystal V, cytotoxic diterpenoidwith a novel C_{19} skeleton from *Isodon eriocalyx* (Dunn.) Hara. Org Lett 2004;6:4327–4330.
119. Han QB, Cheung S, Tai J, Qiao CF, Song JZ, Tso TF, Sun HD, Xu HX. Maoecrystal Z, a cytotoxic diterpene from *Isodon eriocalyx* with a unique skeleton. Org Lett 2006;8:4727–4730.
120. Huang SX, Xiao WL, Li LM, Li SH, Zhou Y, Ding LS, Lou LG, Sun HD. Bisrubescensins A–C: three new dimeric *ent*-kauranoids isolated from *Isodon rubescens*. Org Lett 2006;8:1157–1160.
121. Wang WG, Du X, Li XN, Wu HY, Liu X, Shang SZ, Zhan R, Liang CQ, Kong LM, Li Y, Pu JX, Sun HD. New bicyclo[3.1.0]hexane unit ent-kaurane diterpene and its *seco*-derivative from *Isodon eriocalyx* Var. *laxiflora*. Org Lett 2012;14:302–305.
122. Zou J, Du X, Pang G, Shi YM, Wang WG, Zhan R, Kong LM, Li XN, Li Y, Pu JX, Sun HD. Ternifolide A, a New diterpenoid possessing a rare macrolide motif from *Isodon ternifolius*. Org Lett 2012;14:3210–3213.
123. Wang WG, Li XN, Du X, Wu HY, Liu X, Su J, Li Y, Pu JX, Sun HD. Laxiflorolides A and B, epimeric bishomoditerpene lactones from *Isodon eriocalyx*. J Nat Prod 2012;75:1102–1107.
124. Hu ZX, Xu HC, Hu K, Liu M, Li XN, Li XR, Du X, Zhang YH, Puno PT, Sun HD. Structurally diverse diterpenoids from *Isodon pharicus*. Org Chem Front 2018;5:2379–2389.
125. Yang Q, Hu K, Yan BC, Liu M, Li XN, Sun HD, Puno PT. Maoeriocalysins A–D, four novel *ent*-kaurane diterpenoids from *Isodon eriocalyx* and their structure determination utilizing quantum chemical calculation in conjunction with quantitative interproton distance analysis. Org Chem Front 2019;6:45–53.
126. Huang B, Xiao CJ, Huang ZY, Tian XY, Cheng X, Dong X, Jiang B. Hispidanins A–D: four new asymmetric dimeric diterpenoids from the rhizomes of *Isodon hispida*. Org Lett 2014;16:3552–3555..
127. Xu G, Hou AJ, Zheng YT, Zhao Y, Li XL, Peng LY, Zhao QS. Przewalskin B, a novel diterpenoid with an unprecedented skeleton from *Salvia przewalskii* Maxim. Org Lett 2006;9:291–293.
128. Zou J, Pan L, Li QJ, Zhao JH, Pu JX, Yao P, Gong NB, Lu Y, Kondratyuk TP, Pezzuto JM, Fong HHS, Zhang HJ, Sun HD. Rubesanolides A and B: diterpenoids from *Isodon rubescens*. Org Lett 2011;13:1406–1409.
129. Xia F, Zhang DW, Wu CY, Geng HC, Xu WD, Zhang Y, Yang XW, Qin HB, Xu G. Isolation, structural elucidation, and synthetic study of salviyunnanone a, an abietane derived diterpenoid with a 7/5/6/3 ring system from *Salvia yunnanensis*. Org Chem Front 2018;5:1262–1266.
130. Xia F, Li WY, Yang XW, Yang J, Li X, Nian Y, Xu G. Salpratlactones A and B: a pair of *cis–trans* tautomeric abietanes as Cav3.1 T-type calcium channel agonists from *Salvia prattii*. Org Lett 2019;21:5670–5674.
131. Li LN, Liu XQ, Zhu DR, Chen C, Lin YL, Wang WL, Zhu L, Luo JG, Kong LY. Officinalins A and B, a pair of C_{23} terpenoid epimers with a tetracyclic 6/7/5/5 system from *Salvia officinalis*. Org Chem Front 2019;6:3369–3373.

132. Valdés LJ, Chang HM, Visger DC, Koreeda M. Salvinorin C, a new neoclerodane diterpene from a bioactive fraction of the Hallucinogenic Mexican Mint *Salvia divinorum*. Org Lett 2001;3:3935–3937.
133. Harding WW, Tidgewell K, Schmidt M, Shah K, Dersch CM, Snyder J, Parrish D, Deschamps JR, Rothman RB, Prisinzano TE. Salvinicins A and B, new neoclerodane diterpenes from *Salvia divinorum*. Org Lett 2005;7:3017–3020.
134. Aoyagi Y, Yamazaki A, Nakatsugawa C, Fukaya H, Takeya K, Kawauchi S, Izumi H. Salvileucalin B, a novel diterpenoid with an unprecedented rearranged neoclerodane skeleton from *Salvia Leucantha* Cav. Org Lett 2008;10:4429–4432.
135. Bautista E, Toscano RA, Ortega A. Microphyllandiolide, a new diterpene with an unprecedented skeleton from *Salvia microphylla*. Org Lett 2013;15:3210–3213.
136. Bautista E, Fragoso-Serrano M, Toscano RA, García-Peña MdelR, Ortega A. Teotihuacanin, a diterpene with an unusual spiro-10/6 system from *Salvia amarissima* with potent modulatory activity of multidrug resistance in cancer cells. Org Lett 2015;17:3280–3282.
137. Li XR, Yan BC, Hu K, He SJ, Sun HD, Zuo JP, Puno PT. Spiro *ent*-clerodane dimers: discovery and green approaches for a scalable biomimetic synthesis. Org Lett 2021;23:5647−5651.
138. Zheng GJ, Kadir A, Zheng XF, Jin PF, Liu JJ, Maiwulanjiang M, Yao GM, Aisa HA. Spirodesertols A and B, two highly modified spirocyclic diterpenoids with an unprecedented 6-isopropyl-3H-spiro[benzofuran-2,1′-cyclohexane] motif from *Salvia deserta*. Org Chem Front 2020;7:3137–3145.
139. Xiong L, Zhou QM, Zou Y, Chen MH, Guo L, Hu GY, Liu ZH, Peng C. Leonuketal, a spiroketal diterpenoid from *Leonurus japonicus*. Org Lett 2015;17:6238–6241.
140. Yun YS, Fukaya H, Nakane T, Takano A, Takahashi S, Takahashi Y, Inoue H. A new bis-*seco*-abietane diterpenoid from *Hyptis crenata* Pohlex Benth. Org Lett 2014;16:6188–6191.
141. Wang SJ, Lin S, Zhu CG, Yang YC, Li S, Zhang JJ, Chen XG, Shi JG. Highly acylated diterpenoids with a new 3,4-secograyanane skeleton from the flower buds of *Rhododendron molle*. Org Lett 2010;12:1560–1563.
142. Li Y, Liu YB, Zhang JJ, Li YH, Jiang JD, Yu SS, Ma SG, Qu J, L HN. Mollolide A, a diterpenoid with a new 1,10:2,3-disecograyanane skeleton from the roots of *Rhododendron molle*. Org Lett 2013;15:3074–3077.
143. Zhang MK, Zhu Y, Zhan GQ, Shu PH, Sa RJ, Lei L, Xiang M, Xue YB, Luo ZW, Wan Q, Yao GM, Zhang YH. Micranthanone A, a new diterpene with an unprecedented carbon skeleton from *Rhododendron micranthum*. Org Lett 2013;15:3094–3097.
144. Li Y, Liu YB, Liu YL, Wang C, Wu LQ, Li L, Ma SG, Qu J, Yu SS. Mollanol A, a diterpenoid with a new C-nor-D-homograyanane skeleton from the fruits of *Rhododendron Molle*. Org Lett 2014;16:4320–4323.

145. Zhou JF, Sun N, Zhang HQ, Zheng GJ, Liu JJ, Yao GM. Rhodomollacetals A–C, PTP1B Inhibitory diterpenoids with a 2,3:5,6-di-seco-grayanane skeleton from the leaves of *Rhododendron molle.* Org Lett 2017;19:5352–5355.
146. Zhou JF, Zhan GQ, Zhang HQ, Zhang QH, Li Y, Xue YB, Yao GM. Rhodomollanol A, a highly oxygenated diterpenoid with a 5/7/5/5 tetracyclic carbon skeleton from the leaves of *Rhododendron molle.* Org Lett 2017:19:3935–3938.
147. Zhou JF, Liu JJ, Dang T, Zhou HF, Zhang HQ, Yao GM. Mollebenzylanols A and B, highly modified and functionalized diterpenoids with a 9-benzyl-8,10-dioxatricyclo[5.2.1.$0^{1,5}$]decane core from *Rhododendron molle.* Org Lett 2018;20:2063–2066
148. Zhou JF, Zuo ZL, Liu JJ, Zhang HQ, Zheng GJ, Yao GM. Discovery of highly functionalized 5,6-*seco*-grayanane diterpenoids as potent competitive PTP1B inhibitors. Org Chem Front 2020;7:820–828.
149. Li CH, Niu XM, Luo Q, Xie MJ, Luo SH, Zhou YY, Li SH. Novel polyesterified 3,4-*seco*-grayanane diterpenoids as antifeedants from Pieris formosa. Org Lett 2010;12:2426–2429.
150. Niu CS, Li Y, Liu YB, Ma SG, Liu F, Li L, Xu S, Wang XJ, Wang RB, Qu J, Yu SS. Pierisketolide A and pierisketones B and C, three diterpenes with an unusual carbon skeleton from the roots of *Pieris formosa.* Org Lett 2017;19:906–909.
151. Yin H, Luo JG, Kong LY. Tetracyclic diterpenoids with isomerized isospongian skeleton and labdane diterpenoids from the fruits of *Amomum kravanh.* J Nat Prod 2013;76:237–242.
152. Ji KL, Fan YY, Ge ZP, Sheng L, Xu YK, Gan LS, Li JY, Yue JM. Maximumins A–D, rearranged labdane-type diterpenoids with four different carbon skeletons from *Amomum maximum.* J Org Chem 2019;84:282–288.
153. Yin H, Luo JG, Shan SM, Wang XB, Luo J, Yang MH, Kong LY. Amomaxins A and B, two unprecedented rearranged labdane norditerpenoids with a nine-membered ring from *Amomum maximum.* Org Lett 2013;15:1572–1575.
154. Zhao Q, Gao JJ, Qin XJ, Hao XJ, He HP, Liu HY. Hedychins A and B, 6,7-dinorlabdane diterpenoids with a peroxide bridge from *Hedychium Forrestii.* Org Lett 2018;20:704–707.
155. Ni G, Zhang H, Fan YY, Liu HC, Ding J, Yue JM. Mannolides A–C with an intact diterpenoid skeleton providing insights on the biosynthesis of antitumor *Cephalotaxus troponoids.* Org Lett 2016;18:1880–1883.
156. Chen HL, Wang LW, Su HJ, Wei BL, Yang SZ, Lin CN. New terpenoids from *Amentotaxus Formosana.* Org Lett 2006;8:753–756.
157. Galli B, Gasparrini F, Lanzotti V, Misiti D, Riccio R, Villani C, Guan FH, Zhong WM, Wan FY. Grandione, a new heptacyclic dimeric diterpene from *Torreya grandis* Fort. Tetrahedron 1999;55:11385–11394.
158. Huo CH, Su XH, Wang YF, Zhang XP, Shi QW, Kiyota H. Canataxpropellane, a novel taxane with a unique polycyclic carbon skeleton (tricyclotaxane) from the needles of *Taxus Canadensis.* Tetrahedron Lett 2007;48:2721–2724.

159. Fukuyama Y, Fujii H, Minami H, Takahashi H, Kubo M. Neovibsanin F and its congeners, rearranged vibsane-type diterpenes from *Viburnum suspensum*. J Nat Prod 2006;69:1098–1100.
160. Gao X, Shao LD, Dong LB, Cheng X, Wu XD, Liu F, Jiang WW, Peng LY, He J, Zhao QS. Vibsatins A and B, two new tetranorvibsane-type diterpenoids from *Viburnum tinus* Cv. Variegatus. Org Lett 2014;16:980–983.
161. Liang WJ, Geng CA, Zhang XM, Chen H, Yang CY, Rong GQ, Zhao Y, Xu HB, Wang H, Zhou NJ, Ma YB, Huang XY, Chen JJ. (±)-Paeoveitol, a pair of new norditerpene enantiomers from *Paeonia veitchii*. Org Lett 2014;16:424–427.
162. Lou HY, Zheng S, Li TL, Zhang JX, Fei Y, Hao XJ, Liang GY, Pan WD. Vulgarisin A, a new diterpenoid with a rare 5/6/4/5 ring skeleton from the chinese medicinal plant *Prunella vulgaris*. Org Lett 2014;16:2696–2699.
163. Morikawa T, Xu F, Kashima Y, Matsuda H, Ninomiya K, Yoshikawa M. Novel dolabellane-type diterpene alkaloids with lipid metabolism promoting activities from the seeds of *Nigella sativa*. Org. Lett. 2004;6:869–872.
164. Zhang HJ, Luo J, Shan SM, Wang XB, Luo JG, Yang MH, Kong LY. Aphanamenes A and B, two new acyclic diterpene [4 + 2]-cycloaddition adducts from *Aphanamixis grandifolia*. Org Lett 2013;15:5512–5515.
165. Zhao JX. Yu YY, Wang SS, Huang SL, Shen Y, Gao XH, Sheng L, Li JY, Leng Y, Li J, Yue JM. Structural elucidation and bioinspired total syntheses of ascorbylated diterpenoid hongkonoids A–D. J. Am. Chem. Soc. 2018;140:2485–2492.
166. Hou XF, Yao S, Mándi A, Kurtán T, Tang CP, Ke CQ, Li XQ, Ye Y. Bicunningines A and B, two new dimeric diterpenes from *Cunninghamia lanceolata*. Org Lett 2012;14:460–463.
167. Wu XD, Ding LF, Chen B, Li XN, Peng LY, Zhao QS. Cunlanceloic acids A–D: unprecedented labdane diterpenoid dimers with AChE inhibitory and cytotoxic activities from *Cunninghamia lanceolata*. Org Chem Front 2021;8:5777–5784.
168. Huang XH, Tao LX, Ke CQ, Tang C, Zhang HY, Ye Y, Lin LG, Yao S. Taxodikaloids A and B, two dimeric abietane-type diterpenoids from *Taxodium ascendens* possessing an oxazoline ring linkage. Org Lett 2017;19:556–559.
169. Zhou L, Tuo YL, Hao Y, Guo XM, Tang W, Xue YB, Zeng JF, Zhou Y, Xiang M, Zuo JP, Yao GM, Zhang YH. Cinnamomols A and B, immunostimulative diterpenoids with a new carbon skeleton from the leaves of *Cinnamomum cassia*. Org Lett 2017;19:3029–3032.
170. Zhou HF, Guoruoluo Y DZ; Tuo Y, Zhou J, Zhang H, Wang W, Xiang M, Aisa HA, Yao G. Cassiabudanols A and B, immunostimulative diterpenoids with a cassiabudane carbon skeleton featuring a 3-oxatetracyclo[6.6.1.$0^{2,6}$.$01^{0,14}$]pentadecane scaffold from cassia buds. Org Lett 2019;21:549–553.
171. Li C, Lee D, Graf TN, Phifer SS, Nakanishi Y, Burgess JP, Riswan S, Setyowati FM, Saribi AM, Soejarto DD, Farnsworth NR, Falkinham JO, Kroll DJ, Kinghorn AD, Wani MC, Oberlies NH. A hexacyclic ent-trachylobane

diterpenoid possessing an oxetane ring from *Mitrephora glabra.* Org Lett 2005;7:5709–5712.

172. Chen CH, Hsieh TJ, Liu TZ, Chern CL, Hsieh PY, Chen CY. Annoglabayin, a novel dimeric kaurane diterpenoid, and apoptosis in hepg2 cells of annomontacin from the fruits of *Annona g labra.* J Nat Prod 2004;67:1942–1946.
173. Pan L, Terrazas C, Lezama-Davila CM, Rege N, Gallucci JC, Satoskar AR, Kinghorn AD. Cordifolide A, a sulfur-containing clerodane diterpene glycoside from *Tinospora cordifolia.* Org Lett 2012;14:2118–2121.
174. Guo DX, Zhu RX, Wang XN, Wang LN, Wang SQ, Lin ZM, Lou HX. Scaparvin A, a novel caged cis-clerodane with an unprecedented C-6/C-11 bond, and related diterpenoids from the liverwort *Scapania parva.* Org Lett 2010;12:4404–4407
175. Perry NB, Burgess EJ, Baek SH, Weavers RT. The first atisane diterpenoids from a liverwort: polyols from *Lepidolaena clavigera.* Org Lett 2001;3:4243–4245.
176. Liu KX, Zhu YX, Yan YM, Zeng Y, Jiao YB, Qin FY, Liu JW, Zhang YY, and Cheng YX. Discovery of populusone, a skeletal stimulator of umbilical cord mesenchymal stem cells from *Populus euphratica* exudates. Org Lett 2019;21:1837−1840.
177. Liu YY, Yan YM, Wang DW, Cheng YX. Populusene A, an anti-inflammatory diterpenoid with a bicyclo[8,4,1]pentadecane scaffold from *Populus euphratica* Resins. Org Lett 2021;23:8657−8661.
178. Wang LN, Zhang JZ, Li X, Wang XN, Xie CF, Zhou JC, Lou HX. Pallambins A and B, unprecedented hexacyclic 19-*nor*-secolabdane diterpenoids from the chinese liverwort *Pallavicinia ambigua.* Org Lett 2012;14:1102–1105.
179. Li Y, Xu ZJ, Zhu RX, Zhou JC, Zong Y, Zhang JZ, Zhu MZ, Jin XY, Qiao YY, Zheng HB, Lou HX. Probing the interconversion of labdane lactones from the chinese liverwort *Pallavicinia ambigua.* Org Lett 2020;22:510–514.
180. Zhou JC, Zhang JZ, Li RJ, Liu J, Fan PH, Li Y, Ji M, Dong YW, Yuan HQ, Lou HX. Hapmnioides A–C, rearranged labdane-type diterpenoids from the chinese liverwort *Haplomitrium mnioides.* Org Lett 2016;18:4274–4276.
181. Wu XD, Luo D, Tu WC, Deng ZT, Chen XJ, Su J, Ji X, Zhao QS. Hypophyllins A–D, labdane-type diterpenoids with vasorelaxant activity from *Hypoestes phyllostachya* "Rosea". Org Lett 2016;18:6484–6487.
182. Manríquez-Torres JJ, Torres-Valencia JM, Velázquez-Jiménez R, Valdez-Calderón A, Alvarado-Rodríguez JG, Cerda García Rojas CM, Joseph-Nathan PA. Macrocyclic dimeric diterpene with a C_2 symmetry axis. Org Lett 2013;15:4658–4661.
183. Zheng Y, Zhang SW, Cong HJ, Huang YJ, Xuan LJ. Caesalminaxins A–L, cassane diterpenoids from the seeds of *Caesalpinia minax.* J Nat Prod 2013;76:2210–2218.
184. Wang LY, Wu J, Yang Z, Wang XJ, Fu Y, Liu SZ, Wang HM, Zhu WL, Zhang HY, Zhao WM. (*m*)- and (*p*)-Bicelaphanol A, dimeric trinorditerpenes with

promising neuroprotective activity from *Celastrus orbiculatus*. J Nat Prod 2013;76:745–749.

185. Zhang Y, Lu Y, Mao L, Proksch P, Lin WH. Tagalsins I and J, two novel tetraterpenoids from the Mangrove plant. *Ceriops tagal*. Org Lett 2005;7:3037–3040.
186. Zhang XH, Yang Y, Liu JJ, Shen L, Shi Z, Wu J. Tagalide A and tagalol A, naturally occurring 5/6/6/6- and 5/6/6-fused cyclic dolabrane-type diterpenes: a new insight into the anti-breast cancer activity of the dolabrane scaffold. Org Chem Front 2018;5:1176–1183.
187. Pu DB, Du BW, Chen W, Gao JB, Hu K, Shi N, Li YM, Zhang XJ, Zhang RH, Li XN, Zhang HB, Wang F, Xiao WL. Premnafulvol A: a diterpenoid with a 6/5/7/3-fused tetracyclic core and its biosynthetically related analogues from *Premna fulva*. Org Lett 2018;20:6314–6317.
188. Lin S, Wang SJ, Liu MT, Gan ML, Li S, Yang YC, Wang YH, He WY, Shi JG. Glycosides from the stem bark of *Fraxinus sieboldiana*. J Nat Prod 2007;70:817–823.
189. Zhou J, Wu Z, Guo B, Sun M, Onakpa MM, Yao G, Zhao M, Che CT. Modified diterpenoids from the tuber of *Icacina oliviformis* as protein tyrosine phosphatase 1B inhibitors. Org Chem Front 2020;7:355–367.
190. Xu MM, Zhou JF, Zeng LP, Xu JC, Onakpa MM, Duan JA, Che CT, Bi HK, Zhao M. Pimarane-derived diterpenoids with anti-Helicobacter pylori activity from the tuber of *Icacina trichantha*. Org Chem Front 2021;8:3014–3022.
191. Dong L, Cheng LZ, Yan YM, Wang SM, Cheng YX. Commiphoranes A–D, carbon skeletal terpenoids from *Resina Commiphora*. Org Lett 2017;19:286–289.
192. Acebey L, Sauvain M, Beck S, Moulis C, Gimenez A, Jullian V. Bolivianine, a new sesterpene with an unusual skeleton from *Hedyosmum angustifolium*, and its isomer, isobolivianine. Org Lett 2007;9:4693–4696.
193. Appendino G, Taglialatela Scafati O, Romano A, Pollastro F, Avonto C, Rubiolo P. Genepolide, a sesterpene γ-lactone with a novel carbon skeleton from mountain wormwood (*Artemisia umbelliformis*). J Nat Prod 2009;72:340–344.
194. Choudhary MI, Ranjit R, Atta-ur-Rahman, Shrestha TM, Yasin A, Parvez M. Leucosceptrine–A novel sesterterpene with prolylendopeptidase inhibitory activity from *Leucosceptrum canum*. J Org Chem 2004;69:2906–2909.
195. Li CH, Jing SX, Luo SH, Shi W, Hua J, Liu Y, Li XN, Schneider B, Gershenzon J, Li SH. Peltate glandular trichomes of *Colquhounia coccinea* var. *mollis* harbor a new class of defensive sesterterpenoids. Org Lett 2013;15:1694–1697.
196. Luo SH, Hua J, Li CH, Jing SX, Liu Y, Li XN, Zhao X, Li SH. New antifeedant C*20* terpenoids from *Leucosceptrum canum*. Org Lett 2012;14:5768–5771.
197. Jing SX, Fu R, Li CH, Zhou TT, Liu YC, Liu Y, Luo SH, Li XN, Zeng F, Li S. Immunosuppresive sesterterpenoids and norsesterterpenoids from *Colquhounia coccinea* var. *mollis*. J Org Chem 2021;86:11169−11176
198. Guo K, Liu X, Zhou TT, Liu YC, Liu Y, Shi QM, Li XN, Li SH. Gentianelloids A and B: immunosuppressive 10,11-*seco*-gentianellane sesterterpenoids

from the traditional uighur medicine *Gentianella turkestanorum*. J Org Chem 2020;85;5511–5515.

199. Teng LL, Mu RF, Liu YC, Xiao CJ, Li DS, Gao JX, Guo K, Li XN, Liu Y, Zeng F, Li SH. Immunosuppressive and adipogenesis inhibitory sesterterpenoids with a macrocyclic ether system from *Eurysolen gracilis*. Org Lett 2021;23:2232–2237.
200. Li RT, Zhao QS, Li SH, Han QB, Sun HD, Lu Y, Zhang LL, Zheng QT. Micrandilactone A: a novel triterpene from *Schisandra micrantha*. Org Lett 2003;5;1023–1026.
201. Xiao WL, Li RT, Li SH, Li XL, Sun HD, Zheng YT, Wang RR, Lu Y, Wang C, Zheng QT. Lancifodilactone F: a novel nortriterpenoid possessing a unique skeleton from *Schisandra lancifolia* and its anti-hiv activity. Org Lett 2005;7:1263–1266.
202. Xiao WL, Zhu HJ, Shen YH, Li RT, Li SH, Sun HD, Zheng YT, Wang RR, Lu Y, Wang C, Zheng QT. Lancifodilactone G: a unique nortriterpenoid isolated from *Schisandra lancifolia* and its anti-hiv activity. Org Lett 2005;7:2145–2148.
203. Xiao WL, Yang LM, Gong NB, Wu L, Wang RR, Pu JX, Li XL, Huang SX, Zheng YT, Li RT, Lu Y, Zheng QT, Sun HD. Rubriflordilactones A and B, two novel bisnortriterpenoids from *Schisandra rubriflora* and their biological activities. Org Lett 2006;8:991–994.
204. He F, Pu JX, Huang SX, Wang YY, Xiao WL, Li LM, Liu JP, Zhang HB, Li Y, Sun HD. Schinalactone A, a new cytotoxic triterpenoid from *Schisandra sphenanthera*. Org Lett 2010;12:1208–1211.
205. Xue YB, Yang JH, Li XN, Du X, Pu JX, Xiao WL, Su J, Zhao W, Li Y, Sun HD. Henrischinins A–C: three new triterpenoids from *Schisandra henryi*. Org Lett 2011;13:1564–1567.
206. Xiao WL, Pu JX, Chang Y, Li XL, Huang SX, Yang LM, Li LM, Lu Y, Zheng YT, Li RT, Zheng QT, Sun HD. Sphenadilactones A and B, two novel nortriterpenoids from *Schisandra sphenanthera*. Org Lett 2006;8:1475–1478.
207. Huang SX, Li RT, Liu JP, Lu Y, Chang Y, Lei C, Xiao WL, Yang LB, Zheng QT, Sun HD. Isolation and characterization of biogenetically related highly oxygenated nortriterpenoids from *Schisandra chinensis*. Org Lett 2007;9:2079–2082.
208. Cheng YB, Liao TC, Lo IW, Chen YC, Kuo YC, Chen SY, Chien CT, Shen YC. Arisandilactone A, a new triterpenoid from the fruits of *Schisandra arisanensis*. Org Lett 2010;12:1016–1019.
209. Lin YC, Lo IW, Chen SY, Lin PH, Chien CT, Chang S, Shen YC. Schinarisanlactone A, a new bisnortriterpenoid from *Schisandra arisanensis*. Org Lett 2011;13:446–449.
210. Meng, FY, Sun JX, Li X, Yu HY, Li SM, Ruan HL. Schiglautone A, a new tricyclic triterpenoid with a unique 6/7/9-fused skeleton from the stems of *Schisandra glaucescens*. Org Lett 2011;13:1502–1505.
211. Shi YM, Li XY, Li XN, Luo X, Xue YB, Liang CQ, Zou J, Kong LM, Li Y, Pu JX, Xiao W L, Sun HD. Schicagenins A–C: three cagelike nortriterpenoids from leaves and stems of *Schisandra chinensis*. Org Lett 2011;13:3848–3851.

212. Xiao WL, Huang SX, Zhang L, Tian RR, Wu L, Li L, Pu JX, Zheng YT, Lu Y, Li RT, Zheng QT, Sun HD. Schilancitrilactones A–C: Three unique nortriterpenoids from *Schisandra lancifolia*. Org Lett 2012;14:1286–1289.
213. Shi YM, Wang XB, Li XN, Luo X, Shen ZY, Wang YP, Xiao WL, Sun HD. Lancolides, antiplatelet aggregation nortriterpenoids with tricyclo[6.3.0.0^{2,11}]undecane-bridged system from *Schisandra lancifolia*. Org Lett 2013;15:5068–5071.
214. Shi YM, Yang J, Xu L, Li XN, Shang SZ, Cao P, Xiao WL, Sun HD. Structural characterization and antioxidative activity of lancifonins: unique nortriterpenoids from *Schisandra lancifolia*. Org. Lett. 2014;16:1370–1373.
215. Zhou M, Liu Y, Song J, Peng XG, Cheng Q, Cao H, Xiang M, Ruan HL. Schincalide A, a schinortriterpenoid with a tricyclo[5.2.1.0^{1,6}]decane-bridged system from the stems and leaves of *Schisandra incarnate*. Org Lett 2016;18:4558–4561.
216. Song J, Liu Y, Zhou M, Cao H, Peng XG, Liang JJ, Zhao XY, Xiang M, Ruan HL. Spiroschincarins A–E: five spirocyclic nortriterpenoids from the fruit of *Schisandra incarnata*. Org. Lett. 2017;19:1196–1199.
217. Shi YM, Hu K, Pescitelli G, Liu M, Li XN, Du X, Xiao WL, Sun HD, Puno PT. Schinortriterpenoids with identical configuration but distinct ECD spectra generated by nondegenerate exciton coupling. Org Lett 2018;20:1500–1504.
218. Song J, Zhou M, Zhou J, Liang JJ, Peng XG, Liu J, Ruan HL. Schincalactones A and B, two 5/5/6/11/3 fused schinortriterpenoids with a 13-membered carbon ring system from *Schisandra incarnata*. Org Lett 2018;20:2499–2502.
219. Chen J, Peng XG, Liu J, Gao Y, Zhou M, Zhou J, Song J, Ruan HL. Incarnolides A and B, two schinortriterpenoids with a tricyclo[9.2.1.0^{2,8}]tetradecane-bridged system from *Schisandra incarnata*. Org Lett 2020;22:1071–1075.
220. Pu JX, Xiao WL, Lu Y, Li RT, Li HM, Zhang L, Huang SX, Li X, Zhao QS, Zheng QT, Sun HD. Kadlongilactones A and B, two novel triterpene dilactones from *Kadsura longipedunculata* possessing a unique skeleton. Org Lett 2005;7:5079–5082.
221. Shen YC, Lin YC, Chiang MY, Yeh SF, Cheng YB, Liao CC. Kadsuphilactones A and B, two new triterpene dilactones from *Kadsura philippinensis*. Org Lett 2005;7:3307–3310.
222. Zhang CL, Wang Y, Liu YF, Ni G, Liang D, Luo H, Song XY, Zhang WQ, Chen RY, Chen NH, Yu DQ. Iridal-type triterpenoids with neuroprotective activities from *Iris tectorum*. J Nat Prod 2014;77:411–415.
223. Zhang CL, Liu YF, Wang Y, Liang D, Jiang ZB, Li L, Hao ZY, Luo H, Shi GR, Chen RY, Cao ZY, Yu DQ. Polycycloiridals A–D, four iridal-type triterpenoids with an α-terpineol moiety from *Iris tectorum*. Org Lett 2015;17:5686–5689.
224. Zhang CL, Hao ZY, Liu YF, Wang Y, Shi GR, Jiang ZB, Chen RY, Cao ZY, Yu DQ. Polycycloiridals with a cyclopentane ring from *Iris tectorum*. J Nat Prod 2017;80:156–161.
225. Song ZJ, Xu XM, Deng WL, Peng SL, Ding LS, Xu HH. A new dimeric iridal triterpenoid from *Belamcanda chinensis* with significant molluscicide activity. Org Lett 2011;13:462–465.

226. Song YY, Miao JH, Qin FY, Yan YM, Yang J, Qin DP, Hou FF, Zhou LL, Cheng YX. Belamchinanes A–D from *Belamcanda chinensis*: triterpenoids with an unprecedented carbon skeleton and their activity against age-related renal fibrosis. Org Lett 2018;20:5506–5509.
227. Fan YY, Zhang H, Zhou Y, Liu HB, Tang W, Zhou B, Zuo JP, Yue JM. Phainanoids A–F, a new class of potent immunosuppressive triterpenoids with an unprecedented carbon skeleton from *Phyllanthus hainanensis*. J Am Chem Soc 2015;137:138–141.
228. Fan YY, Gan LS, Liu HC, Li H, Xu CH, Zuo JP, Ding J, Yue JM. Phainanolide A, highly modified and oxygenated triterpenoid from *Phyllanthus hainanensis*. Org Lett 2017;19:4580–4583.
229. Ferreira, RJ, Kincses A, Gajdács M, Spengler G, dos Santos DJVA, Molnár J, Ferreira MJU. Terpenoids from *Euphorbia pedroi* as multidrug-resistance reversers. J Nat Prod 2018;81:2032–2040.
230. Wang C, Yan QS, Wang YF, Huang SS, Ning J, Feng L, Sun CP, Zhang BJ, Li DW, Ma XC. Ebracpenes A and B, Unusual Ring C-*seco* and Ring D-Aromatic Nor-Triterpenoids, from *Euphorbia ebracteolata* and lipase inhibitory evaluation. J Org Chem 2019;84:1624–1629.
231. Yuan FY, Xu F, Fan RZ, Li W, Huang D, Tang GH, Yuan T, Gan LS, Yin S. Structural elucidation of three 9,11-*seco* tetracyclic triterpenoids enables the structural revision of euphorol J. J Org Chem 2021;86:7588−7593.
232. He XF, Wang XN, Gan LS, Yin S, Dong L, Yue JM. Two novel triterpenoids from *Dysoxylum hainanense*. Org Lett 2008;10:4327–4330.
233. Cai XH, Luo XD, Zhou J, Hao XJ. Compound representatives of a new type of triterpenoid from *Aglaia odorata*. Org Lett 2005;7:2877–2879.
234. Zhou ZW, Yin S, Zhang HY, Fu Y, Yang SP, Wang XN, Wu Y, Tang XC, Yue JM. Walsucochins A and B with an unprecedented skeleton isolated from *Walsura cochinchinensis*. Org Lett 2008;10:465–468.
235. Rasamison VE, Rakotondraibe LH, Slebodnick C, Brodie PJ, Ratsimbason M, TenDyke K, Shen Y, Randrianjanaka LM, Kingston DGI. Nitrogen-containing dimeric *nor*-multiflorane triterpene from a *Turraea* sp. Org Lett 2014;16:2626–2629.
236. Zhao QQ, Song QY, Jiang K, Li GD, Wei WJ, Li Y, Gao K. Spirochensilides A and B, two new rearranged triterpenoids from *Abies chensiensis*. Org Lett 2015;17:2760–2763.
237. Li B, Kong DY, Shen YH, Yuan H, Yue RC, He YR, Lu L, Shan L, Li HL, Ye J, Yang XW, Su J, Liu RH, Zhang WD. Pseudolaridimers A and B, hetero-cycloartane–labdane diels-alder adducts from the cone of *Pseudolarix amabilis*. Org Lett 2012;14:5432–5435.
238. Li B, Kong DY, Shen YH, Fu KL, Yue RC, Han ZZ, Yuan H, Liu QX, Shan L, Li HL, Yang XW, Zhang WD. Pseudolarenone, an unusual nortriterpenoid lactone with a fused 5/11/5/6/5 ring system featuring an unprecedented bicyclo[8.2.1]tridecane core from *Pseudolarix amabilis*. Chem Commun 2013;49:1187.

239. Ali Z, Khan SI, Ferreira D, Khan IA. Podocarpaside, a triterpenoid possessing a new backbone from *Actaea podocarpa*. Org Lett 2006;8:5529–5532.
240. Dan C, Zhou Y, Ye D, Peng S, Ding L, Gross ML, Qiu SX. Cimicifugadine from *Cimicifuga foetida*, a new class of triterpene alklaoids with novel reactivity. Org Lett 2007;9:1813–1816.
241. Han QH, Liu X, Yao WQ, Cheng ZB, Lin TT, Song C, Yin S. Unusual 9,19 : 24,32-dicyclotetracyclic triterpenoids from *Lygodium japonicum*. Planta Med 2012;78:1971–1975.
242. Liu JQ, Peng XR, Li XY, Li TZ, Zhang WM, Shi L, Han J, Qiu MH. Norfriedelins A–C with acetylcholinesterase inhibitory activity from acerola tree (*Malpighia emarginata*). Org Lett 2013;15:1580–1583.
243. Xiong L, Zhu M, Zhu CG, Lin S, Yang YC, Shi JG. Structure and bioassay of triterpenoids and steroids isolated from *Sinocalamus affinis*. J Nat Prod 2012;75:1160–1166.
244. Wang F, Ren FC, Liu JK. Alstonic acids A and B, unusual 2,3-secofernane triterpenoids from *Alstonia scholaris*. Phytochemistry 2009;70:650–654.
245. Hu BY, Zhao YL, Xiong DS, He YJ, Zhou ZS, Zhu PF, Wang ZJ, Wang YL, Zhao LX, Luo XD. Potent antihyperuricemic triterpenoids based on two unprecedented scaffolds from the leaves of *Alstonia scholaris*. Org Lett 2021;23:4158–4162.
246. Liu XY, Li CJ, Chen FY, Ma J, Wang S, Yuan YH, Li L, Zhang DM. Nototronesides A–C, three triterpene saponins with a 6/6/9 fused tricyclic tetranordammarane carbon skeleton from the leaves of *Panax notoginseng*. Org Lett 2018;20:4549–4553.
247. Wang C, Huo XK, Luan ZL, Cao F, Tian XG, Zhao XY, Sun CP, Feng L, Ning J, Zhang BJ, Ma XC. Alismanin A, a triterpenoid with a c_{34} skeleton from *Alisma orientale* as a natural agonist of human pregnane X receptor. Org Lett 2017;19:5645–5648.
248. Liang CQ, Shi YM, Luo RH, Li XY, Gao ZH, Li XN, Yang LM, Shang SZ, Li Y, Zheng YT, Zhang HB, Xiao WL, Sun HD. Kadcoccitones A and B, two new 6/6/5/5-fused tetracyclic triterpenoids from *Kadsura coccinea*. Org Lett 2012;14:6362–6365.
249. Wang CQ, Wang L, Fan CL, Zhang DM, Huang XJ, Jiang RW, Bai LL, Shi JM, Wang Y, Ye WC. Ilelic acids A and B, two unusual triterpenes with a seven-membered ring from *Ilex latifolia*. Org Lett 2012;14:4102–4105.
250. Abbet C, Neuburger M, Wagner T, Quitschau M, Hamburger M, Potterat O. Phyteumosides A and B: new saponins with unique triterpenoid aglycons from *Phyteuma orbiculare* L. Org Lett 2011;13:1354–1357.
251. Kamdem RST, Wafo P, Yousuf S, Ali Z, Adhikari A, Rasheed S, Khan IA, Ngadjui BT, Fun HK, Choudhary MI. Canarene: A triterpenoid with a unique carbon skeleton from *Canarium schweinfurthii*. Org Lett 2011;13:5492–5495.
252. Liu MT, Gan ML, Lin S, Zhang YL, Zi JC, Song WX, Fan XN, Liu Y, Yang YC, Shi JG. Machilusides A and B: structurally unprecedented homocucurbitane glycosides from the stem bark of *Machilus Yaoshansis*. Org Lett 2011;13:2856–2859.

253. Mónico A, Ramalhete C, André V, Spengler G, Mulhovo S, Duarte MT, Ferreira MJU. Cucurbalsaminones A–C, rearranged triterpenoids with a 5/6/3/6/5-fused pentacyclic carbon skeleton from *Momordica balsamina*, as multidrug resistance reversers. J Nat Prod 2019;82:2138–2143.
254. Wafo P, Kamdem RST, Ali Z, Anjum S, Khan SN, Begum A, Krohn K, Abegaz BM, Ngadjui BT, Choudhary MI. Duboscic Acid: a potent α-glucosidase inhibitor with an unprecedented triterpenoidal carbon skeleton from *Duboscia macrocarpa*. Org Lett 2010;12:5760–5763.
255. Román LU, Guerra Ramírez D, Morán G, Martínez I, Hernández JD, Cerda García Rojas CM, Torres Valencia JM, Joseph Nathan P. First *seco*-C oleananes from nature. Org Lett 2004;6:173–176.
256. Chiang YM, Kuo YH. Novel triterpenoids from the aerial roots of *Ficus microcarpa*. J Org Chem 2002;67:7656–7661.
257. Ahmad VU, Zahid M, Ali MS, Ali Z, Jassbi AR, Abbas M, Clardy J, Lobkovsky E, Tareen RB, Iqbal MZ. Salvadiones-A and -B: two terpenoids having novel carbon skeleta from *Salvia bucharica*. J Org Chem 1999;64:8465–8467.
258. Li ZY, Qi FM, Zhi DJ, Hu QL, Liu YH, Zhang ZX, Fei DQ. A novel spirocyclic triterpenoid and a new taraxerane triterpenoid from *Teucrium viscidum*. Org Chem Front 2017;4:42–46.
259. Hua J, Liu YC, Luo SH, Liu Y, Xiao CJ, Li XN, Li SH. Immunostimulatory 6/6/6/6 tetracyclic triterpenoid saponins with the methyl-30 incorporated cyclization from the root of *Colquhounia elegans*. Org Lett 2021;23:7462−7466.
260. Kang KB, Kim HW, Kim JW, Oh WK, Kim J, Sung SH. Catechin-bound ceanothane-type triterpenoid derivatives from the roots of *Zizyphus Jujuba*. J Nat Prod 2017;80:1048–1054.
261. Xu J, Xiao D, Lin QH, He JF, Liu WY, Xie N, Feng F, Qu W. Cytotoxic tirucallane and apotirucallane triterpenoids from the stems of *Picrasma quassioides*. J Nat Prod 2016:79:1899–1910.
262. Sun DA, Deng JZ, Starck SR, Hecht SM. Mispyric Acid, a new monocyclic triterpenoid with a novel skeleton from *Mischocarpus pyriformis* that inhibits dna polymerase β. J Am Chem Soc 1999;121:6120–6124.
263. Wang PC, Ran XH, Luo HR, Ma QY, Liu YQ, Dai HF, Zhou J, Zhao YX. Volvalerenol A, a new triterpenoid with a 12-membered ring from *Valeriana hardwickii*. Org Lett 2013;15:2898–2901.
264. Zhang DL, Li M, Han GF, Li SY, Jin DJ, and Tang SA. Longipetalol A: a highly modified triterpenoid from *Dichapetalum longipetalum*. J Nat Prod 2021;84:1556−1562.
265. Yin S, Wang XN, Fan CQ, Liao SG, Yue JM. The first limonoid peroxide in the meliaceae family: walsuronoid A from *Walsura robusta*. Org Lett 2007;9:2353–2356.
266. An FL, Sun DM, Li RJ, Zhou MM, Yang MH, Yin Y, Kong LY, Luo J. Walrobsins A and B, two anti-inflammatory limonoids from root barks of *Walsura robusta*. Org Lett 2017;19:4568–4571.
267. Yang BJ, Fan SR, Cai JY, Wang YT, Jing C, Guo JJ, Chen DZ, Hao XJ. Aphananoid A is an anti-inflammatory limonoid with a new 5/6/5 fused ring

featuring a C24 carbon skeleton from *Aphanamixis polystachya.* J Org Chem 2020;85:8597–8602.

268. Wang XN, Yin S, Fan CQ, Wang FD, Lin LP, Ding J, Yue JM. Turrapubesins A and B, first examples of halogenated and maleimide-bearing limonoids in nature from *Turraea pubescens.* Org Lett 2006;8:3845–3848.
269. Liu CP, Wang GC, Gan LS, Xu CH, Liu QF, Ding J, Yue JM. Ciliatonoids A and B, two limonoids from *Toona ciliata.* Org Lett 2016;18:2894–2897.
270. Luo J, Huang WS, Hu SM, Zhang PP, Zhou XW, Wang XB, Yang MH, Luo JG, Wang C, Liu C, Yao HQ, Zhang C, Sun HB, Chen YJ, Kong LY. Rearranged limonoids with unique 6/5/6/5 tetracarbocyclic skeletons from *Toona ciliata* and biomimetic structure divergence. Org Chem Front 2017;4:2417–2421.
271. Fang X, Di YT, Zhang Y, Xu ZP, Lu Y, Chen QQ, Zheng QT, Hao XJ. Unprecedented quassinoids with promising biological activity from *Harrisonia perforata.* Angew Chem Int Ed 2015;54:5592–5595.
272. Nihei K, Asaka Y, Mine Y, Yamada Y, Iigo M, Yanagisawa T, Kubo I. Musidunin and musiduol, insect antifeedants from *Croton jatrophoides.* J Nat Prod 2006;69:975–977.
273. Yang SP, Chen HD, Liao SG, Xie BJ, Miao ZH, Yue JM. Aphanamolide A, a new limonoid from *Aphanamixis polystachya.* Org Lett 2011;13:150–153.
274. Cai JY, Zhang Y, Luo SH, Chen DZ, Tang GH, Yuan CM, Di YT, Li SH, Hao XJ, He HP. Aphanamixoid A, a potent defensive limonoid, with a new carbon skeleton from *Aphanamixis polystachya.* Org Lett 2012;14:2524–2527.
275. Zhang PP, Xue S, Huang WS, Wang CC, Cui ZR, Luo J, Kong LY. Diverse prieurianin-type limonoids with oxygen-bridged caged skeletons from two *Aphanamixis* Species: discovery and biomimetic conversion. Org Chem Front 2021;8:566–571.
276. Yuan XH, Li BG, Zhou M, Qi HY, Zhang GL. Cipadesins A–C: novel tetranortriterpenoids from *Cipadessa cinerascens.* Org Lett.2005;7:5051–5053.
277. Fang X, Di YT, He HP, Liu HY, Zhang Z, Ren YL, Gao ZL, Gao S, Hao XJ. Cipadonoid A, a novel limonoid with an unprecedented skeleton, from *Cipadessa cinerasecns.* Org Lett 2008;10:1905–1908.
278. Liu CP, Xu JB, Han YS, Wainberg MA, Yue JM. Trichiconins A–C, limonoids with new carbon skeletons from *Trichilia connaroides.* Org Lett 2014;16:5478–5481.
279. Yin S, Fan CQ, Wang XN, Lin LP, Ding J, Yue JM. Xylogranatins A–D: novel tetranortriterpenoids with an unusual 9,10-seco scaffold from marine mangrove *Xylocarpus granatum.* Org Lett 2006;8:4935–4938.
280. Yuan T, Yang SP, Zhang CR, Zhang S, Yue JM. Two limonoids, khayalenoids A and B with an unprecedented 8-oxa-tricyclo[4.3.2.$0^{2,7}$]undecane motif, from *Khaya senegalensis.* Org Lett 2009;11:617–620.
281. Li J, Li MY, Bruhn T, Katele FZ, Xiao Q, Pedpradab P, Wu J, Bringmann G. Thaixylomolins A–C: limonoids featuring two new motifs from the thai *Xylocarpus moluccensis.* Org Lett 2013;15:3682–3685.

282. Wu YB, Wang YZ, Ni ZY, Qing X, Shi QW, Sauriol F, Vavricka CJ, Gu YC, Kiyota H. Xylomexicanins I and J: limonoids with unusual B/C rings from *Xylocarpus granatum*. J Nat Prod 2017;80:2547–2550.
283. Wang GC, Fan YY, Shyaula SL, Yue JM. Triconoids A–D, four limonoids possess two rearranged carbon skeletons from *Trichilia connaroides*. Org Lett 2017;19:2182–2185.
284. Wu J, Xiao Q, Huang JS, Xiao ZH, Qi SH, Li QX, Zhang S. Xyloccensins O and P, unique 8,9,30-phragmalin ortho esters from *Xylocarpus granatum*. Org Lett 2004;6:1841–1844.
285. Zhang CR, Yang SP, Liao SG, Fan CQ, Wu Y, Yue JM. Chuktabularins A–D, four new limonoids with unprecedented carbon skeletons from the stem bark of *Chukrasia tabularis*. Org Lett 2007;9:3383–3386.
286. Zhang CR, Fan CQ, Zhang L, Yang SP, Wu Y, Lu Y, Yue JM. Chuktabrins A and B, two novel limonoids from the twigs and leaves of *Chukrasia tabularis*. Org Lett 2008;10:3183–3186.
287. Luo J, Wang JS, Luo JG, Wang XB, Kong LY. Chukvelutins A–C, 16-norphragmalin limonoids with unprecedented skeletons from *Chukrasia tabularis* var. *Velutina*. Org Lett 2009;11:2281–2284.
288. Liu HB, Zhang H, Li P, Gao ZB, Yue JM. Chukrasones A and B: potential Kv1.2 potassium channel blockers with new skeletons from *Chukrasia tabularis*. Org Lett 2012;14:4438–4441.
289. Hu K, Liu JQ, Li X, Chen JC, Zhang WM, Li Y, Li L, Guo L, Ma W, Qiu MH. Chukfuransins A–D, four new phragmalin limonoids with β-furan ring involved in skeleton reconstruction from *Chukrasia tabularis*. Org Lett 2013;15:3902–3905.
290. Inoue T, Matsui Y, Kikuchi T, In Y, Yamada T, Muraoka O, Matsunaga S, Tanaka R. Guianolides A and B, new carbon skeletal limonoids from the seeds of *Carapa guianensis*. Org Lett 2013;15:3018–3021.
291. Han ML, Zhang H, Yang SP, Yue JM. Walsucochinoids A and B: new rearranged limonoids from *Walsura cochinchinensis*. Org Lett 2012;14:486–489.
292. Zhao JQ, Wang YM, He HP, Li SH, Li XN, Yang CR, Wang D, Zhu HT, Xu M, Zhang YJ. Two new highly oxygenated and rearranged limonoids from *Phyllanthus cochinchinensis*. Org Lett 2013;15:2414–2417.
293. Yuan T, Zhu R X, Zhang H, Odeku OA, Yang SP, Liao SG, Yue JM. Structure determination of grandifotane A from *Khaya grandifoliola* by NMR, X-ray diffraction, and ECD calculation. Org Lett 2010;12:252–255.
294. Liu HB, Zhang H, Li P, Gao ZB, Yue JM. Chukrasones A and B: potential Kv1.2 potassium channel blockers with new skeletons from *Chukrasia tabularis*. Org Lett 2012;14:4438–4441.
295. Lv C, Yan XH, Tu Q, Di YT, Yuan CM, Fang X, Ben-David Y, Xia L, Gong JX, Shen YM, Yang Z, Hao XJ. Isolation and asymmetric total synthesis of perforanoid A. Angew Chem Int Ed 2016;55:7539–7543.
296. Tang XH, Luo RC, Ye MS, Tang HY, Ma YL, Chen YN, Wang XM, Lu QY, Liu S, Li XN, Yan Y, Yang J, Ran XQ, Fang X, Zhou Y, Yao YG, Di YT, Hao XJ. Harpertrioate A, an A,B,D-*seco*-limonoid with promising biological activity

against alzheimer's disease from Twigs of *Harrisonia perforata* (Blanco) Merr. Org Lett 2021;23:262–267.

297. Yu JH, Liu QF, Sheng L, Wang GC, Li J, Yue JM. Cipacinoids A–D, four limonoids with spirocyclic skeletons from *Cipadessa cinerascens*. Org Lett 2016:18:444–447.
298. Wang YL, Ye YS, Fu WW, Wu R, Xiang Q, Lao YZ, Yang JL, Tan HS, Yang XW, Yang BC, Xu HX, Xu G. Garsubelone A, the first dimeric polycyclic polyprenylated acylphloroglucinols with complicated heptacyclic architecture from *Garcinia subelliptica*. Org Lett 2019;21:1534–1537.
299. Lu WJ, Xu WJ, Zhang YQ, Li YR, Zhou X, Li QJ, Zhang H, Luo J, Kong LY. Hyperforones A–C, benzoyl-migrated [5.3.1]-type polycyclic polyprenylated acylphloroglucinols from *Hypericum forrestii*. Org Chem Front 2020;7:1070–1076.
300. Duan YL, Xie SS, Bu PF, Guo Y, Shi ZY, Guo Y, Cao YF, Sun WG, Qi CX, Zhang YH. Hypaluton A, an immunosuppressive 3,4-nor-polycyclic polyprenylated acylphloroglucinol from *Hypericum patulum*. J Org Chem 2021;86:6478–6485.
301. Xie SS, Zhou Y, Tan XS, Sun WG, Duan YL, Feng H, Sun LJ, Guo Y, Shi Z, Hao XC, Chen G, Qi CX, Zhang YH. Norwilsonnol A, an immunosuppressive polycyclic polyprenylated acylphloroglucinol with a spiro[5-oxatricyclo[6.4.0.03,7]dodecane-6',1-1',2'-dioxane] system from *Hypericum wilsonii*. Org Chem Front 2021;8:2280–2286.
302. Yang BY, Huang JZ, Lin S, Tong QY, Yao ZY, Li FL, Ye Y, Hu ZX, Zhang YH. Hyperbeanone A, a 5,6-seco-spirocyclic polycyclic polyprenylated acylphloroglucinol derivative with an unprecedented skeleton from *Hypericum beanii*. Org Chem Front 2021;8:6411–6418.
303. Zeng YR, Yuan CM, Li YN, Huang LJ, Hu ZX, Gu W, Hao XJ. Hymoins A–D: two pairs of polyprenylated acylphloroglucinols from *Hypericum monogynum* and their light-induced transformation. Org Lett 2021;23:3125–3129.
304. Zeng YR, Li YN, Zhang ZZ, Hu ZX, Gu W, Huang LJ, Li YM, Yuan CM, Hao XJ. Hypermoins A–D: rearranged nor-polyprenylated acylphloroglucinols from the flowers of *Hypericum monogynum*. J Org Chem 2021;86:7021–7027.
305. Xu G, Kan WLT, Zhou Y, Song JZ, Han QB, Qiao CF, Cho CH, Rudd JA, Lin G, Xu HX. Cytotoxic acylphloroglucinol derivatives from the twigs of *Garcinia cowa*. J Nat Prod 2010;73:104–108.
306. Zheng D, Zhang H, Zheng CW, Lao YZ, Xu DQ, Xiao LB, Xu HX. Garciyunnanimines A–C, Novel cytotoxic polycyclic polyprenylated acylphloroglucinol imines from *Garcinia yunnanensis*. Org Chem Front 2017;4:2102–2108.
307. Xu WJ, Tang PF, Lu WJ, Zhang YQ, Wang XB, Zhang H, Luo J, Kong LY. Hyperberins A and B, type B polycyclic polyprenylated acylphloroglucinols with bicyclo[5.3.1]hendecane core from *Hypericum beanii*. Org Lett 2019;21:8558–8562.
308. Yang XW, Ding Y, Zhang JJ, Liu X, Yang LX, Li XN, Ferreira D, Walker LA, Xu G. New acylphloroglucinol derivatives with diverse architectures from *Hypericum henryi*. Org Lett 2014;16:2434–2437.

309. Kong LM, Long XW, Yang XW, Xia F, Khan A, Yan H, Deng J, Li X, Xu G. seco-Polycyclic polyprenylated acylphloroglucinols with unusual carbon skeletons from *Hypericum ascyron*. Tetrahedron Lett 2017;58:2113–2117.
310. Zhu HC, Chen CM, Yang J, Li DY, Zhang JW, Guo Y, Wang JP, Luo ZW, Xue YB, Zhang YH. Hyperhexanone A, a crucial intermediate from bicyclo[3.3.1]-to cyclohexanone monocyclic-polycyclic polyprenylated acylphloroglucinols. Tetrahedron 2016;72:4655–4659.
311. Hu YL, Hu K, Kong LM, Xia F, Yang XW, Xu G. Norascyronones A and B, 2,3,4- nor-polycyclic polyprenylated acylphloroglucinols from *Hypericum ascyron*. Org Lett 2019;21:1007–1010.
312. Li QJ, Tang PF, Zhou X, Lu WJ, Xu WJ, Luo J, Kong LY. Elodeoidins A–H, acylphloroglucinol meroterpenoids possessing diverse rearranged skeletons from *Hypericum elodeoides*. Org Chem Front 2021;8:1409–1414.
313. Xie SS, Qi CX, Duan YL, Xu QQ, Liu YP, Huang YY, Yin X, Sun WG, Zhou Y, Zhang YH. Discovery of new polycyclic polyprenylated acylphloroglucinols with diverse architectures as potent cyclooxygenase-2 inhibitors. Org Chem Front 2020;7:1349–1357.
314. Zhu HC, Chen CM, Yang J, Li XN, Liu JJ, Sun B, Huang SX, Li DY, Yao GM, Luo ZW, Li Y, Zhang JW, Xue YB, Zhang YH. Bioactive acylphloroglucinols with adamantyl skeleton from *Hypericum Sampsonii*. Org Lett 2014;16:6322–6325.
315. Liao Y, Liu X, Yang J, Lao YZ, Yang XW, Li XN, Zhang JJ, Ding Z, Xu HX, Xu G. Hypersubones A and B, new polycyclic acylphloroglucinols with intriguing adamantane type cores from *Hypericum subsessile*. Org Lett 2015;17:1172–1175.
316. Tian WJ, Qiu YQ, Yao XJ, Chen HF, Dai Y, Zhang XK, Yao XS. Dioxasampsones A and B, two polycyclic polyprenylated acylphloroglucinols with unusual epoxy-ring-fused skeleton from *Hypericum sampsonii*. Org Lett 2014;16:6346–6349.
317. Liu YY, Ao Z, Xue GM, Wang XB, Luo JG, Kong LY. Hypatulone A, a homoadamantane-type acylphloroglucinol with an intricately caged core from *Hypericum patulum*. Org Lett 2018;20:7953–7956.
318. Tian WJ, Yu Y, Yao XJ, Chen HF, Dai Y, Zhang XK, Yao XS. Norsampsones A–D, four new decarbonyl polycyclic polyprenylated acylphloroglucinols from *Hypericum sampsonii*. Org Lett 2014;16:3448–3451.
319. Tanaka N, Yano Y, Tatano Y, Kashiwada Y. Hypatulins A and B, meroterpenes from *Hypericum patulum*. Org Lett 2016;18:5360–5363.
320. Xie SS, Qi CX, Duan YL, Hao XC, Guo Y, Deng MY, Qiao YB, Shi ZY, Tao L, Cao YF, Gu LH, Zhou Y, Zhang YH. Wilsonglucinols A–C, homoadamantane-type polycyclic polyprenylated acylphloroglucinols with unusual fused epoxy ring skeletons from *Hypericum wilsonii*. Org Chem Front 2020;7:464–471.
321. Chen Y, Xue Q, Teng H, Qin R, Liu H, Xu J, Mei Z, Yang G. Acylphloroglucinol derivatives with a tricyclo-[4.4.1.$1^{1,4}$] dodecane skeleton from *Garcinia bracteata* fruits. J Org Chem 2020;85:6620–6625.

322. Tanaka N, Kashiwada Y, Kim SY, Hashida W, Sekiya M, Ikeshiro Y, Takaishi Y. Acylphloroglucinol, biyouyanagiol, biyouyanagin B, and related spiro-lactones from *Hypericum Chinense.* J Nat Prod 2009;72:1447–1452.
323. Tanaka N, Abe S, Hasegawa K, Shiro M, Kobayashi J. Biyoulactones A–C, new pentacyclic meroterpenoids from *Hypericum chinense.* Org Lett 2011;13:5488–5491.
324. Xu WJ, Luo J, Li RJ, Yang MH, Kong LY. Furanmonogones A and B: two rearranged acylphloroglucinols with a 4,5-*seco*-3(2h)-furanone core from the flowers of *Hypericum monogynum.* Org Chem Front 2017;4:313–317.
325. Zhang N, Shi ZY, Guo Y, Xie SS, Qiao YB, Li XN, Xue YB, Luo ZW, Zhu HC, Chen CM, Hu LZ, Zhang YH. The absolute configurations of hyperilongenols A–C: rare 12,13-*seco*-spirocyclic polycyclic polyprenylated acylphloroglucinols with enolizable β,β'-tricarbonyl systems from *Hypericum longistylum* oliv. Org Chem Front 2019;6:1491–1502.
326. Zhang N, Shi ZY, Xu QQ, Sun WG, Gu LH, Xie SS, Guo Y, Duan YL, Zhang KL, Qi CX, Zhang YH. Longisglucinols A–C, structurally intriguing polycyclic polyprenylated acylphloroglucinols with anti-inflammatory activity from *Hypericum longistylum.* Org Lett 2020;22:7926–7929.
327. Zhang JJ, Yang J, Liao Y, Yang XW, Ma JZ, Xiao QL, Yang LX, Xu G. Hyperuralones A and B, new acylphloroglucinol derivatives with intricately caged cores from *Hypericum uralum.* Org Lett 2014:16:4912–4915.
328. Fan YM, Yi P, Li Y, Yan C, Huang T, Gu W, Ma Y, Huang LJ, Zhang JX, Yang CL, Li Y, Yuan CM, Hao XJ. Two unusual polycyclic polyprenylated acylphloroglucinols, including a pair of enantiomers from *Garcinia multiflora.* Org Lett 2015;17:2066–2069.
329. Tian DS, Yi P, Xia L, Xiao X, Fan YM, Gu W, Huang LJ, Ben David Y, Di YT, Yuan CM, Hao XJ. Garmultins A–G, biogenetically related polycyclic acylphloroglucinols from *Garcinia multiflora.* Org Lett 2016;18:5904–5907.
330. Jian YQ, Huang XJ, Zhang DM, Jiang RW, Chen MF, Zhao BX, Wang Y, Ye WC. Guapsidial A and guadials B and C: three new meroterpenoids with unusual skeletons from the leaves of *Psidium guajava.* Chem Eur J 2015;21:9022–9027.
331. Su JC, Wang S, Cheng W, Huang XJ, Li MM, Jiang RW, Li YL, Wang L, Ye WC, Wang Y. Phloroglucinol derivatives with unusual skeletons from *Cleistocalyx operculatus* and their in vitro antiviral activity. J Org Chem 2018;83:8522–8532.
332. Shang ZC, Yang MH, Jian KL, Wang XB, Kong LY. 1H NMR-guided isolation of formyl-phloroglucinol meroterpenoids from the leaves of *Eucalyptus robusta.* Chem Eur J 2016;22:11778–11784.
333. Qin XJ, Feng MY, Liu H, Ni W, Rauwolf T, Porco JA, Yan H, He L, Liu HY. Eucalyptusdimers A–C, dimeric phloroglucinol–phellandrene meroterpenoids from *Eucalyptus robusta.* Org Lett 2018:20:5066–5070.
334. Ye YS, Li WY, Du SZ, Yang J, Nian Y, Xu G. Congenetic hybrids derived from dearomatized isoprenylated acylphloroglucinol with opposite effects on Cav3.1 low voltage-gated Ca^{2+} channel. J Med Chem 2020:63:1709–1716.

335. Shao M, Wang Y, Jian YQ, Huang XJ, Zhang DM, Tang QF, Jiang RW, Sun XG, Lv ZP, Zhang XQ, Ye WC. Guadial A and psiguadials C and D, three unusual meroterpenoids from *Psidium guajava*. Org. Lett. 2012;14:5262–5265.
336. Tang YX, Fu WW, Xi ZC, Yang JL, Zheng CW, Lu Y, Shen ZW, Xu HX. Xanthone derivatives from the leaves of *Garcinia oligantha*. Eur J Med Chem 2019;181:1–13
337. Song JG, Su JC, Song QY, Huang RL, Tang W, Hu LJ, Huang XJ, Jiang RW, Li YL, Ye WC, Wang Y. Cleistocaltones A and B, antiviral phloroglucinol–terpenoid adducts from *Cleistocalyx operculatus*. Org Lett 2019;21:9579–9583.
338. Yang XL, Hsieh KL, Liu JK. Guajadial: an unusual meroterpenoid from guava leaves *Psidium guajava*. Org Lett 2007;9:5135–5138
339. Shao M, Wang Y, Liu Z, Zhang DM, Cao HH, Jiang RW, Fan CL, Zhang XQ, Chen HR, Yao XS, Ye WC. Psiguadials A and B, two novel meroterpenoids with unusual skeletons from the Leaves of *Psidium guajava*. Org Lett 2010;12:5040–5043.
340. Liu C, Ang S, Huang XJ, Tian HY, Deng YY, Zhang DM, Wang Y, Ye WC, Wang L. Meroterpenoids with new skeletons from *Myrtus communis* and structure revision of myrtucommulone K. Org Lett 2016;18:4004–4007.
341. Zhang YL, Chen C, Wang XB, Wu L, Yang MH, Luo J, Zhang C, Sun HB, Luo JG, Kong LY. Rhodomyrtials A and B, two meroterpenoids with a triketone-sesquiterpene-triketone skeleton from *Rhodomyrtus tomentosa*: structural elucidation and biomimetic synthesis. Org Lett 2016;18:4068–4071.
342. Li CJ, Ma J, Sun H, Zhang D, Zhang DM. Guajavadimer A, a dimeric caryophyllene-derived meroterpenoid with a new carbon skeleton from the leaves of *Psidium guajava*. Org Lett 2016;18:168–171.
343. Yang XW, Li YP, Su J, Ma WG, Xu G. Hyperjapones A–E, terpenoid polymethylated acylphloroglucinols from *Hypericum japonicum*. Org Lett 2016;18:1876–1879.
344. Chen NH, Zhang YB, Huang XJ, Jiang L, Jiang SQ, Li GQ, Li YL, Wang GC. Drychampones A–C: three meroterpenoids from *Dryopteris championii*. J Org Chem 2016;81:9443–9448.
345. Hou JQ, Guo C, Zhao JJ, He QW, Zhang BB, Wang H. Frutescone A–G, tasmanone-based meroterpenoids from the aerial parts of *Baeckea frutescens*. J Org Chem 2017;82:1448–1457.
346. Qin XJ, Rauwolf TJ, Li PP, Liu H, McNeely J, Hua Y, Liu HY, Porco JA. Isolation and synthesis of novel meroterpenoids from *Rhodomyrtus tomentos* a: investigation of a reactive enetrione intermediate. Angew Chem Int Ed 2019;58:4291–4296.
347. Xu J, Zhu HL, Zhang J, Liu WY, Luo JG, Pan K, Cao WY, Bi QR, Feng F, Qu W. Littordials A–E, novel formyl-phloroglucinol-β-caryophyllene meroterpenoids from the leaves of *Psidium littorale*. Org Chem Front 2019;6:1667–1673.

348. Fu HZ, Luo YM, Li CJ, Yang JZ, Zhang DM. Psidials A–C, three unusual meroterpenoids from the leaves of *Psidium guajava* L. Org Lett 2010;12:656–659.
349. Hu LZ, Zhang Y, Zhu HC, Liu JJ, Li H, Li XN, Sun WG, Zeng JF, Xue YB, Zhang YH. Filicinic Acid Based meroterpenoids with anti-epstein–barr virus activities from *Hypericum japonicum*. Org Lett 2016;18:2272–2275.
350. Gu JH, Wang WJ, Chen JZ, Liu JS, Li NP, Cheng MJ, Hu LJ, Li CC, Ye WC, Wang L. Leptosperols A and B, two cinnamoylphloroglucinol–sesquiterpenoid hybrids from *Leptospermum scoparium*: structural elucidation and biomimetic synthesis. Org Lett 2020;22:1796–1800.
351. Ye YS, Du SZ, Jiang NN, Xu HX, Yang J, Fu WW, Nian Y, Xu G. Novel meroterpenoids from *Hypericum patulum*: highly potent late Nav1.5 sodium current inhibitors. Org Lett 2020;22:6339–6343.
352. Yin S, Xue JJ, Fan CQ, Miao ZH, Ding J, Yue JM. Eucalyptals A–C with a new skeleton isolated from *Eucalyptus globulus*. Org Lett 2007;9:5549–5552.
353. Tang ZH, Liu YB, Ma SG, Li L, Li Y, Jiang JD, Qu J, Yu SS. Antiviral spirotriscoumarins A and B: two pairs of oligomeric coumarin enantiomers with a spirodienone–sesquiterpene skeleton from *Toddalia asiatica*. Org Lett 2016;18:5146–5149.
354. Xiong J, Hong ZL, Gao LX, Shen J, Liu ST, Yang GX, Li J, Zeng H, Hu JF. Chlorabietols A–C, phloroglucinol-diterpene adducts from the chloranthaceae plant *Chloranthus oldhamii*. J Org Chem 2015;80:11080–11085.
355. Hu YC, Wu XF, Gao S, Yu SS, Liu Y, Qu J, Liu J, Liu YB. Novel phloroglucinol derivatives from the roots of *Lysidice rhodostegia*. Org Lett 2006;8:2269–2272.
356. Hiranrat A, Mahabusarakam W, Carroll AR, Duffy S, Avery VM. Tomentosones A and B, hexacyclic phloroglucinol derivatives from the thai shrub *Rhodomyrtus tomentosa*. J Org Chem 2012;77:680–683.
357. Choudhary MI, Khan N, Ahmad M, Yousuf S, Fun HK, Soomro S, Asif M, Mesaik MA, Shaheen F. New inhibitors of ROS generation and t-cell proliferation from *Myrtus communis*. Org Lett 2013;15:1862–1865.
358. Yang XW, Yang J, Liao Y, Ye Y, Li YP, Yang SY, Xia F, Xu G. Hypercohin K, a polycyclic polyprenylated acylphloroglucinol with an unusual spiro-fused cyclopropane ring from *Hypericum cohaerens*. Tetrahedron Lett 2015;56:5537–5540.
359. Zong JF, Hu Z, Shao YY, Shi Q, Zhang MM, Zhou YB, Li J, Hou AJ. Hyperprins A and B, two complex meroterpenoids from *Hypericum przewalskii*. Org Lett 2020;22:2797–2800.
360. Deng LM, Hu LJ, Bai Yang TZ, Wang J, Qin GQ, Song QY, Su JC, Huang XJ, Jiang RW, Tang W, Li YL, Li CC, Ye WC, Wang Y. Rhodomentosones A and B: two pairs of enantiomeric phloroglucinol trimers from *rhodomyrtus tomentosa* and their asymmetric biomimetic synthesis. Org Lett 2021;23:4499–4504.
361. Liu F, Tian HY, Huang XL, Wang WJ, Li NP, He J, Ye WC, Wang L. Xanthchrysones A–C: rearranged phenylpropanoyl–phloroglucinol dimers

with unusual skeletons from *Xanthostemon chrysanthus.* J Org Chem 2019;84:15355–15361.

362. Vu VT, Chen XL, Kong LY, Luo JG. Melipatulinones A–C, three lignan–phloroglucinol hybrids from *Melicope patulinervia.* Org Lett 2020;22:1380–1384.
363. Li YS, Matsunaga K, Ishibashi M, Ohizumi Y. Littoralisone, a novel neuritogenic iridolactone having an unprecedented heptacyclic skeleton including four- and nine-membered rings consisting of glucose from *Verbena littoralis.* J Org Chem 2001:66:2165–2167.
364. Liao HB, Lei C, Gao LX, Li JY, Li J, Hou AJ. Two enantiomeric pairs of meroterpenoids from *Rhododendron capitatum.* Org Lett 2015;17:5040–5043.
365. Liao HB, Huang GH, Yu MH, Lei C, Hou AJ. Five pairs of meroterpenoid enantiomers from *Rhododendron capitatum.* J Org Chem 2017;82:1632–1637.
366. Li C, Li CJ, Ma J, Chen FY, Li L, Wang XL, Ye F, Zhang DM. Magterpenoids A–C, three polycyclic meroterpenoids with ptp1b inhibitory activity from the bark of *Magnolia officinalis var. biloba.* Org Lett 2018;20:3682–3686.
367. Xue GM, Zhao CG, Xue JF, Chen H, Zhao ZZ, Si YY, Du K, Zhi Y L, F WS. Fissisternoids A and B, two 2',5'-quinodihydrochalcone-based meroterpenoid enantiomers with unusual carbon skeletons from *Fissistigma bracteolatum.* Org Chem Front 2022;9:190–196.
368. Porzel A, Phuong Lien T, Schmidt J, Drosihn S, Wagner C, Merzweiler K, Van Sung T, Adam G. Fissistigmatins A–D: novel type natural products with flavonoid–sesquiterpene hybrid structure from *Fissistigma bracteolatum.* Tetrahedron 2000;56:865–872.
369. Tanaka N, Okasaka M, Ishimaru Y, Takaishi Y, Sato M, Okamoto M, Oshikawa T, Ahmed SU, Consentino LM, Lee KH. Biyouyanagin A, an anti-HIV agent from *Hypericum chinense* L. var. *salicifolium.* Org Lett 2005;7:2997–2999.
370. Niwa K, Tanaka N, Kim SY, Kojoma M, Kashiwada Y. Hyperdioxane A, a conjugate of dibenzo-1,4-dioxane and sesquiterpene from *Hypericum ascyron.* Org Lett 2018;20:5977–5980
371. Liu ML, Duan YH, Hou YL, Li C, Gao H, Dai Y, Yao XS. Nardoaristolones A and B, two terpenoids with unusual skeletons from *Nardostachys chinensis* batal. Org Lett 2013;15:1000–1003.
372. Liu HX, Chen K, Sun QY, Yang FM, Hu GW, Wang YH, Long CL. Nudibaccatumone, a trimer comprising a phenylpropanoid and two sesquiterpene moieties from *Piper nudibaccatum.* J Nat Prod 2013;76:732–736.
373. Lin XH, Ji S, Qiao X, Hu HB, Chen N, Dong YH, Huang Y, Guo D, Tu PF, Ye M. Density functional theory calculations in stereochemical determination of terpecurcumins J–W, cytotoxic terpene-conjugated curcuminoids from *Curcuma longa* L. J Org Chem 2013;78:11835–11848.
374. Zhou M, Zhou J, Liu J, Liang JJ, Peng XG, Duan FF, Ruan HL. Parasubindoles A–G, seven eremophilanyl indoles from the whole plant of *Parasenecio albus.* J Org Chem 2018;83:12122–12128.
375. Cai YS, Wu Z, Zheng XQ, Wang C, Wang JR, Zhang XX, Qiu G, Zhu K, Cao S, Yu J. Spiroalanpyrroids A and B, sesquiterpene alkaloids with a unique

spiro-eudesmanolide–pyrrolizidine skeleton from *Inula helenium*. Org Chem Front 2020;7:303–309.

376. Fu J, Wang YN, Ma SG, Li L, Wang XJ, Li Y, Liu YB, Qu J, Yu SS. Xanthanoltrimer A–C: three xanthanolide sesquiterpene trimers from the fruits of *Xanthium italicum* Moretti isolated by HPLC-MS-SPE-NMR. Org Chem Front 2021;8:1288–1293.
377. Wang P, Li RJ, Liu RH, Jian KL, Yang MH, Yang L, Kong LY, Luo J. Sarglaperoxides A and B, sesquiterpene–normonoterpene conjugates with a peroxide bridge from the seeds of *Sarcandra glabra*. Org Lett 2016;18:832–835.
378. Feng T, Li XM, He J, Ai HL, Chen HP, Li XN, Li ZH, Liu JK. Nicotabin A, a sesquiterpenoid derivative from *Nicotiana tabacum*. Org Lett 2017;19:5201–5203.
379. Xu G, Hou AJ, Wang RR, Liang GY, Zheng YT, Liu ZY, Li XL, Zhao Y, Huang SX, Peng LY, Zhao QS. Przewalskin A: a new c23 terpenoid with a 6/6/7 carbon ring skeleton from *Salvia przewalskii* Maxim. Org Lett 2006;8:4453–4456.
380. Zhou M, Zhang HB, Wang WG, Gong NB, Zhan R, Li XN, Du X, Li LM, Li Y, Lu Y, Pu JX, Sun HD. Scopariusic acid, a new meroditerpenoid with a unique cyclobutane ring isolated from *Isodon scoparius*. Org Lett 2013;15:4446–4449.
381. Tanaka N, Niwa K, Kajihara S, Tsuji D, Itoh K, Mamadalieva NZ, Kashiwada Y. C28 terpenoids from lamiaceous plant *Perovskia scrophulariifolia*: their structures and anti-neuroinflammatory activity. Org Lett 2020;22:7667–7670.
382. Hayes PY, Chow S, Somerville MJ, De Voss JJ, Fletcher MT. Pimelotides A and B, diterpenoid ketal-lactone orthoesters with an unprecedented skeleton from *Pimelea elongata*. J Nat Prod 2009;72:2081–2083.
383. Zhao JX, Liu CP, Zhang MM, Li J, Yue JM. Dysohonin A, a meroditerpenoid incorporating a 6,15,6-fused heterotricyclic ring system from *Dysoxylum hongkongense*. Org Chem Front 2018;5:2202–2207.
384. Zhang J, He J, Cheng YC, Zhang PC, Yan Y, Zhang HJ, Zhang WK, Xu JK. Fischernolides A–D, four novel diterpene-based meroterpenoid scaffolds with antitumor activities from *Euphorbia fischeriana*. Org Chem Front 2019;6:2312–2318.
385. Fort DM, Ubillas RP, Mendez CD, Jolad SD, Inman WD, Carney JR, Chen JL, Ianiro TT, Hasbun C, Bruening RC, Luo J, Reed M.J, Iwu M, Carlson TJ, King SR, Bierer DE, Cooper R. Novel antihyperglycemic terpenoid-quinones from *Pycnanthus angolensis*. J Org Chem 2000;65:6534–6539.
386. Qin JJ, Fu YF, Chen X, Liu YT, Zhao JH, Zuo JP, Li YM, Zhu WL, Zhao WM. Celamonols A–D, four triterpenoid and catechin conjugates with immunosuppressive activities from the stems of *Celastrus monospermus*. Org Chem Front 2019;6:3786–3792.
387. Farimani MM, Taheri S, Ebrahimi SN, Bahadori MB, Khavasi HR, Zimmermann S, Brun R, Hamburger M. Hydrangenone, a new isoprenoid with an unprecedented skeleton from *Salvia hydrangea*. Org Lett 2012;14:166–169.

388. Chen CC, Wu JH, Yang NS, Chang JY, Kuo CC, Wang SY, Kuo YH. Cytotoxic C35 terpenoid cryptotrione from the bark of *Cryptomeria japonica.* Org Lett 2010;12:2786–2789.
389. Torres A, Molinillo JMG, Varela RM, Casas L, Mantell C, Martínez de la Ossa EJ, Macías FA. Helikaurolides A–D with a diterpene-sesquiterpene skeleton from supercritical fluid extracts of *Helianthus annuus* L. var. *Arianna.* Org Lett 2015;17:4730–4733.
390. Kim SY, Nagashima H, Tanaka N, Kashiwada Y, Kobayashi J, Kojoma M. Hitorins A and B, hexacyclic C_{25} terpenoids from *Chloranthus japonicus.* Org Lett 2016;18:5420–5423.
391. Bueno Pérez L, Pan L, Muñoz Acuña U, Li J, Chai HB, Gallucci JC, Ninh TN.; Carcache de Blanco EJ, Soejarto DD, Kinghorn AD. Caeruleanone A, a rotenoid with a new arrangement of the d-ring from the fruits of *Millettia caerulea.* Org Lett 2014;16:1462–1465.
392. Wang WL, Tang YZ, Liu YX, Yuan L, Wang J, Lin B, Zhou D, Sun L, Huang RB, Chen G, Li N. Isolation, structure elucidation, and synthesis of (±)-millpuline a with a suppressive effect in mir-144 expression. Org Chem Front 2019;6:2850–2859.
393. Zhang LJ, Bi DW, Hu J, Mu WH, Li YP, Xia GH, Yang L, Li XN, Liang XS, Wang LQ. Four hybrid flavan–chalcones, caesalpinnone a possessing a 10,11-dioxatricyclic [5.3.3.$0^{1,6}$]tridecane-bridged system and caesalpinflavans A–C from *Caesalpinia enneaphylla.* Org Lett 2017;19:4315–4318.
394. He QF, Wu ZL, Huang XJ, Zhong YL, Li MM, Jiang RW, Li YL, Ye WC, Wang Y. Cajanusflavanols A–C, three pairs of flavonostilbene enantiomers from *Cajanus cajan.* Org Lett 2018;20:876–879.
395. Jiang JS, He J, Feng ZM, Zhang PC. Two new quinochalcones from the florets of *Carthamus tinctorius.* Org Lett 2010;12:1196–1199.
396. He J, Yang YN, Jiang J, Feng ZM, Zhang PC. Saffloflavonesides A and B, two rearranged derivatives of flavonoid C-glycosides with a furan–tetrahydrofuran ring from *Carthamus tnctorius.* Org Lett 2014;16:5714–5717..
397. Yue SJ, Qu C, Zhang PX, Tang YP, Jin Y, Jiang JS, Yang YN, Zhang PC, Duan JA. Carthorquinosides A and B, quinochalcone C-glycosides with diverse dimeric skeletons from *Carthamus tinctorius.* J Nat Prod 2016;79:2644–2651.
398. Hu X, Wu J W, Wang M, Yu MH, Zhao QS, Wang HY, Hou AJ. 2-Arylbenzofuran, flavonoid, and tyrosinase inhibitory constituents of *Morus yunnanensis.* J Nat Prod 2012;75:82–87
399. Takashima J, Ohsaki A. Acutifolins A−F, a new flavan-derived constituent and five new flavans from *Brosimum acutifolium.* J Nat Prod 2001;64:1493–1496.
400. Qu KJ, Wang B, Jiang CS, Xie BG, Liu AH, Li SW, Guo YW, Li J, Mao SC. Rearranged diels−alder adducts and prenylated flavonoids as potential ptp1b inhibitors from *Morus nigra.* J Nat Prod 2021;84:2303−2311.
401. Chang LC, Chávez D, Song LL, Farnsworth NR, Pezzuto JM, Kinghorn AD. Absolute configuration of novel bioactive flavonoids from *Tephrosia purpurea.* Org Lett 2000;2:515–518.

402. Chen SD, Gao H, Zhu QC, Wang YQ, Li T, Mu ZQ, Wu HL, Peng T, Yao XS. Houttuynoids A–E, anti-herpes simplex virus active flavonoids with novel skeletons from *Houttuynia cordata*. Org. Lett. 2012, 14, 1772–1775.
403. Lu ZY, Van Wagoner RM, Pond CD, Pole AR, Jensen JB, Blankenship D, Grimberg BT, Kiapranis R, Matainaho TK, Barrows LR, Ireland CM. Myristicyclins A and B: antimalarial procyanidins from Horsfieldia spicata from *Papua New Guinea*. Org Lett 2014;16:346–349.
404. Zhang F, Yang Y N, Song XY, Shao SY, Feng ZM, Jiang JS, Li L, Chen NH, Zhang PC. Forsythoneosides A–D, neuroprotective phenethanoid and flavone glycoside heterodimers from the fruits of *Forsythia suspensa*. J Nat Prod 2015;78:2390–2397.
405. Naman CB, Li J, Moser A, Hendrycks JM, Benatrehina PA, Chai H, Yuan C, Keller WJ, Kinghorn AD. Computer-assisted structure elucidation of black chokeberry (*Aronia melanocarpa*) fruit juice isolates with a new fused pentacyclic flavonoid skeleton. Org Lett 2015;17:2988–2991.
406. Li FS, Zhan ZL, Liu F, Yang YN, Li L, Feng ZM, Jiang JS, Zhang PC. Polyflavanostilbene A, a new flavanol-fused stilbene glycoside from *Polygonum cuspidatum*. Org Lett 2013;15:674–677..
407. Yang XW, Li SM, Feng L, Shen YH, Tian JM, Zeng HW, Liu XH, Shan L, Su J, Zhang C, Zhang WD. Abiesanol A, a novel biflavanol with unique six connective hexacyclic rings isolated from *Abies georgei*. Tetrahedron Lett 2008;49:3042–3044.
408. Zhou B, Alania Y, Reis MC, McAlpine JB, Bedran-Russo AK, Pauli GF, Chen SN. Rare A-type, spiro-type, and highly oligomeric proanthocyanidins from *Pinus massoniana*. Org Lett 2020;22:5304–5308.
409. Kuo PC, Hung HY, Hwang TL, Du WK, Ku HC, Lee EJ, Tai SH, Chen FA, Wu TS. Anti-inflammatory flavan-3-ol-dihydroretrochalcones from *Daemonorops draco*. J Nat Prod 2017;80:783–789.
410. Doan TP, Park EJ, Cho HM, Ryu B, Lee BW, Thuong PT, Oh WK. Rugonidines A−F, diastereomeric 1,6-dioxa-7,9-diazaspiro[4.5]dec-7-en-8-amines from the leaves of *Alchornea rugosa*. J Nat Prod 2021;84:3055−3063.
411. Kuo YH, Huang HC, Kuo LMY, Chen, CF. Novel C19 homolignans, taiwanschirin A, B, and cytotoxic taiwanschirin C, and a new C_{18} lignan, schizanrin A, from *Schizandra arisanensis*. J Org Chem 1999;64:7023–7027.
412. Shen YC, Lin YC, Cheng YB, Kuo YH, Liaw CC. Taiwankadsurins A, B, and C, three new C19 Homolignans from *Kadsura philippinensis*. Org Lett 2005;7:5297–5300.
413. Yu HY, Chen ZY, Sun B, Liu J, Meng FY, Liu Y, Tian T, Jin A, Ruan HL. Lignans from the fruit of *Schisandra glaucescens* with antioxidant and neuroprotective properties. J Nat Prod 2014;77:1311−1320.
414. Wang XJ, Liu JB, Pandey P, Fronczek FR, Doerksen RJ, Chen JB, Qi XZ, Zhang PC, Ferreira D, Valeriote FA, Sun H, Li S, Hamann MT. Computationally assisted assignment of the kadsuraols, a class of chemopreventive agents for the control of liver cancer. Org Lett 2018;20:5559–5563.

415. Cheng ZB, Lu X, Bao JM, Han QH, Dong Z, Tang GH, Gan LS, Luo HB.; Yin S. (±)-Torreyunlignans A–D, rare 8–9′ linked neolignan enantiomers as phosphodiesterase-9a inhibitors from *Torreya yunnanensis*. J Nat Prod 2014;77:2651–2657.
416. Lai YJ, Liu TT, Sa RJ, Wei XL, Xue YB, Wu ZD, Luo ZW, Xiang M, Zhang YH, Yao GM. Neolignans with a rare 2-oxaspiro[4.5]deca-6,9-dien-8-one motif from the stem bark of *Cinnamomum subavenium*. J Nat Prod 2015;78:1740–1744.
417. Lee D, Cuendet M, Vigo JS, Graham JG, Cabieses F, Fong HHS, Pezzuto JM, Kinghorn AD. A novel cyclooxygenase-inhibitory stilbenolignan from the seeds of *Aiphanes aculeata*. Org Lett 2001;3:2169–2171.
418. Cardoso CL, Castro Gamboa I, Bergamini GM, Cavalheiro AJ, Silva DHS, Lopes MN, Araújo AR, Furlan M, Verli H, Bolzani VdaS. An unprecedented neolignan skeleton from *Chimarrhis turbinata*. J Nat Prod 2011;74:487–491.
419. do Vale AE, David JM, dos Santos EO, David JP, e Silva LCRC, Bahia MV, Brandão HN. An Unusual caffeic acid derived bicyclic [2.2.2]octane lignan and other constituents from *Cordia rufescens*. Phytochemistry 2012;76:158–161.
420. Shi YS, Liu YB, Li Y, Li L, Qu J, Ma SG, Yu SS. Chiral resolution and absolute configuration of a pair of rare racemic spirodienone sesquineolignans from *Xanthium sibiricum*. Org Lett 2014;16:5406–5409.
421. Zhu GY, Yang J, Yao XJ, Yang X, Fu J, Liu X, Bai LP, Liu L, Jiang ZH. (±)-Sativamides A and B, two pairs of racemic nor-lignanamide enantiomers from the fruits of *Cannabis sativa*. J Org Chem 2018;83:2376–2381.
422. Guo R, Zhao P, Yu X, Yao G, Lin B, Huang XX, Song SJ. (±)-Pinnatifidaones A–D, four pairs of highly modified neolignan enantiomers with a rare spirocyclohexenone skeleton from *Crataegus pinnatifida*. Org Chem Front 2021;8:953–960.
423. Xiao CY, Sun ZL, Huang J, Li RS, He JM, Gibbons S, Ju DW, and Mu Q. Neolignans from Piper betle have synergistic activity against antibiotic-resistant *Staphylococcus aureus*. J Org Chem 2021;86:11072−11085.
424. Yang F, Su BJ, Hu YJ, Liu JL, Li H, Wang YQ, Liao HB, Liang D. Piperhancins A and B, two pairs of antineuroinflammatory cycloneolignane enantiomers from *Piper hancei*. J Org Chem 2021;86:5284−5291.
425. Gan LS, Fan CQ, Yang SP, Wu Y, Lin LP, Ding J, Yue JM. Flueggenines A and B, two novel c,c-linked dimeric indolizidine alkaloids from *Flueggea virosa*. Org Lett 2006:8:2285–2288.
426. Zhao BX, Wang Y, Zhang DM, Jiang RW, Wang GC, Shi JM, Huang XJ, Chen WM, Che CT, Ye WC. Flueggines A and B, two new dimeric indolizidine alkaloids from *Flueggea virosa*. Org Lett 2011:13:3888–3891.
427. Wu ZL, Zhao BX, Huang XJ, Tang GY, Shi L, Jiang RW, Liu X, Wang Y, Ye WC. Suffrutines A and B: a pair of Z/E isomeric indolizidine alkaloids from the roots of *Flueggea suffruticosa*. Angew Chem Int Ed 2014:53:5796–5799.
428. Luo XK, Cai J, Yin ZY, Luo P, Li CJ, Ma H, Seeram NP, Gu Q, Xu J. Fluvirosaones A and B, two indolizidine alkaloids with a pentacyclic skeleton from *Flueggea virosa*. Org Lett 2018;20:991–994.

429. Zhao BX, Wang Y, Zhang DM, Huang XJ, Bai LL, Yan Y, Chen JM, Lu TB, Wang YT, Zhang QW, Ye WC. Virosaines A and B, two new birdcage-shaped securinega alkaloids with an unprecedented skeleton from *Flueggea virosa*. Org Lett 2012:14:3096–3099.
430. Zhang H, Zhang CR, Zhu KK, Gao AH, Luo C, Li J, Yue JM. Fluevirosines A–C: a biogenesis inspired example in the discovery of new bioactive scaffolds from *Flueggea virosa*. Org Lett 2013;15:120–123.
431. Wu ZL, Huang XJ, Xu MT, Ma X, Li L, Shi L, Wang WJ, Jiang RW, Ye WC, Wang Y. Flueggeacosines A–C, dimeric securinine-type alkaloid analogues with neuronal differentiation activity from *Flueggea suffruticosa*. Org Lett 2018;20:7703–7707
432. Tan CJ, Di YT, Wang YH, Zhang Y, Si YK, Zhang Q, Gao S, Hu XJ, Fang X, Li SF, Hao XJ. Three new indole alkaloids from *Trigonostemon lii*. Org Lett 2010;12:2370–2373
433. Ma S.S, Mei WL, Guo ZK, Liu SB, Zhao YX, Yang DL, Zeng YB, Jiang B, Dai HF. Two new types of bisindole alkaloid from *Trigonostemon lutescens*. Org Lett 2013;15:1492–1495.
434. Nge CE, Gan CY, Low YY, Thomas NF, Kam TS. Voatinggine and tabertinggine, pentacyclic indole alkaloids derived from an iboga precursor via a common cleavamine-type intermediate. Org Lett 2013:15:4774–4777.
435. Bracher F, Eisenreich WJ, Mühlbacher J, Dreyer M, Bringmann G. Saludimerines A and B, novel-type dimeric alkaloids with stereogenic centers and configurationally semistable biaryl axes. J Org Chem 2004:69:8602–8608.
436. Gao XH, Xu YS, Fan YY, Gan LS, Zuo JP, Yue JM. Cascarinoids A–C, a class of diterpenoid alkaloids with unpredicted conformations from *Croton cascarilloides*. Org Lett 2018;20:228–231.
437. Morita H, Arisaka M, Yoshida N, Kobayashi J. Serratezomines A–C, new alkaloids from *Lycopodium serratum* Var. s *Erratum*. J Org Chem 2000;65:6241–6245.
438. Kobayashi J, Hirasawa Y, Yoshida N, Morita H. Lyconadin A, a novel alkaloid from *Lycopodium complanatum*. J Org Chem 2001;66:5901–5904.
439. Takayama H, Katakawa K, Kitajima M, Seki H, Yamaguchi K, Aimi N. A new type of lycopodium alkaloid, lycoposerramine-A, from *Lycopodium serratum* Thunb. Org Lett 2001;3:4165–4167.
440. Hirasawa Y, Morita H, Shiro M, Kobayashi J. Sieboldine A, a novel tetracyclic alkaloid from *Lycopodium sieboldii*, inhibiting acetylcholinesterase. Org Lett 2003;5:3991–3993.
441. Morita H, Hirasawa Y, Kobayashi J. Himeradine A, a novel $C_{27}N_3$-type alkaloid from *Lycopodium chinense*. J Org Chem 2003;68:4563–4566.
442. Hirasawa Y, Morita H, Kobayashi J. Nankakurine A, a novel $C_{16}N_2$-type alkaloid from *Lycopodium hamiltonii*. Org Lett 2004;6:3389–3391.
443. Hirasawa Y, Kobayashi J, Morita H. Lycoperine A, a novel $C_{27}N_3$-type pentacyclic alkaloid from *Lycopodium hamiltonii*, inhibiting acetylcholinesterase. Org Lett 2006;8:123–126.

444. He J, Chen XQ, Li MM, Zhao Y, Xu G, Cheng X, Peng LY, Xie MJ, Zheng YT, Wang YP, Zhao QS. Lycojapodine A, a novel alkaloid from *Lycopodium japonicum.* Org Lett 2009;11:1397–1400.
445. Wang XJ, Zhang GJ, Zhuang PY, Zhang Y, Yu SS, Bao XQ, Zhang D, Yuan YH, Chen NH, Ma S, Qu J, Li Y. Lycojaponicumins A–C, three alkaloids with an unprecedented skeleton from *Lycopodium japonicum.* Org Lett 2012;14:2614–2617.
446. Wang XJ, Liu YB, Li L, Yu SS, Lv HN, Ma SG, Bao XQ, Zhang D, Qu J, Li Y. Lycojaponicumins D and E: two new alkaloids from *Lycopodium japonicum.* Org Lett 2012;14:5688–5691.
447. Cheng JT, Liu F, Li XN, Wu XD, Dong LB, Peng LY, Huang SX, He J, Zhao QS. Lycospidine A, a new type of *Lycopodium* alkaloid from *Lycopodium complanatum.* Org Lett 2013;15:2438–2441.
448. Tang Y, Xiong J, Zhang JJ, Wang W, Zhang HY, Hu JF. Annotinolides A–C, three lycopodane-derived 8,5-lactones with polycyclic skeletons from *Lycopodium annotinum.* Org Lett 2016;18:4376–4379.
449. Zhang ZJ, Nian Y, Zhu QF, Li XN, Su J, Wu XD, Yang J, Zhao QS. Lycoplanine A, a $C_{16}N$ *Lycopodium* alkaloid with a 6/9/5 tricyclic skeleton from *Lycopodium complanatum.* Org Lett 2017;19:4668–4671.
450. Yeap JSY, Lim KH, Yong KT, Lim SH, Kam TS, Low YY. *Lycopodium* alkaloids: lycoplatyrine A, an unusual lycodine–piperidine adduct from *Lycopodium platyrhizoma* and the absolute configurations of lycoplanine D and lycogladine H. J Nat Prod 2019;82:324–329.
451. Zhao FW, Sun QY, Yang FM, Hu GW, Luo JF, Tang GH, Wang YH, Long CL. Palhinine A, a novel alkaloid from *Palhinhaea cernua.* Org Lett 2010;12:3922–3925.
452. Dong LB, Yang J, He J, Luo HR, Wu XD, Deng X, Peng LY, Cheng X, Zhao QS. Lycopalhine A, a novel sterically congested lycopodium alkaloid with an unprecedented skeleton from *Palhinhaea cernua.* Chem Commun 2012;48:9038–9040.
453. Dong LB, Gao X, Liu F, He J, Wu XD, Li Y, Zhao QS. Isopalhinine A, a unique pentacyclic Lycopodium alkaloid from *Palhinhaea cernua.* Org Lett 2013;15:3570–3573.
454. Mansoor TA, Ramalhete C, Molnár J, Mulhovo S, Ferreira MJU. Tabernines A−C, *β*-carbolines from the leaves of *Tabernaemontana elegans.* J Nat Prod 2009;72:1147–1150.
455. Carroll AR, Hyde E, Smith J, Quinn RJ, Guymer G, Forster PI. Actinophyllic acid, a potent indole alkaloid inhibitor of the coupled enzyme assay carboxypeptidase u/hippuricase from the leaves of *Alstonia a ctinophylla* (Apocynaceae). J Org Chem 2005;70:1096–1099.
456. Nge CE, Gan CY, Lim KH, Ting KN, Low YY, Kam TS. Criofolinine and vernavosine, new pentacyclic indole alkaloids incorporating pyrroloazepine and pyridopyrimidine moieties derived from a common yohimbine precursor. Org Lett 2014;16:6330–6333.

457. Yu HF, Qin XJ, Ding CF, Wei X, Yang J, Luo JR, Liu L, Khan A, Zhang LC, Xia CF, Luo XD. Nepenthe-like indole alkaloids with antimicrobial activity from *Ervatamia chinensis*. Org Lett 2018;20:4116–4120.
458. Ma K, Wang JS, Luo J, Yang MH, Yao HQ, Sun HB, Kong LY. Bistabercarpamines A and B, first vobasinyl-chippiine-type bisindole alkaloid from *Tabernaemontana corymbosa*. Tetrahedron Lett 2014;55:101–104.
459. Liu ZW, Zhang J, Li ST, Liu MQ, Huang XJ, Ao YL, Fan CL, Zhang DM, Zhang QW, Ye WC, Zhang XQ. Ervadivamines A and B, two unusual trimeric monoterpenoid indole alkaloids from *Ervatamia divaricata*. J Org Chem 2018;83:10613–10618.
460. Hirasawa Y, Miyama S, Hosoya T, Koyama K, Rahman A, Kusumawati I, Zaini NC, Morita H. Alasmontamine A, a first tetrakis monoterpene indole alkaloid from *Tabernaemontana elegans*. Org Lett 2009;11:5718–5721.
461. Yuan YX, Zhang Y, Guo LL, Wang YH, Goto M, Morris Natschke SL, Lee KH, Hao XJ. Tabercorymines A and B, two vobasinyl–ibogan-type bisindole alkaloids from *Tabernaemontana corymbosa*. Org Lett 2017;19:4964–4967.
462. Kam TS, Tan SJ, Ng SW, Komiyama K. Bipleiophylline, an unprecedented cytotoxic bisindole alkaloid constituted from the bridging of two indole moieties by an aromatic spacer unit. Org Lett 2008;10:3749–3752.
463. Koyama K, Hirasawa Y, Nugroho AE, Hosoya T, Hoe TC, Chan KL, Morita H. Alsmaphorazines A and B, novel indole alkaloids from *Alstonia pneumatophora*. Org Lett 2010;12:4188–4191.
464. Lim SH, Tan SJ, Low YY, Kam TS. Lumutinines A–D, linearly fused macroline–macroline and macroline–sarpagine bisindoles from *Alstonia macrophylla*. J Nat Prod 2011;74:2556–2562.
465. Zhu GY, Yao XJ, Liu L, Bai LP, Jiang ZH. Alistonitrine A, a caged monoterpene indole alkaloid from *Alstonia scholaris*. Org Lett 2014;16:1080–1083.
466. Pan ZQ, Qin XJ, Liu YP, Wu T, Luo XD, Xia CF. Alstoscholarisines H–J, indole alkaloids from *Alstonia scholaris*: structural evaluation and bioinspired synthesis of alstoscholarisine H. Org Lett 2016;18:654–657.
467. Yu HF, Ding CF, Zhang LC, Wei X, Cheng GG, Liu YP, Zhang RP, Luo XD. Alstoscholarisine K, an antimicrobial indole from gall-induced leaves of *Alstonia scholaris*. Org Lett 2021;23:5782–5786.
468. Zhu XX, Fan YY, Xu L, Liu QF, Wu JP, Li JY, Li J, Gao K, Yue JM. Alstonlarsines A–D, four rearranged indole alkaloids from *Alstonia Scholaris*. Org Lett 2019;21:1471–1474.
469. Cai XH, Du ZZ, Luo XD. Unique monoterpenoid indole alkaloids from *Alstonia scholaris*. Org Lett 2007:9:1817–1820.
470. Cai XH, Tan QG, Liu YP, Feng T, Du ZZ, Li WQ, Luo XD. A cage-monoterpene indole alkaloid from *Alstonia scholaris*. Org Lett 2008;10:577–580.
471. Zhang J, Song M, Ao YL, Li Y, Zou XY, Xu J, Wang Y, Zhang DM, Zhang XQ, Ye WC. Alstolarines A and B, two unusual monoterpenoid indole alkaloids with an acetal moiety from *Alstonia scholaris*. Org Chem Front 2020;7:3468–3473.

472. Kobayashi J, Sekiguchi M, Shimamoto S, Shigemori H, Ishiyama H, Ohsaki A. Subincanadines A–C, novel quaternary indole alkaloids from *Aspidosperma subincanum*. J Org Chem 2002;67:6449–6455.
473. Feng T, Cai XH, Li Y, Wang YY, Liu YP, Xie MJ, Luo XD. Melohenines A and B, two unprecedented alkaloids from *Melodinus henryi*. Org Lett 2009;11:4834–4837.
474. Feng T, Li Y, Liu YP, Cai XH, Wang YY, Luo XD. Melotenine A, a cytotoxic monoterpenoid indole alkaloid from *Melodinus tenuicaudatus*. Org Lett 2010;12:968–971.
475. Zhang Y, Ding X, Shao S, Guo LL, Zhao Q, Hao XJ, He HP. Melocochines A and B, two alkaloids from the fruits of *Melodinus cochinchinensis*. Org Lett 2019;21:9272–9275.
476. Wu J, Zhao SM, Shi BB, Bao MF, Schinnerl J, Cai XH. Cage-Monoterpenoid quinoline alkaloids with neurite growth promoting effects from the fruits of *Melodinus yunnanensis*. Org Lett 2020;22:7676–7680.
477. Kam TS, Choo YM. Kopsifolines A, B, and C, indole alkaloids with a novel hexacyclic carbon skeleton from *Kopsia*. Tetrahedron Lett 2003;44:1317–1319.
478. Lim KH, Kam TS. Arboflorine, an unusual pentacyclic monoterpenoid indole alkaloid incorporating a third nitrogen atom. Org Lett 2006;8:1733–1735.
479. Zeng J, Zhang DB, Zhou PP, Zhang QL, Zhao L, Chen JJ, Gao, K. Rauvomines A and B, two monoterpenoid indole alkaloids from *Rauvolfia vomitoria*. Org Lett 2017;19:3998–4001.
480. Fu YH, He HP, Di YT, Li SL, Zhang Y, Hao XJ. Mekongenines A and B, two new alkaloids from *Bousigonia mekongensis*. Tetrahedron Lett 2012;53:3642–3646.
481. Hirasawa Y, Hara M, Nugroho AE, Sugai M, Zaima K, Kawahara N, Goda Y, Awang K, Hadi AHA, Litaudon M, Morita H. Bisnicalaterines B and C, atropisomeric bisindole alkaloids from *Hunteria zeylanica*, showing vasorelaxant activity. J Org Chem 2010;75:4218–4223.
482. Beniddir MA, Martin MT, Tran Huu Dau ME, Grellier P, Rasoanaivo P, Guéritte F, Litaudon M. Goniomedines A and B: unprecedented bisindole alkaloids formed through fusion of two indole moieties via a dihydropyran unit. Org Lett 2012;14:4162–4165.
483. N'Nang Obiang EO, Genta Jouve G, Gallard JF, Kumulungui B, Mouray E, Grellier P, Evanno L, Poupon E, Champy P, Beniddir MA. Pleiokomenines A and B: dimeric aspidofractinine alkaloids tethered with a methylene group. Org Lett 2017;19:6180–6183.
484. Li LM, Yang T, Liu Y, Liu J, Li MH, Wang YT, Yang SX, Zou Q, Li GY. Calophyline A, a new rearranged monoterpenoid indole alkaloid from *Winchia calophylla*. Org Lett 2012;14:3450–3453.
485. Gan CY, Robinson WT, Etoh T, Hayashi M, Komiyama K, Kam TS. Leucophyllidine, a cytotoxic bisindole alkaloid constituted from the union of an eburnan and a new vinylquinoline alkaloid. Org Lett 2009;11:3962–3965.

486. Zhang YB, Zhang XL, Chen NH, Wu ZN, Ye WC, Li YL, Wang GC. Four matrine-based alkaloids with antiviral activities against hbv from the seeds of *Sophora alopecuroides*. Org Lett 2017;19:424–427.
487. Zhang YB, Yang L, Luo D, Chen NH, Wu ZN, Ye WC, Li YL, Wang GC. Sophalines E–I, five quinolizidine-based alkaloids with antiviral activities against the hepatitis b virus from the seeds of *Sophora alopecuroides*. Org Lett 2018;20:5942–5946.
488. Zhang YB, Zhan LQ, Li GQ, Wang F, Wang Y, Li YL, Ye WC, Wang GC. Dimeric matrine-type alkaloids from the roots of *Sophora flavescens* and their anti-hepatitis B virus activities. J Org Chem 2016;81:6273–6280.
489. Zhang BJ, Bao MF, Zeng CX, Zhong XH, Ni L, Zeng Y, Cai XH. Dimeric Erythrina alkaloids from the flower of *Erythrina variegata*. Org Lett 2014;16:6400–6403.
490. Tang YT, Wu J, Yu Y, Bao MF, Tan QG, Schinnerl Johann, Cai XH. Colored dimeric alkaloids from the barks of *Erythrina variegata* and their neuroprotective effects. J Org Chem 2021;86:13381−13387.
491. Morita H, Yoshida N, Kobayashi J. Daphnezomines A and B, novel alkaloids with an aza-adamantane core from *Daphniphyllum humile*. J Org Chem 1999;64:7208–7212.
492. Morita H, Yoshida N, Kobayashi J. Daphnezomines F and G: novel alkaloids with 1-azabicyclo[5.2.2]undecane moiety from *Daphniphyllum humile*. J Org Chem 2000;65:3558–3562.
493. Kobayashi J, Inaba Y, Shiro M, Yoshida N, Morita H. Daphnicyclidins A−H, novel hexa- or pentacyclic alkaloids from two species of *Daphniphyllum*. J Am Chem Soc 2001;123:11402–11408.
494. Morita H, Yoshida N, Kobayashi J. Daphnicyclidins J and K, unique polycyclic alkaloids from *Daphniphyllum humile*. J Org Chem 2002;67:2278–2282.
495. Jossang A, Bitar HE, Pham VC, Sévenet T. Daphcalycine, a novel heptacycle fused ring system alkaloid from *Daphniphyllum calycinum*. J Org Chem 2003;68:300–304.
496. Yang SP, Yue JM. Two novel alkaloids with a unique fused hexacyclic skeleton from *Daphniphyllum subverticillatum*. J Org Chem 2003:68:7961–7966.
497. Kobayashi J, Takatsu H, Shen YC, Morita H. Daphniglaucins A and B, novel polycyclic quaternary alkaloids from *Daphniphyllum glaucescens*. Org Lett 2003;5:1733–1736.
498. Morita H, Kobayashi J. Calyciphyllines A and B, two novel hexacyclic alkaloids from *Daphniphyllum calycinum*. Org Lett 2003;5:2895–2898.
499. Yang SP, Yue JM. First diamino Daphniphyllum alkaloid, daphnipaxinin, with an unprecedented heterohexacyclic skeleton from *Daphniphyllum paxianum*. Org Lett 2004:6:1401–1404.
500. Zhan ZJ, Yang SP, Yue JM. Paxdaphnidines A and B, novel penta- and tetracyclic alkaloids from *Daphniphyllum paxianum*. J Org Chem 2004;69:1726–1729.

501. Morita H, Ishioka N, Takatsu H, Shinzato T, Obara Y, Nakahata N, Kobayashi J. Daphmanidins C and D, novel pentacyclic alkaloids from *Daphniphyllum teijsmanii*. Org Lett 2005;7:459–462.
502. Di YT, He HP, Liu HY, Du ZZ, Tian JM, Yang XW, Wang YH, Hao XJ. Calycilactone A, a novel hexacyclic alkaloid from *daphniphyllum calycillum*. Tetrahedron Lett 2006;47:5329–5331.
503. Zhang W, Guo YW, Krohn K. Macropodumines A–C: novel pentacyclic alkaloids with an unusual skeleton or zwitterion moiety from *Daphniphyllum macropodum* Miq. Chem Eur J 2006;12:5122–5127.
504. Saito S, Kubota T, Fukushi E, Kawabata J, Zhang H, Kobayashi J. Calyciphylline C, a novel daphniphyllum alkaloid from *Daphniphyllum calycinum*. Tetrahedron Lett 2007;48:1587–1589.
505. Li CS, He HP, Di YT, Wang YH, Mu SZ, Li SL, Gao S, Gao ZL, Hao XJ. Daphlongeramine A, novel daphniphyllum alkaloid from *Daphniphyllum longeracemosum*. Tetrahedron Lett 2007;48:2737–2740.
506. Saito S, Kubota T, Kobayashi J. Calyciphylline G, a novel alkaloid with an unprecedented fused-hexacyclic skeleton from *Daphniphyllum calycinum*. Tetrahedron Lett 2007;48:5693–5695.
507. Di YT, He HP, Wang YS, Li LB, Lu Y, Gong JB, Fang X, Kong NC, Li SL, Zhu HJ, Hao XJ. Isolation, X-ray crystallography, and computational studies of calydaphninone, a new alkaloid from *Daphniphyllum calycillum*. Org Lett 2007;9:1355–1358.
508. Li CS, Di YT, He HP, Gao S, Wang YH, Lu Y, Zhong JL, Hao XJ. Daphlongeranines A and B, two novel alkaloids possessing unprecedented skeletons from *Daphniphyllum longeracemosum*. Org Lett 2007;9:2509–2512.
509. Li ZY, Chen P, Xu HG, Yang YM, Peng SY, Zhao ZZ, Guo YW. Macropodumines D and E, two new alkaloids with unusual skeletons from *Daphniphyllum macropodum* Miq. Org Lett 2007;9:477–480..
510. Saito S, Kubota T, Fukushi E, Kawabata J, Zhang H, Kobayashi J. Calyciphylline D, a novel alkaloid with an unprecedented fused-pentacyclic skeleton from *Daphniphyllum calycinum*. Org Lett 2007;9:1207–1209.
511. Zhang CR, Liu HB, Dong SH, Zhu JY, Wu Y, Yue JM. Calycinumines A and B, two novel alkaloids from *Daphniphyllum calycinum*. Org Lett 2009;11:4692–4695.
512. Zhang Q, Di YT, Li CS, Fang X, Tan CJ, Zhang Z, Zhang Y, He HP, Li SL, Hao XJ. Daphenylline, a new alkaloid with an unusual skeleton, from *Daphniphyllum longeracemosum*. Org Lett 2009;11:2357–2359.
513. Zhang Y, Di YT, Zhang Q, Mu SZ, Tan CJ, Fang X, He HP, Li SL, Hao XJ. Daphhimalenine A, a new alkaloid with an unprecedented skeleton, from *Daphniphyllum himalense*. Org Lett 2009;11:5414–5417.
514. Zhang CR, Fan CQ, Dong SH, Liu HB, Zhou WB, Wu Y, Yue JM. Angustimine and angustifolimine: two new alkaloids from *Daphniphyllum angustifolium*. Org Lett 2011;13:2440–2443.

515. Wang F, Mao MF, Wei GZ, Gao Y, Ren FC, Liu JK. Hybridaphniphyllines A and B, *Daphniphyllum* alkaloid and iridoid hybrids suggestive of diels–alder cycloaddition in *Daphniphyllum longeracemosum*. Phytochemistry 2013;95:428–435.
516. Xu JB, Zhang H, Gan LS, Han YS, Wainberg MA, Yue JM. Logeracemin A, an anti-HIV *Daphniphyllum* alkaloid dimer with a new carbon skeleton from *Daphniphyllum longeracemosum*. J Am Chem Soc 2014;136:7631–7633.
517. Zhang H, Shyaula SL, Li JY, Li J, Yue JM. Himalensines A and B, alkaloids from *Daphniphyllum himalense*. Org Lett 2016;18:1202–1205.
518. Huang SD, Zhang Y, Cao MM, Di YT, Tang GH, Peng ZG, Jiang JD, He HP, Hao XJ. Myriberine A, a new alkaloid with an unprecedented heteropentacyclic skeleton from *Myrioneuron faberi*. Org Lett 2013;15:590–593.
519. Cao MM, Huang SD, Di YT, Yuan CM, Zuo GY, Gu YC, Zhang Y, Hao XJ. Myrifabine, the first dimeric Myrioneuron alkaloid from *Myrioneuron faberi*. Org Lett 2014;16:528–531.
520. Cao MM, Zhang Y, Li XH, Peng ZG, Jiang JD, Gu YC, Di YT, Li XN, Chen DZ, Xia CF, He HP, Li SL, Hao XJ. Cyclohexane-fused octahydroquinolizine alkaloids from *Myrioneuron faberi* with activity against hepatitis C virus. J Org Chem 2014;79:7945–7950.
521. Wang K, Zhou XY, Wang YY, Li MM, Li YS, Peng LY, Cheng X, Li Y, Wang YP, Zhao QS. Macrophyllionium and macrophyllines A and B, oxindole alkaloids from *Uncaria macrophylla*. J Nat Prod 2011;74:12–15.
522. Geng CA, Huang XY, Ma YB, Hou B, Li TZ, Zhang XM, Chen JJ. (±)-Uncarilins A and B, dimeric isoechinulin-type alkaloids from *Uncaria rhynchophylla*. J Nat Prod 2017:80:959–964.
523. Zhou HF, Li WY, Peng LY, Li XN, Zuo ZL, Zhao QS. Rhynchines A–E: Cav3.1 calcium channel blockers from *Uncaria rhynchophylla*. Org Lett 2021;23:9463–9467.
524. Kerber VA, Gregianini TS, Paranhos JT, Schwambach J, Farias F, Fett JP, Fett-Neto AG, Zuanazzi JÂS, Quirion JC, Elizabetsky E, Henriques AT. Brachycerine, a novel monoterpene indole alkaloid from *Psychotria brachyceras*. J Nat Prod 2001;64:677–679.
525. Li XN, Zhang Y, Cai XH, Feng T, Liu YP, Li Y, Ren J, Zhu HJ, Luo XD. Psychotripine: A new trimeric pyrroloindoline derivative from *Psychotria pilifera*. Org Lett 2011;13:5896–5899.
526. Feng T, Duan KT, He SJ, Wu B, Zheng YS, Ai HL, Li ZH, He J, Zuo JP, Liu JK. Ophiorrhines A and B, two immunosuppressive monoterpenoid indole alkaloids from *Ophiorrhiza japonica*. Org Lett 2018;20:7926–7928.
527. Han JX, Niu ST, Liu YS, Gan LS, Wang TF, Lu CD, Yuan T. Robustanoids A and B, two novel pyrrolo[2,3-b]indole alkaloids from *Coffea canephora*: isolation and total synthesis. Org Chem Front 2018;5:586–589.
528. Fernandez LS, Buchanan MS, Carroll AR, Feng YJ, Quinn RJ, Avery VM. Flinderoles A–C: antimalarial bis-indole alkaloids from *Flindersia* species. Org Lett 2009;11:329–332.

529. Vougogiannopoulou K, Fokialakis N, Aligiannis N, Cantrell C, Skaltsounis AL. The raputindoles: novel cyclopentyl bisindole alkaloids from *Raputia simulans*. Org Lett 2010;12:1908–1911.
530. Yang CH, Cheng MJ, Chiang MY, Kuo YH, Wang CJ, Chen IS. Dihydrobenzo[c]phenanthridine alkaloids from stem bark of *Zanthoxylum nitidum*. J Nat Prod 2008;71:669–673.
531. Shou QY, Banbury LK, Renshaw DE, Smith JE, He XX, Dowell A, Griesser HJ, Heinrich M, Wohlmuth H. Parvifloranines A and B, two 11-carbon alkaloids from *Geijera parviflora*. J Nat Prod 2013;76:1384–1387.
532. Liu YP, Guo JM, Xie Z, Suo XY, Liu ZY, Qiao ZH, Guan RQ, Bian Y, Qiang L, Fu YH. Clausanisumine, a prenylated bicarbazole alkaloid from the fruits of *Clausena anisumolens* and its potential anti-HIV activity. J Org Chem 2021;86:17722–17726.
533. Bringmann G, Lombe BK, Steinert C, Ioset KN, Brun R, Turini F, Heubl G, Mudogo V. Mbandakamines A and B, unsymmetrically coupled dimeric naphthylisoquinoline alkaloids, from a congolese *Ancistrocladus* species. Org Lett 2013;15:2590–2593.
534. Lombe BK, Bruhn T, Feineis D, Mudogo V, Brun R, Bringmann G. Antiprotozoal spirombandakamines A_1 and A_2, fused naphthylisoquinoline dimers from a congolese *Ancistrocladus* plant. Org Lett 2017;19:6740–6743.
535. Lombe BK, Bruhn T, Feineis D, Mudogo V, Brun R, Bringmann G. Cyclombandakamines A_1 and A_2, oxygen-bridged naphthylisoquinoline dimers from a congolese *Ancistrocladus* Liana. Org Lett 2017;19:1342–1345.
536. Zhang W, Huang XJ, Zhang SY, Zhang DM, Jiang RW, Hu JY, Zhang XQ, Wang L, Ye WC. Geleganidines A–C, unusual monoterpenoid indole alkaloids from *Gelsemium elegans*. J Nat Prod 2015;78:2036–2044.
537. Li NP, Liu M, Huang XJ, Gong XY, Zhang W, Cheng MJ, Ye WC, Wang L. Gelsecorydines A–E, five gelsedine–corynanthe-type bisindole alkaloids from the fruits of *Gelsemium elegans*. J Org Chem 2018;83: 5707–5714.
538. Gu JH, Zhang W, Cai WY, Fu XX, Zhou HL, Li NP, Tian HY, Liu JS, Ye WC, Wang L. Gelserancines A–E, monoterpenoid indole alkaloids with unusual skeletons from *Gelsemium elegans*. Org Chem Front 2021;8;1918–1925.
539. Jin PF. Zhan GQ, Zheng GJ, Liu JJ, Peng X, Huang L, Gao B, Yuan XH, and Yao GM. Gelstriamine A, a triamino monoterpene indole alkaloid with a caged 6/5/7/6/6/5 scaffold and analgesic alkaloids from *Gelsemium elegans* stems. J Nat Prod 2021;84:1326–1334.
540. Fu YH, Zhang Y, He HP, Hou L, Di YT, Li SL, Luo XD, Hao XJ. Strynuxlines A and B, alkaloids with an unprecedented carbon skeleton from *Strychnos nux-Vomica*. J Nat Prod 2012;75:1987–1990..
541. Liu JF, Jiang ZY, Wang RR, Zheng YT, Chen JJ, Zhang XM, Ma YB. Isatisine A, a novel alkaloid with an unprecedented skeleton from leaves of *Isatis indigotica*. Org Lett 2007;9:4127–4129.
542. Chen MH, Lin S, Li L, Zhu CG, Wang XL, Wang YN, Jiang BY, Wang SJ, Li YH, Jiang JD, Shi JG. Enantiomers of an indole alkaloid containing unusual

dihydrothiopyran and 1,2,4-thiadiazole rings from the root of *Isatis indigotica*. Org Lett 2012;14:5668–5671

543. Zhang GJ, Li B, Cui HM, Chen L, Tian Y, Liu SJ, Li BW, Li M, Xia ZM, Chen XX, Hou Y, Dong JX. Orychophragines A–C, three biologically active alkaloids from *Orychophragmus violaceus*. Org Lett 2018;20:656–659.
544. Sai CM, Li DH, Xue CM, Wang KB, Hu P, Pei YH, Bai J, Jing YK, Li ZL, Hua HM. Two pairs of enantiomeric alkaloid dimers from *Macleaya cordata*. Org Lett 2015;17:4102–4105
545. Wang B, Yang ZF, Zhao YL, Liu YP, Deng J, Huang WY, Li XN, Wang XH, Luo XD. Anti-inflammatory isoquinoline with bis-*seco*-aporphine skeleton from *Dactylicapnos scandens*. Org Lett 2018;20:1647–1650.
546. Wang B, Zhao YJ, Zhao YL, Liu YP, Li XN, Zhang H, Luo XD. Exploring aporphine as anti-inflammatory and analgesic lead from *Dactylicapnos scandens*. Org Lett 2020;22:257–260.
547. Morita H, Oshimi S, Hirasawa Y, Koyama K, Honda T, Ekasari W, Indrayanto G, Zaini NC. Cassiarins A and B, novel antiplasmodial alkaloids from *Cassia siamea*. Org Lett 2007;9:3691–3693.
548. Zhou M, Zhou K, Gao XM, Jiang ZY, Lv JJ, Liu ZH, Yang GY, Miao MM, Che CT, Hu QF. Fistulains A and B, new bischromones from the bark of *Cassia fistula*, and their activities. Org Lett 2015;17:2638–2641.
549. Mu ZQ, Gao H, Huang ZY, Feng XL, Yao XS. Puberunine and puberudine, two new C_{18}-diterpenoid alkaloids from *Aconitum barbatum* var. *Puberulum*. Org Lett 2012;14:2758–2761.
550. Guo QL, Xia H, Shi GN, Zhang TT, Shi JG. Aconicarmisulfonine A, a sulfonated C_{20}-diterpenoid alkaloid from the lateral roots of *Aconitum carmichaelii*. Org Lett 2018;20:816–819.
551. Zhan ZL, Feng ZM, Yang YN, Li L, Jiang JS, Zhang PC. Ternatusine A, a new pyrrole derivative with an epoxyoxepino ring from *Ranunculus ternatus*. Org Lett 2013;15:1970–1973.
552. Xue JJ, Jiang CY, Zou DL, Li BJ, Lu JC, Li DH, Lin B, Li ZL, Hua HM. Baicalensines A and B, two isoquinoline alkaloids from the roots of *Thalictrum baicalense*. Org Lett 2020;22:7439–7442.
553. Tan CH, Chen GF, Ma XQ, Jiang SH, Zhu DY. Huperzine R, a novel 15-carbon *Lycopodium* alkaloid from *Huperzia serrata*. J Nat Prod 2002;65:1021–1022.
554. Hirasawa Y, Mitsui C, Uchiyama N, Hakamatsuka T, Morita H. Hupercumines A and B, *Lycopodium* alkaloids from *Huperzia cunninghamioides*, inhibiting acetylcholinesterase. Org Lett 2018;20:1384–1387.
555. Meng WJ, Xiong J, Wang WX, Zhang HY, Zeng H, Hu JF. Phlefargesiine A, a $C_{16}N_2$ lycopodium alkaloid with an unprecedented [6/7/6/6]-tetracyclic skeleton from *Phlegmariurus fargesii*. Tetrahedron Lett 2016;57:3218–3221.
556. Zhang ZJ, Wang C, Wu XD, Huang Y, Zhou WX, Zhao QS. Phlegmadine A: a Lycopodium alkaloid with a unique cyclobutane ring from *Phlegmariurus phlegmaria*. J Org Chem 2019;84:11301–11305.
557. Wei K, Li W, Koike K, Chen Y, Nikaido T. Nigramides A–S, dimeric amide alkaloids from the roots of *Piper nigrum*. J Org Chem 2005;70:1164–1176..

558. Zhang XY, Li Y, Wang YY, Cai XH, Feng T, Luo XD. Tirucallane-type alkaloids from the bark of *Dysoxylum laxiracemosum.* J Nat Prod 2010;73:1385–1388.
559. Stærk D, Witt M, Oketch Rabah HA, Jaroszewski JW. A new class of spermidine-derived alkaloids. Org Lett 2003;5:2793–2796.
560. Wang W, Tang X, Hua F, Ling TJ, Wan XC, Bao GH. Camellimidazole A–C, three methylene-bridged dimeric imidazole alkaloids from *Keemun Black* Tea. Org Lett 2018;20:2672–2675.
561. Carroll AR, Arumugan G, Quinn RJ, Redburn J, Guymer G, Grimshaw P. Grandisine A and B, novel indolizidine alkaloids with human δ-opioid receptor binding affinity from the leaves of the australian rainforest tree *Elaeocarpus grandis.* J Org Chem 2005;70:1889–1892.
562. Liu MT, Lin S, Wang YH, He WY, Li S, Wang SJ, Yang YC, Shi JG. Two novel glycosidic triterpene alkaloids from the stem barks of *Machilus yaoshansis.* Org Lett 2007;9:129–132.
563. Toribio A, Bonfils A, Delannay E, Prost E, Harakat D, Henon E, Richard B, Litaudon M, Nuzillard JM, Renault JH. Novel *seco*-dibenzopyrrocoline alkaloid from *Cryptocarya oubatchensis.* Org Lett 2006;8:3825–3828.
564. Tong XG, Zhou LL, Wang YH, Xia C, Wang Y, Liang M, Hou FF, Cheng YX. Acortatarins A and B, two novel antioxidative spiroalkaloids with a naturally unusual morpholine motif from *Acorus tatarinowii.* Org Lett 2010;12:1844–1847.
565. Han JJ, Zhang JZ, Zhu RX, Li Y, Qiao YN, Gao Y, Jin XY, Chen W, Zhou JC, Lou HX. Plagiochianins A and B, two *ent*-2,3-*seco*-aromadendrane derivatives from the liverwort *Plagiochila duthiana.* Org Lett 2018;20:6550–6553.
566. Kobayashi J, Yoshinaga M, Yoshida N, Shiro M, Morita H. Cephalocyclidin A, a novel pentacyclic alkaloid from *Cephalotaxus harringtonia* var. *nana.* J Org Chem 2002;67:2283–2286
567. Wang YH, Gao S, Yang FM, Sun QY, Wang JS, Liu HY, Li CS, Di YT, Li SL, He HP, Hao XJ. Structure elucidation and biomimetic synthesis of hostasinine A, a new benzylphenethylamine alkaloid from *Hosta plantaginea.* Org Lett 2007;9:5279–5281.
568. Matsumoto T, Nakamura S, Ohta T, Fujimoto K, Yoshikawa M, Ogawa K, Matsuda H. A rare glutamine derivative from the flower buds of daylily. Org Lett 2014;16:3076–3078.
569. Wang KB, Di YT, Bao Y, Yuan CM, Chen G, Li DH, Bai J, He HP, Hao XJ, Pei YH, Jing YK, Li ZL, Hua HM. Peganumine A, a β-carboline dimer with a new octacyclic scaffold from *Peganum harmala.* Org Lett 2014;16:4028–4031.
570. Wu ZN, Chen NH, Tang Q, Chen S, Zhan ZC, Zhang YB, Wang GC, Li YL, Ye WC. β-Carboline alkaloids from the seeds of *Peganum harmala* and their anti-HSV-2 virus activities. Org Lett 2020;22:7310–7314.
571. Li SG, Wang YT, Zhang Q, Wang KB, Xue JJ, Li DH, Jing YK, Lin B, Hua HM. Pegaharmols A–B, axially chiral β-carboline-quinazoline dimers from the roots of *Peganum harmala.* Org Lett 2020;22:7522–7525.

572. Chen H, Kong JB, Zhang L, Wang HH, Cao YG, Zeng MN, Li M, Sun YJ, Du K, Xue GM, Wu Y, Zheng XK, Feng WS. Lycibarbarines A–C, three tetrahydroquinoline alkaloids possessing a spiro-heterocycle moiety from the fruits of *Lycium barbarum*. Org Lett 2021;23:858–862.
573. Liu YP, Wang TW, Xie Z, Bian Y, Liu YY, Guan RQ, Liu ZY, Qiang L, Chen GY, Fu YH. Artapilosines A and B, unusual phenanthrene derivatives related to aporphine alkaloids from *Artabotrys pilosus*. J Nat Prod 2021;84:3117–3121.
574. Zhu L, Zhu DR, Zhou WX, Zhu JM, Gong LJ, Li Y, Kong LY, Luo JG. Fortuneicyclidins A and B, pyrrolizidine alkaloids with a 7-azatetracyclo[5.4.3.0.0^{2,8}]tridecane core, from *Cephalotaxus fortunei*. Org Lett 2021;23:2807–2810.
575. Su BN, Park EJ, Nikolic D, Santarsiero BD, Mesecar AD, Vigo JS, Graham JG, Cabieses F, van Breemen RB, Fong HHS, Farnsworth NR, Pezzuto JM, Kinghorn AD. Activity-guided isolation of novel norwithanolides from *Deprea subtriflora* with potential cancer chemopreventive activity. J Org Chem 2003;68:2350–2361.
576. Kuo PC, Kuo TH, Damu AG, Su CR, Lee EJ, Wu TS, Shu R, Chen CM, Bastow KF, Chen TH, Lee KH. Physanolide A, a novel skeleton steroid, and other cytotoxic principles from *Physalis angulata*. Org Lett 2006;8:2953–2956
577. Jin Z, Mashuta MS, Stolowich NJ, Vaisberg AJ, Stivers NS, Bates PJ, Lewis WH, Hammond GB. Physangulidines A, B, and C: three new antiproliferative withanolides from *Physalis angulata* L. Org Lett 2012;14:1230–1233.
578. Ren YH, Liu QF, Chen L, He SJ, Zuo JP, Fan YY, Yue JM. Urceoloids A and B, two C19 steroids with a rearranged spirocyclic carbon skeleton from *Urceola quintaretii*. Org Lett 2019;21:1904–1907.
579. Meng XH, Chai T, Shi YP, Yang JL. Bungsteroid A: one unusual c34 pentacyclic steroid analogue from *Zanthoxylum bungeanum* Maxim. J Org Chem 2020;85:10806–10812.
580. Leverrier A, Dau METH, Retailleau P, Awang K, Guéritte F, Litaudon M. Kingianin A: a new natural pentacyclic compound from *Endiandra kingiana*. Org Lett 2010;12:3638–3641.
581. Saouf A, Guerra FM, Rubal JJ, Moreno Dorado FJ, Akssira M, Mellouki F, López M, Pujadas AJ, Jorge ZD, Massanet GM. Transtaganolides A–D: novel metabolites from *Thapsia transtagana*. Org Lett 2005;7:881–884.
582. Gao YP, Shen YH, Zhang SD, Tian JM, Zeng HW, Ye J, Li HL, Shan L, Zhang WD. Incarvilleatone, a new cyclohexylethanoid dimer from *Incarvillea younghusbandii* and its inhibition against nitric oxide (NO) release. Org Lett 2012;14:1954–1957.
583. Koike K, Jia ZH, Nikaido T, Liu Y, Zhao YY, Guo D. Echinothiophene, a novel benzothiophene glycoside from the roots of *Echinops grijissii*. Org Lett 1999:1:197–198.
584. Zou J, Chen GD, Zhao H, Huang Y, Luo X, Xu W, He RR, Hu D, Yao XS, Gao H. Triligustilides A and B: two pairs of phthalide trimers from *Angelica sinensis* with a complex polycyclic skeleton and their activities. Org Lett 2018;20:884–887.

585. Zou J, Chen GD, Zhao H, Wang XX, Zhang ZJ, Qu YB, He RR, So KF, Yao XS, Gao H. Triangeliphthalides A–D: bioactive phthalide trimers with new skeletons from *Angelica sinensis* and their production mechanism. Chem Commun 2019;55:6221–6224.
586. Liu CP, Xie CY, Zhao JX, Ji KL, Lei XX, Sun H, Lou LG, Yue JM. Dysoxylactam A: a macrocyclolipopeptide reverses p-glycoprotein-mediated multidrug resistance in cancer cells. J Am Chem Soc 2019;141:6812–6816.
587. Wang JX, Zhao YP, Du NN, Han Y, Li H, Wang R, Xu Y, Liu YF, Liang XM. Scocycamides, a pair of macrocyclic dicaffeoylspermidines with butyrylcholinesterase inhibition and antioxidation activity from the roots of *Scopolia tangutica*. Org Lett 2020;22:8240–8244.
588. Zhang B, Wang Y, Yang SP, Zhou Y, Wu WB, Tang W, Zuo JP, LiY, Yue JM. Ivorenolide A, an unprecedented immunosuppressive macrolide from *Khaya ivorensis*: structural elucidation and bioinspired total synthesis. J Am Chem Soc 2012;134:20605–20608.
589. Li XL, Zhao BX, Huang XJ, Zhang DM, Jiang RW, Li YJ, Jian YQ, Wang Y, Li YL, Ye WC. (+)- and (−)-Cajanusine, a pair of new enantiomeric stilbene dimers with a new skeleton from the leaves of *Cajanus cajan*. Org Lett 2014:16:224–227.
590. Li SG, Huang XJ, Li MM, Liu Q, Liu H, Wang Y, Ye WC. Multiflorumisides A–G, dimeric stilbene glucosides with rare coupling patterns from the roots of *Polygonum multiflorum*. J Nat Prod 2018;81:254–263.
591. Gewali MB, Tezuka Y, Banskota AH, Ali MS, Saiki I, Dong H, Kadota S. Epicalyxin F and calyxin I: two novel antiproliferative diarylheptanoids from the seeds of *Alpinia blepharocalyx*. Org Lett 1999;1:1733–1736.
592. Yang Y, Kinoshita K, Koyama K, Takahashi K, Tai T, Nunoura Y, Watanabe K. Two novel anti-emetic principles of *Alpinia katsumadai*. J Nat Prod 1999;62:1672–1674.
593. Ge HM, Huang B, Tan SH, Shi DH, Song YC, Tan RX. Bioactive oligostilbenoids from the stem bark of *Hopea exalata*. J Nat Prod 2006;69:1800–1802.
594. He S, Jiang LY, Wu B, Li C, Pan YJ. Chunganenol: an unusual antioxidative resveratrol hexamer from *Vitis chunganensis*. J Org Chem 2009;74:7966–7969.
595. Li XL, Cheng X, Yang LM, Wang RR, Zheng YT, Xiao WL, Zhao Y, Xu G, Lu Y, Chang Y, Zheng QT, Zhao QS, Sun HD. Dichotomains A and B: two new highly oxygenated phenolic derivatives from *Dicranopteris dichotoma*. Org Lett 2006;8:1937–1940.
596. Liu X, Luo HB, Huang YY, Bao JM, Tang GH, Chen YY, Wang J, Yin S. Selaginpulvilins A–D, new phosphodiesterase-4 inhibitors with an unprecedented skeleton from *Selaginella pulvinata*. Org Lett 2014;16:282–285.
597. Tan X, Han XY, Teng HD, Li QQ, Chen Yu, Lei XX, Yang GD. Structural elucidation of garcipaucinones A and B from *Garcinia paucinervis* using quantum chemical calculations. J Nat Prod 2021;84:972–978.
598. Franzyk H, Jensen SR, Olsen CE, Quiroga JM. A 9-hydroxyiridoid isolated from *Junellia seriphioides* (Verbenaceae). Org Lett 2000;2:699–700.

599. Vongvanich N, Kittakoop P, Kramyu J, Tanticharoen M, Thebtaranonth Y. Phyllanthusols A and B, cytotoxic norbisabolane glycosides from *Phyllanthus acidus* skeels. J Org Chem 2000;65:5420–5423.
600. Sang SM, He K, Liu GM, Zhu NQ, Cheng XF, Wang MF, Zheng QY, Dong ZG, Ghai G, Rosen RT, Ho CT. A new unusual iridoid with inhibition of activator protein-1 (ap-1) from the leaves of *Morinda citrifolia* L. Org Lett 2001;3:1307–1309.
601. Geng CA, Zhang XM, Ma YB, Luo J, Chen JJ. Swerilactones L–O, secoiridoids with C_{12} and C_{13} skeletons from *Swertia mileensis*. J Nat Prod 2011;74:1822–1825.
602. Liu L, Xu GL, Ma XX, Khan A, Tan WH, Yang ZY, Zhou ZH. Sweritranslactones A–C: unusual skeleton secoiridoid dimers via [4+2] cycloaddition from swertiamarin. J Org Chem 2017;82:13263–13267.
603. Hu LH, Sim KY. Cytotoxic polyprenylated benzoylphloroglucinol derivatives with an unusual adamantyl skeleton from *Hypericum sampsonii* (Guttiferae). Org Lett 1999;1:879–882.
604. Xu YJ, Yip SC, Kosela S, Fitri E, Hana M, Goh SH, Sim KY. Novel cytotoxic, polyprenylated heptacyclic xanthonoids from indonesian *Garcinia gaudichaudii* (Guttiferae). Org Lett 2000;2:3945–3948.
605. Cottiglia F, Dhanapal B, Sticher O, Heilmann J. New chromanone acids with antibacterial activity from *Calophyllum brasiliense*. J Nat Prod 2004;67:537–541.
606. Boonnak N, Chantrapromma S, Fun HK, Yuenyongsawad S, Patrick BO, Maneerat W, Williams DE, Andersen RJ. Three types of cytotoxic natural caged-scaffolds: pure enantiomers or partial racemates. J Nat Prod 2014:77:1562–1571.
607. Lin PC, Li S, Wang SJ, Yang YC, Shi JG. A nitrogen-containing 3-alkyl-1,4-benzoquinone and a gomphilactone derivative from *Embelia ribes*. J Nat Prod 2006;69:1629–1632.
608. Tian J, Han C, Guo WH, Yin Y, Wang XB, Sun HB, Yao HQ, Yang Y, Wang C, Liu C, Yang MH, Kong LY. Nigegladines A–C, three thymoquinone dimers from *Nigella glandulifera*. Org Lett 2017;19:6348–6351.
609. Kim SH, Sung SH, Choi SY, Chung YK, Kim J, Kim YC. Idesolide: a new spiro compound from *Idesia polycarpa*. Org Lett 2005;7:3275–3277.
610. Liu MT, Lin S, Gan ML, Chen M, Li L, Wang SJ, Zi JC, Fan XN, Liu Y, Si YK, Yang YC, Chen XG, Shi JG. Yaoshanenolides A and B: new spirolactones from the bark of *Machilus yaoshansis*. Org Lett 2012;14:1004–1007.
611. Itharat A, Plubrukarn A, Kongsaeree P, Bui T, Keawpradub N, Houghton PJ. Dioscorealides and dioscoreanone, novel cytotoxic naphthofuranoxepins, and 1,4-phenanthraquinone from *Dioscorea membranacea* Pierre. Org Lett 2003;5:2879–2882.
612. Zhao SM, Wang Z, Zeng GZ, Song WW, Chen XQ, Li XN, Tan NH. New cytotoxic naphthohydroquinone dimers from *Rubia alata*. Org Lett 2014;16:5576–5579.

613. Wisetsai A, Lekphrom R, Boonmak J, Youngme S, Schevenels FT. Spiroaxillarone A, a symmetric spirobisnaphthalene with an original skeleton from *Cyanotis axillaris*. Org Lett 2019;21:8344–8348.
614. Chen DL, Sun ZC, Liu YY, Li ZY, Liang HQ, Chen L, Xu XD, Yang JS, Ma GX, Huo XW. Eleucanainones A and B: two dimeric structures from the bulbs of *Eleutherine americana* with anti-MRSA activity. Org Lett 2020;22:3449–3453.
615. Geng CA, Jiang ZY, Ma YB, Luo J, Zhang XM, Wang HL, Shen Y, Zuo AX, Zhou J, Chen JJ. Swerilactones A and B, anti-HBV new lactones from a tradtional chinese herb: *Swertia mileensis* as a treatment for viral hepatitis. Org Lett 2009:11:4120–4123.
616. Geng CA, Wang LJ, Zhang XM, Ma YB, Huang XY, Luo J, Guo RH, Zhou J, Shen Y, Zuo A. X, Jiang ZY, Chen JJ. Anti-hepatitis B virus active lactones from the traditional chinese herb: *Swertia mileensis*. Chem Eur J 2011;17:3893–3903.
617. Chen XL, Liu HL, Li J, Xin GR, Guo YW. Paracaseolide A, first α-alkylbutenolide dimer with an unusual tetraquinane oxa-cage bislactone skeleton from Chinese mangrove *Sonneratia paracaseolaris*. Org Lett 2011;13:5032–5035.
618. Yuan T, Ding YQ, Wan CP, Li LY, Xu JL, Liu K, Slitt A, Ferreira D, Khan IA, Seeram NP. Antidiabetic ellagitannins from pomegranate flowers: inhibition of α-glucosidase and lipogenic gene expression. Org Lett 2012;14;5358–5361.
619. Omar M, Matsuo Y, Maeda H, Saito Y, Tanaka T. New metabolites of *C*-glycosidic ellagitannin from Japanese Oak Sapwood. Org Lett 2014;16:1378–1381.
620. Nakano H, Cantrell CL, Mamonov LK, Osbrink WLA, Ross SA. Echinopsacetylenes A and B, new thiophenes from *Echinops transiliensis*. Org Lett 2011;13:6228–6231.
621. Lavoie S, Legault J, Gauthier C, Mshvildadze V, Mercier S, Pichette A. Abibalsamins A and B, two new tetraterpenoids from *Abies balsamea* Oleoresin. Org Lett 2012;14:1504–1507.
622. Kobayashi J, Suzuki H, Shimbo K, Takeya K, Morita H. Celogentins A–C, new antimitotic bicyclic peptides from the seeds of *Celosia argentea*. J Org Chem 2001;66:6626–6633.
623. Tian JM, Shen YH, Yang XW, Liang S, Tang J, Shan, L, Zhang WD. Tunicyclin A, the first plant tricyclic ring cycloheptapeptide from *Psammosilene tunicoides*. Org Lett 2009;11:1131–1133.
624. Wang F, Gao Y, Zhang L, Bai B, Hu YN, Dong ZJ, Zhai QW, Zhu HJ, Liu JK. A pair of windmill-shaped enantiomers from *Lindera aggregata* with activity toward improvement of insulin sensitivity. Org Lett 2010;12:3196–3199.
625. Wang F, Gao Y, Zhang L, Liu JK. Bi-linderone, a highly modified methyl-linderone dimer from *Lindera aggregata* with activity toward improvement of insulin sensitivity in vitro. Org Lett 2010;12:2354–2357.

626. Liu BY, Zhang C, Zeng KW, Li J, Guo XY, Zhao MB, Tu PF, Jiang Y. Exotines A and B, two heterodimers of isopentenyl-substituted indole and coumarin derivatives from *Murraya exotica*. Org Lett 2015:17:4380–4383.
627. Lv HN, Wang S, Zeng KW, Li J, Guo XY, Ferreira D, Zjawiony JK, Tu PF, Jiang Y. Anti-inflammatory coumarin and benzocoumarin derivatives from *Murraya alata*. J Nat Prod 2015;78:279–285.
628. Ma SG, Gao RM, Li YH, Jiang JD, Gong NB, Li L, Lü Y, Tang WZ, Liu YB, Qu J, Lü HN, Li Y, Yu SS. Antiviral spirooliganones A and B with unprecedented skeletons from the roots of *Illicium oligandrum*. Org Lett 2013;15:4450–4453.
629. Xu JF, Zhao HJ, Wang XB, Li ZR, Luo J, Yang MH, Yang L, Yu WY, Yao HQ, Luo JG, Kong LY. (±)-Melicolones A and B, rearranged prenylated acetophenone stereoisomers with an unusual 9-oxatricyclo[3.2.1.1^{3,8}]nonane core from the leaves of *Melicope ptelefolia*. Org Lett 2015;17:146–149.

The manufacturer's authorised representative in the EU is Springer
Nature Customer Service Centre GmbH, Europaplatz 3, 69115 Heidelberg,
Germany. If you have any concerns regarding our products, please
contact ProductSafety@springernature.com

Printed and bound by CPI Group (UK) Ltd, Croydon, CR0 4YY
30/09/2026
02225322-0013